Universitext

Wolfram Pohlers

# Proof Theory

## The First Step into Impredicativity

Springer

Wolfram Pohlers
Universität Münster
Inst. Mathematische Logik
und Grundlagenforschung
Einsteinstr. 62
48149 Münster
Germany

ISBN 978-3-540-69318-5 e-ISBN 978-3-540-69319-2

Library of Congress Control Number: 2008930149

Mathematics Subject Classification (2000): 03F03, 03F05, 03F15, 03F25, 03F30, 03F35



*Cover design*: WMXDesign GmbH, Heidelberg

Printed on acid-free paper

9 8 7 6 5 4 3 2 1

springer.com

To Renate

# Preface

The kernel of this book consists of a series of lectures on infinitary proof theory which I gave during my time at the Westfälische Wilhelms–Universität in Münster. It was planned as a successor of Springer Lecture Notes in Mathematics 1407. However, when preparing it, I decided to also include material which has not been treated in SLN 1407. Since the appearance of SLN 1407 many innovations in the area of ordinal analysis have taken place. Just to mention those of them which are addressed in this book: Buchholz simplified local predicativity by the invention of operator controlled derivations (cf. Chapter 9, Chapter 11); Weiermann detected applications of methods of impredicative proof theory to the characterization of the provable recursive functions of predicative theories (cf. Chapter 10); Beckmann improved Gentzen's boundedness theorem (which appears as Stage Theorem (Theorem 6.6.1) in this book) to Theorem 6.6.9, a theorem which is very satisfying in itself although its real importance lies in the ordinal analysis of systems, weaker than those treated here.

Besides these innovations I also decided to include the analysis of the theory ($\Pi_2$–$\mathsf{REF}$) as an example of a subtheory of set theory whose ordinal analysis only requires a first step into impredicativity. The ordinal analysis of $(\Pi^0_1\text{–}\mathsf{FXP})_0$ of non-monotone $\Pi^0_1$–definable inductive definitions in Chapter 13 is an application of the analysis of ($\Pi_2$–$\mathsf{REF}$).

I have also put more emphasis on the development of the recursion theoretic background. This takes place in Chapters 5 and 6. Chapters 2 and 4 serve as recapitulation of basic facts in general logic. These chapters are indeed very basic. Being aware that this is boring for the more experienced reader the book contains some redundancies. They are intended to avoid the necessity of permanently scrolling back during reading (which is of course unavoidable in most cases).

Chapter 3 on ordinals is misplaced since it is not seriously needed before Chapter 6. However, I found no better place to put it in. In contrast to SLN 4017, where I tried to develop the theory of ordinals on the basis of an (in fact incomplete) axiom system, I decided now to develop the theory on the basis of a not fully axiomatized naive set theory (as it is common in everyday mathematics). According to the well–established custom, ordinals are regarded in the set theoretical sense. Those who regret that as a loss of constructivity are advised to stick to the notation systems developed from the set theoretical study of the ordinals. Notations of ordinals may be viewed as syntactically defined objects as discussed in Section 3.4.3.

As a warm up and a basis for the things to come, Chapter 7 reviews and discusses Gentzen's original results. Chapter 8 is more or less copied from SLN 4017 and treats the boundaries of predicativity in the narrow sense discussed in Section 8.3.

In Chapter 9 we apply Buchholz' technique of operator controlled derivation to obtain an upper bound for the ordinal $\kappa^{\mathsf{ID}_1}$ which coincides with the proof-theoretical ordinal of $\mathsf{ID}_1$.

Weiermann observed that the technique of operator controlled derivations is also applicable to characterize the provably recursive functions of arithmetic. In Chapter 10 we therefore present and discuss an adaption of Weiermann's theory on subrecursive functions to a study of the provably recursive functions of arithmetic.

Chapter 11 starts with a — this time more detailed — introduction to set theory and presents the axiom systems $\mathsf{KP}\omega$ for Kripke–Platek set theory with infinity and the axiom system $(\Pi_2\text{–}\mathsf{REF})$ and computes their proof-theoretical ordinals.

As an application of the ordinal analysis of $(\Pi_2\text{–}\mathsf{REF})$ Chapter 13 computes the proof-theoretical ordinal of non–monotone arithmetical inductive definitions which are $\Pi^0_1$–definable. This analysis is basically built on an observation due to Robin Gandy a couple of decades ago which was never published by him.

Chapter 12 was only inserted after the rest of the book was finished. Gerhard Jäger in his Habilitationsschrift pointed out that the naive view of the impredicativity of the systems $\mathsf{ID}_1$ and $\mathsf{KP}\omega$ is not completely correct. Although the impredicativity of theses system seems to be manifested in the closure axiom $ID^1_1$ or the $\Delta_0$–collection axiom in $\mathsf{KP}\omega$, respectively, their impredicativity comes only in concurrence with foundation. It seemed important to me to include this result. Since Jäger's work is published in [52] the original plan was just to include a few exercises in which Jäger's results could be presented. However, due to the different framework in the present book it turned into a whole chapter in which, however, most of the results are still stated as exercises with extended hints. Here I am especially indebted to Jan Carl Stegert, who checked the exercises there and suggested many improvements.

Chapter 1 displays my personal view of the development of proof theory. Since I am not a historian I cannot guarantee to have checked all sources in a professional way. According to my personal preference and the topic of this book I have put emphasis on ordinal analysis. Of course I am aware of the existence and importance of other parts of proof theory but it was not my aim to get close to a complete history of proof theory.

I want to thank my assistants Christoph Heinatsch and Christoph Duchhardt for proofreading most parts of the book. Besides the correction of a variety of errors they also suggested many improvements. Jan Carl Stegert checked the exercises not only in Chapter 12 but also in the other parts. I also thank Joachim Columbus and Andreas Schlüter who proofread SLN 4017. The parts taken from there still rest on their corrections.

Of course I discussed the topics of the book with many of my colleagues. I am especially indebted to Andreas Weiermann who not only developed the theory presented in Chapter 10 (although he is in no way responsible for any errors which may have sneaked into my adaption of his theory) but also gave many suggestions in discussions.

The counterexample in Exercise 6.7.6 is due to Arnold Beckmann. However, he not only contributed this counterexample but also assisted me on many other occasions during the preparation of this book.

Another colleague who suffered frequently from my inquiring is Wilfried Buchholz to whom we owe the theory of operator controlled derivations.

Many thanks also to the students in our proof theory seminar. Their interest and thorough study helped to improve the book.

Last but not least I want to thank my wife for her patience during the long time it took to finish this work. This book is dedicated to her.

Münster *Wolfram Pohlers*
February 2008

# Contents

# Chapter 1
# Historical Background

The history of "Proof Theory" begins with the foundational crisis of Mathematics in the first decades of the twentieth century. At the turn of the century, in reaction to the explosion of mathematical knowledge, endeavors began, to provide the growing body of mathematics with a firm foundation. Some of the notions used then seemed to be problematic. This was especially true for those notions that embodied "infinities". On the one hand, there was the notion of "infinitesimals" which dealt with "infinity in the small". The elimination of infinitesimals by limit processes meant a big progress in giving existing mathematics a firm fundament.

On the other hand there was the notion of "infinity in the large". Investigations on the uniqueness of representation of functions by trigonometric series forced Georg Cantor to develop a completely new theory of the *infinity in the large*. One of the central points in Cantor's theory was the possibility to collect even infinitely many objects into a new object. These entities – Cantor called them "Mengen", *"sets"* in English – were the mathematical subjects of Cantor's research. Therefore, he called his new theory, *"Mengenlehre"* which is translated as *"set theory"*.

The possibility to form sets unrestrictedly, however, led immediately to contradictions. One example is Russell's paradox about the set of all sets which are not members of themselves. For

$$R := \{x \mid x \notin x\}$$

we obtain $R \in R \Leftrightarrow R \notin R$.

These paradoxes and probably also the seemingly paradoxical fact that the axiom of choice offered the possibility to well-order all sets, created a feeling of uncertainty among the mathematical community. Herman Weyl in his article "Über die neue Grundlagenkrise der Mathematik" [112] pointed out that circular definitions which caused the paradoxes of set theory are also used in analysis (cf. [111]). He introduced the term "new foundational crisis" into the discussion. In his book "Das Kontinuum" [110] he had already proposed a development in mathematics that avoided "circular" definitions.

W. Pohlers, *Proof Theory: The First Step into Impredicativity*, Universitext,

Even more significant was the criticism by Brouwer. He already doubted the logical basis of mathematics. His point of attack was the *law of excluded middle* (or *tertium non datur*) which permits to prove the existence of objects without explicitly constructing them. Brouwer suggested to develop mathematics on intuitive principles which exclude the law of excluded middle. Their formalizations – mostly due to Heyting – is now known as intuitionistic logic.

Both approaches – Brouwer's as well as Weyl's – meant, however, rigid restrictions on mathematics. Hilbert, then one of the most influential mathematicians, was not willing to accept any restriction which could mutilate existing mathematics. Therefore he suggested a programme to save mathematics in its existing form. He started to think about foundations of mathematics very early. Problem number 2 in his famous list of unsolved mathematical problems presented in 1900 at Paris was to find axioms for the theory of reals and to show their consistency. Between 1917 and 1922, he was frequently lecturing on foundational problems (cf. [93]). His ongoing interest in foundational problems was certainly one of the reasons for Hilbert to engage Paul Bernays as assistant. Bernays became one of his most important collaborators in developing his programme, which is nowadays called *Hilbert's programme*. The initial paper in which the phrase *Beweistheorie*, i.e., *proof theory*, is explicitly mentioned is [41] titled "Neubegründung der Mathematik". In this paper, he responds to Weyl's and Brouwer's criticism and sketches his program for the foundation of analysis. He propagates the *axiomatic method* (Axiomatische Methode) which means that mathematical objects and their interrelations should be solely defined by axioms. The remaining task is to show that these axioms are not contradictory. To perform this task a new mathematical discipline is needed – *proof theory*. The objects of research in this new discipline are mathematical proofs. He sketches how this new discipline could work. He points out that a proof is a finite figure which we may think of as a tree whose root is the proved formula, leaves are axioms and nodes are locally correct with respect to rules (Schlußschemata).[1] Therefore, the objects of proof theory are finite objects and it should be possible to show by completely finite reasoning that the root of a proof tree can never be a contradiction.[2] In this way, Hilbert aimed to obtain a justification of the infinite by finite reasoning.[3] This point of view is known as Hilbert's *finitist standpoint* (Finiter Standpunkt). This standpoint was elaborated further in a talk "Die logischen Grundlagen der Mathematik" given in front of the Deutsche Naturforscher–Gesellschaft and printed in [42]. In [41], Hilbert summarized his programme into two steps. The first step is

---

[1] In this paper he has one only single rule, *modus ponens*.

[2] Here Hilbert apparently anticipates the possibility of a calculus for reasoning which produces all the logical consequences of an axiom system. A fact which was proved only in 1930 by K. Gödel in [36].

[3] In a talk "*Über das Unendliche*" given at Münster in June 1925 (printed in [42]) he states literally "*daß das Operieren mit dem Unendlichen nur durch das Endliche gesichert werden kann*", i.e., "*that operating with infinities can only be secured by finite means.*"

to formalize all existing mathematics.[4] The second step is to develop and use proof theory to show that this formal system cannot produce contradictions.[5]

There have been early successes in performing Hilbert's programme. Consistency proofs for weak subsystems of the axioms for number theory have been given by Wilhelm Ackermann [1], John von Neumann [65], Herbrand [39] and [40] and Gerhard Gentzen [28, 29]. But all attempts to extend these proofs to stronger axiom systems which also include an axiom for mathematical induction failed. That this was not an accident followed from Gödels's paper [37] in which he proved his famous incompleteness theorems. His *first incompleteness theorem* shows the impossibility of a (first-order) axiom system which proves all true arithmetical sentences while, even worse, his second incompleteness theorem states that a formal theory cannot prove its own consistency. Gödel's theorems rest on the observation that metamathematical reasoning can be "goedelized", i.e., can be coded by natural numbers. This is especially true for all finite reasoning. As a consequence any finitist consistency proof should already be formalizable within the axioms of number theory (even a weaker axiom system which provides a coding machinery and a bit of mathematical induction suffices). So – according to Gödel's second incompleteness theorem – no finitist consistency proof for the axioms of number theory, let alone for stronger systems, is possible. In 1935, Bernays in his article on "Hilberts Untersuchungen über die Grundlagen der Mathematik" [9] gives an account of the state of proof theory and also discusses the influence of Gödels's incompleteness theorems to Hilbert's proof theory.

---

[4] *"Alles, was bisher die eigentliche Mathematik ausmacht, wird nunmehr streng formalisiert, so daß die* eigentliche Mathematik *oder die Mathematik im engeren Sinne zu einem Bestande an beweisbaren Formeln wird. Die Formeln dieses Bestandes unterscheiden sich von den gewöhnlichen Formeln der Mathematik nur dadurch, daß außer den mathematischen Zeichen noch das Zeichen* $\rightarrow$*, das Allzeichen und die Zeichen für Aussagen darin vorkommen. Dieser Umstand entspricht einer seit langem von mir vertretenen Überzeugung, daß wegen der engen Verknüpfung und Untrennbarkeit arithmetischer und logischer Wahrheiten ein simultaner Aufbau der Arithmetik und formalen Logik notwendig ist."*

I.e., *"Everything, which hitherto has been part of actual mathematics, will be rigidly formalized such that actual mathematics becomes a stock of provable formulas. The formulas in this stock are distinguished from the usual formulas of mathematics in so far that they will, beyond the mathematical symbols, also contain the symbol* $\rightarrow$*, the symbol for generalization and symbols for propositions. This fact reflects my long stated conviction that the strong interrelation and inseparability of arithmetical and logical truth enforce a simultaneous development of arithmetic and formal logic."*

[5] *"Zu dieser eigentlichen Mathematik kommt gewissermaßen eine neue Mathematik, eine Metamathematik, hinzu, die zur Sicherung jener dient, indem sie vor dem Terror der unnötigen Verbote sowie der Not der Paradoxien schützt. In dieser Metamathematik kommt – im Gegensatz zu den rein formalen Schlußweisen der eigentlichen Mathematik – das inhaltliche Schließen zur Anwendung, und zwar zum Nachweis der Widerspruchsfreiheit der Axiome."*

I.e., *"To the real mathematics comes, so to speak, a new mathematics, a metamathematics, which is needed to secure the former by protecting it from the terror of unnecessary prohibitions as well from the disaster of paradoxes. In order to show the consistency of the axioms this metamathematics needs – in contrast to the purely formal reasoning of real mathematics – contentual reasoning".*

While this paper was in print, Gerhard Gentzen [30] succeeded in giving a consistency proof for an axiom system for number theory which included the full scheme of mathematical induction. His proof, however, did not contradict Gödel's second incompleteness theorem. It contained parts which could not be formalized within number theory. Nevertheless, the extensions needed seemed to be so tiny that Bernays (loc. cit.) argued that Gentzen's proof meets the basic requirements of Hilbert's finitist standpoint.[6,7]

In the beginning, Gentzen's first proof was not completely understood. Therefore he withdrew Sects. 14–16 of [30] before printing and replaced them by a version, in which he used Cantor's new transfinite numbers, to prove the termination of his reduction procedure on proof figures.[8] So he succeeded in giving a consistency proof in which the "nonfinitist" means are concentrated in a single principle, an induction along a primitive-recursively definable well-ordering of the natural numbers of transfinite order type $\varepsilon_0$. The remaining parts of his proof were completely finitist.

According to Gödel's second incompleteness theorem, induction along the well-ordering used in Gentzen's proof cannot be proved from the axioms of number theory. Vice versa, Gentzen in [31] showed that induction along any proper segment of this well-ordering is provable from these axioms. The order type $\varepsilon_0$ therefore provides a measure for the "amount of transfiniteness" of the axioms of number theory. Therefore, it is tempting to regard the order type of the shortest primitive recursive well-ordering which is needed in a consistency proof for a theory $T$ as characteristic for $T$. That this idea is malicious was later detected by Georg Kreisel who indicated in [62] that for any reasonable axiom system $Ax$ there is a primitive recursive well-ordering $\prec$ of order type $\omega$ which proves the consistency of $Ax$ by induction along $\prec$ with the other things that only finitist means.[9]

In [31], however, Gentzen showed more. He proved without using Gödel's second incompleteness theorem that any primitive recursive well-ordering whose well-foundedness is provable from the axioms of number theory must have order type less than $\varepsilon_0$. From the previously mentioned result of [31], it follows that conversely for any order type $\alpha < \varepsilon_0$ there is a primitive recursive well-ordering of order type $\alpha$ whose well-foundedness can be proved from the axioms for number theory. This implies that $\varepsilon_0$ is the supremum of the order types of primitive recursive well–orderings whose well–foundedness can be proved from the axioms of number

6 *... ist von G. Gentzen der Nachweis für die Widerspruchsfreiheit des vollen zahlentheoretischen Formalismus erbracht worden, durch eine Methode, die den grundsätzlichen Anforderungen des finiten Standpunktes durchaus entspricht.*

I.e., *... has Gentzen given a consistency proof for the formalism of full number theory by a method which basically meets the fundamental requirements of the finitist standpoint.*

7 In my personal opinion this is – even for number theory – not correct . See [73] §16 and Sect. 7.5 in this book.

8 The galley proofs of the original version of his proof which used a form of the fan theorem have been preserved. It is reprinted in [32]; an English translation is in the collected papers [96].

9 The well–ordering in Kreisel's counterexample is, however, artificially defined (cf. Sect. 7.5.3). For all known "natural well-orderings" the least order type which is needed to prove the consistency of the axioms for arithmetic is $\varepsilon_0$. The problem is that we do not have a definition of "natural well-orderings" and – at least at the moment – have little hope to find one.

theory. Following Gentzen's paper [31], we today define the proof-theoretical ordinal $\|T\|$ of a theory $T$ to be the supremum of the order types of primitive recursive well-orderings on the natural numbers whose well-foundedness can be derived from the axioms in $T$. This is a mathematically well-defined ordinal. For reasons that are explained later this ordinal is now also called the $\Pi^1_1$-ordinal of the theory $T$. But there is also the notion of a $\Pi^0_2$-ordinal of a theory[10] and – most recently – also the notion of the $\Pi^0_1$-ordinal of a theory, which probably is closest to the original intention.

By an ordinal analysis of a theory $T$, we commonly understand the computation of its proof-theoretical ordinal, i.e., $\Pi^1_1$-ordinal. But ordinal analysis includes much more. A good account of the status of ordinal analysis is given by Michael Rathjen in [80].

Ordinal analyzes for systems stronger than pure number theory have been given later by Solomon Feferman in [23] and Kurt Schütte in [87] and [88] who independently characterized the limits of predicativity. Previous work in that direction was done by Paul Lorenzen [63].

An important paper for the further development of impredicative proof theory was Gaisi Takeuti's 1953 paper [101], which showed that the generalization of Gentzen's Hauptsatz for first-order logic to simple type theory entails the consistency of full second-order number theory. Takeuti conjectured that Gentzen's Hauptsatz also holds for simple type theory. In [86] Schütte presented 1960 a semantical equivalent to Takeuti's conjecture. William Tait in [97] proved 1966 Takeuti's conjecture for second-order logic using Schütte's semantical equivalent. Around 1967, Moto-o Takahashi in [100] and Dag Prawitz in [75] independently proved Schütte's semantical equivalent which entails Takeuti's conjecture for full simple type theory. Girard showed 1971 in [33] that simple type theory not only allows cut elimination but also a terminating normalization procedure.[11]

Disappointingly all these results gave neither ordinal analyzes nor essential proof-theoretic reductions, although Girard's results had high impact on structural proof theory and theoretical computer science.

The first results for impredicative subsystems of second-order number theory which also gave ordinal information, were obtained by Takeuti in [102][12], in which he proved the consistency of the system of $\Pi^1_1$-comprehension using transfinite induction on ordinal diagrams. This gave an upper bound for the $\Pi^1_1$ ordinal of

[10] Cf. Sect. 10.

[11] He generalized Gödel's system $T$ of functionals of finite types to a system $F$ of functionals of types possibly containing variables for types. Via the Curry–Howard isomorphism this system corresponds to a natural deduction system for finite types. Then he proved strong normalization for the system $F$ by a generalization of Shoenfield's computability predicates for the system $T$. This generalization, however, was far from obvious since normalization for the system $F$ implies cut elimination for simple type theory which in turn by Takeuti's result entails consistency of full second (and even higher) order number theory. Therefore Girard's computability predicates must not be formalizable in full second-order number theory and it is difficult to invent such predicates. Girard introduced the notion of "candidats de reducibilité" which was perfectly adapted to the situation.

[12] Although Takeuti's emphasis was not on ordinal information but on consistency proofs which were as close as possible to Hilbert's programme.

this system which was later extended by Takeuti and Yasugi in [105] to the system for $\Delta^1_2$–comprehension.

Another way to attack the consistency problem was initiated around 1958 by Gödel [38]. He tried to establish the consistency of the axiom system for pure number theory by translating the provable sentences $F$ of pure number theory into sentences $F^*$ of the form $(\exists\vec{\alpha})(\forall\vec{\beta})\tilde{F}[\alpha,\beta]$, where $\vec{\alpha}$ and $\vec{\beta}$ are strings of variables for functionals of higher types and $\tilde{F}[\alpha,\beta]$ is a quantifier free formula. Then he showed that for every provable sentence $F$ of pure number theory there is a string of terms $\vec{f}$ in his system $T$ of functionals of finite types such that $\tilde{F}[\vec{f},\vec{g}]$ holds for all $\vec{g}$ of appropriate types which means that the quantifier free formula $F[\vec{f},\vec{g}]$ becomes provable in $T$. All that is needed in this consistency proof is the fact that all the functionals in $T$ are computable. Gödel apparently took that for granted. A strict proof, however, again needs a transfinite induction along a well-ordering of order type $\varepsilon_0$ as shown by Howard [46].

There are simpler proofs for the computability of the functionals in $T$ – which probably were completely plain to Gödel and which later have been spelled out by Tait [98][13] – using computability predicates. These computability predicates are locally formalizable in number theory, which means that for every class of terms of restricted type complexity we can define their computability predicate within pure number theory. According to Gödel's second incompleteness theorem, however, the global computability predicate for all terms, which is needed for the consistency proof, is not formalizable.[14] No ordinal information is gained from the proof via computability predicates.

Gödel's paper can be viewed as a paradigmatic example for reductive proof theory. The theory $T$ is quantifier free, based solely on the defining equations for the functionals. No law of excluded middle is needed to show the interpretation theorem. So we have a reduction to an intuitionistic quantifier free theory (but in higher types) which intuitively seems to be much less obscure than the theory PA of pure number theory with the full scheme of mathematical induction. But remember that from a strict mathematical viewpoint the consistency of the theory $T$ is of the same complexity as the consistency of the theory PA. So we have only a reduction, not a real consistency proof.

Building on this way of attack, Clifford Spector in 1962 [95] developed a consistency proof for full second-order number theory by a functional interpretation using functionals defined by bar recursion on all finite types. Of course this proof again did not give any ordinal information. But Kreisel wanted to know whether systems for iterated inductive definitions could serve to model bar recursion of finite types. In 1963 [61], he introduced systems for generalized inductive definitions. Although it turned out that even iterated inductive definitions are much weaker than

[13] Cf. also the proof given in Shoenfield's book [92].

[14] The situation is comparable to the situation when we prove the consistency of number theory directly by truth predicates. The local versions, i.e., truth predicates for formulas of bounded complexity, are formalizable within number theory. However, the global version, which is needed for the consistency proof, cannot be formalized in number theory. Again a consistency proof via truth predicates gives no ordinal information.

full second-order number theory the focus was pointed on systems for iterated generalized inductive definitions.

This topic was blossoming at the 1968 conference on Intuitionism and Proof Theory in Buffalo. The proceedings [60] contain three important papers in this area. Harvey Friedman in [25] showed that the second-order theory with the $\Sigma^1_2$ axiom of choice can be interpreted in the system $(\Pi^1_1\text{–CA})_{<\varepsilon_0}$ of less than $\varepsilon_0$-fold iterated $\Pi^1_1$-comprehensions and Feferman in [24] showed that less than $\nu$-fold iterated $\Pi^1_1$-comprehensions could be interpreted in systems $\mathsf{ID}_{<\nu}$ for less than $\nu$-fold iterated inductive definitions. Tait in [99] used cut elimination for an infinitary propositional logic whose formula complexity is measured by constructive number classes to obtain a consistency proof for second-order number theory with the scheme of $\Sigma^1_2$ dependent choice. No ordinal information could be gained from this proof although it carried already the germs of ideas which later made an ordinal analysis for systems of iterated generalized inductive definitions possible.

By a complicated passage through formal theories for choice sequences, it was known that the theory $\mathsf{ID}_1$ for one generalized inductive definition based on classical logic is reducible to the theory $ID^i_1(\mathcal{O})$ which axiomatizes the first constructive number class based on intuitionistic logic. In $ID^i_1(\mathcal{O})$, only the existence of the accessible part of a computably enumerable ordering is postulated. Accessible parts have a clear constructive meaning. Reduction to theories for accessibility predicates which are based on intuitionistic logic is therefore one of the aims of reductive proof theory. However, Zucker showed [115] that there are definitive obstacles to a straight forward reduction of the theories $\mathsf{ID}_\nu$ for $\nu > 1$ to intuitionistic accessibility theories. Such reductions were later obtained via ordinal analyses.

The first ordinal analysis for the theory $ID^i_1$ was obtained by Willam Howard [47] in 1972. Via the known proof-theoretical reductions this entailed also an ordinal analysis for $\mathsf{ID}_1$. Ordinal analyses for theories for finitely iterated inductive definitions were later obtained by Pohlers in [68] and for transfinitely iterated inductive definitions in [69], using Takeuti's reduction procedure for $\Pi^1_1$–comprehension. Later, more perspicuous methods have been established by Buchholz in [15], using his $\Omega_\nu$-rules, and Pohlers in [71, 70, 72], using the "method of local predicativity". An account to the history of that development by Feferman can be found in [15].

To get more perspicuous ordinal analyzes also for the subtheories of second-order number theories it was an obvious attempt to try to transfer the methods which were successful in the ordinal analyzes of theories for iterated inductive definitions. Generalizing Buchholz' $\Omega_\nu$-rules to second-order number theory worked quite well and led to [56, 16] and the monograph [17] in which the results of Takeuti for $\Pi^1_1$-Comprehension [102] and Takeuti and Yasugi for the theory of $\Delta^1_2$-comprehension [104] could be reobtained by much more perspicuous techniques. But at that time it was not at all clear how this technique could be pushed essentially further.

The obvious generalization of local predicativity to subtheories of second-order number theory meant to extend the ramified analytic hierarchy – the familiar tool in predicative proof theory – to stages beyond $\omega_1^{CK}$ thus leading to proof-theoretic ordinals bigger than $\Gamma_0$, the limiting ordinal of predicativity.[15] This approach, however, needed sets which were not longer sets of natural numbers and therefore outsides the ramified analytical hierarchy. This required a coding machinery which turned out to be unmanageable. The natural remedy was to step outside of the ramified analytic hierarchy – which can be viewed as Gödel's constructible universe intersected with the powerset of the natural numbers – and to work directly in the constructible hierarchy. The pioneering work in this direction has been done by Jäger starting with his Diploma thesis via his Dissertation [48] and finally in [55] and his monograph [52].[16] One of the highlights of this approach was the analysis of the theory KPi by Jäger and Pohlers in [55] which corresponds to $\Delta^1_2$–comprehension plus Bar–induction on the side of subsystems of second-order number theory. Showing the well–foundedness of the notation system needed in the analysis of KPi within Feferman's theory $T_0$ for explicit mathematics Jäger established the lacking direction $(\Delta^1_2\text{–CA})+(\mathrm{BI}) \leq T_0$ in the proof-theoretic equivalence of $T_0$ with $(\Delta^1_2\text{–CA})+(\mathrm{BI})$. The other direction was already established by work of Feferman.

The strongest theories analyzed so far are parameter free $\Pi^1_2$-comprehension by Rathjen [81] which on the side of set theory corresponds to the theory of $\Sigma_1$-separation. He arrived there by sucessively analyzing the theory KPM axiomatizing a recursively Mahlo universe [77], the theory of $\Pi_3$-reflection [78] and [79] and the theory of stability [82]. This analysis still works within the constructible hierarchy but uses methods which by far exceed the method of local predicativity presented in this volume.[17]

This book concentrates on the first step into predicativity. After a short and therefore rather incomplete review of predicative proof theory, we are going to analyze theories on the level of a noniterated inductive definition whose ordinal strength is measured by the Bachmann–Howard ordinal. Stronger theories will need a *second and further steps into impredicativity* which are supposed to become the topic of another monography.

[15] Cf. Chap. 8.

[16] Part of the results presented there are in Chap. 12.

[17] In this connection one should also mention the work of Toshiyasu Arai who, mainly building on Takeuti's approach, also got results for strong axiom systems (e.g. [3]).

# Chapter 2
# Primitive Recursive Functions and Relations

To fix notations and to give a rough overview of some of the recursion theoretic background which we are going to use in this book we start with a chapter on primitive recursive functions and relations. Of course, we cannot be exhaustive here. For further studies we recommend a textbook on computability theory. An old, but still very good source is Rogers' book [84].

## 2.1 Primitive Recursive Functions

The natural numbers can be viewed as generated by a counting process. We start counting with zero. So every natural number is either zero or the immediate successor of another natural number. This counting process is the basis for the definition of the class of primitive recursive functions, a subclass of the functions on the natural numbers which are effectively computable. We obtain the class of primitive recursive functions by first introducing a class of formal terms – the primitive recursive functions terms – and then defining their evaluation on the natural numbers.

**2.1.1 Definition** The primitive recursive function terms are inductively defined by the following clauses.

- The symbol $S$ (for the successor function) is a unary primitive recursive function term.
- The symbol $C_k^n$ (for the function with constant value $k$) is an $n$-ary primitive recursive function term.
- For $1 \leq k \leq n$ the symbol $P_k^n$ (for the projection on the $k$th component) is an $n$-ary primitive recursive function term.

W. Pohlers, *Proof Theory: The First Step into Impredicativity*, Universitext,

- If $h_1,\ldots,h_m$ are $n$-ary primitive recursive function terms and $g$ is an $m$-ary primitive recursive function term then $Sub(g,h_1,\ldots,h_m)$ is an $n$-ary primitive recursive function term (substitution of functions).

- If $g$ is an $n$-ary and $h$ an $n+2$-ary primitive recursive function term then $Rec(g,h)$ is an $n+1$-ary primitive recursive function term (primitive recursion).

**2.1.2 Definition** For an $n$-ary primitive recursive function term $f$ and an $n$-tuple $z_1,\ldots,z_n$ of natural numbers we define the *evaluation* $\mathrm{ev}(f,z_1,\ldots,z_n)$ inductively by the following clauses:

- $\mathrm{ev}(S,z_1) = z$ if $z$ is the successor number of $z_1$.

- $\mathrm{ev}(C^n_k,z_1,\ldots,z_n) = z$ if $z = k$.

- $\mathrm{ev}(P^n_k,z_1,\ldots,z_n) = z$ if $z = z_k$.

- $\mathrm{ev}(Sub(g,h_1,\ldots,h_m),z_1,\ldots,z_n) = z$ holds if there are natural numbers $u_1,\ldots,u_m$ such that $\mathrm{ev}(h_i,z_1,\ldots,z_n) = u_i$ for $i = 1,\ldots,m$ and $\mathrm{ev}(g,u_1,\ldots,u_m) = z$.

- $\mathrm{ev}(Rec(g,h),k,z_1,\ldots,z_n) = z$ holds if there are natural numbers $u_0,\ldots,u_k$ such that $u_k = z$, $\mathrm{ev}(g,z_1,\ldots,z_n) = u_0$ and $\mathrm{ev}(h,i,u_i,z_1,\ldots,z_n) = u_{i+1}$ for $i = 0,\ldots,k-1$.

**2.1.3 Definition** A function $F\colon \mathbb{N}^n \longrightarrow \mathbb{N}$ is *primitive recursive* if and only if there is an $n$-ary primitive recursive function term $f$ such that for every $n$-tuple $z_1,\ldots,z_n$ of natural numbers we have $\mathrm{ev}(f,z_1,\ldots,z_n) = F(z_1,\ldots,z_n)$.

Let $\mathscr{F}$ be a class of number theoretic functions. By $\mathscr{F}^n$, we denote the $n$-ary functions in $\mathscr{F}$. We say that $\mathscr{F}$ is closed by substitution if $H_1,\ldots,H_m \in \mathscr{F}^n$ and $G \in \mathscr{F}^m$ implies that the function $F$ defined by

$$F(z_1,\ldots,z_n) := G(H(z_1,\ldots,z_n),\ldots,H(z_1,\ldots,z_n))$$

also belongs to $\mathscr{F}$. We say that $\mathscr{F}$ is closed under primitive recursion if for $G \in \mathscr{F}^n$ and $H \in \mathscr{F}^{n+2}$ the function $F$, which is uniquely defined by the recursion equations

$$F(0,z_1,\ldots,z_n) = G(z_1,\ldots,z_n)$$

and

$$F(n+1,z_1,\ldots,z_n) = H(n,F(n,z_1,\ldots,z_n),z_1,\ldots,z_n),$$

also belongs to $\mathscr{F}$.

**2.1.4 Theorem** *The primitive recursive functions form the smallest class of functions which contain the* basic functions *"successor", "constant-functions" and "projections" and is closed under substitution and primitive recursion.*

*Proof* Let *PRF* denote the class of primitive recursive functions. By definition *PRF* contains the basic functions. If $H_1,\ldots,H_m$ are $n$-ary primitive recursive functions and $G$ is an $m$-ary primitive recursive function then there are primitive recursive function terms $h_1,\ldots,h_m$ and $g$ such that $H_i(z_1,\ldots,z_n) = \mathrm{ev}(h_i,z_1,\ldots,z_n)$ holds true for $i = 1,\ldots,m$ and $G(u_1,\ldots,u_m) = \mathrm{ev}(g,u_1,\ldots,u_m)$. Then $Sub(g,h_1,\ldots,h_m)$ is an $n$-ary primitive recursive function term satisfying $\mathrm{ev}(Sub(g,h_1,\ldots,h_m),z_1,\ldots,z_n) = G(H_1(z_1,\ldots,z_n),\ldots,H_n(z_1,\ldots,z_n))$. This shows that *PRF* is closed under substitutions.

Now assume that $G$ and $H$ are $n$- and $n+2$-ary primitive recursive functions and $F$ is defined by the recursion equations $F(0,z_1,\ldots,z_n) = G(z_1,\ldots,z_n)$ and $F(i+1,z_1,\ldots,z_n) = H(i,F(i,z_1,\ldots,z_n),z_1,\ldots,z_n)$. There are primitive recursive function terms $g$ and $h$ such that $\mathrm{ev}(g,z_1,\ldots,z_n) = G(z_1,\ldots,z_n)$ and $\mathrm{ev}(h,z_1,\ldots,z_{n+2}) = H(z_1,\ldots,z_{n+2})$. Then $Rec(g,h)$ is an $n+1$-ary primitive recursive function term. We show

$$\mathrm{ev}(Rec(g,h),i,z_1,\ldots,z_n) = F(i,z_1,\ldots,z_n)$$

by induction on $i$. We get the induction begin from $\mathrm{ev}(Rec(g,h),0,z_1,\ldots,z_n) = u_0 = \mathrm{ev}(g,z_1,\ldots,z_n) = G(z_1,\ldots,z_n) = F(0,z_1,\ldots,z_n)$ and the induction step follows from

$$\begin{aligned}&\mathrm{ev}(Rec(g,h),i+1,z_1,\ldots,z_n) = \mathrm{ev}(h,i,u_i,z_1,\ldots,z_n)\\ &= H(i,F(i,z_1,\ldots,z_n),z_1,\ldots,z_n) = F(i+1,z_1,\ldots,z_n).\end{aligned}$$

It remains to show that *PRF* is the least such class. Assume that $\mathscr{F}$ is a class of number theoretic functions having the required closure properties. By induction on the definition of the primitive recursive function term $f$, we show that its evaluation belongs to $\mathscr{F}$. This is obvious if $f$ is one of the terms $S$, $C^n_k$ or $P^n_k$. For a composed term $f = Sub(g,h_1,\ldots,h_m)$ the evaluations of $g$ and $h_j$ belong to $\mathscr{F}$ by induction hypothesis. Since $\mathscr{F}$ is closed under substitutions, the evaluation of $f$ belongs to $\mathscr{F}$, too. If $F$ is the evaluation of the composed term $Rec(g,h)$, then $F$ satisfies the recursion equations $F(0,z_1,\ldots,z_n) = G(z_1,\ldots,z_n)$ and $F(i+1,z_1,\ldots,z_n) = H(i,F(i,z_1,\ldots,z_n),z_1,\ldots,z_n)$ where $G$ and $H$ are the evaluations of the terms $g$ and $h$. By induction hypothesis, $G$ and $H$ belong to $\mathscr{F}$. Since $\mathscr{F}$ is closed under primitive recursion $F$ belongs to $\mathscr{F}$. □

We prove some basic closure properties of the primitive recursive functions which we will need later. For more detailed studies we recommend any textbook on recursion theory. The classical reference is [84].

First we observe that many of the familiar number theoretic functions are primitive recursive (cf. the exercises). Among them is bounded summation satisfying the recursion equations

$$\sum_{i=0}^{0} F(i) = F(0) \text{ and } \sum_{i=0}^{x+1} F(i) = F(x+1) + \sum_{i=0}^{x} F(i)$$

where $F$ is a primitive recursive function. Likewise, we check that the bounded product $\prod_{i=0}^{x} F(i)$ (as a function of its upper bound $x$) is primitive recursive, too. Important primitive recursive functions are the case distinction functions $sg$ and $\overline{sg}$ defined by the recursion equations

$$sg(0) := 0 \text{ and } sg(n+1) := 1$$

and

$$\overline{sg}(0) := 1 \text{ and } \overline{sg}(n+1) := 0.$$

The predecessor function $pred(n)$ satisfies the recursion equations

$$pred(0) = 0 \text{ and } pred(n+1) = n$$

and is thus primitive recursive. Then we obtain the arithmetical difference of two natural numbers as the binary primitive recursive function satisfying the recursion equations

$$n \dot{-} 0 = 0 \text{ and } n \dot{-} (x+1) = pred(n \dot{-} x).$$

The absolute difference between two natural numbers is the primitive recursive function defined by

$$|m-n| := (n \dot{-} m) + (m \dot{-} n).$$

**2.1.5 Exercise** Show that addition, multiplication and exponentiation of natural numbers are primitive recursive.

## 2.2 Primitive Recursive Relations

In the following we identify $n$-ary relations on a set $N$ with subsets of $N^n$.

**2.2.1 Definition** An $n$-ary relation $R \subseteq \mathbb{N}^n$ is *primitive recursive* if and only if its characteristic function

$$\chi_R(z_1, \ldots, z_n) := \begin{cases} 1 & \text{if } (z_1, \ldots, z_n) \in R \\ 0 & \text{otherwise} \end{cases}$$

is primitive recursive.

**2.2.2 Lemma** *The primitive recursive relations are closed under the boolean operations $\neg$, $\wedge$ and $\vee$, bounded quantification and substitution with primitive recursive functions.*

*Proof* Closure under boolean operations is obvious since $\chi_{\neg P} = \overline{sg} \circ \chi_P := Sub(\overline{sg}, \chi_P)$, $\chi_{P \wedge Q} = \chi_P \cdot \chi_Q := Sub(\cdot, \chi_P, \chi_Q)$ and $\chi_{P \vee Q} = sg(\chi_P + \chi_Q) := Sub(sg, Sub(+, \chi_P, \chi_Q))$. To show closure under bounded quantification let $P$ by an $n+1$-ary relation and define

$$Q(x, z_1, \ldots, z_n) \;:\Leftrightarrow\; (\exists y \leq x) P(y, z_1, \ldots, z_n).$$

Then

$$\chi_Q(x, z_1, \ldots, z_n) = sg(\sum_{i=0}^{x} \chi_P(i, z_1, \ldots, z_n))$$

which shows that the primitive recursive relations are closed under bounded existential quantification. Because of the closure under negation this also entails the closure under bounded universal quantification. The closure under substitution with primitive recursive functions follow directly from the fact that the primitive recursive functions are closed under substitutions. □

Using the closure properties of primitive recursive functions we recognize many of the familiar relations on the natural numbers as primitive recursive. Since $x = y \Leftrightarrow |x - y| = 0$ we obtain $\chi_=(m, n) = \overline{sg}(|m - n|)$. Equality is therefore a primitive recursive relation. More primitive recursive relations are listed in Table 2.1.

The next aim is to show that there is a primitive recursive coding machinery for tuples of natural numbers. Let $\mathbb{N}^* := \bigcup_{n \in \mathbb{N}} \mathbb{N}^n$ denote all finite tuples of natural numbers. A coding machinery is an one-to-one mapping

$$\langle\,\rangle : \mathbb{N}^* \xrightarrow{1-1} \mathbb{N}$$

for which we write $\langle x_1, \ldots, x_n \rangle$ instead of $\langle\,\rangle(x_1, \ldots, x_n)$. A coding machinery induces a relation

$$Seq := rng(\langle\,\rangle),$$

the length function satisfying

**Table 2.1** Some primitive recursive relations

| Relation | Notion | Definition |
|---|---|---|
| Equality | $x = y$ | $\chi_=(x, y) = \overline{sg}(\lvert x - y \rvert)$ |
| Less or equal than | $x \leq y$ | $(\exists z \leq y)[y = x + z]$ |
| Less than | $x < y$ | $x \leq y \wedge x \neq y$ |
| $x$ divides $y$ | $x/y$ | $(\exists z \leq y)[y = x \cdot z]$ |
| $p$ is a prime | $Prime(p)$ | $p \neq 0 \wedge p \neq 1 \wedge (\forall z \leq p)[\neg(z/p) \vee z = 1 \vee z = p]$ |

$$Seq(x) \;\Rightarrow\; lh(x) = \min\{n \,|\, \langle\,\rangle^{-1}(x) \in \mathbb{N}^n\}$$

and the decoding functions $(x)_i$ satisfying

$$(\langle x_0,\ldots,x_n\rangle)_i = x_i$$

for all $i \leq n$. We call a coding machinery primitive recursive if all restrictions $\langle\,\rangle \restriction \mathbb{N}^n$ of the encoding function $\langle\,\rangle$, the induced relation *Seq*, the length function *lh* and the decoding functions are all primitive recursive. The elements of *Seq* are the *sequence numbers*.

Let $Pnb(k)$ denote the $k$th prime. Defining

$$\langle\,\rangle = 0 \text{ and } \langle x_0,\ldots,x_n\rangle = \prod_{i=0}^{n} Pnb(i)^{x_i+1} \tag{2.1}$$

we obtain an one-to-one map from $\mathbb{N}^*$ into $\mathbb{N}$ by the unique prime decomposition of natural numbers. This induces a primitive recursive coding machinery (cf. Exercise 2.2.8).

We define the *concatenation* of sequence numbers by

$$\langle x_0,\ldots,x_m\rangle^\frown\langle y_0,\ldots,y_n\rangle := \langle x_0,\ldots,x_m,y_0,\ldots,y_n\rangle.$$

For every function $F\colon \mathbb{N}^{n+1} \longrightarrow \mathbb{N}$ we obtain its *course-of-values* function $[F]$ satisfying the recursion equations

$$[F](0,z_1,\ldots,z_n) := \langle\,\rangle \text{ (the empty sequence)}$$

and

$$[F](i+1,z_1,\ldots,z_n) := [F](i,z_1,\ldots,z_n)^\frown\langle F(i,z_1,\ldots,z_n)\rangle.$$

Observe that $[F](n+1) = \langle F(0),\ldots,F(n)\rangle$.[1] It follows from the recursion equations that $[F]$ is primitive recursive if $F$ is primitive recursive. The converse is also true since $F(x) = ([F](x+1))_x$. So $[F]$ is primitive recursive if and only if $F$ is primitive recursive.

**2.2.3 Theorem** *(Course-of-values recursion) Let G be an $n+2$-ary function. There is a uniquely determined function F satisfying the equation*

$$F(x,z_1,\ldots,z_n) = G(x,[F](x,z_1,\ldots,z_n),z_1,\ldots,z_n).$$

*If G is primitive recursive then F is also primitive recursive.*

*Proof* We obtain the uniqueness by a simple induction on $x$. For the second claim assume that $G$ is primitive recursive. It suffices to show that $[F]$ is primitive recursive. According to the recursion equations for $[F]$ we obtain $[F](i+1,z_1,\ldots,z_n) = [F](i)^\frown\langle G(i,[F](i,z_1,\ldots,z_n),z_1,\ldots,z_n)\rangle$ which entails that $[F]$ is primitive recursive. □

---

[1] If we regard a natural number $n+1$ as the set $\{0,\ldots,n\}$ then the image of $n+1$ under $F$ is the set $F[n+1] = \{F(i) \,|\, i \leq n\}$. Therefore we sometimes use the notions $[F](n)$ and $F[n]$ interchangeable if the ordering of the values of $F$ below $n$ is unimportant.

The main application of course-of-values recursion is the definition of relations.

**2.2.4 Definition** We say that a relation $Q$ is *primitive recursive in* the relations $P_1,\dots,P_n$ if there is a primitive recursive relation $R$ such that

$$\vec{x} \in Q \Leftrightarrow R(\chi_{P_1}(\vec{x}),\dots,\chi_{P_n}(\vec{x}),\vec{x}).$$

For short we sometimes (in abuse of notation) express that by $Q = R(P_1,\dots,P_n)$.

**2.2.5 Observation** *If a relation $Q$ is obtained from relations $P_1,\dots,P_n$ by Boolean operations, bounded quantification and substitution with primitive recursive functions, then $Q$ is primitive recursive in $P_1,\dots,P_n$.*

**2.2.6 Theorem** *(Course-of-values recursion for relations) Assume that $R$ is a primitive recursive $n+2$-ary relation. Then there is a uniquely determined primitive recursive $n+1$-ary relation $Q$ satisfying*

$$Q(k,x_1,\dots,x_n) \Leftrightarrow R([\chi_Q](k,x_1,\dots,x_n),k,x_1,\dots,x_n).$$

*Proof.* We use course-of-values recursion to define the characteristic function of $Q$. We obtain $\chi_Q(k,z_1,\dots,z_n) = \chi_R([\chi_Q](k,z_1,\dots,z_n),k,z_1,\dots,z_n)$. Then $\chi_Q$ is primitive recursive according to Theorem 2.2.3. The uniqueness of $Q$ follows by an easy induction on $k$. □

Theorem 2.2.6 opens the possibility to define primitive recursive relations implicitly with the proviso that only smaller arguments must be used in the defining part. We will use that on many occasions.

We will freely use the $\lambda$-notation for functions. I.e., if $f$ is a function with $n$ arguments then $\lambda x_i \,.\, f(x_1,\dots,x_n)$ denotes the function $x_i \longmapsto f(x_1,\dots,x_n)$. Similarly if $t$ is some term containing a free variable $x$ we denote by $\lambda x \,.\, t(x)$ the function $x \longmapsto t(x)$.

**2.2.7 Exercise** Let $P$ be an $n+1$-ary primitive recursive relation. Show that the function defined by the bounded search operator

$$\mu y \le x \,.\, P(y,z_1,\dots,z_n) := \begin{cases} \min\{y \le x \mid P(y,z_1,\dots,z_n)\} & \text{if this exists} \\ x+1 & \text{otherwise} \end{cases}$$

is primitive recursive.

**2.2.8 Exercise** Show that the encoding function defined in (2.1) induces a primitive recursive coding machinery.

Hint: Use bounded search to show that $Pnb$ is a primitive recursive function.

**2.2.9 Exercise** Show that the concatenation function is primitive recursive.

**2.2.10 Exercise** Prove Observation 2.2.5.

# Chapter 3
# Ordinals

Ordinals play a predominant role in proof-theoretical research. We need ordinals to measure infinitary objects and the run time of infinitary processes. An example of an infinitary process is the stepwise construction of fixed-points in Sect. 6.3, an example for infinitary objects are infinite well-founded trees as introduced in Sect. 5.

This chapter is purposely misplaced. Ordinals are not needed before Sect. 4.4 where we have to measure the complexity of infinite trees. Even there only a superficial knowledge of ordinals is needed. The chapter on inductive definitions (Chap. 6) requires a little more ordinal theoretic background. Only starting from Chap. 7 a more profound knowledge of ordinals and ordinal notations is needed.

However, we found no better place for this chapter. In a first reading this chapter can, therefore, be omitted and revisited later, when ordinals start to play their predominant role.

## 3.1 Heuristics

Ordinals generalize the finite ordinals, first, second, third, ... into the transfinite. A natural number has two aspects. One aspect is that of a cardinal which measures quantity or size. In that sense the natural number $n$ can be viewed as a representative for the finite sets having exactly $n$ elements. The second aspect is that of a finite ordinal, which represents the process of counting. Counting a set means to order its elements according to the way we count them. We take the first, second, etc. For finite sets, however, the difference between the cardinal and ordinal aspects is not immediately visible because there is, up to isomorphism, only one way to order a finite set. Therefore all counting processes for a finite set will lead to the same "ordinal" which is in fact its "cardinality". When passing to infinite sets the situation changes dramatically. As an example, take the set of natural numbers and count according to the canonical ordering of its elements starting with $0, 1, \ldots$ etc. Now change the way of counting and start with $1, 2, \ldots$, and count 0 as the last element. This leads to an ordering that obviously cannot be isomorphic to the canonical one.

The canonical ordering possesses no last element while the new ordering has 0 as its last element. The segment below 0 in the new ordering, which is 1, 2, ..., is an ordering which is apparently isomorphic to the canonical ordering of the natural numbers. The new ordering has therefore "one element more" than the canonical ordering. If we take $\omega$ as a symbol for the canonical ordering of $\mathbb{N}$, the new ordering can be characterized as $\omega + 1$, i.e., we have counted one step behind the canonical ordering of $\mathbb{N}$. This procedure can be iterated by ordering the natural number as

$$2, 3, 4, \ldots, 0, 1,$$

an ordering that can be characterized as $\omega + 2$, or more generally

$$n, n+1, \ldots, 0, 1, \ldots, n-1$$

which can be characterized as $\omega + n$. We may even order the natural numbers by first taking all even numbers in their canonical ordering and then all the odd numbers, i.e.,

$$0, 2, 4, \ldots, 2n, \ldots, 1, 3, 5, \ldots, 2n+1, \ldots,$$

and thus obtain an ordering that can be characterized as $\omega + \omega$ etc.

Observe that we never changed the basic set, which means that the cardinal aspect of all the ordered sets $\omega$, $\omega + 1$, $\omega + \omega$ remains unchanged.

On the other hand not every ordering is suited for counting. Take for example the non-negative rational numbers $\mathbb{Q}^+$ in their canonical ordering. Here we can start with 0, but there is no next element in the canonical ordering because it is a dense ordering, i.e., between 0 and any positive rational number $q > 0$ there is a rational number $q_0$ such that $0 < q_0 < q$. Therefore the canonical ordering of $\mathbb{Q}^+$ is not suited for counting. Only those orderings are suited for counting that have the property that, after having counted many elements arbitrarily, the remaining set of not yet counted elements possesses a least element, provided that it is not empty. Such orderings are called well-orderings. To be more precise, a linear ordering is a pair $(A, \prec)$ where $A$ is a non empty set, the field of the ordering, and $\prec \subseteq A \times A$ is a binary relation satisfying the following conditions:

- $(\forall x \in A)[\neg x \prec x]$ (irreflexivity)
- $(\forall x \in A)(\forall y \in A)(\forall z \in A)[x \prec y \wedge y \prec z \Rightarrow x \prec z]$ (transitivity)
- $(\forall x \in A)(\forall y \in A)[x \prec y \vee y \prec x \vee x = y]$. (linearity)

A linear ordering $(A, \prec)$ is a *well-ordering* if it is also well-founded, i.e., if it also satisfies

- $(\forall X)\big[X \subseteq A \wedge X \neq \emptyset \Rightarrow (\exists y \in X)(\forall x \in A)[x \prec y \Rightarrow x \notin X]\big]$.

The jutting property of well-founded relations is that they admit induction. By induction on a relation $(A, \prec)$ we mean the scheme

$$(\forall x \in A)[(\forall y \prec x)F(y) \Rightarrow F(x)] \quad \Rightarrow \quad (\forall x \in A)F(x). \tag{3.1}$$

The property expressed in (3.1) is obvious. If we assume $(\exists x \in A)\neg F(x)$ we obtain a least element $a \in \{x \in A \mid \neg F(x)\}$. But then $(\forall y \prec a)F(y)$ which, according to the premise of (3.1), would imply $F(a)$ contradicting $a \in \{x \in A \mid \neg F(x)\}$.

A closer look at (3.1) shows that it is essentially the contraposition of well-foundedness.[1] Therefore a relation is well-founded if and only if it allows induction. Another equivalent formulation is the finiteness of descending sequences, which means that a relation $(A, \prec)$ is well-founded if and only if there is no infinite descending sequence $x_0 \succ x_1 \succ \cdots \succ x_n \succ x_{n+1} \succ \cdots$.

This equivalence is again close to trivial. A relation $(A, \prec)$ with an infinite descending sequence is clearly not well-founded. On the other hand, if $(A, \prec)$ is not well-founded we have a non empty subset $B \subseteq A$ which has no $\prec$-least element. Choosing $x_0 \in B$ there is an $x_1 \in B$ such that $x_1 \prec x_0$. Having chosen a descending sequence $x_n \prec \cdots \prec x_0$ of elements in $B$ we still find some $x_{n+1} \in B$ such that $x_{n+1} \prec x_n$. Iterating this procedure we obtain an infinite descending sequence.

We call two orderings $(A_1, \prec_1)$ and $(A_2, \prec_2)$ *equivalent* if there is an order preserving map from $A_1$ onto $A_2$. This is obviously an equivalence relation on orderings. An equivalence class of orderings is called an *order-type*. As we have seen, every counting process well-orders a set and, vice versa, every well-ordering is suited to count the elements of its field. Therefore we call the equivalence class of a well-ordering an *ordinal*. Our previous examples $\omega$, $\omega + n$, $\omega + \omega$ can therefore be understood as ordinals. If $\alpha$ and $\beta$ are ordinals we define an order relation $\alpha < \beta$ iff there is a well-ordering $(A, \prec_A) \in \alpha$ and a well-ordering $(B, \prec_B) \in \beta$ such that $(A, \prec_A)$ is isomorphic to a proper initial segment of $(B, \prec_B)$, i.e., if there is an element $b \in B$ such that $(A, \prec_A)$ is isomorphic to $(B, \prec)\restriction b := (\{x \in B \mid x \prec_B b\}, \prec_B)$. It is not difficult to check that this definition is independent of the representatives $(A, \prec_A) \in \alpha$ and $(B, \prec_B) \in \beta$. Moreover it can be shown that the relation $<$ well-orders the class $On$ of all ordinals. For an ordinal $\alpha \in On$ we see that the well-ordering $(On, <)\restriction \alpha = (\{\xi \mid \xi < \alpha\}, <)$ is a representative of $\alpha$. Therefore we can choose $(\{\xi \mid \xi < \alpha\}, <)$ as a canonical representative for $\alpha$.

By $\omega_1$ we denote the least ordinal which cannot be represented by a well-ordering of the natural numbers. The least ordinal which cannot be represented by a decidable well-ordering of the natural number is commonly denoted by $\omega_1^{CK}$ and pronounced as $\omega_1$–CHURCH–Kleene in honor of these pioneers of computability theory.

From a set-theoretical point of view the just described approach is, however, problematic. The order-type of a well-ordering is, in general, not a set but a proper class in the sense defined below. To avoid this difficulty in a set theoretical framework one represents ordinals by canonical representatives, i.e., defines ordinals in such a way that $\alpha := \{\xi \mid \xi < \alpha\}$. Then ordinals become sets. This can be obtained by requiring that an ordinal is a set that is well-ordered by the membership relation $\in$. Regarding ordinals in the set theoretic sense has become so common in mathematical logic that we will adopt this standpoint. It has many technical advantages. We cannot present the complete set theoretical background. The background needed for this text is, however, very small. It does not exceed the

[1] With $X := \{x \in A \mid \neg F(x)\}$.

usual background set theory which is ubiquitous in mathematics. We will give all proofs on the basis of a naive set theory. There we will talk naively about classes and sets. Every set is also a class. A class is a set if it is a member of another class. Classes which are not sets are called proper classes. The reader who wants to know more about the set theoretical background should consult the first chapters of an introductory text book on set theory, e.g., [58].

## 3.2 Some Basic Facts about Ordinals

As mentioned at the end of the previous section, we will regard ordinals in the set theoretic sense. We assume that an ordinal is a transitive set which is well-ordered by the membership relation $\in$, i.e., we define

$$\alpha \in On \ :\Leftrightarrow\ Tran(\alpha) \wedge (\alpha, \in) \text{ is well-ordered} \tag{3.2}$$

where

$$Tran(M) \ :\Leftrightarrow\ (\forall x \in M)(\forall y \in x)[y \in M]$$

denotes the fact that $M$ is a transitive class, i.e., a class without gaps. By $On$ we denote the class of ordinals. We define

$$\alpha < \beta \ :\Leftrightarrow\ \alpha \in On \wedge \beta \in On \wedge \alpha \in \beta$$

and use mostly lower case Greek letters to denote ordinals. As an immediate consequence of the definition we obtain

$$\alpha \in On \ \Rightarrow\ Tran(\alpha) \wedge (\forall x \in \alpha)[Tran(x)]. \tag{3.3}$$

To see (3.3), we observe that $\alpha \in On$ entails $Tran(\alpha)$ by definition and $z \in y \in x \in \alpha$ implies $z \in x$ because $\alpha$ is well-ordered by $\in$. Hence $Tran(x)$, which proves (3.3). □

For $x \in \alpha \in On$ we get by (3.3) $Tran(x)$ and $x \subseteq \alpha$. But then $x$ is also well-ordered by $\in$ and therefore an ordinal. So we have shown

$$\alpha \in On \wedge x \in \alpha \ \Rightarrow\ x \in On, \text{ i.e., } Tran(On) \tag{3.4}$$

and obtain

$$\alpha \in On \ \Rightarrow\ \alpha = \{\beta \mid \beta < \alpha\} \tag{3.5}$$

as an immediate consequence.

In set theory we commonly assume that the universe is well-founded with respect to the membership relation. This is expressed by the foundation scheme

$$(\exists x)F(x) \ \rightarrow\ (\exists x)\big[F(x) \wedge (\forall y \in x)[\neg F(y)]\big].$$

In the presence of the foundation scheme, also the opposite direction of (3.3) holds true.

**3.2.1 Lemma** *Assume that the membership relation $\in$ well-founded, i.e., that the foundation scheme holds true. Then $\alpha$ is an ordinal if and only if $\alpha$ is a hereditarily transitive set.*

*Proof* One direction of the claim is (3.3). For the opposite direction assume that $\alpha$ is hereditarily transitive, i.e., $Tran(\alpha) \land (\forall x \in \alpha)[Tran(x)]$. By the foundation scheme $\in$ is irreflexive and well-founded on $\alpha$. Since $\alpha$ is hereditarily transitive the membership relation is also transitive on $\alpha$. It remains to check linearity. Assume that $\beta$ is also hereditarily transitive. We show

$$\text{“If } \beta \text{ is well-ordered by} \in \text{and } \alpha \subseteq \beta \text{ then } \alpha = \beta \lor \alpha \in \beta\text{”.} \tag{3.6}$$

To prove (3.6) assume $\alpha \neq \beta$ and let $\xi$ be the $\in$-minimal element in $\beta \setminus \alpha$. Then $\xi \subseteq \alpha$. For $\eta \in \alpha$ we get $\xi \notin \eta$ as well as $\eta \neq \xi$ and thus $\eta \in \xi$ because $\beta$ is linearly ordered by $\in$. Hence $\alpha = \xi \in \beta$. This finishes the proof of (3.6).

Observe that the contraposition of the foundation scheme is the scheme of $\in$-induction

$$(\forall x)[(\forall y \in x)F(y) \to F(x)] \quad \to \quad (\forall x)F(x).$$

We prove

$$\text{“If } \alpha \text{ is hereditarily transitive then } \alpha \text{ is linearly ordered by } \in\text{”}$$

by $\in$-induction. Let $\xi, \eta \in \alpha$ such that $\xi \neq \eta$. Then $\gamma := \xi \cap \eta$ is hereditarily transitive. If $\gamma = \xi \neq \eta$ we get $\xi = \gamma \in \eta$ by the induction hypothesis and (3.6). If $\gamma \neq \xi$ we get $\gamma \in \xi$ by (3.6) and the induction hypothesis. In case of $\gamma = \eta$ we are done and the remaining case $\gamma \in \eta$ is excluded since this would entail $\gamma \in \eta \cap \xi = \gamma$ which contradicts foundation. So we have seen that in the presence of the foundation scheme the ordinals are exactly the hereditarily transitive sets. □

Observe further that (3.6) gives the direction from left to right in the following useful equivalence

$$\alpha \in On \land \beta \in On \;\Rightarrow\; (\alpha \subseteq \beta \;\Leftrightarrow\; \alpha \leq \beta).$$

The other direction holds trivially since $\alpha < \beta$ implies $\alpha \subseteq \beta$ by the transitivity of $\beta$.

But even if we do not assume the foundation scheme we obtain the following fact from the proof of Lemma 3.2.1.

**3.2.2 Lemma** *Assume that $\in$ is well-founded on a set a. Then a is an ordinal if and only if a is hereditarily transitive.*

**3.2.3 Theorem** *The class On is well-ordered by $\in$.*

*Proof* It follows from the definition that the relations $\in$ and $<$ coincide on ordinals. Any infinite descending sequence $\alpha_0 > \alpha_1 \cdots$ of ordinals would also be an infinite descending sequence in $\alpha_0$ contradicting the well-foundedness of $\alpha_0$. Therefore $\in$ is well-founded on $On$. Then $\in$ is also irreflexive on $On$ and transitive because all elements of $On$ are transitive. It remains to check linearity. For $\alpha, \beta \in On$ we obtain

$\gamma := \alpha \cap \beta$ as an $\in$-well-founded hereditarily transitive set, i.e., $\gamma \in On$. By (3.6) we have $\gamma \le \alpha$ and $\gamma \le \beta$. If $\gamma \ne \alpha$ and $\gamma \ne \beta$ we get $\gamma \in \alpha \cap \beta = \gamma$ which is impossible. Therefore $\alpha = \gamma \le \beta$ or $\beta = \gamma \le \alpha$ and we are done. □

Observe that $On$ is a proper class. Since $On$ is hereditarily transitive and well-founded by $\in$ the assumption that $On$ is a set would lead to the contradiction $On \in On$ by Lemma 3.2.2.

By Theorem 3.2.3 we have the principle of induction for $On$. This principle is commonly called *transfinite induction*.

**3.2.4 Lemma** *Let $M \subseteq On$ be transitive. Then $M \in On$ or $M = On$.*

*Proof* Since $M \subseteq On$ the class $M$ is well-ordered by $<$ and all its members are transitive, therefore $M$ is hereditarily transitive. If $M$ is a set we get $M \in On$ by Lemma 3.2.2. Otherwise we show $On \subseteq M$ by transfinite induction. Let $\eta \in On$ and assume by induction hypothesis $\eta \subseteq M$. Since $M$ is well-ordered by $\in$ we obtain $\eta = M$ or $\eta \in M$ by (3.6). But $\eta = M$ is excluded because $M$ is not a set. Hence $\eta \in M$. □

**3.2.5 Lemma** *Let $M \subseteq On$ be a set. Then*

$$\sup M := \min\{\xi \in On \mid (\forall \alpha \in M)[\alpha \le \xi]\}$$

*exists and*

$$\sup M = \bigcup M := \{\xi \mid (\exists \alpha \in M)[\xi \in \alpha]\}.$$

We call $\sup M$ the *supremum* of the set $M$.

*Proof* Let $\beta := \bigcup M$. Then $\beta$ is a hereditarily transitive subset of $On$. Hence $\beta \in On$ by Lemma 3.2.4. For $\alpha \in M$ we obtain $\alpha \subseteq \beta$, which implies $\alpha \le \beta$ by equation (3.6). Hence $\{\xi \mid (\forall \alpha \in M)[\alpha \le \xi]\} \ne \emptyset$. So $\sup M$ exists and $\sup M \le \beta$. For $\xi < \beta$ there is an $\eta \in M$ such that $\xi < \eta$ that also proves $\beta \le \sup M$. □

To gain a feeling for ordinals in the set theoretical setting we enumerate the first ordinals. The least transitive set whose members are also transitive is apparently the empty set $\emptyset$. If $x$ is hereditarily transitive we see immediately that $x \cup \{x\}$ is again a hereditarily transitive set. Therefore we get the first ordinals as

$$0 = \emptyset,\ \ 1 := 0 \cup \{0\} = \{0\},\ \ 2 = 1 \cup \{1\} = \{0,1\},\ \ 3 = \{0,1,2\}, \ldots.$$

**3.2.6 Definition** Let

$$\alpha' := \min\{\beta \mid \alpha < \beta\}.$$

We call $\alpha'$ the *successor* of $\alpha$. It is easily checked that $\alpha' = \alpha \cup \{\alpha\}$. Let

$$Lim := \{\alpha \in On \mid \alpha \ne 0 \land (\forall \beta < \alpha)[\beta' < \alpha]\}$$

denote the class of *limit ordinals*. The existence of limit ordinals is secured by the axiom of infinity. We define

$$\omega := \min Lim$$

and call an ordinal *finite* if it is less than $\omega$.

There are three types of ordinals:

- The ordinal 0
- Successor ordinals of the form $\alpha'$
- Limit ordinals

According to the three types of ordinals we can reformulate transfinite induction on ordinals as

$$F(0) \wedge (\forall\alpha)[F(\alpha) \Rightarrow F(\alpha')] \wedge (\forall\lambda \in Lim)[(\forall\xi < \lambda)F(\xi) \Rightarrow F(\lambda)] \Rightarrow (\forall\alpha)F(\alpha).$$

Another important principle is the principle of *transfinite recursion*, which generalizes primitive recursion into the transfinite. To formulate the principle of transfinite recursion we recall some notions.

A (class) relation (in the set theoretical sense) is a class of ordered pairs. The field of a relation $R$ is defined as

$$field(R) := \{x \mid (\exists y)[(x,y) \in R \vee (y,x) \in R]\}.$$

A (class) function $F$ is a relation in which the second component is uniquely determined by the first. We use the familiar notations:

- $F(a)$ is the uniquely determined element such that $(a,F(a)) \in F$.
- The *domain* $dom(F)$ of a function $F$ is the class $\{a \mid (\exists y)[(a,y) \in F]\}$

and dually

- The *range* $rng(F)$ of $F$ is the class $\{y \mid (\exists x)[(x,y) \in F]\}$.
- The *restriction of $F$ to a set $a$* is the set $F{\restriction}a := \{(x,y) \mid x \in a \wedge (x,y) \in F\}$.

We use the letter $V$ to denote the universe, i.e., the class of all sets. By $F: N \longrightarrow_p M$ we denote that $F$ is a partial function from $N$ into $M$, i.e., $dom(F)$ might be a proper subclass of $N$.

**3.2.7 Theorem** *(Transfinite recursion on ordinals) Let G be a function mapping sets to sets. Then there is a uniquely defined function $F: On \longrightarrow V$ such that*

$$F(\alpha) = G(F{\restriction}\alpha).$$

*Proof* We prove the theorem on the basis of naive set theory. Let

$$M := \{f \mid f\colon On \longrightarrow_p V \wedge dom(f) \in On \wedge (\forall \alpha \in dom(f))[f(\alpha) = G(f{\restriction}\alpha)]\}. \quad \text{(i)}$$

Then we obtain

$$f \in M \wedge g \in M \wedge \alpha \in dom(f) \cap dom(g) \;\Rightarrow\; f(\alpha) = g(\alpha) \quad \text{(ii)}$$

by induction on $\alpha$. By the induction hypothesis we have $f{\restriction}\alpha = g{\restriction}\alpha$ and obtain thus $f(\alpha) = G(f{\restriction}\alpha) = G(g{\restriction}\alpha) = g(\alpha)$. Now we define

$$F := \{(\alpha, b) \mid (\exists f \in M)[f(\alpha) = b].\} \quad \text{(iii)}$$

The class $F$ is a function by (ii). We have to show

$$(\forall \alpha \in On)(\exists b)[F(\alpha) = b] \wedge F{\restriction}\alpha' \in M. \quad \text{(iv)}$$

We prove (iv) by induction on $\alpha$. Let $\alpha \in On$. By the induction hypothesis we have $f_0 := F{\restriction}\alpha \in M$. Now we define

$$f(\beta) := \begin{cases} f_0(\beta) & \text{if } \beta < \alpha \\ G(f_0{\restriction}\alpha) & \text{if } \beta = \alpha. \end{cases}$$

Then $dom(f) = \alpha' \in On$ and for all $\beta \le \alpha$ we have $f(\beta) = G(f{\restriction}\beta)$, i.e., $f \in M$. Therefore $F(\alpha) = f(\alpha)$ and $F{\restriction}\alpha' = f$ and we are done. For the uniqueness of $F$ we assume that $H$ is a second function satisfying the properties of the theorem and show $F(\alpha) = H(\alpha)$ straight forwardly by induction on $\alpha$. □

According to the three different types of ordinals we may again reformulate the principle of transfinite recursion in the following form.

*Let a be a set and S and L functions which map sets to sets. Then there is a uniquely defined function $F\colon On \longrightarrow V$ satisfying*

$$F(0) = a$$
$$F(\alpha') = S(F(\alpha))$$
$$\lambda \in Lim \Rightarrow F(\lambda) = L(F{\restriction}\lambda).$$

The principle of transfinite recursion can be extended to well-founded relations $(A, \prec)$ where we have to require that for $a \in A$ the class $\prec{\restriction}a := \{b \in A \mid b \prec a\}$ is always a set.

Theorem 3.2.7 holds also in the more general setting stated later.

**3.2.8 Theorem** *Let $(A, \prec)$ be a well-founded relation and $G\colon V \longrightarrow V$ a function. Then there is a uniquely determined function $F$ satisfying*

$$(\forall x \in A)[F(x) = G(F{\restriction}\{y \in A \mid y \prec x\})].$$

The proof is essentially the same as that of Theorem 3.2.7. However, since $\prec$ is not supposed to be transitive a rigid proof requires some extra prerequisites such as the definition of the transitive closure of $(A, \prec)$. Spelling out all these details would lead us too far into set theory. Therefore we omit the proof. A rigid proof can be found practically in any elementary textbook on set theory.[2]

**3.2.9 Definition** For a well-founded binary relation $\prec$ and $s \in \mathit{field}(\prec)$ we define

$$\mathrm{otyp}_{\prec}(s) = \sup\{\mathrm{otyp}_{\prec}(t)' \mid t \prec s\}$$

by transfinite recursion. It follows by induction on $\prec$ that $\mathrm{otyp}_{\prec}(s) \in On$ for all $s \in \mathit{field}(\prec)$. We call $\mathrm{otyp}_{\prec}(s)$ the *order-type of s in* $\prec$. The order-type of the relation $\prec$ is

$$\mathrm{otyp}(\prec) := \sup\{\mathrm{otyp}_{\prec}(s)' \mid s \in \mathit{field}(\prec)\}.$$

We define

$$\omega_1^{CK} := \sup\{\mathrm{otyp}(\prec) \mid \prec \subseteq \omega \times \omega \text{ is primitive recursive}\}.$$

Also by transfinite recursion we define for a well-founded relation $\prec$ the *Mostowski collapsing function*

$$\pi_{\prec}: \mathit{field}(\prec) \longrightarrow V$$
$$\pi_{\prec}(s) := \{\pi_{\prec}(t) \mid t \prec s\}.$$

The *Mostowski collapse* of $\prec$

$$\pi_{\prec}[\mathit{field}(\prec)] := \{\pi_{\prec}(s) \mid s \in \mathit{field}(\prec)\}$$

is obviously a transitive set.

**3.2.10 Lemma** *Let $\prec$ be a well-founded and transitive binary relation. Then $\pi_{\prec}(s) = \mathrm{otyp}_{\prec}(s)$ holds true for all $s \in \mathit{field}(\prec)$ and $\mathrm{otyp}(\prec) = \pi_{\prec}[\mathit{field}(\prec)]$.*

*Proof* First we show that $\pi_{\prec}(s)$ is an ordinal. Since $\prec$ is well-founded and $\pi_{\prec}$ is a $\prec$-$\in$ homomorphism the set $\pi_{\prec}(s)$ is well-founded by $\in$. For $x \in y \in \pi_{\prec}(s)$ there is a $t \prec s$ and a $t_0 \prec t$ such that $y = \pi_{\prec}(t)$ and $x = \pi_{\prec}(t_0)$. Since $\prec$ is transitive we have $t_0 \prec s$ and therefore $x \in \pi_{\prec}(s)$. This shows that $\pi_{\prec}(s)$ is transitive for all $s \in \mathit{field}(\prec)$ which implies that $\pi_{\prec}(s)$ is hereditarily transitive and thus an ordinal by Lemma 3.2.2. Next we show

$$\pi_{\prec}(s) = \mathrm{otyp}_{\prec}(s) \tag{i}$$

by induction along $\prec$. For $\alpha \in \mathrm{otyp}_{\prec}(s)$ we find a $t \prec s$ such that $\alpha \leq \mathrm{otyp}_{\prec}(t) = \pi_{\prec}(t) < \pi_{\prec}(s)$. Hence $\alpha \in \pi_{\prec}(s)$ and thus $\mathrm{otyp}_{\prec}(s) \leq \pi_{\prec}(s)$. Conversely $\alpha \in \pi_{\prec}(s)$ implies $\alpha = \pi_{\prec}(t) = \mathrm{otyp}_{\prec}(t) < \mathrm{otyp}_{\prec}(s)$ for some $t \prec s$ which shows $\pi_{\prec}(s) \leq \mathrm{otyp}_{\prec}(s)$. For the second claim we obviously have $\pi_{\prec}[\mathit{field}(\prec)] \subseteq \mathrm{otyp}(\prec)$. To obtain also the opposite inclusion assume $\alpha \in \mathrm{otyp}(\prec)$. Then there is an $s \in \mathit{field}(\prec)$ such that $\alpha \leq \mathrm{otyp}_{\prec}(s) = \pi_{\prec}(s)$. Hence $\alpha \in \pi_{\prec}[\mathit{field}(\prec)]$. □

For a well-ordering $\prec$ and $S \subseteq \mathit{field}(\prec)$ we denote by $\mu_{\prec}S$ the $\prec$-least element in $S$.

---

[2] Cf. also Theorem 11.4.10 in Chap. 11.

**3.2.11 Lemma** *Let $\prec$ be a well-ordering. Then the functions* $\text{otyp}_{\prec}$ *and* $\pi_{\prec}$ *coincide, are order preserving, hence one-one, and also onto on* $\text{otyp}(\prec)$. *The inverse function* $en_{\prec} := \text{otyp}_{\prec}^{-1}$ *satisfies*

$$en_{\prec}\colon \text{otyp}(\prec) \longrightarrow \mathit{field}(\prec)$$
$$en_{\prec}(s) := \mu_{\prec}\{t \in \mathit{field}(\prec) \mid (\forall x \prec s)[en_{\prec}(x) \prec t]\}$$

*and thus enumerates the elements in the field of $\prec$ increasingly. We call $en_{\prec}$ the* enumerating function *of $\prec$.*

The proof is obvious from Lemma 3.2.10 and the definitions of $\text{otyp}(\prec)$ and $\pi_{\prec}$, respectively. □

Let $M \subseteq On$. Then $(M,\in)$ is a well-ordering. Therefore all previous definitions apply to $(M,\in)$. In this case we do not mention the relation $\in$, but just talk about $\text{otyp}(M)$ and denote its enumerating function by $en_M$.

**3.2.12 Lemma** *For a class $M \subseteq On$ we have either* $\text{otyp}(M) = On$ *or* $\text{otyp}(M) \in On$. *It is* $\text{otyp}(M) \in On$ *if and only if $M$ is a set.*

*Proof* This follows from Lemma 3.2.4 since $\text{otyp}(M)$ is hereditarily transitive. □

**3.2.13 Lemma** *Let $f\colon On \longrightarrow_p On$ be an order preserving function such that $dom(f)$ is transitive. Then $a \leq f(a)$ for all $a \in dom(f)$.*

*Proof* Assume that there is an $\alpha \in dom(f)$ such that $f(\alpha) < \alpha$. Then there is a least such $\alpha$. But then $f(\alpha) \in dom(f)$ and we obtain $f(f(\alpha)) < f(\alpha) < \alpha$, contradicting the minimality of $\alpha$. □

Observe that, as a special case of Lemma 3.2.13, we always have $\alpha \leq en_M(\alpha)$ for a class $M \subseteq On$.

Let us now turn to the cardinal aspect of sets. The cardinality of a set is the number of its elements. This is difficult to grasp in an absolute way. It is easier to compare the size of sets by bringing their elements in a one-one correspondence. Therefore we call two sets $a$ and $b$ are *equivalent* if there is a one-one map from $a$ onto $b$, i.e.,

$$a \sim b \;:\Leftrightarrow\; (\exists f)\left[f\colon a \xrightarrow[1-1]{\mathit{onto}} b\right].$$

If we assume that every set can be well-ordered – a fact which is equivalent to the axiom of choice – there is a possibility to measure the size of a set by an ordinal.

**3.2.14 Definition** For a set $a$ we define

$$\overline{\overline{a}} := \min\{\beta \in On \mid (\exists f)[f\colon a \xrightarrow[1-1]{\mathit{onto}} \beta]\}$$

and call $\overline{\overline{a}}$ the cardinality of $a$.

An ordinal $\alpha$ is a cardinal iff $\overline{\overline{\alpha}} = \alpha$.

Let

$$Card := \{\kappa \mid \kappa \text{ is a cardinal.}\}$$

Since

$$a \sim b \;\Leftrightarrow\; \overline{\overline{a}} = \overline{\overline{b}}$$

we can measure the size of a set by determining its cardinal which – in some sense – corresponds to "counting its elements".

Cardinals are "initial ordinals" in the sense that there is no smaller ordinal of the same size. All finite ordinals are cardinals. Also $\omega$ is a cardinal. The cardinal $\omega^+$, i.e., the first cardinal bigger than $\omega$ is commonly denoted by $\omega_1$.

Of special interest are initial ordinals $\kappa$, which satisfy the additional closure condition that they can only be approximated by a set of smaller ordinals if this set has at least the size of $\kappa$. Such ordinals will be called regular.

**3.2.15 Definition** The class of regular ordinals is defined by

$$Reg := \{\alpha \in On \mid (\forall x)[x \subseteq \alpha \wedge \overline{\overline{x}} < \alpha \Rightarrow \sup x < \alpha]\}.$$

Observe that all regular ordinals are cardinals. It is a folklore result of set theory that the class of all cardinals as well as the class of the regular cardinals are unbounded in $On$ and thus proper classes. For any ordinal $\alpha$ there is a least cardinal $\alpha^+$ that is bigger than $\alpha$. All cardinals of the form $\alpha^+$ are regular.

**3.2.16 Definition** Let $\kappa$ be a regular ordinal. A class $M \subseteq On$ is *unbounded in* $\kappa$ if for all $\alpha < \kappa$ there is a $\beta \in M \cap \kappa$ such that $\alpha < \beta$.

**3.2.17 Definition** Let $\kappa$ be a regular ordinal. A class $M \subseteq On$ is *closed in* $\kappa$ if $\sup U \in M$ holds for all $U \subseteq M \cap \kappa$ such that $U \neq \emptyset$ and $\overline{\overline{U}} < \kappa$.

We denote by "M is $\kappa$-club" that $M$ is closed and unbounded in $\kappa$.

**3.2.18 Definition** Let $\kappa$ be a regular ordinal and $f\colon On \longrightarrow_p On$ be order preserving. We call $f$ *$\kappa$-continuous* if $dom(f)$ is closed in $\kappa$ and $\sup f[U] = f(\sup U)$ holds for all $U \subseteq dom(f) \cap \kappa$ such that $\overline{\overline{U}} < \kappa$.

An order preserving function $f\colon On \longrightarrow_p On$ is a *$\kappa$-normal function* if $f$ is $\kappa$-continuous and $\kappa \subseteq dom(f)$.

**3.2.19 Theorem** *Let $\kappa$ be a regular ordinal. A class $M \subseteq On$ is $\kappa$-club iff $en_M$ is a $\kappa$-normal function.*

*Proof* "$\Rightarrow$:" Let $M$ be $\kappa$-club. Then $M$ is unbounded in $\kappa$, which implies otyp$(M \cap \kappa) = \kappa$. Hence $\kappa \subseteq dom(en_M)$. Let $U \subseteq dom(f) \cap \kappa$ such that $\overline{\overline{U}} < \kappa$. Then we also have $\overline{\overline{en_M[U]}} = \overline{\overline{U}} < \kappa$. Since $\kappa$ is regular this implies $\sup U < \kappa$ and $\sup(en_M[U]) < \kappa$. Therefore we find an $\alpha < \kappa$ such that $\sup(en_M[U]) = en_M(\alpha)$.

For $\xi \in U$ we have $en_M(\xi) < en_M(\alpha)$ which implies $\xi < \alpha$. Hence $\sup U \leq \alpha$. Assume $\sup U < \alpha$. Then $en_M(\sup U) < en_M(\alpha) = \sup en_M[U]$ and we obtain a $\xi \in U$ such that $en_M(\sup U) < en_M(\xi)$. But then $\sup U < \xi \in U$ which is absurd.

"$\Leftarrow$:" Because of $\kappa \subseteq dom(en_M)$ we obtain $\text{otyp}(M \cap \kappa) = \kappa$. Therefore $M$ is unbounded in $\kappa$. Let $U \subseteq M \cap \kappa$ such that $\overline{\overline{U}} < \kappa$ and $A = \text{otyp}(U)$. Then $\overline{\overline{A}} = \overline{\overline{U}} < \kappa$ and, since $\kappa$ is regular, $\sup A < \kappa$. We obtain $en_M(\sup A) = \sup en_M[A] = \sup U$ which shows $\sup U \in M$. □

**3.2.20 Exercise** The open intervals $(\alpha, \beta) := \{\gamma \mid \alpha < \gamma \wedge \gamma < \beta\}$ together with the class $On$ form a basis of a topology on $On$, the *order-topology*. The order-topology induces a topology on every regular ordinal $\kappa$. Prove the following claims:

(a) A set $M \subseteq \kappa$ is closed in the sense of Definition 3.2.17 iff it is closed in the order-topology on $\kappa$.

(b) Let $M \subseteq \kappa$ be closed in $\kappa$. An order preserving function $f : M \longrightarrow \kappa$ is continuous in the sense of Definition 3.2.18 iff it is continuous in the order-topology on $\kappa$.

(c) Let $M \subseteq \kappa$ be closed in $\kappa$. Characterize the class of functions satisfying $f(sup\ U) = \sup\{f(\xi) \mid \xi \in U\}$ for all non void sets $U \subseteq M$ that are bounded in $M$.

**3.2.21 Exercise** Prove the following facts:

(a) Every regular ordinal is a cardinal.

(b) The class of cardinals is closed and unbounded.

(c) If $\lambda$ and $\kappa$ are cardinals and there is no cardinal in the interval $(\lambda, \kappa)$ then $\kappa$ is regular.

(d) The class of regular ordinals is unbounded.

## 3.3 Fundamentals of Ordinal Arithmetic

**3.3.1 Definition** For an ordinal $\alpha$ let $On_\alpha := \{\beta \in On \mid \alpha \leq \beta\}$ denote the class of ordinals $\geq \alpha$. Let $\alpha + \xi := en_{On_\alpha}(\xi)$ and call $\alpha + \beta$ the *ordinal sum* of $\alpha$ and $\beta$. Since $On_\alpha$ is obviously club in any regular $\kappa > \alpha$, the function $\lambda\xi \,.\, \alpha + \xi$ is a $\kappa$-normal function by Theorem 3.2.19.

**3.3.2 Observation** *The function $\lambda\xi \,.\, \alpha + \xi$ satisfies the following recursion equations:*

$$\alpha + 0 = \alpha$$
$$\alpha + \xi' = (\alpha + \xi)'$$
$$\alpha + \lambda = \sup_{\xi < \lambda} (\alpha + \xi) \ \textit{for} \ \lambda \in \textit{Lim}$$

We see from Observation 3.3.2 that $\alpha + \xi$ extends the addition of natural numbers into the transfinite. We easily check the following properties of ordinal addition.

**3.3.3 Theorem** *The ordinal sum satisfies the following properties:*

- $\xi < \eta \Rightarrow \alpha + \xi < \alpha + \eta$
- $\alpha \leq \alpha + \xi$ *and* $\xi \leq \alpha + \xi$
- $\alpha + (\beta + \gamma) = (\alpha + \beta) + \gamma$

*Proof* The first two properties follow directly from the fact that $\lambda\xi \,.\, \alpha + \xi$ is an enumerating function. The last property follows straight forwardly by induction on $\gamma$. □

We obtain $\alpha + 1 = (\alpha + 0') = \alpha'$ and will therefore mostly write $\alpha + 1$ instead of $\alpha'$.

**3.3.4 Definition** Put

$$\mathbb{H} := \{\alpha \in On \,|\, \alpha \neq 0 \wedge (\forall \xi < \alpha)(\forall \eta < \alpha)[\xi + \eta < \alpha]\}.$$

We call the ordinals in $\mathbb{H}$ *principal* or *additively indecomposable*.

**3.3.5 Lemma** *Principal ordinals have the following properties:*

*(1)* $\alpha \notin \mathbb{H} \ \Leftrightarrow \ (\exists \xi < \alpha)(\exists \eta < \alpha)[\alpha = \xi + \eta]$.

*(2)* $\mathbb{H} \subseteq Lim \cup \{0'\}$.

*(3)* $\{0', \omega\} \subseteq \mathbb{H}$ *and* $(0', \omega) \cap \mathbb{H} = \emptyset$.

*(4) Any infinite cardinal* $\kappa$ *is principal.*

*(5)* $\mathbb{H}$ *is* $\kappa$*-club for all regular ordinals* $\kappa > \omega$.

*Proof* By the definition of $\mathbb{H}$ we get

$$\alpha \notin \mathbb{H} \ \Leftrightarrow \ (\exists \xi \in \alpha)(\exists \eta \in \alpha)[\alpha \leq \xi + \eta]. \tag{i}$$

To obtain (1) we have to show that for $\alpha \notin \mathbb{H}$ and $\xi < \alpha$ there is an $\eta < \alpha$ such that $\alpha = \xi + \eta$. Since $\alpha \in On_\xi$ there is an $\eta_0$ such that $\alpha = \xi + \eta_0$. Together with (i) we obtain $\xi + \eta_0 = \alpha \leq \xi + \eta$ which implies $\eta_0 \leq \eta < \alpha$.

Properties (2) and (3) are obvious.

We prove (4). Let $\xi < \kappa$. Then $\overline{\overline{On_\xi \cap \kappa}} = \kappa$. Therefore we have $\xi + \eta \in On_\xi \cap \kappa$ for all $\eta < \kappa$.

To prove (5) we first show that $\mathbb{H}$ is unbounded in $\kappa$. Let $\alpha < \kappa$. Put $\alpha_0 := \alpha + 1$ and $\alpha_{n+1} = \alpha_n + \alpha_n$ and $M := \{\alpha_n \mid n \in \omega\}$. By (4) we have $M \subseteq \kappa$. Since $\omega < \kappa$ this implies $\sup M < \kappa$. For $\xi, \eta < \sup M$ there is an $n \in \omega$ such that $\xi, \eta < \alpha_n$. Hence $\xi + \eta < \alpha_n + \alpha_n = \alpha_{n+1} \leq \sup M$. Therefore we have $\alpha < \sup M \in \mathbb{H}$ and $\sup M < \kappa$.

To show that $\mathbb{H}$ is closed pick $U \subseteq \mathbb{H} \cap \kappa$ such that $\overline{\overline{U}} < \kappa$. Then $\sup U < \kappa$. If $\sup U \in U$ we are done. Otherwise $\sup U \in Lim$ and for $\xi, \eta < \sup U$ we find an $\alpha \in U \subseteq \mathbb{H}$ such that $\xi, \eta < \alpha$. Hence $\xi + \eta < \alpha \leq \sup U$ that shows $\sup U \in \mathbb{H}$. □

Let

$$\omega^\xi := en_{\mathbb{H}}(\xi). \tag{3.7}$$

Then we obtain

$$\omega^0 = 1, \quad \omega^1 = \omega \quad \text{and} \quad \lambda \in Lim \Rightarrow \omega^\lambda = \sup\{\omega^\xi \mid \xi < \lambda\}$$

and

$$\alpha < \beta \;\Rightarrow\; \omega^\alpha < \omega^\beta.$$ [3]

**3.3.6 Observation** *An ordinal $\alpha$ is additively indecomposable if and only if $\xi + \alpha = \alpha$ holds true for all $\xi < \alpha$.*

*Proof* Let $\alpha \in \mathbb{H}$. If $\alpha = 1$ the only ordinal less than 1 is 0 and $0 + 1 = 1$. Otherwise we have $\alpha \in Lim$ and obtain $\xi + \alpha = \sup\{\xi + \eta \mid \eta < \alpha\} \leq \alpha \leq \xi + \alpha$.

The opposite direction follows because for $\xi, \eta < \alpha$ we get $\xi + \eta < \xi + \alpha = \alpha$. □

The next lemma is a direct consequence of Observation 3.3.6.

**3.3.7 Lemma** *Let $\{\alpha_1, \ldots, \alpha_n\} \subseteq \mathbb{H}$. Then there are $\{k_1, \ldots, k_m\} \subseteq \{1, \ldots, n\}$ such that $k_i < k_{i+1}$, $\alpha_{k_i} \geq \alpha_{k_{i+1}}$ for $i = 1, \ldots, m$ and $\alpha_1 + \cdots + \alpha_n = \alpha_{k_1} + \cdots + \alpha_{k_m}$.*

We define

$$\begin{aligned} \alpha =_{NF} \alpha_1 + \cdots + \alpha_n \;\; :\Leftrightarrow \;\; & \alpha = \alpha_1 + \cdots + \alpha_n \wedge \{\alpha_1, \ldots, \alpha_n\} \subseteq \mathbb{H} \\ & \wedge\, \alpha_i \geq \alpha_{i+1} \text{ for } i = 1, \ldots, n-1. \end{aligned} \tag{3.8}$$

**3.3.8 Theorem** *(Cantor normal-form) For all ordinals $\alpha \neq 0$ there are uniquely determined ordinals $\alpha_1, \ldots, \alpha_n$ such that $\alpha =_{NF} \alpha_1 + \cdots + \alpha_n$.*

---

[3] Writing $\omega^\xi$ as an exponential is not by accident. (cf. Exercise 3.3.11)

*Proof* We prove the existence by induction on $\alpha$. If $\alpha \in \mathbb{H}$ then $\alpha =_{NF} \alpha$. Otherwise we have $\alpha = \xi + \eta$ with $\xi, \eta < \alpha$. By induction hypothesis we get $\xi =_{NF} \xi_1 + \cdots + \xi_m$ and $\eta =_{NF} \eta_1 + \cdots + \eta_n$. Then $\alpha =_{NF} \xi_1 + \cdots + \xi_j + \eta_1 + \cdots + \eta_n$ where $1 \leq j \leq m$ is the biggest index such that $\xi_j \geq \eta_1$.

For uniqueness we assume $\alpha =_{NF} \beta_1 + \cdots + \beta_m$ and $\alpha =_{NF} \alpha_1 + \cdots + \alpha_n$ and show $m = n$ and $\alpha_i = \beta_i$ for $i = 1, \ldots, n$ by induction on $m$. Let $\alpha_1 = \omega^\xi$ and $\beta_1 = \omega^\eta$. Then $\omega^\xi = \alpha_1 \leq \alpha < \omega^{\eta+1}$ which entails $\xi \leq \eta$. Dually we also obtain $\eta \leq \xi$. Hence $\xi = \eta$, i.e., $\alpha_1 = \beta_1$. But then $\alpha_2 + \cdots + \alpha_m = \beta_2 + \cdots + \beta_n$ which by induction hypothesis implies $m = n$ and $\alpha_i = \beta_i$ for $i = 2, \ldots, n$. □

Using the Cantor normal-form the following observation is obvious.

**3.3.9 Observation** *If $\alpha =_{NF} \alpha_1 + \cdots + \alpha_m$ and $\beta =_{NF} \beta_1 + \cdots + \beta_n$ then $\alpha < \beta$ holds iff one of the following conditions is satisfied:*

- *$m < n$ and $\alpha_i = \beta_i$ for all $i \leq m$,*
- *There is a $j < m$ such that $\alpha_i = \beta_i$ for all $i < j$ and $\alpha_j < \beta_j$.*

Besides the $\omega$-power $\omega^\alpha$ (which is in fact a power) we will later also need the power to the basis 2. This, however, is easily defined by the recursion equations

$$2^0 := 1$$
$$2^{\alpha+1} := 2^\alpha + 2^\alpha$$
$$\lambda \in Lim \;\Rightarrow\; 2^\lambda := \sup\{2^\xi \mid \xi < \lambda\}.$$

**3.3.10 Exercise** We define multiplication and exponentiation of ordinals by transfinite recursion:

$$\alpha \cdot 0 := 0$$
$$\alpha \cdot \beta' := \alpha \cdot \beta + \alpha$$
$$\alpha \cdot \lambda := \sup\{\alpha \cdot \xi \mid \xi < \lambda\} \text{ for } \lambda \in Lim$$

and

$$exp(\alpha, 0) := 1$$
$$exp(\alpha, \beta') := exp(\alpha, \beta) \cdot \alpha$$
$$exp(\alpha, \lambda) := \sup\{exp(\alpha, \xi) \mid \xi < \lambda\} \text{ for } \lambda \in Lim.$$

Proof or refute the following statements:

(a) $\alpha < \beta \wedge \gamma > 0 \;\Leftrightarrow\; \gamma \cdot \alpha < \gamma \cdot \beta$

(b) $\alpha < \beta \Rightarrow \alpha \cdot \gamma < \beta \cdot \gamma$

(c) $\alpha \cdot (\beta + \gamma) = \alpha \cdot \beta + \alpha \cdot \gamma$

(d) $(\alpha+\beta)\cdot\gamma = \alpha\cdot\gamma+\beta\cdot\gamma$

(e) $\alpha\cdot(\beta\cdot\gamma) = (\alpha\cdot\beta)\cdot\gamma$

(f) $\beta \neq 0 \Rightarrow (\forall\alpha)(\exists!\gamma)(\exists!\delta)[\alpha = \beta\cdot\gamma+\delta \wedge \delta < \beta]$

(g) $1 < \gamma \wedge \alpha < \beta \Rightarrow exp(\gamma,\alpha) < exp(\gamma,\beta)$

(h) $exp(\alpha,\beta)\cdot exp(\alpha,\gamma) = exp(\alpha,\beta+\gamma)$

(i) $exp(exp(\alpha,\beta),\gamma) = exp(\alpha,\beta\cdot\gamma)$

(j) $\alpha < \beta \Rightarrow exp(\alpha,\gamma) \leq exp(\beta,\gamma)$

(k) $\alpha > 0 \wedge \beta > 1 \Rightarrow (\exists!\delta)[exp(\beta,\delta) \leq \alpha < exp(\beta,\delta+1)]$

**3.3.11 Exercise** Show

(a) $exp(\omega,\alpha) = \omega^{\alpha}$

(b) $\alpha \in Lim \Rightarrow exp(2,\alpha) \in \mathbb{H}$

(c) $2^{\alpha} = exp(2,\alpha)$

for all ordinals $\alpha$. Use these results to prove $\omega^{\alpha} = 2^{\omega\cdot\alpha}$.

Hint: To prove (a) start by showing $exp(\omega,\alpha) \in \mathbb{H}$.

**3.3.12 Exercise** Define the enumerating function of all ordinals which are not successor ordinals.

**3.3.13 Exercise** *(Cantor normal-form with basis $\beta$)* Let $\beta > 1$. Show that for all $\alpha > 0$ there are uniquely determined ordinals $\alpha_1 > \cdots > \alpha_n$ and $\gamma_1,\ldots,\gamma_n$ with $0 < \gamma_i < \beta$ for $i = 1,\ldots,n$ such that

$$\alpha = \beta^{\alpha_1}\cdot\gamma_1 + \cdots + \beta^{\alpha_n}\cdot\gamma_n,$$

where $\beta^{\alpha_i}$ stands for $exp(\beta,\alpha_i)$ (cf. Exercise 3.3.10).

**3.3.14 Exercise** Let $\alpha \# 0 = 0 \# \alpha = \alpha$. For $\alpha =_{NF} \alpha_1 + \cdots + \alpha_m$ and $\beta =_{NF} \alpha_{m+1} + \cdots + \alpha_n$ let $\alpha \# \beta := \alpha_{\pi(1)} + \cdots + \alpha_{\pi(n)}$ where $\pi$ is a permutation of the numbers $\{1,\ldots,n\}$ such that $\alpha_{\pi(i)} \geq \alpha_{\pi(i+1)}$ for $i \in \{1,\ldots,n-1\}$. We call $\alpha \# \beta$ the *symmetric sum* of $\alpha$ and $\beta$.

Prove the following properties of the symmetric sum:

- $\alpha \# \beta = \beta \# \alpha$.

- $\alpha \# \xi < \alpha \# \eta$ for all ordinals $\xi < \eta$.

**3.3.15 Exercise** *(Multiplicative indecomposable ordinals)* An ordinal $\alpha \neq 0$ is *multiplicatively indecomposable* if

$$(\forall \xi < \alpha)(\forall \eta < \alpha)[\xi \cdot \eta < \alpha].$$

Prove the following statements:

(a) An ordinal $\alpha > 1$ is multiplicatively indecomposable iff $\xi \cdot \alpha = \alpha$ for all $\xi < \alpha$.

(b) An ordinal $\alpha$ is multiplicatively indecomposable iff $\alpha \in \{1,2\}$ or there is an ordinal $\delta$ such that $\alpha = \omega^{\omega^\delta}$.

**3.3.16 Exercise** Prove the equation in (a) and characterize the classes in (b)–(d) similarly.

(a) $M_1^= := \{\alpha \in On \mid \omega \cdot \alpha = \alpha \cdot \omega\} = (\mathbb{H} \cap \omega^\omega) \cup \{0\}$

(b) $M_2^= := \{\alpha \in On \mid \omega^\alpha = \alpha^\omega\}$

(c) $M_1^< := \{\alpha \in On \mid \omega \cdot \alpha < \alpha \cdot \omega\}$

(d) $M_2^< := \{\alpha \in On \mid \omega^\alpha < \alpha^\omega\}$

### *3.3.1 A Notation System for the Ordinals below $\varepsilon_0$*

We will later prove that all ($\kappa$-) normal functions possess fixed-points. The fixed-points of $\lambda \xi \,.\, \omega^\xi$ are commonly called $\varepsilon$-numbers. Let

$$\varepsilon_0 := \min\{\alpha \mid \omega^\alpha = \alpha\}.$$

For ordinals less than $\varepsilon_0$ we get a normal-form $\alpha =_{NF} \omega^{\alpha_1} + \cdots + \omega^{\alpha_n}$ with $\alpha_i < \alpha$ for $i = 1, \ldots, n$. This opens the possibility to denote every ordinal $< \varepsilon_0$ by a word formed from the alphabet $\{0, +, \omega\}$. By goedelizing these words, i.e., coding them by natural numbers, we obtain codes for the ordinals below $\varepsilon_0$ in the natural numbers. We do that by simultaneously defining a set $OT$ of ordinal codes and a mapping $|\ | : OT \longrightarrow On$ that assigns the ordinal value $|a|$ to every code $a \in OT$.

First we define

(O1) $0 \in OT$ and $|0| = 0$

(O2) If $a_1, \ldots, a_n \in OT$ and $|a_1| \geq |a_2| \geq \cdots \geq |a_n|$ then $\langle a_1, \ldots, a_n \rangle \in OT$ and $|\langle a_1, \ldots, a_n \rangle| := \omega^{|a_1|} + \cdots + \omega^{|a_n|}$.

Second we define a relation

$$a \prec b \;:\Leftrightarrow\; a \in OT \wedge b \in OT \wedge |a| < |b|$$

and claim that the set $OT$ as well as the binary relation $\prec$ are both primitive recursive. To check that we show that both notions can be simultaneously defined by course-of-values recursion.

We get

$$
\begin{aligned}
x \in OT \;\Leftrightarrow\; & Seq(x) \wedge [x = 0 \vee (\forall i < lh(x))[(x)_i \in OT \wedge \\
& (i+1 < lh(x) \to (a)_{i+1} \preceq (a)_i)]],
\end{aligned}
$$

where $a \preceq b$ stands for $a \prec b \vee a = b$, and

$$
\begin{aligned}
a \prec b \;\Leftrightarrow\; & a \in OT \wedge b \in OT \wedge [(a = 0 \wedge b \neq 0) \vee (lh(b) = 1 \wedge a \preceq (b)_0) \vee \\
& (\exists j < \min\{lh(a), lh(b)\})(\forall i < j)[(a)_i = (b)_i \wedge (a)_j \prec (b)_j] \vee \\
& (lh(a) < lh(b) \wedge (\forall i < lh(a))[(a)_i = (b)_i])].
\end{aligned}
$$

**3.3.17 Theorem** *The set OT and the binary relation $\prec$ are primitive recursive. For $a \in OT$ it is $|a| < \varepsilon_0$ and for every ordinal $\alpha < \varepsilon_0$ there is an $a \in OT$ such that $|a| = \alpha$.*

*Proof* We have just seen that $OT$ and $\prec$ are simultaneously definable by course-of-values recursion and are thus primitive recursive. Since $\varepsilon_0$ is in $\mathbb{H}$ and closed under $\lambda\xi\,.\,\omega^{\xi}$ we obtain $|a| < \varepsilon_0$ easily by induction on $a$. Conversely we show by induction on $\alpha < \varepsilon_0$ that there is an $a \in OT$ such that $|a| = \alpha$. This is obvious for $\alpha = 0$. If $\alpha \neq 0$ we have $\alpha =_{NF} \omega^{\alpha_1} + \cdots + \omega^{\alpha_n}$ with $\alpha_i < \alpha$ for $i = 1, \ldots, n$. By induction hypothesis there are ordinal notations $a_1, \ldots, a_n$ such that $|a_i| = \alpha_i$. Then $|\alpha_1| \geq |\alpha_2| \geq \cdots \geq |\alpha_n|$ which implies $a := \langle a_1, \ldots, a_n \rangle \in OT$ and $|a| = \omega^{|a_1|} + \cdots + \omega^{|a_n|} = \omega^{\alpha_1} + \cdots + \omega^{\alpha_n} = \alpha$. □

**3.3.18 Lemma** *For $a \in OT$ it is* $\mathrm{otyp}_{\prec}(a) = |a|$. *Hence* $\varepsilon_0 = \mathrm{otyp}(\prec) < \omega_1^{CK}$.

*Proof* Since $\prec$ is a well-ordering we have $\mathrm{otyp}_{\prec}(a) = \pi_{\prec}(a)$ for all $a \in OT$. We show $|a| = \pi_{\prec}(a)$ by induction on $\prec$. By Theorem 3.3.17 we have $|a| = \{|b| \mid |b| < |a|\} = \{|b| \mid b \prec a\} = \{\pi_{\prec}(b) \mid b \prec a\} = \pi_{\prec}(a)$. □

**3.3.19 Remark** After having defined the notation system $(OT, \prec)$ we may forget the set theoretical background we used to develop the system. We may consider $OT$ as a syntactically defined set of natural numbers together with a syntactically defined relation $\prec$.[4] This is of importance when ordinal notations are used to obtain consistency proofs as finitist as possible. In a finitist consistency proof we must not rely on the set theoretic background but have to argue purely combinatorially. The notation system can be viewed as purely combinatorially given. But without the set theoretical background we need to prove that $(OT, \prec)$ is a well-ordering. The proof that it is a linear ordering is tedious but completely elementary. The proof of its well-foundedness, however, needs means that exceed the strength of Peano Arithmetic (cf. Sect. 7.1). The well-foundedness proof is given in Sect. 7.4. A discussion about its foundational status is in Sect. 7.5.4.

[4] This will also be true for all the notation systems we are going to develop.

**3.3.20 Exercise** Let $\mathscr{F}$ be the least set of functions on the natural numbers that contains the function $\lambda x.0$ and satisfies:

- *If* $f_1, \ldots, f_n$ *are in* $\mathscr{F}$ *then* $\lambda x. x^{f_1(x)} + \cdots + x^{f_n(x)}$ *is in* $\mathscr{F}$.

For $f, g \in \mathscr{F}$ let

$$f \prec g :\Leftrightarrow (\exists k)(\forall x \geq k)[f(x) < g(x)].$$

Show that $(\mathscr{F}, \prec)$ is a well-ordering of order-type $\varepsilon_0$.

What is the order-type of the polynomials in one variable with respect to $\prec$?

## 3.4 The Veblen Hierarchy

To obtain notions for ordinals beyond $\varepsilon_0$ we need decompositions also for additively indecomposable ordinals. Thereto we have to study ordinals more profoundly. The main tool will be a hierarchy of closed and unbounded classes based on the class of additively indecomposable ordinals.

### *3.4.1 Preliminaries*

**3.4.1 Definition** Let $f\colon On \longrightarrow_p On$ be a partial function. Put

$$Fix(f) := \{\eta \in dom(f) \mid f(\eta) = \eta\}.$$

The *derivative* of the function $f$ is defined by

$$f' := en_{Fix(f)}.$$

For a class $M \subseteq On$ we define its derivative by

$$M' := Fix(en_M) = \{\alpha \in M \mid en_M(\alpha) = \alpha\}.$$

**3.4.2 Lemma** *Let* $\kappa > \omega$ *be a regular ordinal and* $f\colon On \longrightarrow_p On$ *a* $\kappa$*-normal function. Then* $Fix(f)$ *is club in* $\kappa$ *and therefore* $f'$ *again a* $\kappa$*-normal function.*

*Proof* First we show that $Fix(f)$ is unbounded in $\kappa$. Let $\alpha < \kappa$ and put $\beta_0 := \alpha + 1$ and $\beta_{n+1} := f(\beta_n)$. Then $\beta := \sup_{n\in\omega} \beta_n < \kappa$ and $f(\beta) = \sup_{n\in\omega} f(\beta_n) = \sup_{n\in\omega} \beta_{n+1} = \beta$. Hence $\alpha < \beta \in Fix(f) \cap \kappa$ and $Fix(f)$ is unbounded in $\kappa$. Next we show that $Fix(f)$ is closed in $\kappa$. Let $U \subseteq Fix(f) \cap \kappa$ such that $\overline{\overline{U}} < \kappa$. We obtain $f(\sup U) = \sup f[U] = \sup U$, i.e., $\sup U \in Fix(f)$. □

**3.4.3 Corollary** *The derivative* $M'$ *of a* $\kappa$*-club class* $M$ *is again* $\kappa$*-club.*

*Proof* Since $M' = Fix(en_M)$ we obtain the claim immediately from Lemma 3.4.2 and Theorem 3.2.19. □

**3.4.4 Lemma** *Let $\kappa > \omega$ be a regular ordinal and $\bar{\bar{I}} < \kappa$. If $\{M_\iota \mid \iota \in I\}$ is a collection of classes which are club in $\kappa$ then $\bigcap_{\iota \in I} M_\iota$ is also club in $\kappa$.*

*Proof* The intersection $\bigcap_{\iota \in I} M_\iota$ is obviously closed in $\kappa$. To visualize the fact that it is also unbounded in $\kappa$ we choose $\alpha < \kappa$ and $\alpha < m_{0,0} \in M_0$, arrange the classes $M_\iota$ in a row and choose elements $m_{k,\iota} \in M_\iota$ increasingly as shown in the following figure:

$$\begin{array}{ccccccc}
M_0 & & M_1 & \cdots & M_\iota & \cdots & \iota \in I \\
m_{0,0} & \leq & m_{0,1} & \leq \cdots \leq & m_{0,\iota} & \leq \cdots \leq & \sup_{\iota \in I} m_{0,\iota} \leq \\
\vdots & & \vdots & & \vdots & & \vdots \\
m_{k,0} & \leq & m_{k,1} & \leq \cdots \leq & m_{k,\iota} & \leq \cdots \leq & \sup_{\iota \in I} m_{k,\iota} \leq \\
\vdots & & \vdots & & \vdots & & \vdots \\
\sup_k m_{k,0} & = & \sup_k m_{k,1} & = \cdots = & \sup_k m_{k,\iota} & = \cdots = &
\end{array}$$

Since all the classes $M_\iota$ are unbounded in $\kappa$ and $\bar{\bar{I}} < \kappa$ we can choose the $m_{k,\iota} < \kappa$ in such a way that the ordinals in the figure are increasing from left to right and from top to down. But then $\sup_k m_{k,\iota} \in M_\iota$ for all $\iota \in I$ and since all these suprema coincide we see that $\sup_k m_{k,\iota} \in \bigcap_{\iota \in I} M_\iota$. □

**3.4.5 Exercise** Let $\kappa > \omega$ be a regular ordinal and $\{X_\alpha \mid \alpha < \kappa\}$ be a family of sets which are club in $\kappa$. Show that their *diagonal intersection*

$$\Delta_{\alpha<\kappa} X_\alpha := \{\xi \in \kappa \mid \xi \in \bigcap_{\alpha<\xi} X_\alpha\}$$

is club in $\kappa$.

### 3.4.2 The Veblen Hierarchy

Based on the class of additively indecomposable ordinal we will now establish a hierarchy of closed and unbounded classes using derivatives.

**3.4.6 Definition** We define the Veblen hierarchy of $\alpha$-critical ordinals by

$$Cr(0) = \mathbb{H}$$
$$Cr(\alpha+1) = Cr(\alpha)'$$
$$\lambda \in Lim \Rightarrow Cr(\lambda) = \bigcap_{\xi<\lambda} Cr(\xi)$$

and put

$$\varphi_\alpha := en_{Cr\,(\alpha)}.$$

**3.4.7 Theorem** *The classes $Cr\,(\alpha)$ are club in all regular ordinals $\kappa > \max\{\alpha, \omega\}$. The functions $\varphi_\alpha$ are,therefore, $\kappa$-normal functions for all regular $\kappa > \max\{\alpha, \omega\}$.*

*Proof* This follows immediately from Corollary 3.4.3, Lemma 3.4.4 and Theorem 3.2.19. □

**3.4.8 Lemma** *The Veblen functions $\varphi_\alpha$ have the following basic properties:*

*(1)* $\varphi_0(\alpha) = \omega^\alpha$

*(2)* $\varphi_1(0) = \varepsilon_0$

*(3)* $\beta < \alpha \;\Rightarrow\; \varphi_\xi(\alpha) < \varphi_\xi(\beta)$

*(4)* $\beta \leq \varphi_\alpha(\beta)$

*(5)* $\alpha < \beta \;\Rightarrow\; Cr\,(\beta) \subsetneqq Cr\,(\alpha) \wedge \varphi_\alpha(\gamma) \leq \varphi_\beta(\gamma) \wedge \varphi_\alpha(\varphi_\beta(\gamma)) = \varphi_\beta(\gamma)$

*Proof* Properties (1) through (4) follow directly from the definition and Lemma 3.2.13. Only (5) needs a proof. By definition we have

$$\alpha < \beta \;\Rightarrow\; Cr\,(\beta) \subseteq Cr\,(\alpha) \tag{i}$$

that already implies

$$\varphi_\alpha(\gamma) \leq \varphi_\beta(\gamma). \tag{ii}$$

For $\alpha < \beta$ we get $\varphi_\beta(\gamma) \in Cr\,(\beta) \subseteq Cr\,(\alpha + 1) = Cr\,(\alpha)'$ which shows

$$\varphi_\alpha(\varphi_\beta(\gamma)) = \varphi_\beta(\gamma). \tag{iii}$$

Since $0 < \beta$ we obtain $0 < \varphi_\beta(0)$ and thus $\varphi_\alpha(0) < \varphi_\alpha(\varphi_\beta(0)) = \varphi_\beta(0)$. Hence $\varphi_\alpha(0) \in Cr\,(\alpha) \setminus Cr\,(\beta)$, and the inclusion is proper. □

**3.4.9 Theorem** *(A) We have $\varphi_{\alpha_1}(\beta_1) = \varphi_{\alpha_2}(\beta_2)$ if and only if one of the following conditions is satisfied:*

*(1)* $\alpha_1 < \alpha_2 \wedge \beta_1 = \varphi_{\alpha_2}(\beta_2)$

*(2)* $\alpha_1 = \alpha_2 \wedge \beta_1 = \beta_2$

*(3)* $\alpha_2 < \alpha_1 \wedge \varphi_{\alpha_1}(\beta_1) = \beta_2.$

*(B) We have $\varphi_{\alpha_1}(\beta_1) < \varphi_{\alpha_2}(\beta_2)$ if and only if one of the following conditions is satisfied:*

*(1)* $\alpha_1 < \alpha_2 \wedge \beta_1 < \varphi_{\alpha_2}(\beta_2)$

*(2)* $\alpha_1 = \alpha_2 \wedge \beta_1 < \beta_2$

*(3)* $\alpha_2 < \alpha_1 \wedge \varphi_{\alpha_1}(\beta_1) < \beta_2$.

*Proof* We prove (A) and (B) simultaneously and distinguish the following cases:
1. $\alpha_1 < \alpha_2$. Then $\varphi_{\alpha_1}(\varphi_{\alpha_2}(\beta_2)) = \varphi_{\alpha_2}(\beta_2)$ and we obtain

$$\varphi_{\alpha_1}(\beta_1) = \varphi_{\alpha_2}(\beta_2) = \varphi_{\alpha_1}(\varphi_{\alpha_2}(\beta_2)) \;\Leftrightarrow\; \beta_1 = \varphi_{\alpha_2}(\beta_2)$$

as well as

$$\varphi_{\alpha_1}(\beta_1) < \varphi_{\alpha_2}(\beta_2) = \varphi_{\alpha_1}(\varphi_{\alpha_2}(\beta_2)) \;\Leftrightarrow\; \beta_1 < \varphi_{\alpha_2}(\beta_2).$$

2. $\alpha_1 = \alpha_2$. Then

$$\varphi_{\alpha_1}(\beta_1) = \varphi_{\alpha_2}(\beta_2) \;\Leftrightarrow\; \beta_1 = \beta_2$$

and

$$\varphi_{\alpha_1}(\beta_1) < \varphi_{\alpha_2}(\beta_2) \;\Leftrightarrow\; \beta_1 < \beta_2.$$

3. $\alpha_1 > \alpha_2$. Then we obtain by 1.

$$\varphi_{\alpha_1}(\beta_1) = \varphi_{\alpha_2}(\beta_2) \;\Leftrightarrow\; \beta_2 = \varphi_{\alpha_1}(\beta_1)$$

and

$$\varphi_{\alpha_1}(\beta_1) > \varphi_{\alpha_2}(\beta_2) \;\Leftrightarrow\; \beta_2 < \varphi_{\alpha_1}(\beta_1).$$

Hence

$$\varphi_{\alpha_1}(\beta_1) < \varphi_{\alpha_2}(\beta_2) \;\Leftrightarrow\; \varphi_{\alpha_1}(\beta_1) < \beta_2.$$ □

From Theorem 3.4.9 we obtain especially

$$\varphi_\alpha(0) < \varphi_\beta(0) \;\Leftrightarrow\; \alpha < \beta \tag{3.9}$$

which entails the following corollary.

**3.4.10 Corollary** *The function $\lambda\xi\,.\,\varphi_\xi(0)$ is order preserving. Hence $\alpha \leq \varphi_\alpha(0) \leq \varphi_\alpha(\beta)$ for all $\beta$.*

**3.4.11 Theorem** *For all principal ordinals $\alpha \in \mathbb{H}$ there are uniquely determined ordinals $\xi$ and $\eta$ such that $\alpha = \varphi_\xi(\eta)$ and $\eta < \alpha$.*

*Proof* We show first uniqueness. Let $\alpha = \varphi_\xi(\eta)$ and $\alpha = \varphi_\nu(\mu)$ such that $\eta < \alpha$ and $\mu < \alpha$. If $\xi < \nu$ we get the contradiction $\alpha = \varphi_\xi(\eta) < \varphi_\xi(\alpha) = \varphi_\xi(\varphi_\nu(\mu)) = \varphi_\nu(\mu) = \alpha$. Hence $\xi = \nu$ which immediately also implies $\eta = \mu$.

To show also the existence let $\xi := \min\{\mu \mid \alpha < \varphi_\mu(\alpha)\}$. This minimum exists because $\alpha \leq \varphi_\alpha(0) < \varphi_\alpha(\alpha)$. For $\xi = 0$ we have $\alpha < \varphi_0(\alpha)$ and because of $\alpha \in \mathbb{H}$ there is an $\eta$ such that $\varphi_0(\eta) = \alpha < \varphi_0(\alpha)$. Hence $\eta < \alpha$. If $\xi \neq 0$ we have $\alpha = \varphi_\rho(\alpha)$ for all $\rho < \xi$. Hence $\alpha \in \bigcap_{\rho<\xi} Cr(\rho+1) = Cr(\xi)$. Therefore there is an $\eta$ such that $\varphi_\xi(\eta) = \alpha < \varphi_\xi(\alpha)$ that implies $\eta < \alpha$. □

An ordinal $\beta \in Cr(\alpha)$ is closed under ordinal addition (since $Cr(\alpha) \subseteq \mathbb{H}$) and closed under all functions $\varphi_\xi$ with $\xi < \alpha$ since $\beta = \varphi_\alpha \rho$ for some $\rho$ and $\eta < \beta = \varphi_\alpha(\rho)$ implies $\varphi_\xi(\eta) < \varphi_\alpha(\rho) = \beta$ for all $\xi < \alpha$ by Theorem 3.4.9 (B)(1). The ordinals in $Cr(\alpha)$ are thus inaccessible by ordinal addition and the functions $\varphi_\xi$ for $\xi < \alpha$. Therefore we call the ordinals in $Cr(\alpha)$ *$\alpha$-critical*. Ordinals $\alpha$, which are themselves $\alpha$-critical are therefore closed under the function $\varphi := \lambda\xi\eta\,.\,\varphi_\xi(\eta)$ viewed as a binary function. We call these ordinals *strongly critical* and define

$$SC := \{\alpha \mid \alpha \in Cr(\alpha)\} \tag{3.10}$$

and

$$\Gamma_\xi := en_{SC}(\xi).$$

**3.4.12 Lemma** *We have*

*(1)* $\quad \alpha \in SC \wedge \xi, \eta < \alpha \;\Rightarrow\; \varphi_\xi(\eta) < \alpha$

*and*

*(2)* $\quad \alpha \in SC \;\Leftrightarrow\; \varphi_\alpha(0) = \alpha.$

*Proof* Property (1) is already clear. Property (2) holds because $\alpha \in SC$ implies $\alpha = \varphi_\alpha(\xi)$ for some $\xi$ and $\alpha \leq \varphi_\alpha(0) \leq \varphi_\alpha(\xi) = \alpha$ implies $\alpha = \varphi_\alpha(0)$. The opposite direction is obvious. □

**3.4.13 Lemma** *The ordinals in SC are exactly the ordinals which are closed under $\varphi$ viewed as a binary function.*

*Proof* In Lemma 3.4.12 (1) we have already seen that strongly critical ordinals are closed under $\varphi$. For the opposite inclusion let $\alpha$ be closed under $\varphi$. We show

$$\xi < \varphi_\alpha(0) \;\Rightarrow\; \xi < \alpha \tag{i}$$

by induction on $\xi$. If $\xi \notin \mathbb{H}$ we obtain (i) directly by the induction hypothesis and the fact that $\alpha \in \mathbb{H}$. So assume $\xi = \varphi_{\xi_0}(\xi_1)$ with $\xi_1 < \xi$. From the premise of (i) and Theorem 3.4.9 we then obtain $\xi_0 < \alpha$ and $\xi_1 < \varphi_\alpha(0)$. But then $\xi_1 < \alpha$ by induction hypothesis and we obtain $\xi = \varphi_{\xi_0}(\xi_1) < \alpha$ because $\alpha$ is closed under $\varphi$.

From (i) we get $\varphi_\alpha(0) = \alpha$ which entails $\alpha \in SC$. □

**3.4.14 Theorem** *The class SC is club in all regular ordinals $\kappa > \omega$.*

*Proof* We show unboundedness first. Let $\alpha < \kappa$ and put $\alpha_0 := \alpha + 1$ and $\alpha_{n+1} := \varphi_{\alpha_n}(0)$. Then $\beta := \sup_{n\in\omega} \alpha_n < \kappa$ and for $\xi < \beta$ there is a $k$ such that $\xi < \alpha_k \le \alpha_m$ for all $m \ge k$. This implies $\varphi_\xi(\alpha_{m+1}) = \varphi_\xi(\varphi_{\alpha_m}(0)) = \varphi_{\alpha_m}(0) = \alpha_{m+1}$ for all $m \ge k$. So $\beta \le \varphi_\xi(\beta) = \varphi_\xi(\sup_{k<m<\omega} \alpha_m) = \sup_{k<m<\omega} \varphi_\xi(\alpha_m) = \sup_{k<m<\omega} \alpha_m \le \beta$. Hence $\beta \in \bigcap_{\xi<\beta} Cr(\xi) = Cr(\beta)$ which shows $\alpha < \beta \in SC \cap \kappa$.

To prove that $SC$ is also closed let $M \subseteq SC \cap \kappa$ such that $\overline{\overline{M}} < \kappa$. Then $\alpha := \sup M < \kappa$. For $\xi, \eta < \alpha$ there is an ordinal $\beta \in M \cap \kappa$ such that $\xi, \eta < \beta$. Hence $\varphi_\xi(\eta) < \beta \le \alpha$ and $\alpha \in SC$ by Lemma 3.4.13. □

**3.4.15 Exercise** Let $\kappa > \omega$ be a regular ordinal. Show that $SC \cap \kappa = \Delta_{\alpha<\kappa} Cr(\alpha)$.

**3.4.16 Exercise** Define $Cr_\xi(\alpha)$ by the following clauses:

- $Cr_0(0) = \mathbb{H}$.
- $Cr_{\alpha+1}(0) := \Delta_{\lambda<\kappa} Cr_\alpha(\lambda)$.
- $Cr_\lambda(0) = \bigcap_{\xi<\lambda} Cr_\xi(0)$ for $\lambda \in Lim$.
- $Cr_\xi(\alpha+1) := Cr_\xi(\alpha)'$

and

- $Cr_\xi(\lambda) := \bigcap_{\eta<\lambda} Cr_\xi(\eta)$ for $\lambda \in Lim$

and define

$$\varphi_{\xi,\eta} := en_{Cr_\xi(\eta)}.$$

Show that $\varphi_{\xi,\eta}$ are all $\kappa$-normal-functions and prove

(a) $\varphi_{0,\xi}(\alpha) = \varphi_\xi(\alpha)$.

(b) $\varphi_{1,0}(\alpha) = \Gamma_\alpha$.

(c) $\alpha := \varphi_{\alpha_1,\alpha_2}(\alpha_3) < \varphi_{\beta_1,\beta_2}(\beta_3) =: \beta$ holds true iff one of the following conditions is satisfied:

- $\alpha_1 = \beta_1 \wedge \alpha_2 < \beta_2 \wedge \alpha_3 < \beta$.
- $\alpha_1 = \beta_1 \wedge \alpha_2 = \beta_2 \wedge \alpha_3 < \beta_3$.
- $\alpha_1 = \beta_1 \wedge \beta_2 < \alpha_2 \wedge \alpha < \beta_3$.
- $\alpha_1 < \beta_1 \wedge \alpha_2 < \beta \wedge \alpha_3 < \beta$.
- $\beta_1 < \alpha_1 \wedge (\alpha < \beta_2 \vee (\alpha = \beta_2 \wedge 0 < \beta_3) \vee \alpha < \beta_3)$.

### 3.4.3 A Notation System for the Ordinals below $\Gamma_0$

The first strongly critical ordinal is $\Gamma_0$. We are going to use the Veblen hierarchy to develop a system of notations for the ordinals below $\Gamma_0$. Besides the Cantor normal-form for ordinals, which we already used in the development of the notations for the ordinals below $\varepsilon_0$, the key in the development is the following lemma.

**3.4.17 Lemma** *For every ordinal $\alpha \in \mathbb{H} \setminus SC$ there are uniquely determined ordinals $\xi$ and $\eta$ such that $\alpha = \varphi_\xi(\eta)$ and $\xi, \eta < \alpha$.*

*Proof* In Theorem 3.4.11 we have already seen that there are uniquely determined ordinals $\xi$ and $\eta$ such that $\alpha = \varphi_\xi(\eta)$ and $\eta < \alpha$. Of course we have $\xi \le \alpha$. So assume $\xi = \alpha$. But then $\alpha \le \varphi_\alpha(0) \le \varphi_\alpha(\eta) = \alpha$ in contradiction to $\alpha \notin SC$. □

By Lemma 3.4.17 and the Cantor normal-form we obtain for the ordinals below $\Gamma_0$ a uniquely determined normal-form

$$\alpha =_{NF} \varphi_{\xi_1}(\eta_1) + \cdots + \varphi_{\xi_n}(\eta_n)$$

such that $\xi_i, \eta_i < \alpha$ for $i = 1, \ldots, n$. This shows that we can represent every ordinal $<$ $\Gamma_0$ as a word over the alphabet $\{0, +, \varphi_\cdot(\cdot)\}$. This opens the possibility for a notation system. Again we define a set $OT$ of ordinal notations and a subset $PT$ of notations for principal ordinals together with an evaluation function $|\ |$.

(O1) $0 \in OT$ and $|0| = 0$

(O2) If $a_1, \ldots, a_n \in PT$ and $a_1 \succeq \cdots \succeq a_n$ then $\langle 1, a_1, \ldots, a_n \rangle \in OT$ and $|\langle 1, a_1, \ldots, a_n \rangle| = |a_1| + \cdots + |a_n|$

(O3) If $a_1, a_2 \in OT$ then $\langle 2, a_1, a_2 \rangle \in PT$ and $|\langle 2, a_1, a_2 \rangle| = \varphi_{|a_1|}(|a_2|)$.

(O4) $PT \subseteq OT$.

Moreover we define

$$a \prec b \ :\Leftrightarrow\ a \in OT \wedge b \in OT \wedge |a| < |b|$$

and

$$a \equiv b \ :\Leftrightarrow\ a \in OT \wedge b \in OT \wedge |a| = |b|.$$

By Theorem 3.4.9 and Lemma 3.4.17 it is easy to check that $OT$ and the relations $\prec$ and $\equiv$ are simultaneously definable by course-of-values recursion and thence primitive recursive. We leave the details and the following two claims as (easy) exercises.

**3.4.18 Theorem** *The sets $OT$, $PT$ and the relations $\prec$ and $\equiv$ are primitive recursive. For every notation $a \in OT$ we have $|a| < \Gamma_0$ and for every ordinal $\alpha < \Gamma_0$ there is conversely an $a \in OT$ such that $|a| = \alpha$.*

This is all we need about ordinals for predicative proof theory. Later, however, before stepping into impredicative proof theory, we will have to return to the theory of ordinals.

**3.4.19 Exercise** Show that every cardinal is strongly critical.

**3.4.20 Exercise** We define a fixed-point-free version $\bar{\varphi}$ of the Veblen function $\varphi$.

$$\bar{\varphi}_\alpha(\beta) := \begin{cases} \varphi_\alpha(\beta+1) & \text{if } \beta = \gamma+n \text{ for some ordinal } \gamma \text{ such that } \varphi_\alpha(\gamma)=\gamma \\ & \text{or } \varphi_\alpha(\gamma)=\alpha \\ \varphi_\alpha(\beta) & \text{otherwise.} \end{cases}$$

Show:

(a) $\bar{\varphi}_\alpha(\beta) < \bar{\varphi}_\gamma(\delta) \;\Leftrightarrow\; \begin{cases} \alpha<\gamma \wedge \beta<\bar{\varphi}_\gamma(\delta) & \text{or} \\ \alpha=\gamma \wedge \beta<\delta & \text{or} \\ \gamma<\alpha \wedge \bar{\varphi}_\alpha(\beta) \le \delta \end{cases}$

(b) $\alpha < \bar{\varphi}_\alpha(\beta) \wedge \beta < \bar{\varphi}_\alpha(\beta)$.

(c) $\bar{\varphi}_{\alpha_1}(\beta_1) = \bar{\varphi}_{\alpha_2}(\beta_2) \;\Leftrightarrow\; \alpha_1=\alpha_2 \wedge \beta_1=\beta_2$.

**3.4.21 Exercise** Prove Theorem 3.4.18.

# Chapter 4
# Pure Logic

To fix the formal framework we recall the notions of first- and second-order languages and introduce calculi for first- and weak second-order predicate logic.

## 4.1 Heuristics

To begin with, we will follow HILBERT's programme and, in a first step, try to formalize, if not the whole of mathematics, at least parts of it. To get a feeling about how this could be done, we start with some heuristic remarks.

The object(ive)s of mathematical research are, broadly speaking, structures. The "working mathematician" wants to figure out which theorem will hold in the structure of his or her interest. Therefore (s)he needs a language in which (s)he can formulate the theorems. This is usually done in English (or in another language) augmented with technical terms that are characteristic for the structure. But as we know from mathematical logic only the technical terms do matter; the use of English or any other colloquial language can (in principle) be dispensed with. We can do with the formal language of logic. From a heuristical point of view the basic ingredients for a formal language are:

- symbols for the elements of the structure; fixed elements symbolized by *constants* and arbitrary elements symbolized by *variables*
- *symbols for functions*
- *predicate symbols* which describe relations between terms (e.g., equality, less than, ...) and thus form primitive propositions
- *logical connectives* by which we can compose propositions

- *quantifiers* ranging over the elements of a structure that allow us to express properties which are satisfied by some or all elements of the structure. Here, we can also imagine quantifiers which range over the functions, or the subsets, of the structure or even over sets of subsets etc. In these cases we talk about higher order languages. If the range is restricted to the elements of the structure alone we talk about a *first-order language*. *First-order logic* is the logic that deals with first-order languages.

From the basic symbols we build two types of well-formed expressions.

- *Terms* which are built up recursively from constants and variables by function symbols

and

- *formulas* which are built up recursively from primitive propositions by logical connectives and quantifiers.

Once we have established a formal language which is adequate for a structure, we can formulate *sentences* in this language. The problem is now to figure out which sentences hold in the structure. This could be done by pure intuition. But it may happen that our intuition about the truth of a sentence is erroneous or may not be shared by our colleagues. Therefore we have to *prove* the truth of a sentence. The next problem we have to deal with is therefore to settle the notion of *"proof"*. Experience tells us that a proof consists of a series of *inferences*. But how can we characterize an inference? Inferences have to preserve the truth of sentences in a structure. Let $\mathfrak{S} \models F$ denote that the sentence $F$ holds in the structure $\mathfrak{S}$. Call an inference $A_1,\ldots,A_n \models_{\mathfrak{S}} F$ *adequate* for the structure $\mathfrak{S}$ if $\mathfrak{S} \models A_i$ for $i = 1,\ldots,n$ implies $\mathfrak{S} \models F$. Adequateness in a special structure doesnot, however, provide any progress. To secure that an inference $A_1,\ldots,A_n \models_{\mathfrak{S}} F$ is adequate for the structure $\mathfrak{S}$ we have to ensure $\mathfrak{S} \models F$. But that is what we aimed for and therefore it is completely superfluous to check also $\mathfrak{S} \models A_i$ for the premises $A_i$.

Therefore we have to widen the notion of inference such that it becomes independent of a particular structure, i.e., we regard only those inferences which are adequate for all possible structures. We define

$$A_1,\ldots,A_n \models_L F \quad :\Leftrightarrow \quad (\forall \mathfrak{S})\,[A_1,\ldots,A_n \models_{\mathfrak{S}} F]$$

and call that a *logical inference*. Logical inferences preserve truth in *all possible structures*.[1] It is one of the big achievements of mathematical logic to have shown that in the case of first-order languages, the notion $\models_L$ of a logical inference can be replaced by a set of *formal rules*. A formal rule is syntactically given by a figure of the form

$$F_1,\ldots,F_n \vdash G$$

[1] By narrowing the range of possible test-structures $\mathfrak{S}$ we obtain stronger logics but lose the possibility to replace $\models_L$ by a strictly formal rule. (Cf. Exercise 5.4.13).

where $F_1,\dots,F_n,G$ is a finite set of formulas in the formal language. Let $F_1,\dots,F_n \vdash G$ denote the fact that $G$ is *deducible* from $F_1,\dots,F_n$, i.e., derivable from $F_1,\dots,F_n$ by a finite number of applications of formal rules. The theorems which connect logical inferences and deductibility are the Soundness- and Completeness-Theorem for first-order logic which state that there are formal rules such that

$$A_1,\dots,A_n \models_L F \;\Leftrightarrow\; A_1,\dots,A_n \vdash F.$$

The direction from left to right is known as the Completeness Theorem while the opposite direction is called Soundness- or Correctness-Theorem.

Since formal rules are syntactically defined, it is decidable whether a formal rule is correctly applied. So it becomes decidable whether a finite series of formal inferences is correct. If we start from a set $A_1,\dots,A_n$ of sentences, which are obviously true in the structure $\mathfrak{S}$ and infer $A_1,\dots,A_n \vdash F$ we can replace $\vdash$ by $\models_L$ and obtain $\mathfrak{S} \models F$, i.e., that $F$ is a theorem of $\mathfrak{S}$. It remains decidable whether the presented proof is correct, i.e., whether the rules have been applied correctly. This is one of the most important features of mathematics. The truth of its proved theorems is (in principle) machine-checkable on the basis of logically legitimated rules and does not depend on the knowledge of conventions and regulations which are only admissible to a selected group of insiders. This is probably one of the reasons for the broad applicability of mathematics.

Of course no "working mathematician" works formally. Finding proofs is a matter of high intuition. But in writing up a proof she or he is (at least unconsciously) aware that mathematical proofs are in principle syntactically formalizable and thus decidable. Otherwise it would be impossible for the mathematical community to check the correctness of a proof. Apparently this has been common knowledge to the mathematical community long before GÖDEL was able to prove the completeness theorem which states that there are calculi, i.e., sets of formal rules, which are complete for logical reasoning.[2]

The notion of logical inference is closely connected to that of logical validity. A sentence $F$ is *logically valid* – denoted by $\models F$ – if it holds in all structures. According to the definition of a logical inference, a sentence is logically valid if and only if it is derivable from an empty set of premises, i.e., $\models F \;\Leftrightarrow \models_L F$. An immediate consequence of the definition of a logical inference is the *deduction theorem* stating

$$(A_1,\dots,A_n,G \models_L F) \;\Leftrightarrow\; (A_1,\dots,A_n \models_L G \to F).$$

Therefore we obtain

$$A_1,\dots,A_n \models_L F \;\Leftrightarrow\; \models A_1 \land \cdots \land A_n \to F.$$

Since logical validity and thence also logical deductibility rely on *all* suited structures we cannot expect to obtain information about a particular structure purely logically. If we want to prove something about a special structure we have to anticipate

[2] Hilbert in [41] apparently already anticipated the existence of such calculi.

basic facts which we consider to be characteristic of that structure. These facts form the axioms. We know, however, that for sufficiently complex structures it is impossible to specify axioms which characterize the structure completely. This incompleteness is twofold. First, it is impossible to characterize an infinite structure by a set of first-order axioms up to isomorphisms. Even if we take as axioms all the sentences which are true in a structure $\mathfrak{S}$ there are still structures which are not isomorphic to $\mathfrak{S}$ but satisfy exactly the same first-order sentences. This is a consequence of the compactness theorem for first-order logic which says that an infinite set $M$ of first-order sentences is satisfiable in a structure if and only if every finite subset of $M$ is satisfiable. In full second-order logic there is an axiom system which characterizes the structure of natural numbers up to isomorphism but there is no complete deduction formalism for second-order logic. The correctness of a proof in second-order logic is therefore not machine-checkable. There are, however, formal calculi which are at least sound for second-order languages. From a logical point of view, these calculi are not really second-order calculi but rather two sorted first-order calculi. This distinction is a topic of an introductory text of Mathematical Logic and will not be discussed further here. We will, however, introduce a second-order language for arithmetic and set theory and (especially in the exercises) sometimes also work with formal calculi for second-order logic.

If we insist in machine-verifiability of mathematical proofs it makes no sense to take all valid first-order sentences of a structure as axioms because it is in general not decidable whether a sentence is valid in a structure. To check whether $A_1, \ldots, A_n \vdash F$ is a correct proof, requires us also to check that $A_1, \ldots, A_n$ are axioms. The property of being an axiom must therefore be decidable. Here comes another incompleteness which is known as GÖDEL's first incompleteness theorem. It is generally impossible to derive all first-order theorems of a structure from a decidable set of axioms. Exploring the limits of the deductibility power of decidable axioms systems is one of the aims of this book.

The following sections fix the formal logical framework for this book and recall some general background and results of Mathematical Logic.

## 4.2 First-Order and Second-Order Logics

Logical languages describe structures. In a formal language for logic we distinguish between logical and non logical symbols. The logical symbols are common to all languages while the non logical symbols are characteristic for the intended structures.

**4.2.1 Definition** (Logical symbols) The logical symbols are:

- Countably many free object variables, denoted by $u, v, w, u_0, \ldots$.
- Countably many bounded object variables, denoted by $x, y, z, x_0, \ldots$.

- For every non zero natural number $n$ countably many $n$-ary free relation variables, denoted by $U, V, U_0, \ldots$.
- For every non zero natural number $n$ countably many $n$-ary bounded relation variables, denoted by $X, Y, Z, X_0, \ldots$.
- The propositional connectives $\neg, \vee, \wedge, \rightarrow$.
- The quantifiers $\forall$ and $\exists$.
- Auxiliary symbols such as parentheses, square brackets etc.

**4.2.2 Definition** The non logical symbols of a language comprise:

- A set $\mathscr{C}$ of *constants*.
- A set $\mathscr{F}$ of *function symbols*. Every function symbol $f \in \mathscr{F}$ is equipped with an arity $\#f$ which is a non zero natural number.
- A set $\mathscr{R}$ of *relation symbols*. Every relation symbol $R \in \mathscr{R}$ comes with an arity $\#R$ which is a non zero natural number.

**4.2.3 Definition** (Inductive definition of the $\mathscr{L}$-terms)

- Every constant $c$ is an $\mathscr{L}$-term with $FV(c) = \emptyset$.
- Every free object variable $u$ is an $\mathscr{L}$-term with $FV(u) = \{u\}$.
- If $t_1,\ldots,t_n$ are $\mathscr{L}$-terms and $f$ is an $n$-ary function symbol then $(ft_1\ldots t_n)$ is an $\mathscr{L}$-term with $FV(ft_1\ldots t_n) = FV(t_1) \cup \cdots \cup FV(t_n)$.

We call $FV(t)$ the set of variables which occur freely in the term $t$.

**4.2.4 Definition** (Inductive definition of the $\mathscr{L}$-formulas)

- If $t_1,\ldots,t_n$ are $\mathscr{L}$-terms and $R$ is an $n$-ary relation symbol then $(Rt_1\ldots t_n)$ is an atomic formula with $FV(Rt_1\ldots t_n) = FV(t_1) \cup \cdots \cup FV(t_n)$, $BV(Rt_1\ldots t_n) = \emptyset$ and $FV_2(Rt_1\ldots t_n) = BV_2(Rt_1\ldots t_n) = \emptyset$.
- If $U$ is an $n$-ary free relation variable and $t_1,\ldots,t_n$ are $\mathscr{L}$-terms then $(Ut_1\ldots t_n)$ is an atomic formula with $FV(Ut_1\ldots t_n) = FV(t_1) \cup \cdots \cup FV(t_n)$, $BV(Ut_1\ldots t_n) = BV_2(Ut_1\ldots t_n) = \emptyset$ and $FV_2(Ut_1\ldots t_n) = \{U\}$. Instead of $(Ut_1\ldots t_n)$ we often also write $(t_1,\ldots,t_n) \,\varepsilon\, U$.
- Every atomic formula is a formula.

- If $F$ is a formula then $(\neg F)$ is formula with $FV(\neg F) = FV(F)$, $BV(\neg F) = BV(F)$, $FV_2(\neg F) = FV_2(F)$ and $BV_2(\neg F) = BV_2(F)$.

- If $F$ and $G$ are formulas then $(F \wedge G)$, $(F \vee G)$ and $(F \to G)$ are formulas with $FV(F \circ G) = FV(F) \cup FV(G)$, $BV(F \circ G) = BV(F) \cup BV(G)$, $FV_2(F \circ G) = FV_2(F) \cup FV_2(G)$ and $BV_2(F \circ G) = BV_2(F) \cup BV_2(G)$ for $\circ \in \{\wedge, \vee, \to\}$.

- If $F$ is a formula and $x \notin BV(F)$ then $(\forall x)F_u(x)$ and $(\exists x)F_u(x)$ are formulas with $FV((\mathsf{Q}x)F_u(x)) = FV(F) \setminus \{u\}$, $BV((\mathsf{Q}x)F_u(x)) = BV(F) \cup \{x\}$, $FV_2((\mathsf{Q}x)F_u(x)) = FV_2(F)$ and $BV_2((\mathsf{Q}x)F_u(x)) = BV_2(F)$ for $\mathsf{Q} \in \{\forall, \exists\}$.

- If $F$ is a formula and $X \notin BV_2(F)$ then $(\forall X)F_U(X)$ and $(\exists X)F_U(X)$ are formulas with $FV((\mathsf{Q}X)F_U(X)) = FV(F)$, $BV((\mathsf{Q}X)F_U(X)) = BV(F)$, $FV_2((\mathsf{Q}X)F_U(X)) = FV_2(F) \setminus \{U\}$ and $BV_2((\mathsf{Q}X)F_U(X)) = BV_2(F) \cup \{X\}$ for $\mathsf{Q} \in \{\forall, \exists\}$.[3]

A language is characterized by its non logical symbols. We call $(\mathscr{C}, \mathscr{F}, \mathscr{R})$ together with the arities of the function- and predicate symbols the *signature* of the language. To interpret a formal language we need a structure which matches the signature of the language. This is the following definition.

**4.2.5 Definition** Let $\mathscr{L}(\mathscr{C}, \mathscr{F}, \mathscr{R})$ be a formal language. An $\mathscr{L}$-structure $\mathfrak{S} = (S, {}^{\mathfrak{S}})$ is a non void set $S$ (the domain of the structure) together with a mapping ${}^{\mathfrak{S}}$ which assigns an element $c^{\mathfrak{S}} \in S$ to every constant $c \in \mathscr{C}$, a function $f^{\mathfrak{S}} : S^{\# f} \longrightarrow S$ to every function symbol $f \in \mathscr{F}$ and a relation $R^{\mathfrak{S}} \subseteq S^{\# R}$ to every relation symbol $R \in \mathscr{R}$.

Let $\mathfrak{S} = (S, {}^{\mathfrak{S}})$ be an $\mathscr{L}$-structure. An $\mathfrak{S}$-assignment is a family $\Phi = (\Phi^1, \Phi^2_n)_{n \in \omega \setminus \{0\}}$ of mappings $\Phi^1 : FV \longrightarrow S$ and $\Phi^2_n : FV^n_2 \longrightarrow \mathrm{Pow}(S^n)$ for all positive natural numbers $n$ where $FV$ denotes the set of all free object variables and $FV^n_2$ the set of all $n$-ary relation variables. Usually we write shortly $\Phi(u)$ instead of $\Phi^1(u)$ and $\Phi(U)$ instead of $\Phi^2_n(U)$ since it is mostly obvious from the context which mapping applies.

**4.2.6 Definition** (Inductive definition of the value $t^{\mathfrak{S}}[\Phi]$ for an $\mathscr{L}$-term $t$ in a structure $\mathfrak{S}$ with the $\mathfrak{S}$-assignment $\Phi$)

- $c^{\mathfrak{S}}[\Phi] := c^{\mathfrak{S}}$.

- $u^{\mathfrak{S}}[\Phi] := \Phi(u)$.

- $(ft_1 \ldots t_n)^{\mathfrak{S}}[\Phi] := f^{\mathfrak{S}}(t_1^{\mathfrak{S}}[\Phi], \ldots, t_n^{\mathfrak{S}}[\Phi])$.

[3] Here $F_u(x)$ and $F_U(X)$ stand for the symbol string that is obtained from $F$ replacing all occurrences of $u$ or $U$ by $x$ or $X$, respectively.

Observe that the interpretation $t^{\mathfrak{S}}[\Phi]$ of an $\mathscr{L}$-term is an element of $S$. To define that a structure satisfies a formula with an assignment we introduce the notation

$$\Phi \sim_u \Psi \quad :\Leftrightarrow \quad \Phi(v) = \Psi(v) \text{ for all } v \neq u$$

for $\mathfrak{S}$-assignments $\Phi$ and $\Psi$. We use the analogous notation also for the second-order variables.

**4.2.7 Definition** (Inductive definition of the satisfaction relation $\mathfrak{S} \models F[\Phi]$ for an $\mathscr{L}$-formula $F$ in an $\mathscr{L}$-structure $\mathfrak{S}$ with an $\mathfrak{S}$-assignment $\Phi$)

- $\mathfrak{S} \models (Rt_1 \dots t_n)$ iff $(t_1^{\mathfrak{S}}[\Phi], \dots, t_n^{\mathfrak{S}}[\Phi]) \in R^{\mathfrak{S}}$.
- $\mathfrak{S} \models (Ut_1, \dots, t_n)[\Phi]$ iff $(t_1^{\mathfrak{S}}[\Phi], \dots, t_n^{\mathfrak{S}}[\Phi]) \in \Phi(U)$.
- $\mathfrak{S} \models \neg F[\Phi]$ iff $\mathfrak{S} \not\models F[\Phi]$.
- $\mathfrak{S} \models (F \wedge G)[\Phi]$ iff $\mathfrak{S} \models F[\Phi]$ and $\mathfrak{S} \models G[\Phi]$.
- $\mathfrak{S} \models (F \vee G)[\Phi]$ iff $\mathfrak{S} \models F[\Phi]$ or $\mathfrak{S} \models G[\Phi]$.
- $\mathfrak{S} \models (F \to G)[\Phi]$ iff $\mathfrak{S} \not\models F[\Phi]$ or $\mathfrak{S} \models G[\Phi]$.
- $\mathfrak{S} \models (\forall x) F_u(x)[\Phi]$ iff $\mathfrak{S} \models F[\Psi]$ for all assignments $\Psi \sim_u \Phi$.
- $\mathfrak{S} \models (\exists x) F_u(x)[\Phi]$ iff $\mathfrak{S} \models F[\Psi]$ for some assignment $\Psi \sim_u \Phi$.
- $\mathfrak{S} \models (\forall X) F_U(X)[\Phi]$ iff $\mathfrak{S} \models F[\Psi]$ for all assignments $\Psi \sim_U \Phi$.
- $\mathfrak{S} \models (\exists X) F_U(X)[\Phi]$ iff $\mathfrak{S} \models F[\Psi]$ for some assignment $\Psi \sim_U \Phi$.

We read $\mathfrak{S} \models F[\Phi]$ as "*$\mathfrak{S}$ satisfies $F$ with the assignment $\Phi$*" or as "*$F$ is true in $\mathfrak{S}$ with the assignment $\Phi$*".

In the above definition we used $\mathfrak{S} \not\models F[\Phi]$ to denote that $\mathfrak{S}$ does not satisfy $F$ with the assignment $\Phi$. We will often denote by $F(v_1, \dots, v_n)$ that $FV(F) \subseteq \{v_1, \dots, v_n\}$. In general, however, it does not mean $FV(F) = \{v_1, \dots, v_n\}$. The same notation is also used for second-order variables.

Class terms of the form $\{x_1, \dots, x_n \mid A_{u_1,\dots,u_n}(x_1, \dots, x_n)\}$ are not constituents of the formal language. We will, however, use them in the form

$$(t_1, \dots, t_n) \, \varepsilon \, \{x_1, \dots, x_n \mid A_{u_1,\dots,u_n}(x_1, \dots, x_n)\}$$

as an "abbreviation" for the formula $A_{u_1,\dots,u_n}(t_1, \dots, t_n)$ (which in fact sometimes is easier to write and read). If $F$ and $A$ are formulas we define,

$$F_U(\{x_1, \dots, x_n \mid A_{u_1,\dots,u_n}(x_1, \dots, x_n)\})$$

as the formula which is obtained from $F$ by replacing all occurrences $(Ut_1 \dots t_n)$ in $F$ by $A_{u_1,\dots,u_n}(t_1,\dots,t_n)$. Sometimes we write just $F_U(A)$ or even shorter $F(A)$ if it is clear which variable is to be replaced.

We have introduced $\mathscr{L}$ as a second-order language. We call $FV(F)$ the *free first-order variables of* $F$, $BV(F)$ the *bounded first-order variables of* $F$ and analogously $FV_2(F)$ and $BV_2(F)$ the free or bounded second-order variables of $F$.

A term $t$ with $FV(t) = \emptyset$ is *closed.* A formula $F$ is a *sentence* if $FV(F) = FV_2(F) = \emptyset$. A formula $F$ is called *first-order* if $BV_2(F) = \emptyset$.

Observe that the value $t^{\mathfrak{S}}[\Phi]$ and and the satisfaction relation $\mathfrak{S} \models F[\Phi]$ only depend on the values of $\Phi$ on the free variables which occur in $t$ or $F$, respectively. i.e., we have

$$\Phi \restriction FV(t) = \Psi \restriction FV(t) \;\Rightarrow\; t^{\mathfrak{S}}[\Phi] = t^{\mathfrak{S}}[\Psi] \tag{4.1}$$

and

$$\Phi \restriction FV(F) \cup FV_2(F) = \Psi \restriction FV(F) \cup FV_2(F) \;\Rightarrow\; (\mathfrak{S} \models F[\Phi] \;\Leftrightarrow\; \mathfrak{S} \models F[\Psi]). \tag{4.2}$$

We say that an $\mathscr{L}$-formula $F$ is *satisfiable* if there is a structure $\mathfrak{S}$ and an $\mathfrak{S}$-assignment $\Phi$ such that $\mathfrak{S} \models F[\Phi]$. We call $F$ *valid (or true) in a structure* $\mathfrak{S}$ if $\mathfrak{S} \models F[\Phi]$ for all $\mathfrak{S}$-assignments $\Phi$.

It follows from (4.2) that sentences have a truth value in a structure which is independent of the choice of assignments.

**4.2.8 Definition** (Logical consequence) Let $M$ be a set of $\mathscr{L}$-formulas. We say that $F$ is a logical consequence of $M$, denoted by $M \models_L F$, if for every $\mathscr{L}$-structure $\mathfrak{S}$ and every $\mathfrak{S}$-assignment $\Phi$ which satisfies $\mathfrak{S} \models G[\Phi]$ for all $G \in M$ we also have $\mathfrak{S} \models F[\Phi]$.

A formula $F$ is logically valid , denoted by $\models F$, if it is a logical consequence of the empty set, i.e., if $\mathfrak{S} \models F[\Phi]$ for all $\mathscr{L}$-structures $\mathfrak{S}$ and all $\mathfrak{S}$-assignments $\Phi$.

Two formulas $F$ and $G$ are *logically equivalent*, denoted by $F \equiv_L G$, if $F \models_L G$ and $G \models_L F$.

We define $F \leftrightarrow G \;:\Leftrightarrow\; (F \to G) \land (G \to F)$ and obtain $F \equiv_L G$ if and only if $\models F \leftrightarrow G$.

The notion of logical consequence can be reformulated in terms of satisfiability.

**4.2.9 Theorem** *A formula $F$ is a logical consequence of a set $M$ of formulas if and only if the set $M \cup \{\neg F\}$ is unsatisfiable.*

*Proof* If $M \models_L F$ then $M \cup \{\neg F\}$ is unsatisfiable because every structure $\mathfrak{S}$ and every $\mathfrak{S}$-assignment $\Phi$ which satisfies all formulas in $M$ also satisfies $F$. Thus $\mathfrak{S} \not\models \neg F[\Phi]$. On the other hand if $M \cup \{\neg F\}$ is unsatisfiable then every structure $\mathfrak{S}$ and every $\mathfrak{S}$-assignment $\Phi$ which satisfies all formulas in $M$ has to falsify $\neg F$ and thus satisfies $F$. $\square$

An immediate consequence of Theorem 4.2.9 is the Deduction Theorem.

**4.2.10 Theorem** *(Deduction Theorem) A formula $F$ is a logical consequence of a set $M \cup \{A_1,\dots,A_n\}$ if and only if $A_1 \wedge \cdots \wedge A_n \to F$ is a logical consequence of $M$.*

*Proof* This is obvious by Theorem 4.2.9 since $M \cup \{A_1,\dots,A_n\} \cup \{\neg F\}$ is unsatisfiable iff $M \cup \{\neg(A_1 \wedge \cdots \wedge A_n \to F)\}$ is unsatisfiable. □

**4.2.11 Exercise** Let $\mathscr{L}$ be a first-order language.

(a) Define $t_u(s)$ and $F_u(s)$ inductively for $\mathscr{L}$-terms $s$ and $t$ and $\mathscr{L}$-formulas $F$

and prove that

(b) $s_u(t)$ is again an $\mathscr{L}$-term and $F_u(s)$ an $\mathscr{L}$-formula.

**4.2.12 Exercise** Give an inductive definition of $F_U(A)$ for $\mathscr{L}$-formulas $F$ and $A$ and prove that $F_U(A)$ is again a formula.

**4.2.13 Exercise** Prove the following claims.

(a) $\models (\forall x)F_u(x) \to F_x(t)$ for any $\mathscr{L}$-term $t$.

(b) $\models (\forall X)F_U(X) \to F_U(G)$ for any $\mathscr{L}$-formula $G$ such that $F_U(G)$ is again a formula.

Assume that $M$ is a set of $\mathscr{L}$-formulas. Show

(c) $M \models_L A \to F$ implies $M \models_L A \to (\forall x)F_u(x)$ if $u \notin FV(M \cup \{A\})$,

(d) $M \models_L A \to F$ implies $M \models_L A \to (\forall X)F_U(X)$ if $U \notin FV_2(M \cup \{A\})$.

**4.2.14 Exercise** Let $\mathscr{L}_{\mathbb{N}} = (\{\underline{0},\underline{1}\},\{\underline{+},\underline{\cdot}\},\{\underline{\leq},\underline{=}\})$ be the language of number theory and $\mathscr{N} := (\mathbb{N},\{0,1\},\{+,\cdot\},\{<,=\})$ be the $\mathscr{L}_{\mathbb{N}}$-structure of the natural numbers. Define formulas $F(u)$ and $G$ such that

(a) $\mathscr{N} \models F[\Phi]$ iff $\Phi(u)$ is a prime number.

(b) $\mathscr{N} \models G$ iff there are infinitely many natural numbers $n$ such that $n$ and $n+2$ are prime.

**4.2.15 Exercise** Let $F(u,v)$ be an $\mathscr{L}$-formula. Decide whether the following formulas are logically valid:

(a) $(\forall x)(\exists y)F_{u,v}(x,y) \to (\exists x)(\forall y)F_{u,v}(x,y)$

(b) $(\exists x)(\forall y)F_{u,v}(x,y) \to (\forall x)(\exists y)F_{u,v}(x,y)$

**4.2.16 Exercise**

(a) Let $F$ and $G$ be $\mathscr{L}$-formulas. Show

$$\models (\forall x)(F \land G)_u(x) \to (\forall x)F_u(x) \land (\forall x)G_u(x)$$

(b) Let $\mathscr{L}$ be a first-order language which contains the constant $\underline{0}$ and the relation symbol $=$. Find $\mathscr{L}$-formulas $F$ and $G$ such that

$$\not\models (\forall x)(F \lor G)_u(x) \to (\forall x)F_u(x) \lor (\forall x)G_u(x).$$

## 4.3 The TAIT-Calculus

As mentioned in the previous section, there are formal calculi which are sound and complete for $M \models_L F$ if $M \cup \{F\}$ is a set of first-order formulas. This section will introduce such a calculus which is especially adapted for proof-theoretic studies. In his pioneering papers [28] and [29] G. GENTZEN formalized logical reasoning by his *sequent calculus*. A sequent has the form $A_1,\dots,A_m \vdash B_1,\dots,B_n$, where $A_i$ and $B_j$ are first-order formulas. The intended meaning of this sequent is $\{A_1,\dots,A_m\} \models_L B_1 \lor \cdots \lor B_n$. The sequence $(A_1,\dots,A_m)$ is the *antecedent*, the sequence $(B_1,\dots,B_n)$ the *succedent* of the sequent. GENTZEN's striking result is his Hauptsatz which states that there is a sound and complete set of rules for the sequent calculus which is deviation free, i.e., every formula occurring in the conclusion of an inference already occurs in its premises. The paradigmatic example for an inference which incorporates a deviation is the cut rule

$$\frac{A_1,\dots,A_m \vdash B_1,\dots,B_n,F \qquad F,A_1,\dots,A_m \vdash B_1,\dots,B_n}{A_1,\dots,A_m \vdash B_1,\dots,B_n},$$

where the "deviation" via the cut-formula $F$ is unrecoverable from the conclusion. Examples for the rules in the sequent calculus are

($\land$-rules) $$\frac{\Delta,A \vdash \Gamma}{\Delta,A \land B \vdash \Gamma} \qquad\qquad \frac{\Delta_1 \vdash \Gamma_1,A \qquad \Delta_2 \vdash \Gamma_2,B}{\Delta_1,\Delta_2 \vdash \Gamma_1,\Gamma_2,A \land B}$$

and

($\lor$-rules) $$\frac{\Delta_1,A \vdash \Gamma_1 \qquad \Delta_2,B \vdash \Gamma_2}{\Delta_1,\Delta_2,(A \lor B) \vdash \Gamma_1,\Gamma_2} \qquad\qquad \frac{\Delta \vdash \Gamma,A}{\Delta \vdash \Gamma,A \lor B}.$$

These examples show that we need two inferences for every logical symbol, one for the antecedent and one for the succedent. Antecedent and succedent are connected via the rules for the negation symbol which are,

$$\frac{\Delta,A \vdash \Gamma}{\Delta \vdash \Gamma,\neg A} \qquad\qquad \frac{\Delta \vdash \Gamma,A}{\Delta,\neg A \vdash \Gamma}.$$

It was W. TAIT's observation that by removing the negation symbol from the basic propositional connectives, the distinction between antecedent and succedent in sequents becomes dispensable. This reduces the number of required inferences by half.[4]

To fix the syntactical framework for this book, we introduce a formal system for first-order logic (without identity) which is based on a one-sided sequent calculus à la TAIT. Let $\mathcal{L}$ be a first-order language. We define the Tait language $\mathcal{L}_T$ of $\mathcal{L}$ by introducing for every relation variable $U$ a new relation variable $\neg U$ and for every relation symbol $R$ a new relation symbol $\neg R$ of the same arity as $U$ or $R$, respectively. As logical symbols we only allow the positive boolean connectives $\wedge$ and $\vee$ and the quantifiers $\forall$ and $\exists$. Terms and formulas are defined according to Definitions 4.2.3 and 4.2.4 with respect to the extended number of relation variables and -symbols and the restricted number of logical connectives.

We expand an $\mathcal{L}$-structure $\mathfrak{M}$ to an $\mathcal{L}_T$-structure $\mathfrak{M}_T$ by interpreting $(\neg R)^{\mathfrak{M}}$ as the complement of $R^{\mathfrak{M}}$. An $\mathfrak{M}$-assignment $\Phi$ is extended to an $\mathfrak{M}_T$-assignment $\Phi_T$ by defining $\Phi_T(\neg U) := \{(s_1,\dots,s_n) \,|\, (s_1,\dots,s_n) \notin \Phi(U)\}$.

Negation is not among the logical symbols but becomes definable via de Morgan's laws.

**4.3.1 Definition** For a formula $F$ in the TAIT-language $\mathcal{L}_T$ we define ${\sim}F$ by the following clauses.

- ${\sim}(Rt_1,\dots,t_n) :\equiv (\neg Rt_1,\dots,t_n)$; ${\sim}(\neg Rt_1,\dots,t_n) :\equiv (Rt_1,\dots,t_n)$.
- ${\sim}(Us_1,\dots,s_n) :\equiv (\neg Us_1,\dots,s_n)$; ${\sim}(\neg Us_1,\dots,s_n) :\equiv (Us_1,\dots,s_n)$.
- ${\sim}(A \wedge B) :\equiv {\sim}A \vee {\sim}B$; ${\sim}(A \vee B) :\equiv {\sim}A \wedge {\sim}B$.
- ${\sim}(\forall x)F(x) :\equiv (\exists x){\sim}F(x)$; ${\sim}(\exists x)F(x) :\equiv (\forall x){\sim}F(x)$.
- ${\sim}(\forall X)F(X) :\equiv (\exists X){\sim}F(X)$; ${\sim}(\exists X)F(X) :\equiv (\forall X){\sim}F(X)$.

For any $\mathcal{L}$-structure $\mathfrak{M}$, any $\mathfrak{M}$-assignment $\Phi$ and any $\mathcal{L}_T$-formula $F$ we then obtain,

$$\mathfrak{M}_T \models {\sim}F[\Phi_T] \;\Leftrightarrow\; \mathfrak{M} \models \neg F[\Phi]. \tag{4.3}$$

Therefore we mostly identify $\neg F$ and ${\sim}F$ although $\neg$ is not among the basic connectives of $\mathcal{L}_T$.

Replacing $\neg$ by $\sim$ we obtain a translation $F \mapsto F^T$ from a language $\mathcal{L}$ into its TAIT-language $\mathcal{L}_T$. We commonly will, however, identify an $\mathcal{L}$-formula $F$ and its TAIT translation $F^T$. It will always be clear from the context whether we talk about $F$ or $F^T$.

---

[4] In a private talk William Tait told me that one of his objectives was to reduce the number of cases in proof theory.

The TAIT-calculus for first-order logic derives finite sets (not sequences) of first-order formulas in the TAIT-language which are to be interpreted disjunctively. Unless otherwise mentioned we will only talk about first-order formulas in this and the following section. We use upper case Greek letters $\Gamma$, $\Delta$, $\Lambda$, $\Gamma_1, \ldots$ as syntactical variables for finite sets of formulas. Instead of $\Delta \cup \Gamma$ we commonly write $\Delta, \Gamma$. Instead of $\Delta \cup \{F\}$ we write $\Delta, F$.

**4.3.2 Definition** (The TAIT-calculus) We define the derivation relation $\underset{\mathsf{T}}{\overset{m}{\models}} \Delta$ inductively by the following clauses.

(AxL) If $A$ is an atomic formula then $\underset{\mathsf{T}}{\overset{m}{\models}} \Delta, A, {\sim}A$ for all natural numbers $m$.

($\vee$) If $\underset{\mathsf{T}}{\overset{m_0}{\models}} \Delta, A_i$ for some $i \in \{1,2\}$, then $\underset{\mathsf{T}}{\overset{m}{\models}} \Delta, A_1 \vee A_2$ for all $m > m_0$.

($\wedge$) If $\underset{\mathsf{T}}{\overset{m_i}{\models}} \Delta, A_i$ and $m_i < m$ for all $i \in \{1,2\}$, then $\underset{\mathsf{T}}{\overset{m}{\models}} \Delta, A_1 \wedge A_2$.

($\exists$) If $\underset{\mathsf{T}}{\overset{m_0}{\models}} \Delta, A_v(t)$, then $\underset{\mathsf{T}}{\overset{m}{\models}} \Delta, (\exists x)A_v(x)$ for all $m > m_0$.

($\forall$) If $\underset{\mathsf{T}}{\overset{m_0}{\models}} \Delta, A(u)$ and $u$ not free in $\Delta$, then $\underset{\mathsf{T}}{\overset{m}{\models}} \Delta, (\forall x)A_u(x)$ for all $m > m_0$.

The derivation relation $\underset{\mathsf{T}}{\overset{m}{\models}} \Delta$ should be read as: *"There is a derivation of length $\leq m$ of the finite set $\Delta$"*.

As an immediate consequence of Definition 4.3.2 we obtain the *structural rule*

(STR) If $\underset{\mathsf{T}}{\overset{m}{\models}} \Delta$, $m \leq n$ and $\Delta \subseteq \Gamma$ then $\underset{\mathsf{T}}{\overset{n}{\models}} \Gamma$,

which is proved by induction on $m$. We leave the straightforward proof by induction on $m$ as an exercise.

First we observe that this calculus is sound. Let $\bigvee \Delta := \bigvee \{F \mid F \in \Delta\}$.

**4.3.3 Theorem** *(Soundness Theorem) If $\underset{\mathsf{T}}{\overset{m}{\models}} \Delta$ then $\models \bigvee \Delta$.*

*Proof* Let $\mathfrak{M}$ be an $\mathscr{L}$-structure and $\Phi$ an $\mathfrak{M}$-assignment. We show

$$\underset{\mathsf{T}}{\overset{m}{\models}} \Delta \text{ implies } \mathfrak{M} \models \bigvee \Delta[\Phi] \tag{i}$$

by induction on $m$.

If the premise of (i) holds by (AxL) then $\Delta$ contains a formula $A$ and its dual ${\sim}A$. According to (4.3) on p. 53 we obtain $\mathfrak{M} \models (A \vee {\sim}A)[\Phi]$ which implies $\mathfrak{M} \models \bigvee \Delta[\Phi]$.

If the premise of (i) holds by a rule we obtain the claim from the induction hypothesis because the rules preserve validity. Let us just check the case of an inference $(\forall)$ to pinpoint the role of the variable condition in this rule. If $\mathfrak{M} \not\models (\forall x)A_u(x)[\Phi]$ then there is an assignment $\Psi \sim_u \Phi$ such that $\mathfrak{M} \not\models A[\Psi]$. From the premise $\overset{m_0}{\underset{\mathsf{T}}{\models}} \Delta, A$ we obtain $\mathfrak{M} \models (\bigvee\Delta \vee A)[\Psi]$ by the induction hypothesis. Hence $\mathfrak{M} \models \bigvee\Delta[\Psi]$. But, since $u$ does not occur in $\Delta$, we have $\mathfrak{M} \models \bigvee\Delta[\Psi]$ if and only if $\mathfrak{M} \models \bigvee\Delta[\Phi]$ and thus also $\mathfrak{M} \models (\bigvee\Delta \vee (\forall x)A_u(x))[\Phi]$. □

**4.3.4 Exercise** Give an inductive definition of the translation $F \mapsto F^T$.

## 4.4 Trees and the Completeness Theorem

We are now going to show that the TAIT-calculus is also complete.

**4.4.1 Theorem** *(Completeness Theorem) Assume $M \models_L \bigvee\Delta$ for a countable set $M$ and a finite set $\Delta$ of $\mathcal{L}_T$-formulas. Then there is a finite subset $\Gamma \subseteq M$ and an $m \in \omega$ such that $\overset{m}{\underset{\mathsf{T}}{\models}} \neg\Gamma, \Delta$, where $\neg\Gamma := \{\neg G \mid G \in \Gamma\}$.*

We will prove Theorem 4.4.1 by the method of *search trees* introduced by Schütte. Therefore we have to describe trees mathematically. A tree can be visualized as shown in Fig. 4.1. A tree grows out of its root. We enumerate the immediate

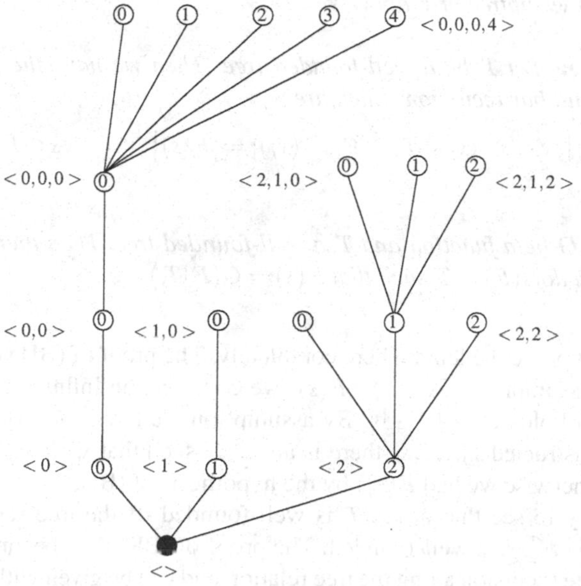


**Fig. 4.1** Visualization of a (mathematical) tree

successors of a node from left to right. Then every node in a tree can be addressed by a finite sequence of numbers as indicated in Fig. 4.1. This leads to the following definition.

**4.4.2 Definition** A *tree* is a set of sequence numbers which is closed under initial segments. i.e.,

$$Tree(T) \;:\Leftrightarrow\; T \subseteq Seq \wedge (\forall s \in T)[t \subseteq s \rightarrow t \in T],$$

where for sequence numbers $s$ and $t$ we define,

$$t \subseteq s \;:\Leftrightarrow\; lh(t) \leq lh(s) \wedge (\forall i < lh(t))[(t)_i = (s)_i].$$

A node $s \in T$ is a *leaf* if it possesses no successors in $T$, i.e., if $(\forall x)[s^\frown\langle x\rangle \notin T]$.

A *path* in a tree $T$ is a subset $P \subseteq T$ which is linearly ordered by $\subseteq$ and closed under initial sequences, i.e.,

$$P \text{ is a path in } T \;:\Leftrightarrow\; P \subseteq T \wedge (\forall s \in P)(\forall t \in P)[s \subseteq t \vee t \subseteq s] \wedge (\forall s)(\forall t \in P)[s \subseteq t \rightarrow s \in P].$$

Observe that an infinite path $P$ can be identified with a function $f\colon \mathbb{N} \longrightarrow \mathbb{N}$ such that $P = \{[f](m) \mid m \in \omega\}$, where $[f](m) := \langle f(0), \ldots, f(m-1)\rangle$ codes the course-of-values of $f$ below $m$.

A tree $T$ is well-founded if and only if it possesses no infinite paths.

For $s \in T$ we define,

$$T_s := \{t \mid s^\frown t \in T\}$$

and call $T_s$ the *subtree of $T$ above $s$*.

**4.4.3 Lemma** *Let $T$ be a well-founded tree. Then we have the principles of bar induction and bar recursion which are*

*(BI)* $\quad (\forall s)\big[(\forall x)[s^\frown\langle x\rangle \in T \Rightarrow F(s^\frown\langle x\rangle)] \Rightarrow F(s)\big] \;\Rightarrow\; (\forall s \in T)[F(s)]$

*and*

*(BR)* $\quad$ *Let $G$ be a function and $T$ a well-founded tree. Then there is a function $F$ with $dom(F) = T$ such that $F(s) = G(F{\restriction}T_s)$.*

We will not prove the lemma here completely. The proof of (BI) is straight forward. From the assumption $(\exists s \in T)[\neg F(s)]$ we construct an infinite path $s_0, s_1, \ldots$ such that $\neg F(s_i)$ holds for all $i \in \mathbb{N}$. By assumption we have $\neg F(s_0)$ for some $s_0 \in \mathbb{N}$. Having constructed $s_0, \ldots, s_n$ there is an $i \in \mathbb{N}$ such that $s_n^\frown\langle i\rangle \in T$ and $\neg F(s_n^\frown\langle i\rangle)$ because otherwise we had $F(s_n)$ by the hypothesis of (BI).

It is easy to see that a tree $T$ is well-founded iff the tree relation, defined by $s \prec_T t \;:\Leftrightarrow\; t \subsetneq s$, is well-founded. The proof of (BR) then becomes a special case of transfinite recursion along the tree relation and can be given within the framework of set theory (cf. Theorem 3.2.8 and Exercise 4.4.13). □

Since it is easy to visualize how a function defined by bar recursion is constructed, bar recursion is sometimes regarded as a constructive principle.[5]

Using bar recursion we define the order type of a node $s$ in a well-founded tree $T$ by

$$\mathrm{otyp}_T(s) := \sup\{\mathrm{otyp}_T(s^\frown\langle x\rangle)+1 \mid s^\frown\langle x\rangle \in T\}$$

and

$$\mathrm{otyp}(T) := \mathrm{otyp}_T(\langle\,\rangle).$$

Bar induction on a well-founded tree $T$ then corresponds to induction on $\mathrm{otyp}_T(s)$.

These are the preliminaries which we need in the definition of the search tree. To define the search tree for a finite set $\Delta$ of $\mathscr{L}_T$-formulas we order $\Delta$ arbitrarily to obtain a finite sequence $\langle\Delta\rangle$. Instead of $\langle\Delta,F\rangle$ we write shortly $\langle\Delta\rangle,F$. The first formula in $\langle\Delta\rangle$, i.e., the formula with lowest index in the sequence $\langle\Delta\rangle$, which is not atomic, is its *redex*, denoted by $R(\langle\Delta\rangle)$. We put $R(\langle\Delta\rangle) := \emptyset$ if $\langle\Delta\rangle$ possesses no redex. The sequence $\langle\Delta\rangle^r$ is obtained by discarding the redex $R(\langle\Delta\rangle)$ in $\langle\Delta\rangle$. Assume that $M$ is a countable set of $\mathscr{L}_T$-formulas and fix an enumeration for the formulas in $M$. Observe that there are only countably many terms which can be built from the variables, constants and function symbols occurring in $M \cup \Delta$. Let $\{t_i \mid i \in \mathbb{N}\}$ be an enumeration of these terms.

We define the search tree $S^M_{\langle\Delta\rangle}$ together with the label function $\delta$ by the following clauses.

$(S_{\langle\rangle})$ $\langle\rangle \in S^M_{\langle\Delta\rangle} \wedge \delta(\langle\rangle) = \langle\Delta\rangle$

For the coming clauses assume $s \in S^M_{\langle\Delta\rangle}$ and that $\delta(s)$, viewed as a finite set, is not an axiom according to (AxL).

$(S_{Id})$ If $R(\delta(s)) = \emptyset$ then $s^\frown\langle 0\rangle \in S^M_{\langle\Delta\rangle}$ and $\delta(s^\frown\langle 0\rangle) = \delta(s), \neg G_i$, where $G_i$ is the first formula in the enumeration of the set $M$ such that $\neg G_i \notin \bigcup_{s_0 \subseteq s} \delta(s_0)$. If there is no such $G_i$ then $\delta(s^\frown\langle 0\rangle) = \delta(s)$.

$(S_\wedge)$ If $R(\delta(s))$ is a formula $(A_0 \wedge A_1)$ then $s^\frown\langle i\rangle \in S_{\langle\Delta\rangle}$ and $\delta(s^\frown\langle i\rangle) = \delta(s)^r, A_i$ for $i = 0, 1$

$(S_\forall)$ If $R(\delta(s))$ is a formula $((\forall x)A_v(x))$ then $s^\frown\langle 0\rangle \in S_{\langle\Delta\rangle}$ and $\delta(s^\frown\langle 0\rangle) = \delta(s)^r, A_v(u)$, where $u$ is the first free variable in a given enumeration of the free variables such that $u$ does not occur in $\delta(s)$ and $M$.

$(S_\vee)$ If the redex $R(\delta(s))$ is a formula $(A_0 \vee A_1)$ then $s^\frown\langle 0\rangle \in S^M_{\langle\Delta\rangle}$ and $\delta(s^\frown\langle 0\rangle) = \delta(s)^r, A_i, \neg G_j, R(\delta(s))$, where $A_i$ is the first formula in $\langle A_0, A_1\rangle$ which does not occur in $\bigcup_{s_0 \subseteq s}\delta(s_0)$ and $G_j$ the first formula in the enumeration of $M$ such that $\neg G_j$ does not occur $\bigcup_{s_0 \subseteq s}\delta(s_0)$. If such a formula $A_i$ or $G_j$ does not exist we put $A_i = \emptyset$ or $G_j = \emptyset$, respectively.

[5] The constructiveness of a function defined by bar recursion depends of course on the complexity of the well-founded tree.

($S_\exists$) If $R(\delta(s))$ is a formula $(\exists x)A_v(x)$ then $s^\frown\langle 0\rangle \in S^M_{\langle\Delta\rangle}$ and we define $\delta(s^\frown\langle 0\rangle) = \delta(s)^r, A_v(t), \neg G_i, R(\delta(s))$, where $G_i$ is the first formula in the enumeration of $M$ such that $\neg G_i \notin \bigcup_{s_0 \subseteq s} \delta(s_0)$ and $G_i = \emptyset$ if such a formula does not exist, and $t$ is the first term in the enumeration of the terms such that $A_v(t)$ does not occur in $\bigcup_{s_0 \subseteq s} \delta(s_0)$. If no such term exists then $\delta(s^\frown\langle 0\rangle) = \delta(s)^r, R(\delta(s)), \neg G_i$.

We observe that the search tree is a binary tree. If $S^M_{\langle\Delta\rangle}$ is well-founded then $S^M_{\langle\Delta\rangle}$ is finite by König's lemma and therefore all nodes in $S^M_{\langle\Delta\rangle}$ have finite order type. We first show,

**4.4.4 Lemma** *(Syntactical Main Lemma) If the search tree $S^M_{\langle\Delta\rangle}$ is well-founded then for every $s \in S^M_{\langle\Delta\rangle}$ there is a finite set $\Gamma_s \subseteq M$ such that $\vdash_{\mathsf{T}}^{\mathrm{otyp}(s)} \neg\Gamma_s, \delta(s)$, where* otyp($s$) *denotes the order type of s in the well-founded trees $S^M_{\langle\Delta\rangle}$.*

*Proof* We show the lemma by bar induction on $S^M_{\langle\Delta\rangle}$. If $s$ is a leaf in $S^M_{\langle\Delta\rangle}$ then none of the clauses ($S_{Id}$) through ($S_\exists$) apply. Therefore $\delta(s)$, viewed as a set, is an axiom according to (AxL) and we obtain $\vdash_{\mathsf{T}}^{0} \delta(s)$ and choose $\Gamma_s := \emptyset$. So assume $s \in S^M_{\langle\Delta\rangle}$ such that $\delta(s)$ is not an axiom according to (AxL). If $R(\delta(s)) = \emptyset$ then $s^\frown\langle 0\rangle \in S^M_{\langle\Delta\rangle}$ and by induction hypothesis there is a finite set $\Gamma_{s^\frown\langle 0\rangle}$ such that $\vdash_{\mathsf{T}}^{\mathrm{otyp}(s^\frown\langle 0\rangle)} \neg\Gamma_{s^\frown\langle 0\rangle}, \delta(s^\frown\langle 0\rangle)$. But either $\delta(s^\frown\langle 0\rangle) = \delta(s)$ or $\delta(s^\frown\langle 0\rangle) = \delta(s), \neg G_i$ for some formula $G_i \in M$ and we obtain $\vdash_{\mathsf{T}}^{\mathrm{otyp}(s)} \neg\Gamma_s, \delta(s)$ by putting $\Gamma_s := \Gamma_{s^\frown\langle 0\rangle}$ in the first case and $\Gamma_s := \Gamma_{s^\frown\langle 0\rangle} \cup \{G_i\}$ in the second.

For $R(\delta(s)) \neq \emptyset$ we have to distinguish cases according to the shape of $R(\delta(s))$.

If $R(\delta(s))$ is a formula $A_0 \wedge A_1$ we are in the case of ($S_\wedge$). Then $s^\frown\langle i\rangle \in S^M_{\langle\Delta\rangle}$ for $i = 0, 1$ and we obtain $\vdash_{\mathsf{T}}^{\mathrm{otyp}(s^\frown\langle i\rangle)} \neg\Gamma_{s^\frown\langle i\rangle}, \delta(s^\frown\langle i\rangle)$ by the induction hypothesis, where $\delta(s^\frown\langle i\rangle) = \delta(s)^r, A_i$. By an inference ($\wedge$) we then obtain $\vdash_{\mathsf{T}}^{\mathrm{otyp}(s)} \neg\Gamma_s, \delta(s)$ for $\Gamma_s := \Gamma_{s^\frown\langle 0\rangle} \cup \Gamma_{s^\frown\langle 1\rangle}$.

In the case that $R(\delta(s))$ is a formula $(\forall x)A_v(x)$ we have $s^\frown\langle 0\rangle \in S^M_{\langle\Delta\rangle}$ and obtain $\vdash_{\mathsf{T}}^{\mathrm{otyp}(s^\frown\langle 0\rangle)} \neg\Gamma_{s^\frown\langle 0\rangle}, \delta(s)^r, A_v(u)$ by the induction hypothesis where $u$ does not occur in $\delta(s)$ and $\Gamma_{s^\frown\langle 0\rangle}$. Putting $\Gamma_s := \Gamma_{s^\frown\langle 0\rangle}$ we obtain $\vdash_{\mathsf{T}}^{\mathrm{otyp}(s)} \neg\Gamma_s, \delta(s)$ by an inference ($\forall$).

Now assume that $R(\delta(s))$ is a formula $A_0 \vee A_1$. Then $s^\frown\langle 0\rangle \in S^M_{\langle\Delta\rangle}$ and we obtain $\vdash_{\mathsf{T}}^{\mathrm{otyp}(s^\frown\langle 0\rangle)} \neg\Gamma_{s^\frown\langle 0\rangle}, \delta(s), \neg G_j$ or $\vdash_{\mathsf{T}}^{\mathrm{otyp}(s^\frown\langle 0\rangle)} \neg\Gamma_{s^\frown\langle 0\rangle}, \delta(s), A_i, \neg G_j$ for some $i \in \{0, 1\}$ and some formula $G_j \in M$ by the induction hypothesis. Put

$\Gamma_s := \Gamma_{s^\frown\langle 0\rangle} \cup \{G_j\}$. In the first case we get by the structural rule $\frac{|\text{otyp}(s)}{\mathsf{T}} \neg\Gamma_s, \delta(s)$, in the second case we need an inference $(\vee)$ to get $\frac{|\text{otyp}(s)}{\mathsf{T}} \neg\Gamma_s, \delta(s)$.

The last case is that $R(\delta(s))$ is a formula $(\exists x)A_u(x)$. Then $s^\frown\langle 0\rangle \in S^M_{\langle\Delta\rangle}$ and we get $\frac{|\text{otyp}(s^\frown\langle 0\rangle)}{\mathsf{T}} \neg\Gamma_{s^\frown\langle 0\rangle}, \delta(s), \neg G_i$ or $\frac{|\text{otyp}(s^\frown\langle 0\rangle)}{\mathsf{T}} \neg\Gamma_{s^\frown\langle 0\rangle}, \delta(s), A_u(t), \neg G_i$ for some formula $G_i \in M$ and some $\mathscr{L}$-term $t$ by the induction hypothesis. Letting $\Gamma_s := \Gamma_{s^\frown\langle 0\rangle} \cup \{G_i\}$ we get $\frac{|\text{otyp}(s)}{\mathsf{T}} \neg\Gamma_s, \delta(s)$ either by the structural rule or by an application of an inference $(\exists)$. □

**4.4.5 Lemma** *(Semantical Main Lemma) If the search tree $S^M_{\langle\Delta\rangle}$ is not well-founded then there is a model $\mathfrak{M}$ and an assignment $\Phi$ over $\mathfrak{M}$ such that $\mathfrak{M} \models G[\Phi]$ for all formulas $G \in M$ but $\mathfrak{M} \not\models F[\Phi]$ for all formulas in $\Delta$.*

*Proof* Assume that $S^M_{\langle\Delta\rangle}$ is not well-founded. Then there is an infinite path in $S^M_{\langle\Delta\rangle}$. Such a path can be viewed as a function $f{:}\mathbb{N} \longrightarrow \mathbb{N}$ such that $[f](n) := \langle f(0), \ldots, f(n-1)\rangle \in S^M_{\langle\Delta\rangle}$ for all $n \in \mathbb{N}$. Let $\delta(f) := \bigcup_{n\in\mathbb{N}} \delta([f](n))$. We pick an infinite path $f$ in $S^M_{\langle\Delta\rangle}$ and check the following properties.

(Atm) If an atomic formula $A$ occurs in $\delta([f](n))$ then $A$ occurs in all $\delta([f](m))$ for $m \geq n$.

$(M)$ It is $\neg G \in \delta(f)$ for all $G \in M$.

(Red) If a non atomic formula $F$ occurs in $\delta([f](n))$ then there is a $m \geq n$ such that $F = R(\delta([f](m)))$.

$(\wedge)$ If a formula $(A_0 \wedge A_1)$ occurs in $\delta(f)$ then there is an $i \in \{0,1\}$ such that $A_i \in \delta(f)$.

$(\forall)$ If a formula $((\forall x)A_v(x))$ occurs in $\delta(f)$ then there is a variable $u$ such that $A_v(u)$ occurs in $\delta(f)$.

$(\vee)$ If a formula $(A_0 \vee A_1)$ occurs in $\delta(f)$ then $A_i \in \delta(f)$ for every $i \in \{0,1\}$.

$(\exists)$ If a formula $((\exists x)A_v(x))$ occurs in $\delta(f)$ then $A_v(t) \in \delta(f)$ for every term $t$.

Property (Atm) is obvious because atomic formulas are never discarded.

Property $(M)$ follows because either rule $S_\emptyset$, $S_\vee$ or rule $S_\exists$ is infinitely often applied in $f$.

We prove property (Red) by induction on the number of non atomic formulas in $\delta([f](n))$ with lower index than $F$. If there is no such formula then $F$ is the redex in $\delta([f](n))$. If $\delta([f](n))$ possesses a redex with lower index than $F$ then $F$ is discharged in $\delta([f](m+1))$ and by induction hypothesis there is an $m \geq n+1$ such that $F = R(\delta([f](m)))$.

Property ($\wedge$) follows because by property (Red) there is an $m$ such that $R(\delta([f](m))) = (A_0 \wedge A_1)$. Then $[f](m+1) = [f](m)^\frown\langle i\rangle$ and $A_i \in \delta([f](m+1))$ for some $i \in \{0,1\}$.

Similarly we show property ($\forall$). If $(\forall x)A_v(x)$ occurs in $\delta(f(n))$ there is an an $m$ such that $R(\delta([f](m))) = (\forall x)A_v(x)$. Then $A_v(u) \in \delta([f](m+1))$ which proves ($\forall$).

To prove property ($\vee$) assume first $A_0 \notin \delta(f)$. There is an $m$ such that $R(\delta([f](m))) = (A_0 \vee A_1)$. Then we have $[f](m+1) = [f](m)^\frown\langle 0\rangle$ and obtain $\delta([f](m+1)) = \delta([f](m))^r, A_0, \neg G_j, A_0 \vee A_1$ because $A_0 \notin \bigcup_{k \leq m} \delta([f](k))$. Therefore the assumption $A_0 \notin \delta(f)$ is false. Now assume $A_0 \in \delta(f)$ but $A_1 \notin \delta(f)$. Then there is an $n_0$ such that $A_0 \in \delta([f](n_0))$. Since also $(A_0 \vee A_1) \in \delta([f](n_0))$ there is by (Red) an $n \geq n_0$ such that $(A_0 \vee A_1) = R(\delta([f](n)))$. But then $A_1 \in \delta([f](n+1))$ showing that also the assumption $A_1 \notin \delta(f)$ is false.

We finally regard property ($\exists$) and show $A_v(t_i) \in \delta(f)$ by induction on the index $i$ in our fixed enumeration of the terms. Assume $\{A_v(t_j), (\exists x)A_v(x)\} \subseteq \bigcup_{l<m_0} \delta([f](l))$ for all $j < i$ but $A_v(t_i) \notin \bigcup_{l<m_0} \delta([f](l))$. Observe that $(\exists x)A_v(x)$ is never discarded in $\delta(f)$. By (Red) there is therefore an $m \geq m_0$ such that $R(\delta([f](m))) = (\exists x)A_v(x)$. Then $[f](m+1) = [f](m)^\frown\langle 0\rangle$ and $\{A_v(t_i), (\exists x)A_v(x)\} \subseteq \delta([f](m+1))$.

We define a model $\mathfrak{M}$ together with an $\mathfrak{M}$-assignment $\Phi$. The domain $M$ of the model $\mathfrak{M}$ is given by,

$$M := \{t_i \mid i \in \mathbb{N}\}$$

and the assignment

$$\Phi(u) := u \text{ for all free variables } u.$$

Any constant $c$ is interpreted by $c^{\mathfrak{M}} := c$ and the interpretation of a function symbol $f$ is defined by,

$$f^{\mathfrak{M}}(t_1,\ldots,t_n) := (f(t_1,\ldots,t_n)).$$

Then we obtain $t^{\mathfrak{M}}[\Phi] = t$ for all terms $t$ by a simple induction on the the complexity of the term $t$ and define the interpretation of an $n$-ary relation symbol by,

$$R^{\mathfrak{M}} := \{t_1,\ldots,t_n \mid (\neg Rt_1,\ldots,t_n) \in \delta(f)\}.$$

From the above properties we then obtain,

$$F \in \delta(f) \;\Rightarrow\; \mathfrak{M} \not\models F[\Phi] \tag{i}$$

by induction on the complexity of the formula $F$. If $F$ is an atomic formula $Rt_1,\ldots,t_n$ then $(\neg Rt_1,\ldots,t_n) \notin \delta(f)$ by (Atm) since otherwise there existed an $m$ such that

$$\{(Rt_1,\ldots,t_n), (\neg Rt_1,\ldots,t_n)\} \subseteq \delta([f](m))$$

contradicting the infinity of the path $f$. Hence $(t_1^{\mathfrak{M}}[\Phi],\ldots,t_n^{\mathfrak{M}}[\Phi]) \notin R^{\mathfrak{M}}$, i.e., $\mathfrak{M} \not\models F[\Phi]$. If $F$ is a formula $(\neg Rt_1,\ldots,t_n)$ we obtain $(t_1^{\mathfrak{M}}[\Phi],\ldots,t_n^{\mathfrak{M}}[\Phi]) \in R^{\mathfrak{M}}$ by definition which in turn entails $\mathfrak{M} \not\models (\neg Rt_1,\ldots,t_n)[\Phi]$. In case that $F$ is not atomic we obtain the claim immediately from the induction hypothesis using properties $(\wedge)$ through $(\exists)$.

Since $\neg G \in \delta(f)$ for all $G \in M$ we obtain from (i) $\mathfrak{M} \models G[\Phi]$ for all $G \in M$. Since $\langle \Delta \rangle = \delta([f](0))$ we finally get $\mathfrak{M} \not\models G[\Phi]$ for all $G \in \Delta$. □

We can now prove Theorem 4.4.1. Let $M$ be countable and $\Delta$ finite and assume $\not\models^{m}_{\mathsf{T}} \neg\Gamma,\Delta$ for all $m \in \omega$ and all finite subsets $\Gamma \subseteq M$. Then $S^M_{\langle\Delta\rangle}$ is not well-founded by the Syntactical Main Lemma. By the Semantical Main Lemma we therefore find a model $\mathfrak{M}$ and an $\mathfrak{M}$-assignment $\Phi$ such that $\mathfrak{M} \models M[\Phi]$ but $\mathfrak{M} \not\models \bigvee\Delta[\Phi]$. Hence $M \not\models_L \bigvee\Delta$. □

The next theorem is the combination of the Soundness and the Completeness Theorem.

**4.4.6 Theorem** *Let $M$ be a countable set of $\mathscr{L}_T$-formulas. Then we have $M \models_L F$ if and only if there is a finite subset $\Gamma \subseteq M$ and an $m \in \omega$ such that $\models^{m}_{\mathsf{T}} \neg\Gamma, F$.*

**4.4.7 Corollary** *Let $F$ be a first-order formula. Then $\models F$ if and only if there is an $m \in \omega$ such that $\models^{m}_{\mathsf{T}} F$.*

*Proof* Let $M = \emptyset$ and $\Delta = \{F\}$ in Theorem 4.4.6. □

By a theory we understand a set of sentences. We call a theory $T$ also an axiom system if $A \in T$ is decidable. A formula $F$ is provable from a theory $T$ if $T \models_L F$. We obtain from Theorem 4.4.6 that a formula is provable from a first-order theory $T$ if and only if there are finitely many sentences $A_1,\ldots,A_n$ in $T$ such that $\models^{m}_{\mathsf{T}} \neg A_1,\ldots,\neg A_n, F$ for some $m < \omega$.

**4.4.8 Definition** For a first-order theory $T$, we define,

$$T \vdash_{\mathsf{T}} F \;:\Leftrightarrow\; \text{there is an } m \text{ and there are sentences } A_1,\ldots,A_n \text{ in } T \text{ such that } \models^{m}_{\mathsf{T}} \neg A_1,\ldots,\neg A_n, F. \tag{4.4}$$

(Remark: Don't confuse the $T$ standing for the theory $T$ with the subscript $\vdash_{\mathsf{T}}$ standing for "Tait").

**4.4.9 Remark** We have introduced the notion of logical consequence for pure predicate logic without identity. However, we will always assume that a language contains the binary identity symbol $=$, whose intended meaning is equality. Therefore we assume that any theory, formulated in a language $\mathscr{L} = \mathscr{L}(\mathscr{C},\mathscr{F},\mathscr{R})$, contains the defining axioms for $=$, i.e., the identity axioms $\mathsf{IDEN}$ which are

- $(\forall x)[x = x]$
- $(\forall x)(\forall y)[x = y \rightarrow y = x]$

- $(\forall x)(\forall y)(\forall z)[x = y \land y = z \to x = y]$
- $(\forall x_1)\dots(\forall x_n)(\forall y_1)\dots(\forall y_n)[\bigwedge_{i=1}^n x_i = y_i \to f(x_1,\dots,x_n) = f(y_1,\dots,y_n)]$ for all $f \in \mathscr{F}$

and

- $(\forall x_1)\dots(\forall x_n)(\forall y_1)\dots(\forall y_n)[\bigwedge_{i=1}^n x_i = y_i \to (Rx_1,\dots,x_n) \to (Ry_1,\dots,y_n)]$ for all $R \in \mathscr{R}$.

Sometimes we also write $T \overset{m}{\underset{\mathsf{T}}{\vdash}} \Delta$ to denote that there are finitely many axioms $A_1,\dots,A_n$ of $T$ (including the identity axioms) such that $\overset{m}{\underset{\mathsf{T}}{\vdash}} \neg A_1,\dots,\neg A_n,\Delta$.

As another corollary of the Completeness Theorem we obtain the Compactness Theorem for countable sets $M$.

**4.4.10 Corollary** *(Compactness Theorem) Let $M$ be a countable set of $\mathscr{L}$-formulas. If $M \models_L F$ then there is already a finite subset $N \subseteq M$ such that $N \models_L F$.*

*Proof* If $M \models_L F$ we get $\underset{\mathsf{T}}{\vdash} \neg N, F$ for a finite subset $N \subseteq M$ by Theorem 4.4.6. Then $\models \bigvee \neg N, F$ by the Soundness Theorem (Theorem 4.3.3). Using the Deduction Theorem (Theorem 4.2.10) we finally obtain $N \models_L F$. □

For a more familiar formulation of the Compactness Theorem we go back to Theorem 4.2.9 to obtain:

$$\begin{aligned} &\textit{The theory } T \cup \{\neg F\} \textit{ has no model} \Leftrightarrow \\ &T \models_L F \Leftrightarrow \\ &N \models_L F \textit{ for some finite set } N \subseteq T \Leftrightarrow \end{aligned} \tag{4.5}$$

*There is a finite subset $N \subseteq T$ such that $N \cup \{\neg F\}$ has no model.*

Choosing $F$ as a true sentence and taking contrapositions in (4.5), we obtain the Compactness Theorem in its more familiar formulation.

**4.4.11 Theorem** *(Compactness Theorem) A countable theory $T$ possesses a model if and only if every finite subset of $T$ possesses a model.*

We just want to remark that the Compactness Theorem is also true for uncountable theories. This needs, however, a modified proof which we omitted since only countable (even recursive) theories are proof-theoretically interesting.

**4.4.12 Exercise** Let $M$ be a countable set of formulas. Extend the definition of the TAIT-calculus to a calculus $M \overset{m}{\underset{\mathsf{T}}{\vdash}} \Delta$ by replacing $\overset{m}{\underset{\mathsf{T}}{\vdash}} \Delta$ everywhere in Definition 4.3.2 by $M \overset{m}{\underset{\mathsf{T}}{\vdash}} \Delta$ and adding a *theory rule*

(MR) If $M \overset{m_0}{\underset{\mathsf{T}}{\models}} \neg A, \Delta$ and $A \in M$ then $M \overset{m}{\underset{\mathsf{T}}{\models}} \Delta$ for all $m \geq m_0$.

Show that this definition coincides with the definition given in (4.4).

**4.4.13 Exercise** Let $T \subseteq Seq$ be a tree. The *tree-ordering* $\prec_T$ is defined on $T$ by

$$s \prec_T t :\Leftrightarrow (\exists u \in Seq)[u \neq \langle\rangle \wedge t^\frown u = s].$$

Prove:

(a) The tree $T$ is well-founded iff the tree ordering $\prec_T$ is well-founded.

For a well-founded tree $T$ define $\operatorname{depth}(T)$ as the least ordinal $\alpha$ such that there is a function

$$f{:}T \longrightarrow \alpha + 1 \text{ such that } (\forall s,t \in T)\big[s \prec_T t \Rightarrow f(s) < f(t)\big].$$

(b) Show $\operatorname{depth}(T) = \operatorname{otyp}(T)$.

Hint: Show first

$$\operatorname{depth}(T) = \sup\{\operatorname{depth}(T_{\langle a\rangle}) + 1 \mid \langle a\rangle \in T\}.$$

The Kleene–Brouwer-ordering on a tree $T$ is defined by,

$$s <_{T_{\mathrm{KB}}} t :\Leftrightarrow s \in T \wedge t \in T \wedge \big[s \prec_T t \vee (\exists b)[b \subseteq s \wedge b \subseteq t \wedge (s)_{lh(b)} < (t)_{lh(b)}]\big]$$

(c) Show that $<_{T_{\mathrm{KB}}}$ is a linear ordering.

(d) Show that $T$ is well-founded iff $<_{T_{\mathrm{KB}}}$ is well-founded.

(e) Compute and upper bound for $\operatorname{otyp}(<_{T_{\mathrm{KB}}})$ in terms of $\operatorname{depth}(T)$.

(f) Show that $\operatorname{otyp}(T) < \omega_1^{CK}$ for every primitive recursive tree $T$.

Hint: Use the Kleene–Brouwer-ordering.

**4.4.14 Exercise** A hydra is a finite tree $H$ (c.f. Definition 3.3.5). The heads of the hydra are the leaves of $H$ different from $\langle\ \rangle$. A hydra without heads is dead. A battle between Hercules and the hydra $H$ runs as follows. Hercules consecutively chops off one of the heads $s$ of the Hydra. If the neck of $s$ was sufficiently long the hydra $H$ regrows $m$ many new necks each of them carrying the remaining number of heads, i.e., it turns into the hydra $H(s,m)$ according to the following rules:

- If $lh(s) = 1$ then $H(s,m) := H \setminus \{s\}$.
- If $lh(s) = n+1 > 1$ then

$$H(s,m) := H \setminus \{s\} \cup$$
$$\{\langle (s)_0, \ldots, (s)_{n-2}\rangle^\frown \langle l\rangle^\frown b \mid b \in H_{\langle (s)_0,\ldots,(s)_{n-1}\rangle} \setminus \{\langle (s)_n\rangle\} \wedge k < l \leq k+m\},$$

where $k := \max\{i \mid \langle (s)_0, \ldots, (s)_{n-2}, i\rangle \in H\}$.

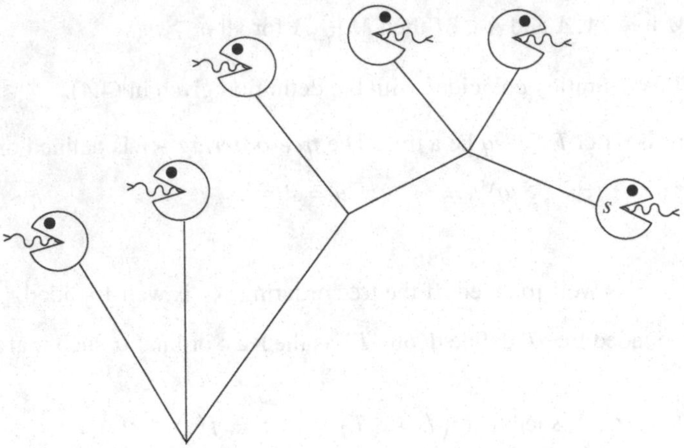


**Fig. 4.2** A hydra

A run in the battle between Hercules and a hydra $H$ is a sequence

$$\{H_i \mid i \in \omega\}$$

such that

- $H_0 = H$,
- $H_{i+1} = \begin{cases} H_i(s_i, m_i) & \text{for some head } s_i \in H_i \text{ and some } m_i \in \mathbb{N} \text{ if } H_i \text{ is alive} \\ \{\langle\,\rangle\} & \text{if } H_i \text{ is dead.} \end{cases}$

Hercules wins the battle if the hydra eventually dies.

(a) Let $H$ be the hydra in Fig. 4.2. Draw a picture of $H(s,2)$.

(b) Show that Hercules wins any battle against any hydra.

Hint: Assign an ordinal notation $o(H) < \varepsilon_0$ from Sect. 3.3.1 to any hydra $H$ and show that $o(H(s,m)) < o(H)$ holds true for any hydra $H$. Use the symmetric sum (cf. Exercise 3.3.14) instead of the ordinary ordinal sum.

## 4.5 GENTZEN's Hauptsatz for Pure First-Order Logic

One of the consequences of the completeness theorem for the first-order TAIT-calculus is the admissibility of the cut rule.

**4.5.1 Theorem** *(Weak* GENTZEN*'s Hauptsatz) If* $\overset{m}{\underset{\mathsf{T}}{\models}} \Delta, F$ *and* $\overset{n}{\underset{\mathsf{T}}{\models}} \Delta, \neg F$ *then there is a k such that* $\overset{k}{\underset{\mathsf{T}}{\models}} \Delta$.

The *proof* is simple. From $\overset{m}{\underset{\mathsf{T}}{\vdash}} \Delta, F$ and $\overset{n}{\underset{\mathsf{T}}{\vdash}} \Delta, \neg F$ we get by the soundness of the calculus that $\bigvee\Delta \vee F$ as well as $\bigvee\Delta \vee \neg F$ are valid in any model with any assignment. But then $\bigvee\Delta$ has to be valid in all models with all assignments. Hence $\overset{k}{\underset{\mathsf{T}}{\vdash}} \bigvee\Delta$ for some $k$ by the completeness of the calculus. It is an easy exercise to show by induction on $k$ that this entails $\overset{k}{\underset{\mathsf{T}}{\vdash}} \Delta$. □

Theorem 4.5.1 gives no information about the size of $k$. This information, however, might be crucial. To compute $k$ we define the *rank* $\text{rnk}(F)$ of a formula $F$ as the number of logical symbols occurring in $F$ and augment the clauses in Definition 4.3.2 by an additional rule, the cut rule

(Cut) If $\text{rnk}(F) < r$, $\overset{m}{\underset{r}{\vdash}} \Delta, F$ and $\overset{m}{\underset{r}{\vdash}} ,\Delta \neg F$ then $\overset{n}{\underset{r}{\vdash}} \Delta$ for all $n > m$.

To make this definition correct we have to replace $\overset{m}{\underset{\mathsf{T}}{\vdash}} \Delta, \dots$ in all clauses in Definition 4.3.2 by $\overset{m}{\underset{r}{\vdash}} \Delta, \dots$. The subscript $r$ is thus a measure for the complexity of all cut formulas occurring in the derivation. Obviously we have $\overset{m}{\underset{\mathsf{T}}{\vdash}} \Delta \;\Leftrightarrow\; \overset{m}{\underset{0}{\vdash}} \Delta$. We can now formulate GENTZEN's Hauptsatz in a version which gives information about the size of the cut free derivation.

**4.5.2 Theorem** *(GENTZEN's Hauptsatz) If* $\overset{m}{\underset{r}{\vdash}} \Delta$ *then* $\overset{2_r(m)}{\underset{0}{\vdash}} \Delta$ *where* $2_r(x)$ *is defined by* $2_0(x) = x$ *and* $2_{n+1}(x) = 2^{2_n(x)}$.

We need GENTZEN's Hauptsatz for pure logic, in fact only in its weak version (Theorem 4.5.1). We will, however, prove modifications of the strong version (Theorem 4.5.2) for different infinitary systems. Therefore we leave the proof of GENTZEN's Hauptsatz as an exercise which should be solved after having seen the cut-elimination for the semi-formal calculus which we are going to introduce in Definition 7.3.5.

**4.5.3 Exercise** Prove Theorem 4.5.2.

For the following exercises assume that the first-order language $\mathcal{L}$ contains at least one unary predicate symbol $C$ and a distinguished constant $c$. Put $\bot \;:\Leftrightarrow\; Cc \wedge \neg Cc$. Then $\mathfrak{S} \not\models \bot$ holds in all $\mathcal{L}$-structures $\mathfrak{S}$. Therefore $\bot$ represents the false formula. Dually $\top \;:\Leftrightarrow\; \sim\bot$ stands for the true sentence.

**4.5.4 Exercise** (a) Show that $\overset{2}{\underset{0}{\vdash}} \Delta$ holds for all finite sets $\Delta$ of $\mathcal{L}$-formulas containing $\top$.

(b) Show that $\overset{m}{\underset{n}{\vdash}} \Delta, \bot$ implies that there is a natural number $k$ such that $\overset{k}{\underset{0}{\vdash}} \Delta$. Compute an upper bound for $k$.

(c) Conclude that $\bot \models_L F$ holds for all $\mathcal{L}$-formulas $F$. This rule is known as "*ex falso quodlibet*".

**4.5.5 Exercise** Let $\Delta$ and $\Gamma$ be finite sets of formulas such that $\frac{n}{r}\, \Delta, \Gamma$. Show that there is a formula $F$ with $FV(F) \subseteq FV(\Delta) \cap FV(\Gamma)$ containing only those relation variables and -constants, with the possible exception of $C$ and $\neg C$, which occur in all formulas of $\Gamma$ and in all formulas of $\neg\Delta := \{\sim A \mid A \in \Delta\}$ and that there are natural numbers $m_0$ and $m_1$ such that $\frac{m_0}{0}\, \Delta, F$ and $\frac{m_1}{0}\, \neg F, \Gamma$.

Hint: Use GENTZEN's Hauptsatz.

**4.5.6 Exercise** (Craig's interpolation theorem). An interpolation formula for two $\mathscr{L}$-formulas $A$ and $B$ is a formula $F$ satisfying the following conditions.

- $FV(F) \subseteq FV(A) \cap FV(B)$ and $FV_2(F) \subseteq FV_2(A) \cap FV_2(B)$.
- All relation constants occurring in $F$ occur in $A$ as well as in $B$.
- $\models A \to F$ and $\models F \to B$ hold true.

Assume $\models A \to B$. Show that there is an interpolation formula for $A$ and $B$.

Hint: Use the Completeness Theorem and Exercise 4.5.5.

**4.5.7 Exercise** (Interpolation theorem of Craig and Lyndon) We say that a relation symbol $R$ or a relation variable $X$ occurs *positively* in a formula $F$ if $R$ or $X$, respectively, occurs in $F^T$. We say that $R$ or $X$ occurs *negatively* in $F$ if $\neg R$ or $\neg X$, respectively, occurs in $F^T$. Strengthen Craig's interpolation theorem (Exercise 4.5.6) in so far that the interpolation formula only contains positive (negative) occurrences of relation variables and -predicates which occur positively (negatively) as well in $A$ as in $B$.

## 4.6 Second-Order Logic

We have already mentioned that there is no calculus for second-order logic which is complete. This is due to the fact that there is a categorical axiom system for arithmetic in second-order logic (cf. Exercise 7.1.1). Let $T$ denote this axiom system and $c$ be a new constant. Then every finite subset of $T' := T \cup \{c \neq \underline{n} \mid n \in \mathbb{N}\}$ has a model (just take $\mathbb{N}$, which is a model of $T$, and interprete the new constant $c$ by a natural number which is different form all $\underline{n}^{\mathbb{N}}$ occurring in the finite subset). If there were a compactness theorem for second-order logic we would also obtain a second-order model $\mathscr{N} \models T'$ which cannot be isomorphic to $\mathbb{N}$ because $c^{\mathscr{N}} \neq \underline{n}^{\mathbb{N}} = n$ for all $n \in \mathbb{N}$.

As we have just seen, a completeness theorem entails a compactness theorem. Therefore there cannot be a completeness theorem for second-order logic.

Nevertheless there are sound calculi for second-order logic. We obtain such a calculus if we extend Definition 4.3.2 by the rules for second-order quantifiers which are,

($\exists^2$) If $\underset{\mathsf{T}}{\overset{m_0}{\vert\!\!-}} \Delta, A_U(V)$, then $\underset{\mathsf{T}}{\overset{m}{\vert\!\!-}} \Delta, (\exists X)A_U(X)$ for all $m > m_0$

and

($\forall^2$) If $\underset{\mathsf{T}}{\overset{m_0}{\vert\!\!-}} \Delta, A(U)$ and $U$ not free in $\Delta$, then $\underset{\mathsf{T}}{\overset{m}{\vert\!\!-}} \Delta, (\forall X)A_U(X)$ for all $m > m_0$.

We often write $\underset{\mathsf{T}^2}{\overset{m}{\vert\!\!-}} \Delta$ to emphasize that we talk about a derivation in the sense of second-order logic.

We transfer the definition of $T \vdash \Delta$ given in (4.4) on p. 61 also to theories $T$ in second-order language.

**4.6.1 Theorem** *Let* $\underset{\mathsf{T}^2}{\overset{m}{\vert\!\!-}} \Delta$. *Then* $\models \bigvee \Delta$ *in the sense of second-order logic.*

*Proof* The proof is by induction on $m$ and continues the proof of the soundness theorem for first-order logic (Theorem 4.3.3). The new additional cases are inferences according to ($\forall^2$) and ($\exists^2$). We start with the case of an inference ($\forall^2$). Let $\mathfrak{M}$ be a structure and $\Phi$ an assignment. By induction hypothesis we have $\mathfrak{M} \models (\bigvee \Delta \vee A(U))[\Psi]$ for all assignments $\Psi$. This is especially true for all assignments $\Psi \sim_U \Phi$. Because of the variable condition $U \notin FV_2(\Delta)$ we obtain $\mathfrak{M} \models \bigvee \Delta[\Phi] \;\Leftrightarrow\; \mathfrak{M} \models \bigvee \Delta[\Psi]$. If $\mathfrak{M} \not\models \bigvee \Delta[\Phi]$ we therefore obtain $\mathfrak{M} \not\models \bigvee \Delta[\Psi]$ and thus $\mathfrak{M} \models A(U)[\Psi]$ for all assignments $\Psi \sim_U \Phi$. Hence $\mathfrak{M} \models (\forall X)A_U(X)[\Phi]$.

In case of an inference ($\exists^2$) assume $\mathfrak{M} \not\models \bigvee \Delta[\Phi]$. By the induction hypothesis we then obtain $\mathfrak{M} \models A_U(V)[\Phi]$. Now define

$$\Psi(W) := \begin{cases} \Phi(W) & \text{for } W \neq U \\ \Phi(V) & \text{for } W = U. \end{cases}$$

Then $\Psi \sim_U \Phi$ and $\mathfrak{M} \models A(U)[\Psi]$. Hence $\mathfrak{M} \models (\exists X)A_U(X)[\Phi]$. □

**4.6.2 Note** We introduced only a very weak form of a second-order calculus. Commonly much stronger calculi are regarded. They differ in so far that class terms become well-formed expressions of the second-order language. Whenever $F(u_1, \ldots, u_n)$ is a second-order formula the expression $\{(x_1, \ldots, x_n) | F_{u_1,\ldots,u_n}(x_1, \ldots, x_n)\}$ is a second-order term. The defining axiom for second-order terms then is

$$(t_1, \ldots, t_n) \in \{(x_1, \ldots, x_n) \,|\, F_{u_1,\ldots,u_n}(x_1, \ldots, x_n)\} \leftrightarrow F_{u_1,\ldots,u_n}(t_1, \ldots, t_n).$$

So far there is no real difference to our convention to use class terms as "abbreviations". The difference comes by extending the ($\exists^2$) rule to

($\exists^2$) If $\underset{SO}{\overset{m_0}{\vert\!\!-}} \Delta, A_U(S)$ for a second-order term $S$, then $\underset{SO}{\overset{m}{\vert\!\!-}} \Delta, (\exists X)A_U(X)$ for all $m > m_0$.

Adapting the remaining clauses in Definition 4.3.2 and clause $(\forall)^2$ above we get the calculus $\underset{SO}{\vert\!\!-} \Delta$ for simple second-order logic. By rule $(\exists)^2$ we obtain obtain full comprehension

(CA) $(\exists X)(\forall x_1) \ldots (\forall x_n)[(x_1, \ldots, x_n) \in X \leftrightarrow F_{u_1,\ldots,u_n}(x_1, \ldots, x_n)]$

for arbitrary formulas $F(u_1,\dots,u_n)$ in simple second-order logic, which shows that already a bit of mathematics sneaks into pure logic. Commonly the calculus is also equipped with a cut rule

(cut) If $\vdash_{\overline{SO}} \Delta, F$ and $\vdash_{\overline{SO}} \Gamma, \neg F$ then $\vdash_{\overline{SO}} \Delta, \Gamma$.

It has been shown by Takeuti [101] that the eliminability of the cut rule in simple second-order logic already implies the consistency of full second-order arithmetic (cf. Sect. 1). Therefore the proof of the Hauptsatz for the calculus $\vdash_{\overline{SO}}$ cannot be as simple as for the weak second-order calculus $\vdash_{\overline{T^2}}$ introduced above (cf. Exercise 4.6.4). The first proof of the Hauptsatz for a calculus similar to $\vdash_{\overline{SO}}$ has been given by Tait in [97].

**4.6.3 Exercise** Prove the soundness of the simple second-order logic, i.e., show

$$\vdash_{\overline{SO}} \Delta \;\Rightarrow\; \mathfrak{S} \models \bigvee \Delta[\Phi]$$

for any structure $\mathfrak{S}$ and any $\mathfrak{S}$ assignment $\Phi$.

**4.6.4 Exercise** Extend the definition of $\models^{n}_{r} \Delta$ to weak second-order logic, i.e., second-order logic without second-order terms. Prove

$$\models^{m}_{r} \Delta \;\Rightarrow\; \models^{2_r(m)}_{0} \Delta.$$

Hint: Postpone this exercise until you have seen the cut elimination procedure for the infinitary calculus in Sect. 7.3

# Chapter 5
# Truth Complexity for $\Pi^1_1$-Sentences

We will now turn to the language of arithmetic. The first-order sentences of the language of arithmetic can be arranged in a hierarchy, the arithmetical hierarchy. Arranging the second-order sentences in the language of arithmetic into a hierarchy leads to the analytical hierarchy, which continues the arithmetical hierarchy. Of special interest in proof theory are the sentences at the $\Pi^1_1$-level of the analytical hierarchy.

## 5.1 The Language $\mathscr{L}(\mathsf{NT})$ of Arithmetic

To fix the language $\mathscr{L}(\mathsf{NT}) := \mathscr{L}(\mathscr{C}, \mathscr{F}, \mathscr{R})$ of arithmetic it suffices to define its signature. We put:

- $\mathscr{C} := \{\underline{0}\}$.
- $\mathscr{F} := \{\underline{f} \mid f$ is a symbol for a primitive recursive function term$\}$. The arity $\#\underline{f}$ is the arity of the denoted primitive recursive function term.
- $\mathscr{R} := \{=\}$. The arity of $=$ is 2.

**5.1.1 Remark** Since we have a primitive recursive coding machinery we lose no expressive power by restricting the relation variables to unary ones. Any atomic formula $(t_1, \ldots, t_n) \ \varepsilon \ U^n$ for an $n$-ary relation variable $U^n$ can be expressed as $\langle t_1, \ldots, t_n \rangle \ \varepsilon \ U^1$ with a unary "set variable" $U^1$ and vice versa. Sometimes it is more convenient to allow only set variables; sometimes it is more convenient to work with $n$-ary relation variables. We will switch between both concepts without always emphasizing it. This, however, means no loss of generality.

We call $\mathscr{L}(\mathsf{NT})$-terms also *arithmetical terms*. An $\mathscr{L}(\mathsf{NT})$-formula $F$ is called *arithmetical* if $FV_2(F) = BV_2(F) = \emptyset$. The first-order formulas of $\mathscr{L}(\mathsf{NT})$, i.e., the formulas that also allow occurrences of free relation variables, are sometimes called *bold face arithmetical formulas*. They will play an important role in the ordinal analysis of axiom systems for arithmetic.

We are interested in the structure of natural numbers which form the *standard model* of $\mathcal{L}(\mathsf{NT})$. In abuse of notation we put $\mathbb{N} := (\mathbb{N}, {}^{\mathbb{N}})$, i.e., we denote the structure and its domain by the same symbol. We define

- $\underline{0}^{\mathbb{N}} := 0$.
- $\underline{f}^{\mathbb{N}}$ is the function represented by $f$, i.e., $\underline{f}^{\mathbb{N}}(z_1,\dots,z_n) := \mathrm{ev}(f, z_1,\dots,z_n)$ according to Definition 2.1.2.
- $z_1 =^{\mathbb{N}} z_2 \;:\Leftrightarrow\; z_1 = z_2$, i.e., $=^{\mathbb{N}}$ is interpreted as the identity on the natural numbers.

Observe that for every natural number $z$ there is a closed term $\underline{z}$ such that $\underline{z}^{\mathbb{N}} = z$. Just put $\underline{z} := \underbrace{(\underline{S}\cdots(\underline{S}\underline{0})}_{z\text{-times}}\cdots)$.

**5.1.2 Lemma** *Let $t$ be an arithmetical term and $\Phi$ an $\mathbb{N}$ assignment such that $\Phi(v) = z$. Then $t^{\mathbb{N}}[\Phi] = t_v(\underline{z})^{\mathbb{N}}[\Phi]$.*

*Proof* We induct on the length of the term $t$. If $t = \underline{0}$ there is nothing to show. If $t$ is a free variable different from $v$ we get $t_v(\underline{z}) = t$ and are done. If $t$ is the free variable $v$ we get $t^{\mathbb{N}}[\Phi] = \Phi(v) = z = t_v(\underline{z})^{\mathbb{N}}$. Now assume that $t$ is a term $(\underline{f}t_1\dots t_n)$. Then we obtain $t^{\mathbb{N}}[\Phi] = \underline{f}^{\mathbb{N}}(t_1^{\mathbb{N}}[\Phi],\dots,t_n^{\mathbb{N}}[\Phi]) = \underline{f}^{\mathbb{N}}(t_{1,v}(\underline{z})^{\mathbb{N}}[\Phi],\dots,t_{n,v}(\underline{z})^{\mathbb{N}}[\Phi]) = (\underline{f}t_1\dots t_n)_v(\underline{z})^{\mathbb{N}}[\Phi] = t_v(\underline{z})^{\mathbb{N}}[\Phi]$ using the induction hypothesis. □

**5.1.3 Lemma** *Let $F$ be an $\mathcal{L}(\mathsf{NT})$-formula and $\Phi$ an $\mathbb{N}$-assignment such that $\Phi(v) = z$. Then $\mathbb{N} \models F[\Phi]$ if and only if $\mathbb{N} \models F_v(\underline{z})[\Phi]$.*

*Proof* Induction is on the length of the formula $\Phi$. The claim is obvious if $v \notin FV(F)$. If $F$ is an atomic formula $s = t$ or $(t_1,\dots,t_n)\,\varepsilon\,U$ we obtain $\mathbb{N} \models F[\Phi]$ iff $s^{\mathbb{N}}[\Phi] = t^{\mathbb{N}}[\Phi]$ or $(t_1^{\mathbb{N}}[\Phi],\dots,t_n^{\mathbb{N}}[\Phi]) \in \Phi(U)$, respectively. By the previous lemma (Lemma 5.1.2) we obtain either $\mathbb{N} \models F_v(\underline{z})[\Phi]$ iff $t_v(\underline{z})^{\mathbb{N}}[\Phi] = s_v(\underline{z})^{\mathbb{N}}[\Phi]$ iff $t^{\mathbb{N}}[\Phi] = s^{\mathbb{N}}[\Phi]$ or $\mathbb{N} \models F_v(\underline{z})[\Phi]$ if and only if $(t_{1,v}^{\mathbb{N}}(\underline{z}),\dots,t_{n,v}^{\mathbb{N}}(\underline{z})) \in \Phi(U)$ iff $(t_1^{\mathbb{N}}[\Phi],\dots,t_n^{\mathbb{N}}[\Phi]) \in \Phi(U)$, respectively.

In case that $F$ is a composed formula we get the claim immediately from the induction hypothesis. □

The next theorem shows that the somewhat weird looking definition of the satisfaction relation in case of first-order quantification matches in fact the intuitive meaning of the universal and existential quantifiers.

**5.1.4 Theorem** *Let $F(v)$ be an $\mathcal{L}(\mathsf{NT})$-formula. Then*

$$\mathbb{N} \models (\forall x)F_v(x)[\Phi] \;\Leftrightarrow\; \mathbb{N} \models F_v(\underline{z})[\Phi] \text{ for all } z \in \mathbb{N}$$

*and*

$$\mathbb{N} \models (\exists x)F_v(x)[\Phi] \;\Leftrightarrow\; \mathbb{N} \models F_v(\underline{z})[\Phi] \text{ for some } z \in \mathbb{N}$$

*Proof* By Definition 4.2.7 we have $\mathbb{N} \models (\forall x)F_v(x)[\Phi]$ if and only if $\mathbb{N} \models F[\Psi]$ for all assignments $\Psi \sim_v \Phi$. Let $z \in \mathbb{N}$ and define $\Psi \sim_v \Phi$ such that $\Psi(v) = z$. Then we obtain by Lemma 5.1.3 $\mathbb{N} \models F[\Psi] \;\Leftrightarrow\; \mathbb{N} \models F_v(\underline{z})[\Psi]$ which by (4.2) is equivalent to $\mathbb{N} \models F_v(\underline{z})[\Phi]$ because $v \notin FV(F_v(\underline{z}))$. This proves the direction from left to right. For the opposite direction let $\Phi \sim_v \Psi$ and put $z := \Psi(v)$. By hypothesis we have $\mathbb{N} \models F_v(\underline{z})[\Phi]$ that (4.2) is equivalent to $\mathbb{N} \models F_v(\underline{z})[\Psi]$. Again by Lemma 5.1.3 this entails $\mathbb{N} \models F[\Psi]$.

The proof for the existential quantifier follows the same pattern and is left as an exercise. □

We just want to remark that an analogous theorem holds also for second-order quantifiers if we enrich the language by constants for subsets of $\mathbb{N}^n$.

A subset $A \subseteq \mathbb{N}$ is $\mathcal{L}(\mathsf{NT})$-*definable* if there are an $\mathcal{L}(\mathsf{NT})$ formula $F(u, v_1, \ldots, v_n)$ and natural numbers $z_1, \ldots, z_n$ such that

$$A = \{x \in \mathbb{N} \,|\, \mathbb{N} \models F(x, \underline{z}_1, \ldots, \underline{z}_n)\}.$$

The definable subsets of $\mathbb{N}$ can be hierarchically ordered by the complexity of their defining formulas. To do that we first introduce a syntactically defined hierarchy for $\mathcal{L}$-formulas:

- The class of $\forall^0_0$-formulas comprises the quantifier free $\mathcal{L}$-formulas. We put $\exists^0_0 := \forall^0_0$.

- If $G$ is a $\exists^0_n$-formula then $(\exists x)G_u(x)$ is a $\exists^0_n$ and $(\forall x)G_u(x)$ is a $\forall^0_{n+1}$-formula.

- If $G$ is a $\forall^0_n$-formula then $(\forall x)G_u(x)$ is a $\forall^0_n$-formula and $(\exists x)G_u(x)$ is a $\exists^0_{n+1}$-formula.

A $\exists^0_n$-formula is therefore a first-order formula in prenex form which has $n$ alternating blocks of quantifiers starting with a block of existential quantifiers. Dually a $\forall^0_n$-formula is a first-order formula in the prenex form with $n$ alternating quantifiers starting with a block of universal quantifiers. It is a theorem of Mathematical Logic that every first-order formula is logically equivalent to a $\forall^0_n$- or $\exists^0_n$-formula. We call a formula a $\Delta^0_n$-formula if it is logically equivalent to a $\exists^0_n$-formula and simultaneously also logically equivalent to a $\forall^0_n$-formula.

We extend the hierarchy to second-order formulas by adding the following clauses.

- The class of $\forall^1_0$-formulas comprises the union of all $\forall^0_n$ formulas. We put $\Delta^1_0 := \exists^1_0 := \forall^1_0$.

- If $G$ is a $\exists^1_n$-formula then $(\exists x)G_u(x)$, $(\forall x)G_u(x)$ and $(\exists X)G_U(X)$ are $\exists^1_n$-formulas and $(\forall X)G_U(X)$ is a $\forall^1_{n+1}$-formula.

- If $G$ is a $\forall^1_n$-formula then $(\exists x)G_u(x)$, $(\forall x)G_u(x)$ and $(\forall X)G_U(X)$ are $\forall^1_n$-formulas and $(\exists X)G_U(X)$ is a $\exists^1_{n+1}$-formula.

The hierarchy of second-order formulas follows the same pattern as the hierarchy of first-order formulas but we count only blocks of second-order quantifiers and neglect first-order quantifiers.

The basis of the hierarchy of definable sets of natural numbers form the primitive recursively definable sets. Due to the richness of the language $\mathscr{L}(\mathsf{NT})$ every primitive recursively definable set is already definable by a $\forall^0_0$ formula. We put:

- $\Pi^0_0 := \Sigma^0_0 := \Delta^0_0 = \{A \subseteq \mathbb{N} \mid A$ is definable by an arithmetical $\forall^0_0$-formula$\}$

as well as

- $\Pi^0_n := \{A \subseteq \mathbb{N} \mid A$ is definable by an arithmetical $\forall^0_n$-formula$\}$

and dually

- $\Sigma^0_n := \{A \subseteq \mathbb{N} \mid A$ is definable by an arithmetical $\exists^0_n$-formula$\}$.

We put:

- $\Delta^0_n := \Sigma^0_n \cap \Pi^0_n$.

A set $A \subseteq \mathbb{N}$ is called *arithmetical* if there is a $n$ such that $A \in \Delta^0_n$. The hierarchy of arithmetical sets is cumulative and strict in the sense that for all $n$ we have

$$\Sigma^0_n \cup \Pi^0_n \subsetneq \Delta^0_{n+1} \subsetneq \Sigma^0_{n+1} \cup \Pi^0_{n+1}.$$

These facts are known as the *Arithmetical Hierarchy Theorem,* which is a theorem of abstract recursion theory. Without proof we cite that for $n > 0$ the levels $\Sigma^0_n$ are closed under the positive boolean operations $\wedge$ and $\vee$, bounded first-order $\forall$-quantification and unbounded $\exists$-quantification. Dually the classes $\Pi^0_n$ are closed under the positive boolean operations, bounded $\exists$-quantification and unbounded $\forall$-quantification.

The classes $\Delta^0_n$ are closed under the propositional operations $\neg$, $\wedge$ and $\vee$ and bounded $\forall$- and $\exists$-quantification.

It is an easy observation that all subsets of $\mathbb{N}$ that are first-order definable in the language of arithmetic are members of the arithmetical hierarchy, i.e., they are definable by an arithmetical $\exists^0_n$ or $\forall^0_n$ formula for some finite $n$.

The arithmetical hierarchy can be expanded into the *analytical hierarchy* that comprises sets of natural numbers which can be defined by second-order formulas. We put:

- $\Sigma^1_0 := \Pi^1_0 := \Delta^1_0$ is the collection of all arithmetical sets

For $n \geq 1$ we define:

- $\Pi^1_n$ is the collection of sets of natural numbers which are definable by a $\forall^1_n$-formula.

and dually

- $\Sigma_n^1$ is the collection of sets of natural numbers which are definable by a $\exists_n^1$-formula.

Again we put:

- $\Delta_n^1 := \Pi_n^1 \cap \Sigma_n^1$.

The sets $\Delta_n^1$ form the levels of the analytical hierarchy. It is easy to see that the collection of $\Pi_n^1$-sets is closed under the positive boolean operations, first-order quantification and second-order $\forall$-quantification. Dually the $\Sigma_n^1$ sets are closed under the positive boolean operations, first-order quantifications and second-order $\exists$-quantification. The class of $\Delta_n^1$-sets is closed under all boolean operations and first-order quantifications.

The analytical hierarchy, too, is cumulative and strict. The study of this hierarchy is the subject of descriptive set theory. In this book we are only concerned with $\Pi_1^1$-sets.

In the arithmetic language, it is common to talk about $\Sigma_n^i$-formulas or $\Pi_n^i$-formulas instead of $\exists_n^i$-formulas or $\forall_n^i$-formulas, respectively. Due to the existence of a primitive recursive coding machinery we obtain for every $\Pi_n^0$ or $\Sigma_n^0$-formula $F$ a $\Pi_n^0$ or $\Sigma_n^0$-formula $F'$ in that there are exactly $n$ alternating first-order quantifiers followed by a quantifier-free kernel such that:

$$\mathbb{N} \models F \leftrightarrow F'.$$

Similarly, we obtain for every $\Pi_n^1$ or $\Sigma_n^1$-formula $F$ a $\Pi_n^1$ or $\Sigma_n^1$-formula $F'$ with exactly $n$ alternating second-order quantifiers followed by an arithmetical kernel such that:

$$\mathbb{N} \models F \leftrightarrow F'.$$

**5.1.5 Exercise** Let $A \subseteq \mathbb{N}$ be a $\Pi_1^1$-set. Show that there is a first-order formula $F(X,x)$ such that $A = \{x \in \mathbb{N} \mid \mathbb{N} \models (\forall X)F(X,x)\}$.

Hint: Let $z \in (X)_y :\Leftrightarrow \langle z,y\rangle \in X$. Show that $(\exists y)(\forall X)G(X,y) \leftrightarrow (\forall Z)(\exists y)G((Z)_y,y)$ holds true in $\mathbb{N}$ and use these facts to obtain a block of universal second quantifers followed by a first-order formula. Then show that $(\forall X)(\forall Y)G(X,Y) \leftrightarrow (\forall Z)G((Z)_{\underline{0}},(Z)_{\underline{1}})$ is true in $\mathbb{N}$ and use that to contract the universal second-order quantifers.

## 5.2 The TAIT-Language for Second-Order Arithmetic

We adapt the definition of the TAIT-language of Sect. 4.3 to the language $\mathcal{L}(\mathsf{NT})$.

**5.2.1 Definition** The TAIT-language for arithmetic contains the following symbols:

- Bounded number variables $x, y, z, x_0, \ldots$.
- Set variables $X, Y, X_0, \ldots$.

- The logical symbols $\land, \lor, \forall, \exists$.
- The binary relation symbols $\varepsilon, \not\varepsilon, =, \neq$.
- The constant $\underline{0}$.
- Symbols for all primitive recursive functions.

Terms are constructed from the constant $\underline{0}$ using function symbols in the usual way. Atomic formulas are equations $s = t$, inequalities $s \neq t$ between terms and the formulas $t \,\varepsilon\, X$ and $t \not\varepsilon X$ where $t$ is a term and $X$ a set variable. Formulas are obtained from atomic formulas closing them under the logical operations.

The definition of $\sim F$ specializes as follows:

- $\sim(s = t) :\equiv s \neq t; \quad \sim(s \neq t) :\equiv s = t;$
- $\sim(s \,\varepsilon\, X) :\equiv s \not\varepsilon X; \quad \sim(s \not\varepsilon X) :\equiv s \,\varepsilon\, X;$
- $\sim(A \land B) :\equiv \sim A \lor \sim B; \quad \sim(A \lor B) :\equiv \sim A \land \sim B;$
- $\sim(\forall x)F(x) :\equiv (\exists x)\sim F(x); \quad \sim(\exists x)F(x) :\equiv (\forall x)\sim F(x).$

For any assignment $\Phi$ of subsets of $\mathbb{N}$ to the set variables occurring in $F$ we again obtain

$$\mathbb{N} \models \sim F[\Phi] \;\Leftrightarrow\; \mathbb{N} \models \neg F[\Phi]. \tag{5.1}$$

Therefore, we commonly write $\neg F$ instead of $\sim F$.

Observe that we do not allow free number variables in the formulas of the TAIT-language of $\mathcal{L}(\mathsf{NT})$. Nevertheless, any formula in the language of arithmetic can be canonically translated into its *Tait-form*. This is done by replacing $\neg$ by $\sim$. However, we have to replace number variables in an $\mathcal{L}(\mathsf{NT})$-formula by closed number terms to obtain a well-formed formula of the TAIT-language in the sense of Definition 5.2.1. This could be avoided by also allowing free number variables in the TAIT-language. To dispense with free number variables in the TAIT-language of $\mathcal{L}(\mathsf{NT})$ has, however, good technical advantages.

## 5.3 Truth Complexities for Arithmetical Sentences

As a heuristic preparation we study first the truth complexities of arithmetical sentences, i.e., of formulas in the first-order TAIT-language of arithmetic that must not contain free set variables.

Let *Diag*($\mathbb{N}$) be the *diagram* of $\mathbb{N}$, i.e., the set of true atomic sentences in the Tait-language.

**5.3.1 Observation** *The true arithmetical sentences can be characterized by the following types:*

- *The sentences in* $Diag(\mathbb{N})$,
- *The sentences of the form* $(F_0 \lor F_1)$ *or* $(\exists x)F(x)$ *where* $F_i$ *and* $F(\underline{k})$ *is true for* some $i \in \{0,1\}$ *or* $k \in \omega$*, respectively,*
- *The sentences of the form* $(F_0 \land F_1)$ *or* $(\forall x)F(x)$ *where* $F_i$ *and* $F(\underline{k})$ *is true for* all $i \in \{0,1\}$ *or* $k \in \omega$*, respectively.*

According to Observation 5.3.1 we divide the arithmetical sentences into two types.

**5.3.2 Definition** Let

$$\bigwedge\text{-type} := Diag(\mathbb{N}) \cup \{\text{sentences of the form } (F_0 \land F_1)\} \cup \{\text{sentences of the form } (\forall x)F(x)\}.$$

We say that the sentences in $\bigwedge$-type are of *conjunctive type*. Dually put

$$\begin{aligned}\bigvee\text{-type} &:= \{\neg F \mid F \in \bigwedge\text{-type}\}\\ &= \neg Diag(\mathbb{N}) \cup \{\text{sentences of the form } (F_0 \lor F_1)\} \cup \\ &\quad \{\text{sentences of the form } (\exists x)F(x)\}.\end{aligned}$$

The sentences in $\bigvee$-type are of *disjunctive type.*

Then we define the *characteristic sequence* $\mathrm{CS}(F)$ of sub-sentences of $F$.

**5.3.3 Definition** Let

$$\mathrm{CS}(F) := \begin{cases} \emptyset & \text{if } F \text{ is atomic} \\ \langle F_0, F_1 \rangle & \text{if } F \equiv (F_0 \circ F_1) \\ \langle G(\underline{k}) \mid k \in \omega \rangle & \text{if } F \equiv (\mathrm{Q}x)G(x) \end{cases}$$

for $\circ \in \{\land, \lor\}$ and $\mathrm{Q} \in \{\forall, \exists\}$. The *length of the type* of a sentence $F$ is the length of its characteristic sequence $\mathrm{CS}(F)$.

From Observation 5.3.1 and Definition 5.3.2 we get immediately

**5.3.4 Observation**

$$F \in \bigwedge\text{-type} \;\Rightarrow\; [\mathbb{N} \models F \;\Leftrightarrow\; (\forall G \in \mathrm{CS}(F))(\mathbb{N} \models G)]$$

*and*

$$F \in \bigvee\text{-type} \;\Rightarrow\; [\mathbb{N} \models F \;\Leftrightarrow\; (\exists G \in \mathrm{CS}(F))(\mathbb{N} \models G)]$$

We use Observation 5.3.4 to define the *truth complexity* of a sentence $F$.

**5.3.5 Definition** The infinitary verification calculus $\models^{\alpha} F$ is inductively defined by the following clauses:

$$(\bigwedge) \quad \text{If } F \in \bigwedge\text{–type and } (\forall G \in \mathrm{CS}(F))(\exists \alpha_G < \alpha)\left[\models^{\alpha_G} G\right] \text{ then } \models^{\alpha} F.$$

$$(\bigvee) \quad \text{If } F \in \bigvee\text{–type and } (\exists G \in \mathrm{CS}(F))(\exists \alpha_G < \alpha)\left[\models^{\alpha_G} G\right] \text{ then } \models^{\alpha} F.$$

We read $\models^{\alpha} F$ as "$F$ is *$\alpha$-verifiable*" and put

$$\mathrm{tc}(F) := \min(\{\alpha \mid \models^{\alpha} F\} \cup \{\omega_1\}).$$

We call $\mathrm{tc}(F)$ the *truth complexity* of the sentence $F$.

The next theorem is obvious from Observation 5.3.4 and Definition 5.3.5.

**5.3.6 Theorem** $\models^{\alpha} F$ *implies* $\mathbb{N} \models F$.

**5.3.7 Observation** *Let* $\mathrm{rnk}(F)$ *be the number of logical symbols occurring in* $F$. *Then*

$$\mathbb{N} \models F \;\Rightarrow\; \mathrm{tc}(F) \leq \mathrm{rnk}(F)$$

*and*

$$\mathbb{N} \models F \;\Leftrightarrow\; \mathrm{tc}(F) < \omega.$$

*Proof* For the first claim we prove

$$\mathbb{N} \models F \;\Rightarrow\; \models^{\mathrm{rnk}(F)} F \tag{i}$$

by induction on $\mathrm{rnk}(F)$. If $\mathrm{rnk}(F) = 0$ then $F$ is atomic. Hence, $F \in \mathit{Diag}(\mathbb{N})$ and we obtain $\models^{0} F$ by a clause $(\bigwedge)$ with empty premise. If $\mathrm{rnk}(F) > 0$ and $F \in \bigwedge$–type then $\mathbb{N} \models G$ for all $G \in \mathrm{CS}(F)$. Since $\mathrm{rnk}(G) < \mathrm{rnk}(F)$ we obtain $\models^{\mathrm{rnk}(G)} G$ for all $G \in \mathrm{CS}(F)$ by induction hypothesis and infer $\models^{\mathrm{rnk}(F)} F$ by a clause $(\bigwedge)$. If $F \in \bigvee$–type then $\mathbb{N} \models G$ for some $G \in \mathrm{CS}(F)$ and $\models^{\mathrm{rnk}(G)} G$ by induction hypothesis. But then $\models^{\mathrm{rnk}(F)} F$ by clause $(\bigvee)$.

The second claim follows from the fact that $\mathrm{rnk}(F) < \omega$ together with the first claim and Theorem 5.3.6. □

According to Observation 5.3.7 the notion of truth complexity is not very exciting for arithmetical sentences. This, however, will change if we extend it to the class of formulas containing free set variables.

## 5.4 Truth Complexity for $\Pi^1_1$-Sentences

The aim of this section is to extend the notion of truth complexity to $\Pi^1_1$-sentences, the first level of the analytic hierarchy. This is made possible by the $\omega$-completeness theorem (Theorem 5.4.9 below). The $\omega$-completeness theorem allows for an infinitary syntactical verification calculus that is $\mathbb{N}$-complete for $\Pi^1_1$-sentences. In the form as we are going to present this verification calculus it derives finite sets of first-order formulas of arithmetic that may contain free set parameters but must not contain free number variables.

**5.4.1 Definition** We call an arithmetical formula that does not contain free number variables, but may contain free set parameters, a *pseudo $\Pi^1_1$-sentence.* For pseudo $\Pi^1_1$-sentences $F(\vec{X})$ we define

$$\mathbb{N} \models F(\vec{X}) \;:\Leftrightarrow\; \mathbb{N} \models (\forall \vec{X})F(\vec{X}) \;\Leftrightarrow\; \mathbb{N} \models F(\vec{X})[\Phi] \text{ for all assignments } \Phi.$$

We adopt the definition of $\bigwedge$–type and $\bigvee$–type for pseudo $\Pi^1_1$-sentences. Observe, however, that open atomic pseudo sentences, i.e., sentences of the form

$$(t \,\varepsilon\, X) \quad \text{and} \quad (s \,\not\varepsilon\, X),$$

do not belong to any type. Observation 5.3.4 fails for pseudo $\Pi^1_1$-sentences. But it can be weakened to the following observation.

**5.4.2 Observation** *For any assignment $\Phi$ to the free set variables in a pseudo $\Pi^1_1$-sentence $F$ we obtain*

$$F \in \bigwedge\text{–type} \;\Rightarrow\; (\mathbb{N} \models F[\Phi] \;\Leftrightarrow\; (\forall G \in \mathrm{CS}(F))[\mathbb{N} \models G[\Phi]])$$
$$F \in \bigvee\text{–type} \;\Rightarrow\; (\mathbb{N} \models F[\Phi] \;\Leftrightarrow\; (\exists G \in \mathrm{CS}(F))[\mathbb{N} \models G[\Phi]])$$

The *proof* is a simple induction on the rank of formula $F$.

Our aim is to use Observation 5.4.2 to obtain a more syntactic verification calculus for pseudo $\Pi^1_1$-sentences. However, Observation 5.4.2 says nothing about the verification of pseudo $\Pi^1_1$-sentence that have no type. Here we observe that a pseudo $\Pi^1_1$-sentence of the form $t \,\varepsilon\, X \vee s \,\not\varepsilon\, X$ is verifiable if the terms $t$ and $s$ yield the same value. In interpreting finite sets of $\Pi^1_1$-sentence as finite disjunctions we can use these observations to define a "syntactic" verification calculus that is similar to the Tait calculus for pure logic but uses inferences with infinitely many premises. This is made precise in the following definition.

**5.4.3 Definition** For a finite set $\Delta$ of pseudo $\Pi^1_1$-sentences we define the verification calculus $\models^{\alpha} \Delta$ inductively by the following clauses:

(*Ax*) If $s^{\mathbb{N}} = t^{\mathbb{N}}$ then $\models^{\alpha} \Delta, s \,\varepsilon\, X, t \,\not\varepsilon\, X$ holds true for all ordinals $\alpha$.

($\bigwedge$) If $F \in (\Delta \cap \bigwedge\text{–type})$ and $(\forall G \in \mathrm{CS}(F))(\exists \alpha_G < \alpha)[\models^{\alpha_G} \Delta, G]$ then $\models^{\alpha} \Delta$.

($\bigvee$) If $F \in (\Delta \cap \bigvee\text{–type})$ and $(\exists G \in \mathrm{CS}(F))(\exists \alpha_G < \alpha)[\,\models^{\alpha_G} \Delta, G]$ then $\models^{\alpha} \Delta$.

We call the formulas $s \,\varepsilon\, X$ and $t \,\not\varepsilon\, X$ in a clause $(Ax)$ and the formula $F$ in a clauses ($\bigwedge$) and ($\bigvee$) the *critical formula(s)* of the clause.

As previously remarked, the finite set $\Delta$ should be read as a finite disjunction. Again we write $F_1,\ldots,F_n$ instead of $\{F_1,\ldots,F_n\}$, $\Delta,\Gamma$ instead of $\Delta \cup \Gamma$ and $\Delta, F$ for $\Delta \cup \{F\}$.

It is a simple observation that there is a "structural rule" also for the verification calculus.

(STR) If $\models^{\alpha} \Delta$ then $\models^{\beta} \Gamma$ for all $\beta \geq \alpha$ and all $\Gamma \supseteq \Delta$.

*Proof* We induct on $\alpha$. If $\models^{\alpha} \Delta$ holds by $(Ax)$ then there are terms $s$ and $t$ such that $s^{\mathbb{N}} = t^{\mathbb{N}}$ and $\{t \,\varepsilon\, X, s \,\not\varepsilon\, X\} \subseteq \Delta \subseteq \Gamma$. Hence, $\models^{\beta} \Gamma$ by $(Ax)$. In case of an inference according to ($\bigwedge$) (or ($\bigvee$)) with critical formula $F$ we have $F \in \Delta \subseteq \Gamma$ and the premise(s) $\models^{\alpha_G} \Delta, G$ for all (or some) $G \in \mathrm{CS}(F)$. If $\mathrm{CS}(F) = \emptyset$ then $F \in \bigwedge\text{–type}$ and we obtain $\models^{\beta} \Gamma$ by an inference ($\bigwedge$). If $\mathrm{CS}(F) \neq \emptyset$ we obtain $\models^{\alpha_G} \Gamma, G$ for all (or some) $G \in \mathrm{CS}(F)$ by the induction hypothesis. Since $\alpha_G < \alpha \leq \beta$ we obtain $\models^{\beta} \Gamma$ by an inference ($\bigwedge$) or ($\bigvee$), respectively. □

**5.4.4 Lemma** *The verification calculus* $\models^{\alpha} \Delta$ *is sound for pseudo* $\Pi^1_1$*-sentences, i.e., if we assume that all free set variables occurring in* $\Delta$ *are among* $\vec{X} := (X_1,\ldots,X_n)$ *then*

$$\models^{\alpha} \Delta \;\Rightarrow\; \mathbb{N} \models (\forall \vec{X})[\bigvee \Delta(\vec{X})].$$

*Proof* Let $\Phi$ be any assignment. We show $\mathbb{N} \models (\bigvee \Delta)[\Phi]$ by induction on $\alpha$. If $\alpha = 0$ we are either in the situation of a rule according to $(Ax)$ or there is an $F \in \Delta \cap \bigwedge\text{–type}$ such that $\mathrm{CS}(F) = \emptyset$, hence $F \in Diag(\mathbb{N})$, which makes the claim trivial. In the first case we get $\mathbb{N} \models (\bigvee \Delta)[\Phi]$ because $\{t \,\varepsilon\, X_i, s \,\not\varepsilon\, X_i\} \subseteq \Delta$ and $s^{\mathbb{N}} \in \Phi(X_i) \vee t^{\mathbb{N}} \notin \Phi(X_i)$ since $s^{\mathbb{N}} = t^{\mathbb{N}}$.

Now assume $\alpha > 0$ and $\not\models^{\beta} (\bigvee \Delta)[\Phi]$ for all $\beta < \alpha$. Let $F$ be the critical formula of the last inference in $\models^{\alpha} \Delta$. Then either $F \in \Delta \cap \bigvee\text{–type}$ or $F \in \Delta \cap \bigwedge\text{–type}$ and $\mathrm{CS}(F) \neq \emptyset$ and we have the premise(s) $\models^{\alpha_G} \Delta, G$ for some or for all $G \in \mathrm{CS}(F)$, respectively. By the inductive hypothesis we, therefore, get $\mathbb{N} \models (\bigvee \Delta \vee G)[\Phi]$ for some or all $G \in \mathrm{CS}(F)$. But this implies $\mathbb{N} \models G[\Phi]$ for some or all $G \in \mathrm{CS}(F)$. Hence, $\mathbb{N} \models F[\Phi]$ and thus also, $\mathbb{N} \models (\bigvee \Delta)[\Phi]$ by Observation 5.4.2. □

The next task is to show that the verification calculus is also complete. Similar to the completeness proof for pure logic this will be done by defining search trees. Let $\Delta$ be a finite set of pseudo $\Pi^1_1$-sentences. We order the formulas in $\Delta$ arbitrarily to transform them into a finite sequence $\langle \Delta \rangle$ of pseudo $\Pi^1_1$-sentences. The leftmost formula in the sequence $\langle \Delta \rangle$ that is in $\bigwedge\text{–type}$ or in $\bigvee\text{–type}$ is the *redex* $R(\langle \Delta \rangle)$ of $\langle \Delta \rangle$. The sequence $\langle \Delta \rangle^r$ is obtained from $\langle \Delta \rangle$ by canceling its redex $R(\langle \Delta \rangle)$. We put:

$$Ax(\Delta) \;:\Leftrightarrow\; \exists s,t,X[s^{\mathbb{N}} = t^{\mathbb{N}} \wedge \{t \,\varepsilon\, X, s \,\not\varepsilon\, X\} \subseteq \Delta].$$

Two pseudo $\Pi^1_1$-sentences are *numerically equivalent* if they only differ in terms whose evaluations yield the same value. To simplify notations we mostly identify numerically equivalent pseudo $\Pi^1_1$-sentences.

**5.4.5 Definition** For a finite sequence $\langle\Delta\rangle$ of pseudo $\Pi^1_1$-sentences, we define its *search tree* $S^{\omega}_{\langle\Delta\rangle}$ together with a *label function* $\delta$ that assigns finite sequences of $\Pi^1_1$-sentences to the nodes of $S^{\omega}_{\langle\Delta\rangle}$ inductively by the following clauses:

($S_{\langle\rangle}$) Let $\langle\rangle \in S^{\omega}_{\langle\Delta\rangle}$ and $\delta(\langle\rangle) = \langle\Delta\rangle$.

For the following clauses assume $s \in S^{\omega}_{\langle\Delta\rangle}$ and $\neg Ax(\delta(s))$.

($S_{Id}$) If $\delta(s)$ has no redex then $s^\frown\langle 0\rangle \in S^{\omega}_{\langle\Delta\rangle}$ and $\delta(s^\frown\langle 0\rangle) = \delta(s)$.

($S_{\bigwedge}$) If the redex $R(\delta(s))$ belongs to $\bigwedge$–type then $s^\frown\langle i\rangle \in S_{\langle\Delta\rangle}$ for every $F_i \in \mathrm{CS}(R(\delta(s)))$ and $\delta(s^\frown\langle i\rangle) = \delta(s)^r, F_i$.

($S_{\bigvee}$) If the redex $R(\delta(s))$ is in $\bigvee$–type then $s^\frown\langle 0\rangle \in S^{\omega}_{\langle\Delta\rangle}$ and $\delta(s^\frown\langle 0\rangle) = \delta(s)^r, F_i, R(\delta(s))$, where $F_i$ is the first formula in $\mathrm{CS}(F)$ that is not numerically equivalent to a formula in $\bigcup_{s_0\subseteq s}\delta(s_0)$. If no such formula exists then $\delta(s^\frown\langle 0\rangle) = \delta(s)^r, R(\delta(s))$.

**5.4.6 Remark** The search tree $S^{\omega}_{\langle\Delta\rangle}$ and the label function $\delta$ are defined by course-of-values recursion. Therefore, $S^{\omega}_{\langle\Delta\rangle}$ as well as $\delta$ are primitive recursive. The order type of $S^{\omega}_{\langle\Delta\rangle}$, in case that it is well-founded, is therefore an ordinal $< \omega^{CK}_1$.

**5.4.7 Lemma** *(Syntactical Main Lemma) If $S^{\omega}_{\langle\Delta\rangle}$ is well-founded then $\models^{\mathrm{otyp}(s)} \delta(s)$ for all $s \in S^{\omega}_{\langle\Delta\rangle}$, where $\delta(s)$ is viewed as a finite set.*

*Proof* We induct on $\mathrm{otyp}(s)$. If $\mathrm{otyp}(s) = 0$ then $s$ is a leaf. By definition of $S^{\omega}_{\langle\Delta\rangle}$ the only possibilities for $s \in S^{\omega}_{\langle\Delta\rangle}$ to become a leaf are that either $Ax(\delta(s))$ holds true or $R(\delta(s)) \in \bigwedge$–type and $\mathrm{CS}(\delta(s)) = \emptyset$. In the first case we get $\models^{0} \delta(s)$ according to ($Ax$). In the second case we obtain $\models^{0} \delta(s)$ according to $\bigwedge$.

If $\mathrm{otyp}(s) > 0$ we are in the case $S_{\bigvee}$ or $S_{\bigwedge}$. For the redex $F := R(\delta(s)$ we get either $F \in \bigvee$–type or $F \in \bigwedge$–type, and obtain by the induction hypothesis and the structural rule

$$\models^{\mathrm{otyp}(s^\frown\langle k\rangle)} \delta(s), G_k \tag{i}$$

for some or all $G_k \in \mathrm{CS}(F)$, respectively. From (i), however, we get the claim by an inference according to ($\bigvee$) or ($\bigwedge$), respectively. □

**5.4.8 Lemma** *(Semantical Main Lemma) If $S^{\omega}_{\langle\Delta\rangle}$ is not well-founded then there is an assignment $S_1,\ldots,S_n$ to the set variables occurring in $\Delta$ such that $\mathbb{N} \not\models F[S_1,\ldots,S_n]$ for all $F \in \Delta$.*

*Proof* The proof is very similar to the proof of Lemma 4.4.5 in Sect. 4.4. Pick an infinite path $f$ in $S^{\omega}_{\langle\Delta\rangle}$. Let $\delta(f) := \bigcup_{m\in\omega} \delta([f](m))$. Observe

$$F \notin (\bigvee\text{–type} \cup \bigwedge\text{–type}) \wedge F \in \delta([f](n)) \;\Rightarrow\; (\forall m \geq n)[F \in \delta([f](m))] \quad \text{(i)}$$

$$F \in \delta([f](n)) \cap (\bigvee\text{–type} \cup \bigwedge\text{–type}) \;\Rightarrow\; (\exists m \geq n)[F = R(\delta([f](m)))] \quad \text{(ii)}$$

$$F \in \delta(f) \cap \bigwedge\text{–type} \;\Rightarrow\; (\exists G \in \mathrm{CS}(F))[G \in \delta(f)] \quad \text{(iii)}$$

$$F \in \delta(f) \cap \bigvee\text{–type} \;\Rightarrow\; (\forall G \in \mathrm{CS}(F))[G \in \delta(f)]. \quad \text{(iv)}$$

Recall that we identify numerically equivalent formulas.

The proof of (i) is obvious because there are no rules that change or discard formulas of the form $t \,\varepsilon\, X$ or $t \,\not\varepsilon\, X$.

The proof of (ii) is by induction on the index of $F$ in $\delta([f](n))$. If there are no formulas in $\bigvee$–type $\cup \bigwedge$–type that have a smaller index in $\delta([f](n))$ then $F = R(\delta([f](n)))$. Otherwise, the redex $G := R(\delta([f](n)))$ has a smaller index than $F$. But $f$ is infinite, so $\mathrm{CS}(G) \neq \emptyset$. Thus $\delta([f](n+1)) = \delta([f](n))^r{}^\frown\langle G_i\rangle$ for some $G_i \in \mathrm{CS}(G)$ and $F$ gets a smaller index in $\delta([f](n+1))$. By induction hypothesis there is an $m$ such that $F = R(\delta([f](m)))$.

For the proof of (iii) we may by (ii) assume that $F = R(\delta([f](n)))$. Since $f$ is infinite, $\mathrm{CS}(F)$ cannot be empty. Thus, by clause $\bigwedge$, we find a $G \in \mathrm{CS}(F)$ such that $G \in \delta([f](n+1))$.

To show (iv) let $G \in \mathrm{CS}(f)$ and denote by $i(G)$ its index in the sequence $\mathrm{CS}(F)$. By (ii) we may again assume $F = R(\delta([f](n)))$. Then $[f](n+1) = [f](n)^\frown\langle 0\rangle$ and $\delta([f](n+1)) = \delta([f](n))^r{}^\frown\langle G_0\rangle^\frown\langle F\rangle$ for some $G_0 \in \mathrm{CS}(F)$. If $i(G) \leq i(G_0)$ we are done. Otherwise, we proceed by induction on $i(G) - i(G_0)$ and obtain by (ii) an $m$ such that $F = R(\delta([f](m)))$. Then $\delta([f](m+1)) = \delta([f](m))^r{}^\frown\langle G_1\rangle^\frown\langle F\rangle$ for some $G_1 \in \mathrm{CS}(F)$. Since $i(G) - i(G_1) < i(G) - i(G_0)$ the claim follows by induction hypothesis.

We define an assignment

$$\Phi(X) := \{t^{\mathbb{N}} \mid (t \,\not\varepsilon\, X) \in \delta(f)\}$$

and show by induction on $\mathrm{rnk}(F)$

$$(\forall F \in \delta(f))[\mathbb{N} \not\models F[\Phi]]. \quad \text{(v)}$$

If $F \equiv (t \,\varepsilon\, X) \in \delta([f](n))$ then $(t \,\varepsilon\, X) \in \delta([f](m))$ for all $m \geq n$ by (i). This, however, implies $(s \,\not\varepsilon\, X) \notin \delta(f)$ for all $s$ such that $s^{\mathbb{N}} = t^{\mathbb{N}}$ because, otherwise, we had by (i) $Ax(\delta([f](m)))$ for some $m$ contradicting the infinity of the path $f$. Hence $t^{\mathbb{N}} \notin \Phi(X)$, which means $\mathbb{N} \not\models (t \,\varepsilon\, X)[\Phi]$.

If $(t \not\varepsilon X) \in \delta([f](n))$ we get by definition of $\Phi$ directly $\mathbb{N} \models (t \,\varepsilon\, X)[\Phi]$, i.e., $\mathbb{N} \not\models (t \not\varepsilon X)[\Phi]$.

If $F \in \bigwedge$–type we get by (iii) an $m$ such that $G \in \delta([f](m))$ for some $G \in \mathrm{CS}(F)$. By the induction hypothesis this implies $\mathbb{N} \not\models G[\Phi]$ which, in turn, shows $\mathbb{N} \not\models F[\Phi]$.

If $F \in \bigvee$–type we get for all $G \in \mathrm{CS}(F)$ a $m$ such that $G \in \delta([f](m))$ by (iv). By the induction hypothesis this implies $\mathbb{N} \not\models G[\Phi]$ for all $G \in \mathrm{CS}(F)$. Hence $\mathbb{N} \not\models F[\Phi]$.

Since $\Delta = \delta([f](0))$ the claim follows from (v). □

**5.4.9 Theorem** *($\omega$-completeness Theorem) For all $\Pi^1_1$-sentences of the form $(\forall X_1)\dots(\forall X_n)F(X_1,\dots,X_n)$ we have*

$$\mathbb{N} \models (\forall X_1)\dots(\forall X_n)F(X_1,\dots,X_n) \;\Leftrightarrow\; (\exists a < \omega_1^{CK}) \models^{\alpha} F(X_1,\dots,X_n).$$

*Proof* The direction from right to left is Lemma 5.4.4. For the opposite direction we assume

$$\not\models^{\alpha} F(X_1,\dots,X_n) \tag{i}$$

for all $\alpha < \omega_1^{CK}$. Then $S^{\omega}_{(F(X_1,\dots,X_n))}$ cannot be well-founded by the Syntactical Main Lemma (Lemma 5.4.7). Thus, by the Semantical Main Lemma, we obtain an assignment $\Phi$ to the set variables $X_1,\dots,X_n$ such that $\mathbb{N} \not\models F(X_1,\dots,X_n)[\Phi]$. □

It is an easy exercise to prove that the verification calculus has the property of $\vee$-exportation, i.e., to prove

$$\models^{\alpha} \Delta, A \vee B \;\Rightarrow\; \models^{\alpha} \Delta, A, B$$

by induction on $\alpha$. For a formula $F$ let $F^{\nabla}$ denote the finite set that is obtained from $F$ by exporting all disjunction , i.e.,

$$F^{\nabla} := \begin{cases} \{F\} & \text{if } F \notin \bigvee\text{–type or } F \equiv (\exists y)G_v(y) \\ A^{\nabla} \cup B^{\nabla} & \text{if } F \equiv (A \vee B). \end{cases}$$

Then we obtain[1]

$$\models^{\alpha} F \;\Rightarrow\; \models^{\alpha} F^{\nabla}.$$

**5.4.10 Definition** Let $(\forall \vec{X})F(\vec{X})$ be a $\Pi^1_1$ sentence. We put

$$\mathrm{tc}\big((\forall \vec{X})F(\vec{X})\big) := \min(\{\alpha \mid \models^{\alpha} F(\vec{X})^{\nabla}\} \cup \{\omega_1^{CK}\})$$

[1] There is no profoundness behind the definition of $F^{\nabla}$. We just wanted to avoid that certain formulas, e.g., the scheme for Mathematical Induction, get weird looking truth complexities, caused by additional applications of $\bigvee$-rules.

and call $\text{tc}(F)$ the *truth complexity* of $F$. For a pseudo $\Pi_1^1$-sentence $G(\vec{X})$ containing the free set parameters $\vec{X}$ we define

$$\text{tc}\big(G(\vec{X})\big) := \text{tc}\big((\forall\vec{X})G(\vec{X})\big).$$

We can now reformulate Theorem 5.4.10 in the following form.

**5.4.11 Theorem** *For any (pseudo) $\Pi_1^1$-sentence $F$ we have*

$$\mathbb{N} \models F \;\Leftrightarrow\; \text{tc}(F) < \omega_1^{CK}.$$

**5.4.12 Exercise** Let $F$ be a pseudo $\Pi_1^1$- sentence. Show that the search tree $S_{\{F\}}$ and the label function $\delta$ are primitive recursive.

**5.4.13 Exercise** Let $\mathfrak{S} = (S, \ldots)$ be a structure for $\mathcal{L}(\mathsf{NT})$. We call $\mathfrak{S}$ a weak second-order structure if we only allow $\mathfrak{S}$-assignments to second-order variables that lie in some subset $\mathscr{A}$ of the full power-set of $S$. This implies that the range of second-order quantifiers is restricted to $\mathscr{A}$.[2] We denote weak second-order structures by $\mathfrak{S} = (S, \mathscr{A}, \ldots)$. Call a weak second-order structure $\mathfrak{S}$ a *$\omega$-structure* if $(S, \ldots)$, i.e., its first-order part, is isomorphic to the standard model $\mathbb{N}$. Define $M \models_{\omega} F$ iff $\mathfrak{S} \models G[\Phi]$ for all $G \in M$ implies $\mathfrak{S} \models F[\Phi]$ for all $\omega$-structures $\mathfrak{S}$ and all $\mathfrak{S}$-assignments $\Phi$. Show that for a $\Pi_1^1$-sentence $F$ and a countable set $M$ of $\Sigma_1^1$-sentences we have $M \models_{\omega} F$ iff there is a $\alpha < \omega_1^{CK}$ and a finite set $\Gamma \subseteq M$ such that $\models^{\alpha} \neg\Gamma, F$.

---

[2] That means that we treat second-order variables as a new sort of first-order variables, the range of which is restricted to $\mathscr{A}$. Therefore, we are in fact in the realm of a two sorted first-order logic.

# Chapter 6
# Inductive Definitions

Inductive definitions are ubiquitous in mathematics, especially in mathematical logic. In the following section, we give a summary of the basic theory of inductive definitions and link it with the the theory of truth complexity for $\Pi^1_1$-sentences of the previous Chapter.

## 6.1 Motivation

We already introduced a sample of inductively defined notions in this book. Recall, for instance, the inductive definition of a term by the following clauses:

1. Every free variable and constant is a term.
2. If $t_1,\ldots,t_n$ are terms and $f$ is an $n$-ary function symbol then $(ft_1,\ldots,t_n)$ is a term.

Generally speaking, inductive definitions consist of a set of clauses. A clause has the form:

*If* $a \in X$ *for all* $a \in A$ *then* $b \in X$.

Put in abstract terms, a clause is a pair $(A,b)$ where $A$ is the set of premises and $b$ the conclusion of the clause. Sometimes we write briefly

$$A \Rightarrow b$$

to denote a clause. The set of premises may well be infinite.

**6.1.1 Definition** An inductive definition is a collection $\mathscr{A}$ of clauses. Let $\mathscr{A}$ be an inductive definition. We say that, a set $X$ satisfies $\mathscr{A}$ if for $(A \Rightarrow b) \in \mathscr{A}$ and $A \subseteq X$ we also have $b \in X$. Synonymously, we say that, the set $X$ is closed under the clauses in $\mathscr{A}$ (or just $\mathscr{A}$-closed).

**6.1.2 Definition** A set $X$ is *inductively generated* by an inductive definition $\mathscr{A}$ if $X$ is the least set, with respect to set inclusion, which is closed under $\mathscr{A}$. We denote this set sometimes by $I(\mathscr{A})$.

Observe that the set $C_{\mathscr{A}} := \{b \mid (A \Rightarrow b) \in \mathscr{A}\}$ of conclusions of an inductive definition $A$ is always $\mathscr{A}$-closed. Moreover the intersection of $\mathscr{A}$-closed sets is again $\mathscr{A}$-closed. We therefore obtain

$$I(\mathscr{A}) = \bigcap\{X \mid X \text{ is } \mathscr{A}\text{-closed}\}.$$

An inductive definition $\mathscr{A}$ is *deterministic* if $(A \Rightarrow b) \in \mathscr{A}$ and $(B \Rightarrow b) \in \mathscr{A}$ imply $A = B$, i.e., if there is only one way in which $b$ can get into $I(\mathscr{A})$.

## 6.2 Inductive Definitions as Monotone Operators

In this section we will show that inductive definitions can be viewed as *monotone operators*.

**6.2.1 Definition** Let an inductive definition $\mathscr{A}$ operate on a set $X$, i.e., $A \cup \{a\} \subseteq X$ holds for all $(A \Rightarrow a) \in \mathscr{A}$. We define an operator,

$$\Phi_{\mathscr{A}}: \text{Pow}(X) \longrightarrow \text{Pow}(X)$$
$$\Phi_{\mathscr{A}}(S) := \{b \in X \mid (\exists (A \Rightarrow b) \in \mathscr{A})[A \subseteq S]\}.$$

For $S_0 \subseteq S_1 \subseteq X$ we then obviously have

$$\Phi_{\mathscr{A}}(S_0) \subseteq \Phi_{\mathscr{A}}(S_1)$$

which means that the operator $\Phi_{\mathscr{A}}$ is *monotone*. Vice versa every monotone operator

$$\Phi: \text{Pow}(X) \longrightarrow \text{Pow}(X)$$

can be viewed as the collection

$$\{(A \Rightarrow a) \mid A \subseteq X \wedge a \in \Phi(A)\}$$

of clauses. Inductive definitions and monotone operators are thus two sides of the same medal.

So we simplify Definition 6.1.1 in the following way.

**6.2.2 Definition** A generalized inductive definition on a set $A$ is a monotone operator $\Phi: \text{Pow}(A^n) \longrightarrow \text{Pow}(A^n)$.

**6.2.3 Definition** Let

$$\Phi: \text{Pow}(A^n) \longrightarrow \text{Pow}(A^n)$$

be an inductive definition. A set $S \subseteq A^n$ is *$\Phi$-closed* iff $\Phi(S) \subseteq S$.

We define

$$I(\Phi) = \bigcap \{S \mid S \subseteq A^n \wedge \Phi(S) \subseteq S\}.$$

**6.2.4 Observation** *Let* $\Phi\colon \mathrm{Pow}(A^n) \longrightarrow \mathrm{Pow}(A^n)$ *be an inductive definition. Then* $I(\Phi)$ *is the least fixed-point of the operator* $\Phi$.

*Proof* Let

$$\mathfrak{M}_\Phi := \{S \mid \Phi(S) \subseteq S\}.$$

Then

$$I(\Phi) = \bigcap \mathfrak{M}_\Phi.$$

For $S \in \mathfrak{M}_\Phi$ we have $I(\Phi) \subseteq S$ and thus $\Phi(I(\Phi)) \subseteq \Phi(S) \subseteq S$ by monotonicity. Thus

$$\Phi(I(\Phi)) \subseteq \bigcap \mathfrak{M}_\Phi = I(\Phi). \tag{i}$$

From (i) we obtain

$$\Phi(\Phi(I(\Phi))) \subseteq \Phi(I(\Phi)) \tag{ii}$$

again by monotonicity. Hence $\Phi(I(\Phi)) \in \mathfrak{M}_\Phi$, which entails

$$I(\Phi) \subseteq \Phi(I(\Phi)). \tag{iii}$$

But (i) and (iii) show that $I(\Phi)$ is a fixed-point and by definition of $I(\Phi)$ this has to be the least one. □

We call, $I(\Phi)$ *the fixed-point* of the inductive definition $\Phi$.

## 6.3 The Stages of an Inductive Definition

The definition of a fixed-point as the intersection of all $\Phi$-closed sets – which is sometimes called the definition of $I(\Phi)$ "from above" – does, however, not really reflect the inductive nature of the fixed-point. An inductively defined set should be obtained step by step from below but not as the intersection of $\Phi$-closed sets. This step by step construction of $I(\Phi)$ becomes visible regarding the *stages* of an inductive definition $\Phi$. The idea is to apply $\Phi$ successively starting with the empty set. In general it is not guaranteed that the fixed-point will be constructed in finitely many steps but might well need transfinite iterations. Therefore we need ordinals to describe the stages of an inductive definition $\Phi$. For a set $A$ we denote by $\overline{\overline{A}}$ its cardinality. It is a folklore result of set theory that $(\overline{\overline{A}})^+$, the first cardinal bigger than $\overline{\overline{A}}$, is always a regular cardinal. We call a set $a$ countable if $\overline{\overline{a}} \leq \omega$. Observe that the first uncountable cardinal $\omega_1 := \omega^+$ is regular. Since $\sup\{\alpha \mid \alpha < (\overline{\overline{A}})^+\} = (\overline{\overline{A}})^+$ we obtain that $\overline{\overline{\{\alpha \mid \overline{\overline{\alpha}} \leq \overline{\overline{A}}\}}} = \overline{\overline{\{\alpha \mid \alpha < (\overline{\overline{A}})^+\}}} = (\overline{\overline{A}})^+$, i.e., there are $(\overline{\overline{A}})^+$-many ordinals of cardinality less or equal to $\overline{\overline{A}}$. Especially there are $\omega_1$-many countable ordinals.

**6.3.1 Definition** By transfinite recursion we define

$$\Phi^{\alpha} := \Phi(\Phi^{<\alpha})$$

where we used the abbreviation

$$\Phi^{<\alpha} := \bigcup_{\xi<\alpha} \Phi^{\xi}.$$

Then $\Phi^0 = \Phi(\emptyset)$, $\Phi^1 = \Phi(\Phi(\emptyset))$, $\Phi^2 = \Phi(\Phi(\Phi(\emptyset)) \cup (\Phi(0)))$, $\ldots$.

**6.3.2 Lemma** *Let $\Phi$ be an inductive definition on a set A. Then there is an ordinal $\sigma < (\overline{\overline{A}})^+$ such that $\Phi^{<\sigma} = \Phi^{\sigma}$.*

*Proof* By definition and the monotonicity of $\Phi$, we get

$$\xi < \eta \;\Rightarrow\; \Phi^{\xi} \subseteq \Phi^{\eta}.$$

All the sets $\Phi^{\xi}$ are subsets of $A$, hence of cardinality $\leq \overline{\overline{A}}$. Since there are $(\overline{\overline{A}})^+$-many ordinals below $(\overline{\overline{A}})^+$, the hierarchy of stages cannot be strict for all ordinals $\leq \overline{\overline{A}}$. Therefore there exist an ordinal $\sigma < (\overline{\overline{A}})^+$ such that $\Phi^{<\sigma} = \Phi^{\sigma}$. □

**6.3.3 Definition** We define

$$|\Phi| := \min\{\xi \mid \Phi^{\xi} = \Phi^{<\xi}\}$$

and call $|\Phi|$ the *closure ordinal* of the inductive definition $\Phi$.

**6.3.4 Theorem** *For an inductive definition $\Phi$ we have*

$$I(\Phi) = \Phi^{<|\Phi|}.$$

*Proof* We obtain

$$\Phi^{\xi} \subseteq I(\Phi) \tag{i}$$

easily by induction on $\xi$. The induction hypothesis yields $\Phi^{<\xi} \subseteq I(\Phi)$ and by the monotonicity of $\Phi$ we obtain $\Phi^{\xi} = \Phi(\Phi^{<\xi}) \subseteq \Phi(I(\Phi)) = I(\Phi)$.

Since $\Phi(\Phi^{<|\Phi|}) = \Phi^{|\Phi|} = \Phi^{<|\Phi|}$ and $I(\Phi)$ is the least $\Phi$-closed set we also have

$$I(\Phi) \subseteq \Phi^{<|\Phi|}. \tag{ii}$$

Hence

$$I(\Phi) = \Phi^{<|\Phi|}.$$

By (i) and (ii) and we have a construction of the fixed-point $I(\Phi)$ from below. □

**6.3.5 Definition** For an element $\vec{n} \in I(\Phi)$ we define its *inductive norm* by

$$|\vec{n}|_\Phi := \min\{\xi \mid \vec{n} \in \Phi^\xi\}.$$

**6.3.6 Lemma** *For the closure ordinal of an inductive definition* $\Phi$ *we get*

$$|\Phi| = \sup\{|\vec{n}|_\Phi + 1 \mid \vec{n} \in I(\Phi)\}.$$

*Proof* Since $I(\Phi) = \Phi^{<|\Phi|}$ we have $\eta := \sup\{|\vec{n}|_\Phi + 1 \mid \vec{n} \in I(\Phi)\} \leq |\Phi|$. Assuming $\eta < |\Phi|$ we get $\Phi^\eta \subsetneqq \Phi^{\eta+1}$. But then, there is an $\vec{n} \in \Phi^{\eta+1} \setminus \Phi^\eta$ which contradicts the definition of $\eta$. □

**6.3.7 Exercise** Let $A$ be an arbitrary set and $M \subseteq \mathrm{Pow}(A)$. Show that there is an inductive definition $\Phi_M$ whose fixed-point is the sigma-algebra induced by $M$. Show that $|\Phi_M| \leq \omega_1$.

## 6.4 Arithmetically Definable Inductive Definitions

The computation of the closure ordinals of inductive definitions is one of the objectives of generalized recursion theory. In the general case, however, all we know is that these ordinals are below $(\overline{\overline{A}})^+$. To say more, we need more information about the operator. In proof theory, we are primarily interested in inductively generated sets of natural numbers and will therefore concentrate on inductive definitions on natural numbers. Therefore we concentrate on operators which are *arithmetically definable*, i.e., inductive definitions whose clauses can be described in the language of arithmetic.

**6.4.1 Definition** An operator

$$\Phi\colon \mathrm{Pow}(\mathbb{N}^n) \longrightarrow \mathrm{Pow}(\mathbb{N}^n)$$

is *arithmetically definable* if there is a formula $F(X,\vec{x},y_1,\ldots,y_n)$ in the first-order language of number theory, in which only the shown variables occur freely, together with number parameters $a_1,\ldots,a_m$ such that

$$\Phi(S) = \{k \in \mathbb{N} \mid \mathbb{N} \models F(X,\vec{x},y_1,\ldots,y_m)[S,k,a_1,\ldots,a_m]\}.$$

Given a formula $F(X,\vec{x},y_1,\ldots,y_n)$ and a tuple $a_1,\ldots,a_m$ of number constants, we denote the induced operator by $\Phi_F[a_1,\ldots,a_m]$. Usually we suppress mentioning the additional parameters $a_1,\ldots,a_m$.

**6.4.2 Remark** Since there is a primitive recursive coding machinery in the language of arithmetic, it means no loss of generality to restrict inductive definitions to operators $\Phi\colon \mathrm{Pow}(\mathbb{N}) \longrightarrow \mathrm{Pow}(\mathbb{N})$. For an $n$-ary relation variable $X$, we can define an unary relation variable $\langle X\rangle$ and replace $(u_1,\ldots,u_n) \; \varepsilon \; X$ by $\langle u_1,\ldots,u_n\rangle \; \varepsilon \; \langle X\rangle$.

To simplify notation we will therefore mostly work with set variables, i.e., unary relation variables.

**6.4.3 Definition** Let $F(X)$ be a formula in the language of arithmetic. We say that $F(X)$ is $X$-positive, if, after translating $F(X)$ into the TAIT-language, there are no occurrences of $t \not\varepsilon X$ in $F(X)$.

**6.4.4 Lemma** *The operator* $\Phi_F$ *which is induced by an* $X$*-positive formula* $F(X,x)$ *is monotone. In this case we talk about* positive *operators or* positive inductive definitions.

*Proof* We show

$$S \subseteq T \;\Rightarrow\; (\forall x)[F(S,x) \to F(T,x)] \tag{i}$$

by induction on the rank of the $X$-positive formula $F(X,x)$. If $X$ does not occur in $F(X,x)$ then (i) holds trivially. If $F(X,x)$ is a formula $x \;\varepsilon\; X$ we obtain $(\forall x)[F(S,x) \to F(T,x)]$ from the hypothesis $S \subseteq T$. In case that $F(X,x)$ possesses sub-formulas we obtain the claim immediately from the induction hypothesis. □

**6.4.5 Definition** If $\Phi_F$ is an inductive definition defined by the $X$-positive formula $F(X,x,a_1,\dots,a_n)$ we write $I_F[a_1,\dots,a_n]$ instead of $I(\Phi_F[a_1,\dots,a_n])$ and $|F|$ instead of $|\Phi_F|$ as well as $I_F^{<\xi}[a_1,\dots,a_n]$ and $I_F^{\xi}[a_1,\dots,a_n]$ instead of $\Phi_F[a_1,\dots,a_n]^{<\xi}$ and $\Phi_F[a_1,\dots,a_n]^{\xi}$, respectively. Again we suppress the extra parameters $a_1,\dots,a_n$ whenever they are inessential or obvious from the context.

**6.4.6 Remark** It follows from the Craig–Lyndon interpolation theorem that every definable operator whose monotonicity is provable in first-order logic is positive. This may not be the case if the proof of the monotonicity needs non logical axioms. All relevant monotone inductive definitions, however, are positively definable. (Cf. Exercise 6.4.8.)

For a set $S$ and a natural number $a$, the *a-slice* is the set

$$S_a := \{x \mid \langle a,x\rangle \in S\}.$$

**6.4.7 Definition** A set $S \subseteq \mathbb{N}$ is *positive-inductively definable* if there is an $X$-positive formula $F(X,x)$, possibly with further numerical parameters, and a natural number $a$ such that $S$ is the $a$ slice of the fixed-point $I_F$.

By Observation 6.2.4 we see that the fixed-point of an arithmetically definable inductive definition is definable by a $\Pi^1_1$-formula. For an $X$-positive arithmetical formula $F(X,x,y_1,\dots,y_n)$ we obtain

$$x \in I_F[a_1,\dots,a_n] \;\Leftrightarrow\; (\forall X)\big[(\forall y)[F(X,y,a_1,\dots,a_n) \to y \;\varepsilon\; X] \;\to\; x \;\varepsilon\; X\big]. \tag{6.1}$$

Therefore we have the following lemma;

**6.4.8 Lemma** *All positive-inductively definable sets are in $\Pi^1_1$.*

*Proof* This is obvious from (6.1) and Definition 6.4.7. □

We will see later that vice versa all $\Pi^1_1$-set are positive-inductively definable.

An ordinal which is characteristic for the structure $\mathbb{N}$ of natural numbers is the ordinal

$$\kappa^{\mathbb{N}} := \sup\{|\Phi_F| \mid F(X,x,a_1,\ldots,a_n) \text{ is an } X\text{-positive arithmetical formula and } a_1,\ldots,a_n \in \mathbb{N}.\}$$

It is a folklore result of abstract recursion theory that $\kappa^{\mathbb{N}} = \omega_1^{CK}$, where $\omega_1^{CK}$ is the least ordinal which cannot be represented as the order-type of a primitive recursive (even $\Sigma^1_1$-definable) well-ordering, a result which we will reprove in Theorem 6.6.4. We will later even get a finer calibration regarding not only definability but also provability within certain axiom systems.

**6.4.9 Exercise** Let $\Phi\colon \mathrm{Pow}(\mathbb{N}) \longrightarrow \mathrm{Pow}(\mathbb{N})$ be an arithmetically definable operator such that $\models (\forall X)(\forall Y)\big[X \subseteq Y \to (\forall x)[F(X,x) \to F(Y,x)]\big]$ holds true for its defining formula $F(X,x)$. Show that there is an $X$-positive formula which defines $\Phi$.

Hint: Use the CRAIG–LYNDON interpolation theorem (Exercise 4.5.7).

**6.4.10 Exercise** *(Simultaneous inductive definitions).* Assume that $F(X,Y,x)$ and $G(X,Y,y)$ are $X$- and $Y$-positive formulas. We define

$$\Phi_F^\alpha := \{n \in \mathbb{N} \mid \mathbb{N} \models F[\Phi_F^{<\alpha},\Phi_G^{<\alpha},n]\}$$

and

$$\Phi_G^\alpha := \{n \in \mathbb{N} \mid \mathbb{N} \models G[\Phi_F^{<\alpha},\Phi_G^{<\alpha},n]\}$$

where we again abbreviate $\Phi_F^{<\alpha} := \bigcup_{\xi<\alpha}\Phi_F^\xi$ and $\Phi_G^{<\alpha} := \bigcup_{\xi<\alpha}\Phi_G^\xi$. Show that $\bigcup_{\xi\in On}\Phi_F^\xi$ as well as $\bigcup_{\xi\in On}\Phi_G^\xi$ are positive-inductively definable.

**6.4.11 Exercise** *(Transitivity theorem).* Let $F(X,Y,x)$ be an $X,Y$-positive and $G(Z,z)$ a $Z$-positive formula. Show that the fixed-point of the operator $\Phi_{F(I_G)}$ defined by

$$\Phi_{F(I_G)}(S) := \{n \in \mathbb{N} \mid \mathbb{N} \models F[I_G,S,n]\}$$

is positive-inductively definable.

**6.4.12 Exercise** *(Stage comparison theorem).* Let $F(X,x)$ and $G(X,x)$ be $X$-positive formulas. Show that the stage comparison relations

$$m \le^*_{F,G} n \;:\Leftrightarrow\; m \in I_F \wedge |m|_F \le |n|_G$$

and

$$m <^{*}_{F,G} n \;:\Leftrightarrow\; |m|_F < |n|_G$$

are positive-inductively definable. Here we put $|n|_F := \omega_1$ for $n \notin I_F$.

## 6.5 Inductive Definitions, Well-Orderings and Well-Founded Trees

An important special case of inductive definitions are *accessible parts* which will be studied in this section.

**6.5.1 Definition** Let $\prec$ be an arithmetically definable binary relation on the natural numbers and regard the operator defined by the formula

$$\mathrm{Acc}(X,x,\prec) \;:\Leftrightarrow\; x \in \mathit{field}(\prec) \wedge (\forall y)[y \prec x \rightarrow y \,\varepsilon\, X].$$

This is obviously a $X$-positive formula inducing a monotone operator $\mathsf{Acc}_{\prec}$. Its fixed-point is the *accessible part* of $\prec$, usually denoted by $\mathrm{Acc}_{\prec}$. By $\mathrm{Acc}^{\alpha}_{\prec}$ we denote the $\alpha$-th stage of the operator $\mathsf{Acc}_{\prec}$. For $s \in \mathrm{Acc}_{\prec}$ we denote by $|s|_{\mathrm{Acc}_{\prec}}$ the inductive norm of $s$.

**6.5.2 Lemma** *Let $\prec$ be an arithmetically definable binary well-founded relation on the natural numbers. Then $\mathrm{Acc}_{\prec} = \mathit{field}(\prec)$ and for all $s \in \mathit{field}(\prec)$ we have $\mathrm{otyp}_{\prec}(s) = |s|_{\mathrm{Acc}_{\prec}}$.*

*Proof* First we show

$$s \in \mathit{field}(\prec) \wedge \mathrm{otyp}_{\prec}(s) = \alpha \;\Rightarrow\; s \in \mathrm{Acc}^{\alpha}_{\prec} \tag{i}$$

by induction on $\alpha$. By the induction hypothesis we have

$$(\forall t)[t \prec s \rightarrow t \in \mathrm{Acc}^{<\alpha}_{\prec}] \tag{ii}$$

which immediately implies $s \in \mathrm{Acc}^{\alpha}_{\prec}$. From (i) we obtain $\mathit{field}(\prec) \subseteq \mathrm{Acc}_{\prec}$ and $|s|_{\mathrm{Acc}_{\prec}} \leq \mathrm{otyp}_{\prec}(s)$. The opposite inclusion $\mathrm{Acc}_{\prec} \subseteq \mathit{field}(\prec)$ holds anyway. Therefore it suffices to show

$$s \in \mathrm{Acc}^{\alpha}_{\prec} \;\Rightarrow\; \mathrm{otyp}_{\prec}(s) \leq \alpha. \tag{iii}$$

We prove (iii) by induction on $\alpha$. By the induction hypothesis we get

$$\mathrm{Acc}^{<\alpha}_{\prec} \subseteq \{x \in \mathit{field}(\prec) \mid \mathrm{otyp}_{\prec}(x) < \alpha\}. \tag{iv}$$

For $s \in \mathrm{Acc}^{\alpha}_{\prec}$ and $t \prec s$ we obtain $t \in \mathrm{Acc}^{<\alpha}_{\prec}$ and thence $\mathrm{otyp}_{\prec}(t) < \alpha$. Hence $\mathrm{otyp}_{\prec}(s) \leq \alpha$. □

We see from Lemma 6.5.2 that induction along a well-ordering or induction on the definition of an inductively defined set are two sides of the same medal. We shall see that Bar Induction on well-founded trees belongs to the same category. Well-founded trees have been introduced in Definition 4.4.2. Let $T$ be a tree which is arithmetically definable, i.e., we have an arithmetical formula $T(x)$ without further free variables such that

$$s \in T \;\Leftrightarrow\; T(s).$$

Define

$$F_T(X,x) \;:\Leftrightarrow\; T(x) \wedge (\forall y)[T(x^\frown\langle y\rangle) \;\rightarrow\; x^\frown\langle y\rangle \,\varepsilon\, X].$$

Then $F_T(X,x)$ is an $X$-positive formula. Let $I_T$ denote its fixed-point and $I_T^\alpha$ the stages of the fixed-point.

**6.5.3 Lemma** *Let $T$ be a well-founded arithmetically definable tree. Then $s \in I_T^{\mathrm{otyp}_T(s)}$ for all $s \in T$.*

*Proof* We induct on $\mathrm{otyp}_T(s)$. By the induction hypothesis we have $s^\frown\langle x\rangle \in I_T^{<\mathrm{otyp}_T(s)}$ for all $x$ such that $s^\frown\langle x\rangle \in T$, i.e $(\forall x)[T(s^\frown\langle x\rangle) \rightarrow s^\frown\langle x\rangle \,\varepsilon\, I_T^{<\mathrm{otyp}_T(s)}]$. But this immediately implies $s \in I_T^{\mathrm{otyp}_T(s)}$. ☐

**6.5.4 Lemma** *Let $T$ be an arithmetically definable tree. Then $T_s$ is well-founded for all $s \in I_T$ and* $\mathrm{otyp}(T_s) \leq |s|_{I_T}$.

*Proof* We induct on $|s|_{I_T}$. By the induction hypothesis $T_{s^\frown\langle x\rangle}$ is well-founded of order type $< |s|_{I_T}$ for all $x$ such that $s^\frown\langle x\rangle \in T$. But then $T_s$ is well-founded and we easily compute

$$\begin{aligned}\mathrm{otyp}(T_s) &= \mathrm{otyp}_{T_s}(\langle\rangle) = \sup\{\mathrm{otyp}_{T_s}(\langle x\rangle)+1 \mid \langle x\rangle \in T_s\}\\ &= \sup\{\mathrm{otyp}(T_{s^\frown\langle x\rangle})+1 \mid s^\frown\langle x\rangle \in T\} \leq |s|_{I_T}.\end{aligned}$$

☐

Pulling Lemmata 6.5.3 and 6.5.4 together we obtain the following theorem.

**6.5.5 Theorem** *An arithmetically definable tree $T$ is well-founded if and only if $\langle\,\rangle \in I_T$. If $T$ is well-founded then*

$$(\forall s \in T)[\mathrm{otyp}_T(s) = |s|_{I_T}] \;\; \textit{and} \;\; |I_T| = \mathrm{otyp}(T)+1.$$

*Proof* Let $T$ be arithmetically definable. If $T$ is well-founded we obtain $\langle\,\rangle \in I_T$ by Lemma 6.5.3. If conversely $\langle\,\rangle \in I_T$ then we obtain the well-foundedness of $T_{\langle\,\rangle} = T$ by Lemma 6.5.4. By Lemma 6.5.3 we obtain $|s|_{I_T} \leq \mathrm{otyp}_T(s)$ and by Lemma 6.5.4 we get $\mathrm{otyp}_T(s) = \mathrm{otyp}(T_s) \leq |s|_{I_T}$. Finally we observe that $|t|_{I_T} \leq |s|_{I_T}$ for all $s \subseteq t \in T$. Hence $|I_T| = |\langle\,\rangle|_{I_T} + 1 = \mathrm{otyp}_T(\langle\,\rangle) + 1 = \mathrm{otyp}(T) + 1$ by Lemma 6.3.6. ☐

**6.5.6 Corollary** *The class of $\Pi^1_1$-sets and the class of positive-inductively definable sets coincide.*

*Proof* We have already seen that all positive-inductively definable sets are in $\Pi^1_1$. If $P \in \Pi^1_1$, say $P = \{x \mid F_u(x)\}$, we obtain $n \in P$ iff the search tree for $F_u(\underline{n})$ is well-founded from the $\omega$-Completeness Theorem (Theorem 5.4.9). By Theorem 6.5.5 this is the case iff $\langle\rangle \in I_{S^{\omega}_{F_u(\underline{n})}}$. This shows that $P$ is positive-inductively definable.

□

## 6.6 Inductive Definitions and Truth Complexities

The aim of this section is to connect the closure ordinal of a positive inductive definition to the truth complexity of its defining $\Pi^1_1$-sentence. To improve readability we introduce some abbreviations. Let

$$Cl_F(X) \;:\Leftrightarrow\; (\forall x)[F(X,x) \to x \,\varepsilon\, X]$$

denote the fact that the set $X$ is closed under the operator induced by the formula $F(X,x)$. The $\Pi^1_1$-definition of $n \in I_F$ is then obtained as

$$n \in I_F \;\Leftrightarrow\; (\forall X)[Cl_F(X) \to n \,\varepsilon\, X].$$

The first step is to show a stage theorem.

**6.6.1 Theorem** *(Stage Theorem) Let $F(X,x)$ be an $X$-positive arithmetical formula. Then*

$$(\forall n \in I_F)\big[|n|_F < 2^{\mathrm{tc}(n \in I_F)}\big].$$

To prove the Stage Theorem we need the more general Stage Lemma. To formulate the lemma let $F(X,x)$ be an $X$-positive arithmetical formula and put

$$F[t_1,\dots,t_k] \;:\Leftrightarrow\; F(X,x) \vee x = t_1 \vee \cdots \vee x = t_k.$$

Then

$$s \in I^{\alpha}_F \;\Rightarrow\; I^{\beta}_{F[s]} \subseteq I^{\alpha+\beta}_F. \tag{6.2}$$

Claim (6.2) is shown by induction on $\beta$. If $x \in I^{\beta}_{F[s]}$ then $F(I^{<\beta}_{F[s]},x) \vee x = s$. By induction hypothesis and $X$-positivity, this implies $F(I^{<\alpha+\beta}_F,x) \vee x = s$. Hence $x \in I^{\alpha+\beta}_F$ or $x = s \in I^{\alpha}_F \subseteq I^{\alpha+\beta}_F$ and (6.2) is shown.

The proof of the Stage Lemma uses the following Inversion Lemma.

**6.6.2 Lemma** *(Inversion Lemma) Let $F \in \bigwedge$-type and $\mathrm{CS}(F) \neq \emptyset$. Then $\models^{\alpha} \Delta, F$ implies $\models^{\alpha} \Delta, G$ for all $G \in \mathrm{CS}(F)$.*

*Proof* We induct on $\alpha$. Assume first that the last inference in verifying $\models^{\alpha} \Delta, F$ has a critical formula $A$ different from $F$. Then $A \in \Delta$ and we have the premise(s)

$\models^{\alpha_B} \Delta, F, B$ for some (or all) $B \in \mathrm{CS}(A)$. By the induction hypothesis we get $\models^{\alpha_B} \Delta, G, B$ for some (or all) $B \in \mathrm{CS}(A)$ and infer $\models^{\alpha} \Delta, G$ by the same inference.

If the last clause was $(\bigwedge)$ with critical formula $F$ then we have the premises

$$\models^{\alpha_G} \Delta, F, G \tag{i}$$

for all $G \in \mathrm{CS}(F)$. From the induction hypothesis we then get $\models^{\alpha_G} \Delta, G$ for all $G \in \mathrm{CS}(F)$ and obtain the claim by the structural rule (STR) (cf. p. 78). □

**6.6.3 Lemma** *(Stage Lemma) Let $F(X,x)$ be an $X$-positive arithmetical formula and $\Delta(X,\vec{Y})$ a finite set of pseudo $\Pi^1_1$-sentences in which $X$ occurs at most positively. If*

$$\models^{\alpha} \neg Cl_F(X),\ t_1 \notin X, \ldots, t_k \notin X,\ \Delta(X,\vec{Y}), \tag{i}$$

*then*

$$\mathbb{N} \models \bigvee \Delta[I^{<2^{\alpha}}_{F[t_1,\ldots,t_k]}, S_1, \ldots, S_n] \tag{ii}$$

*holds true for all assignments $S_1, \ldots, S_n$ to the set variables $Y_1, \ldots, Y_n$ in $\Delta$.*

*Proof* We prove the lemma by induction on $\alpha$. These are the following cases to distinguish.

1) (i) holds by (Ax). Then there is a formula $s \, \varepsilon \, X$ in $\Delta$ such that $s^{\mathbb{N}} = t_i^{\mathbb{N}}$ for some $i \in \{1, \ldots, k\}$. Then $s \in I^0_{F[t_1,\ldots,t_k]}$ and we obtain $\mathbb{N} \models \bigvee \Delta[I^{<1}_{F[t_1,\ldots,t_k]}, S_1, \ldots, S_n]$.

2) (i) holds by an application of $(\bigwedge)$. Then there is an $A \in \bigwedge\text{–type} \cap \Delta$. If $\mathrm{CS}(A) = \emptyset$ then $A \in Diag(\mathbb{N})$ and we are done. Otherwise we have the premises

$$\models^{\alpha_G} \neg Cl_F(X),\ t_1 \notin X, \ldots, t_k \notin X,\ \Delta(X,\vec{Y}),\ G(X,\vec{Y})$$

for all $G \in \mathrm{CS}(A)$. All $G(X,\vec{Y})$ are again $X$-positive and we obtain by the induction hypothesis

$$\mathbb{N} \models \bigvee \Delta[I^{<2^{\alpha_G}}_{F[t_1,\ldots,t_k]}, S_1, \ldots, S_n] \vee G[I^{<2^{\alpha_G}}_{F[t_1,\ldots,t_k]}, S_1, \ldots, S_n] \tag{iii}$$

for all $G(X,\vec{Y}) \in \mathrm{CS}(A)$. Since $I^{<2^{\alpha_G}}_{F[t_1,\ldots,t_k]} \subseteq I^{<2^{\alpha}}_{F[t_1,\ldots,t_k]}$ we obtain (ii) from (iii).

3) (i) holds by an application of $(\bigvee)$. Then there is a formula $A \in \bigvee\text{–type}$ belonging also to $\{\neg Cl_F(X), t_1 \notin X, \ldots, t_n \notin X\} \cup \Delta(X,\vec{Y})$ such that

$$\models^{\alpha_G} \neg Cl_F(X), t_1 \notin X, \ldots, t_n \notin X, \Delta(X,\vec{Y}), G \tag{iv}$$

for some $G \in \mathrm{CS}(A)$ and $\alpha_G < \alpha$. If $A \in \Delta(X,\vec{Y})$ we obtain the claim from the induction hypothesis as in case 2. Therefore suppose that $A \equiv \neg Cl_F(X) \equiv (\exists x)[F(X,x) \wedge x \notin X]$. Then $G \equiv F(X,t) \wedge t \notin X$ for some term $t$ and we obtain from (iv)

$$\models^{\alpha_G} \neg Cl_F(X), t_1 \notin X, \ldots, t_n \notin X, \Delta(X,\vec{Y}), F(X,t) \tag{v}$$

and

$$\models^{\alpha_G} \neg Cl_F(X), t_1 \notin X, \ldots, t_n \notin X, t \notin X, \Delta(X, \vec{Y}) \tag{vi}$$

by inversion. From (v) we get

$$\mathbb{N} \models \bigvee \Delta[I^{<2^{\alpha_G}}_{F[t_1,\ldots,t_k]}, S_1, \ldots, S_n] \vee F[I^{<2^{\alpha_G}}_{F[t_1,\ldots,t_k]}, t] \tag{vii}$$

by the induction hypothesis. If

$$\mathbb{N} \models \bigvee \Delta[I^{<2^{\alpha_G}}_{F[t_1,\ldots,t_k]}, S_1, \ldots, S_n]$$

we are done because $I^{<2^{\alpha_G}}_{F[t_1,\ldots,t_k]} \subseteq I^{<2^{\alpha}}_{F[t_1,\ldots,t_k]}$. Otherwise, we get

$$\mathbb{N} \models F[I^{<2^{\alpha_G}}_{F[t_1,\ldots,t_k]}, t], \text{ i.e., } t \in I^{2^{\alpha_G}}_{F[t_1,\ldots,t_k]}. \tag{viii}$$

From (viii) and assertion (6.2), we obtain

$$I^{<2^{\alpha_G}}_{F[t_1,\ldots,t_k,t]} \subseteq I^{<2^{\alpha_G}+2^{\alpha_G}}_{F[t_1,\ldots,t_k]} \subseteq I^{<2^{\alpha}}_{F[t_1,\ldots,t_k]}. \tag{ix}$$

The induction hypothesis applied to (vi) yields

$$\mathbb{N} \models \bigvee \Delta[I^{<2^{\alpha_G}}_{F[t_1,\ldots,t_k,t]}, S_1, \ldots, S_n] \tag{x}$$

and we obtain the claim from (x) and (ix). □

To infer the Stage Theorem from the Stage Lemma let $\alpha := \mathrm{tc}\big(n \in I_F\big)$. Then $\models^{\alpha} \neg Cl_F(X), \underline{n} \,\varepsilon\, X$ and we obtain $\mathbb{N} \models n \in I_F^{<2^{\alpha}}$ by the Stage Lemma. Hence $|n|_F < 2^{\alpha}$. □

**6.6.4 Theorem** $\kappa^{\mathbb{N}} = \omega_1^{CK}$.

*Proof* By Lemma 6.5.2 we have $\mathrm{otyp}(\prec) < \kappa^{\mathbb{N}}$ for every recursive well-ordering $\prec$. Hence $\omega_1^{CK} \leq \kappa^{\mathbb{N}}$. If $F(X,x)$ is an $X$-positive arithmetical formula and $n \in I_F$ we obtain

$$\alpha := \mathrm{tc}\big((\forall X)[(\forall x)[F(X,x) \to x \,\varepsilon\, X] \to \underline{n} \,\varepsilon\, X]\big) < \omega_1^{CK}$$

because the search tree $S_{(\forall x)[F(X,x) \to x\varepsilon X] \to \underline{n}\varepsilon X}$ is primitive recursive. Since $\omega_1^{CK}$ is closed under ordinal addition and exponentiation we obtain

$$|n|_F \leq 2^{\alpha} < \omega_1^{CK} \tag{i}$$

for all formulas $F$ and all $n \in I_F$. Hence $\kappa^{\mathbb{N}} \leq \omega_1^{CK}$. □

**6.6.5 Note** The Stage Theorem is a theorem of Gentzen (cf. [31]) in disguise. Gentzen showed (of course using a different terminology) $\mathrm{otyp}(\prec) \leq \omega^{\mathrm{tc}\left(TI(\prec)\right)}$ for a well-ordering $\prec$. In case of a well-ordering, however, there is a strengthening of Gentzen's theorem due to Beckmann [8] which we are going to prove next.

To sharpen the Stage Theorem we need some additional notations. Assume that $\prec$ is a binary relation. We extend the accessibility operator $\mathsf{Acc}_\prec$ defined by the formula

$$\mathrm{Acc}(X,x,\prec) \;:\Leftrightarrow\; x \in \mathit{field}(\prec) \wedge (\forall y)[y \prec x \rightarrow y \,\varepsilon\, X]$$

to its inflation

$$\overline{\mathsf{Acc}_\prec}(X) := X \cup \mathsf{Acc}_\prec(X) = X \cup \{n \in \mathit{field}(\prec) \,|\, (\forall y \prec n)[y \in X]\}.$$

The operator $\overline{\mathsf{Acc}_\prec}$ is called inflationary because of $X \subseteq \overline{\mathsf{Acc}_\prec}(X)$, which is in general not true for $\mathsf{Acc}_\prec$. The $\alpha$-th iterate of the operator $\overline{\mathsf{Acc}_\prec}$ is defined by

$$\overline{\mathsf{Acc}_\prec}^{\alpha}(X) := \overline{\mathsf{Acc}_\prec}(X \cup \bigcup_{\xi<\alpha} \overline{\mathsf{Acc}_\prec}^{\xi}(X)).$$

It is then plain that, starting with the empty set, all iterations of $\overline{\mathsf{Acc}_\prec}$ and $\mathsf{Acc}_\prec$ coincide. i.e., we have

$$\overline{\mathsf{Acc}_\prec}^{\alpha}(\emptyset) = \mathrm{Acc}^{\alpha}_\prec. \tag{6.3}$$

Let $X \subseteq \mathrm{Acc}_\prec$. We want to enumerate the stages of the elements in $\mathrm{Acc}_\prec$ which are different from the stages of elements in $X$. Therefore we define

$$\overline{X} := \{n \,|\, (\exists y \in X)[|y|_{\mathrm{Acc}_\prec} = |n|_{\mathrm{Acc}_\prec}]\},$$

$$O(X) := \{|n|_{\mathrm{Acc}_\prec} \,|\, n \in X\} \text{ and } \overline{en}_X := en_{On \setminus O(X)}.$$

We use the enumerating function $\overline{en}_X$ in the explicit definition of a new operator.

$$\mathrm{R}^{\alpha}_{\prec}(X) := X \cup \{x \in \mathrm{Acc}_\prec \,|\, |x|_{\mathrm{Acc}_\prec} \leq \overline{en}_X(\alpha)\} = X \cup \mathrm{Acc}_\prec^{\overline{en}_X(\alpha)}.$$

Since $\overline{en}_{X \cup \{x\}}(\alpha) \leq \overline{en}_X(\alpha+1)$ holds trivially, we obtain the innocently looking, but crucial property

$$\mathrm{R}^{\alpha}_{\prec}(\overline{X \cup \{x\}}) \subseteq \mathrm{R}^{\alpha+1}_{\prec}(\overline{X}) \cup \overline{\{x\}}. \tag{6.4}$$

Put

$$\mathrm{R}^{<\alpha}_{\prec}(\overline{X}) := \overline{X} \cup \bigcup_{\xi<\alpha} \mathrm{R}^{\xi}_{\prec}(\overline{X}).$$

There is a close connection between the operators $\mathrm{R}^{\alpha}_{\prec}(\overline{X})$ and the iterations $\overline{\mathsf{Acc}_\prec}^{\alpha}(\overline{X})$, which is expressed by the next lemma.

**6.6.6 Lemma** *Assume that* $\prec$ *is a* transitive *binary relation and* $X \subseteq \mathrm{Acc}_\prec$. *Then*

$$\mathrm{R}^{\alpha}_{\prec}(\overline{X}) = \overline{\mathsf{Acc}_\prec}(\mathrm{R}^{<\alpha}_{\prec}(\overline{X})).$$

*Proof* For the inclusion from left to right, let $s \in \mathrm{R}^{\alpha}_{\prec}(\overline{X})$. Then either $s \in \overline{X} \subseteq \overline{\mathsf{Acc}_\prec}(\mathrm{R}^{<\alpha}_{\prec}(\overline{X}))$ and we are done, or $|s|_{\mathrm{Acc}_\prec} \leq \overline{en}_{\overline{X}}(\alpha)$. Pick $t \prec s$. If $\overline{en}_{\overline{X}}(\beta) <$

$|t|_{\mathrm{Acc}_\prec} < |s|_{\mathrm{Acc}_\prec} \le \overline{en}_{\overline{X}}(\alpha)$ for all $\beta < \alpha$ then $|t|_{\mathrm{Acc}_\prec} \notin rng(\overline{en}_{\overline{X}})$, which implies $t \in \overline{X} \subseteq \mathrm{R}_\prec^{<\alpha}(\overline{X})$. Otherwise, there is a $\beta < \alpha$ such that $|t|_{\mathrm{Acc}_\prec} \le \overline{en}_{\overline{X}}(\beta)$. Therefore, we have

$$(\forall t \prec s)[t \in \mathrm{R}_\prec^{<\alpha}(\overline{X})]$$

which in turn implies $s \in \overline{\mathsf{Acc}_\prec(\mathrm{R}_\prec^{<\alpha}(\overline{X}))}$.

For the opposite inclusion, let $s \in \overline{\mathsf{Acc}_\prec(\mathrm{R}_\prec^{<\alpha}(\overline{X}))}$. Then $s \in \mathrm{R}_\prec^{<\alpha}(\overline{X})$ and we are done, or

$$(\forall t \prec s)[t \in \overline{X} \vee |t|_{\mathrm{Acc}_\prec} < \overline{en}_{\overline{X}}(\alpha)]. \tag{i}$$

We show

$$(\forall t \prec s)[t \in \overline{X} \vee |t|_{\mathrm{Acc}_\prec} < \overline{en}_{\overline{X}}(\alpha)] \;\Rightarrow\; |s|_{\mathrm{Acc}_\prec} \le \overline{en}_{\overline{X}}(\alpha) \tag{ii}$$

by induction on $|s|_{\mathrm{Acc}_\prec}$. Let $p \prec s$. If $p \notin \overline{X}$ we obtain $|p|_{\mathrm{Acc}_\prec} < \overline{en}_{\overline{X}}(\alpha)$ by the hypothesis of (ii). Therefore assume $p \in \overline{X}$. If $t' \prec p$ and $t' \notin \overline{X}$ we obtain $t' \prec s$ by the transitivity of $\prec$. Hence $|t'|_{\mathrm{Acc}_\prec} < \overline{en}_{\overline{X}}(\alpha)$ by (i). Using the induction hypothesis, we obtain $|p|_{\mathrm{Acc}_\prec} \le \overline{en}_{\overline{X}}(\alpha)$. Since $p \in \overline{X}$ we cannot have $|p|_{\mathrm{Acc}_\prec} = \overline{en}_{\overline{X}}(\alpha)$. Hence $|p|_{\mathrm{Acc}_\prec} < \overline{en}_{\overline{X}}(\alpha)$. Therefore $|p|_{\mathrm{Acc}_\prec} < \overline{en}_{\overline{X}}(\alpha)$ holds true for all $p \prec s$ which entails $|s|_{\mathrm{Acc}_\prec} \le \overline{en}_{\overline{X}}(\alpha)$.

From (i) and (ii) we finally get $|s|_{\mathrm{Acc}_\prec} \le \overline{en}_{\overline{X}}(\alpha)$ and thus $s \in \mathrm{R}_\prec^{\alpha}(\overline{X})$. □

**6.6.7 Lemma** *Let $\prec$ be a transitive binary relation. For any set $X \subseteq \mathrm{Acc}_\prec$ and any ordinal $\alpha$ we have $\mathrm{R}_\prec^{\alpha}(\overline{X}) = \overline{\mathsf{Acc}_\prec}^{\alpha}(\overline{X})$.*

*Proof* Proving the lemma by induction on $\alpha$ we have the induction hypothesis

$$\mathrm{R}_\prec^{<\alpha}(\overline{X}) = \overline{X} \cup \bigcup_{\xi<\alpha} \overline{\mathsf{Acc}_\prec}^{\xi}(\overline{X}) =: \overline{X} \cup \overline{\mathsf{Acc}_\prec}^{<\alpha}(\overline{X}). \tag{i}$$

Together with Lemma 6.6.6 we obtain

$$\begin{aligned}\mathrm{R}_\prec^{\alpha}(\overline{X}) &= \overline{\mathsf{Acc}_\prec}(\mathrm{R}_\prec^{<\alpha}(\overline{X})) = \\ &= \overline{\mathsf{Acc}_\prec}(\overline{X} \cup \overline{\mathsf{Acc}_\prec}^{<\alpha}(\overline{X})) = \overline{\mathsf{Acc}_\prec}^{\alpha}(\overline{X}).\end{aligned}$$ □

**6.6.8 Lemma** *(Boundedness Lemma) Let $\prec$ be an arithmetical definable binary transitive relation and $\Delta(X,\vec{Y})$ a finite set of X-positive pseudo $\Pi_1^1$-sentences such that*

$$\models^{\alpha} \neg(\forall x)[(\forall y \prec x)[y \,\varepsilon\, X] \to x \,\varepsilon\, X],\; t_1 \not\varepsilon X, \ldots, t_k \not\varepsilon X,\; \Delta(X,\vec{Y}). \tag{i}$$

*Then*

$$\mathbb{N} \models \bigvee \Delta[\mathrm{R}_\prec^{\alpha}(\overline{\{t_1^{\mathbb{N}}, \ldots, t_k^{\mathbb{N}}\}}), \vec{S}] \tag{ii}$$

*holds true for all assignments $\vec{S}$ of sets to the variables in $\vec{Y}$.*

*Proof* Let $Prog(\prec,X) \;:\Leftrightarrow\; (\forall x)\big[(\forall y \prec x)[y \,\varepsilon\, X] \to x \,\varepsilon\, X\big]$ express the progressiveness of the set $X$ relative to $\prec$. We show the lemma by induction on $\alpha$. There are the following cases.

1) Hypothesis (i) holds by (Ax). Then there is a formula $(s \,\varepsilon\, X) \in \Delta(X,\vec{Y})$ with $s^{\mathbb{N}} = t_i^{\mathbb{N}}$ for some $i \in \{1,\dots,k\}$. But $t_i^{\mathbb{N}} \in \mathrm{R}^0_\prec(\{t_1^{\mathbb{N}},\dots,t_k^{\mathbb{N}}\})$ which implies

$$\mathbb{N} \models \bigvee \Delta[\mathrm{R}^0_\prec(\overline{\{t_1^{\mathbb{N}},\dots,t_k^{\mathbb{N}}\}}),\vec{S}].$$

2) Hypothesis (i) holds by an inference whose critical formula $F$ belongs to $\Delta(X,\vec{Y})$. If $\mathrm{CS}(F) = \emptyset$ then $F \in Diag(\mathbb{N})$ which implies

$$\mathbb{N} \models \bigvee \Delta[\mathrm{R}^0_\prec(\overline{\{t_1^{\mathbb{N}},\dots,t_k^{\mathbb{N}}\}}),\vec{S}].$$

Otherwise all $G(X,\vec{Y}) \in \mathrm{CS}(F)$ are again $X$-positive and we have the premise(s)

$$\models^{\alpha_G} \neg Prog(\prec,X), t_1 \not\varepsilon\, X,\dots,t_k \not\varepsilon\, X, \Delta(X,\vec{Y}), G(X,\vec{Y}) \tag{iii}$$

for some (or all) $G(X,\vec{Y}) \in \mathrm{CS}(F)$. If $\mathbb{N} \not\models \bigvee \Delta[\mathrm{R}^\alpha_\prec(\overline{\{t_1^{\mathbb{N}},\dots,t_k^{\mathbb{N}}\}}),\vec{S}]$ then also

$$\mathbb{N} \not\models \bigvee \Delta[\mathrm{R}^{\alpha_G}_\prec(\overline{\{t_1^{\mathbb{N}},\dots,t_k^{\mathbb{N}}\}}),\vec{S}]$$

for all $\alpha_G < \alpha$. Therefore we get $\mathbb{N} \models G[\mathrm{R}^{\alpha_G}_\prec(\overline{\{t_1^{\mathbb{N}},\dots,t_k^{\mathbb{N}}\}}),\vec{S}]$ for some (or all) $G \in \mathrm{CS}(F)$ from (iii) by the induction hypothesis. Since $F$ is $X$-positive this implies

$$\mathbb{N} \models F[\mathrm{R}^\alpha_\prec(\overline{\{t_1^{\mathbb{N}},\dots,t_k^{\mathbb{N}}\}}),\vec{S}]$$

and this in turn entails $\mathbb{N} \models \bigvee \Delta[\mathrm{R}^\alpha_\prec(\overline{\{t_1^{\mathbb{N}},\dots,t_k^{\mathbb{N}}\}}),\vec{S}]$.

3) Hypothesis (i) holds by an inference $(\bigvee)$ with premise

$$\models^{\alpha_0} \neg Prog(\prec,X), t_1 \not\varepsilon\, X,\dots,t_k \not\varepsilon\, X, \Delta(X,\vec{Y}), (\forall y)[y \prec t \to y \,\varepsilon\, X] \wedge t \not\varepsilon\, X. \tag{iv}$$

By inversion we obtain from (iv)

$$\models^{\alpha_0} \neg Prog(\prec,X), t_1 \not\varepsilon\, X,\dots,t_k \not\varepsilon\, X, \Delta(X,\vec{Y}), (\forall y)[y \prec t \to y \,\varepsilon\, X] \tag{v}$$

and

$$\models^{\alpha_0} \neg Prog(\prec,X), t_1 \not\varepsilon\, X,\dots,t_k \not\varepsilon\, X, t \not\varepsilon\, X, \Delta(X,\vec{Y}). \tag{vi}$$

Assume $\mathbb{N} \not\models \bigvee \Delta[\mathrm{R}^\alpha_\prec(\overline{\{t_1^{\mathbb{N}},\dots,t_k^{\mathbb{N}}\}}),\vec{S}]$. Since $\alpha_0 < \alpha$ this implies also

$$\mathbb{N} \not\models \bigvee \Delta[\mathrm{R}^{\alpha_0}_\prec(\overline{\{t_1^{\mathbb{N}},\dots,t_k^{\mathbb{N}}\}}),\vec{S}]$$

which by the induction hypothesis for (v) yields

$$\mathbb{N} \models (\forall y)[y \prec t^{\mathbb{N}} \Rightarrow y \in \mathrm{R}^{\alpha_0}_\prec(\{t_1^{\mathbb{N}},\dots,t_k^{\mathbb{N}}\})]$$

and therefore by Lemma 6.6.6

$$t^{\mathbb{N}} \in \mathrm{R}^\alpha_\prec(\overline{\{t_1^{\mathbb{N}},\dots,t_k^{\mathbb{N}}\}}) \tag{vii}$$

which in turn implies

$$\overline{\{t^{\mathbb{N}}\}} \subseteq \mathrm{R}^{\alpha}_{\prec}(\overline{\{t_1^{\mathbb{N}},\dots,t_k^{\mathbb{N}}\}}). \tag{viii}$$

From the induction hypothesis applied to (vi) we obtain

$$\mathbb{N} \models \bigvee \Delta[\mathrm{R}^{\alpha_0}_{\prec}(\overline{\{t_1^{\mathbb{N}},\dots,t_k^{\mathbb{N}},t^{\mathbb{N}}\}}),\vec{S}]. \tag{ix}$$

By the assertion (6.4) (on p. 95) and (viii) we obtain

$$\mathrm{R}^{\alpha_0}_{\prec}(\overline{\{t_1^{\mathbb{N}},\dots,t_k^{\mathbb{N}},t^{\mathbb{N}}\}}) \subseteq \mathrm{R}^{\alpha}_{\prec}(\overline{\{t_1^{\mathbb{N}},\dots,t_k^{\mathbb{N}}\}})$$

which together (ix) implies

$$\mathbb{N} \models \bigvee \Delta[\mathrm{R}^{\alpha}_{\prec}(\overline{\{t_1^{\mathbb{N}},\dots,t_k^{\mathbb{N}}\}}),\vec{S}]. \qquad \square$$

From the Boundedness Lemma together with (6.3) and Lemma 6.6.7 we obtain the Boundedness Theorem which strengthens the Stage Theorem in case of accessible parts for *transitive* binary relations.

**6.6.9 Theorem** *(Boundedness Theorem for accessible parts of transitive relations) Let $\prec$ be an arithmetically definable and transitive binary relation and $\Delta(X,\vec{Y})$ a finite set of X-positive pseudo-$\Pi^1_1$-sentences which at most contain the shown free set variables. Then*

$$\models^{\alpha} \neg Prog(\prec,X),\Delta(X,\vec{Y})$$

*implies*

$$\mathbb{N} \models (\forall \vec{Y})\big[\bigvee \Delta[\mathrm{Acc}^{\alpha}_{\prec},\vec{Y}]\big].$$

*Proof* From

$$\models^{\alpha} \neg Prog(\prec,X),\Delta(X,\vec{Y})$$

we obtain by the Boundedness Lemma

$$\mathbb{N} \models (\forall \vec{Y})\big[\bigvee \Delta[\mathrm{R}^{\alpha}_{\prec}(\emptyset),\vec{Y}]\big]$$

which by Lemma 6.6.7 implies

$$\mathbb{N} \models (\forall \vec{Y})\big[\bigvee \Delta[\overline{\mathsf{Acc}_{\prec}}^{\alpha}(\emptyset),\vec{Y}]\big].$$

By (6.3) we therefore obtain

$$\mathbb{N} \models (\forall \vec{Y})\big[\bigvee \Delta[\mathrm{Acc}^{\alpha}_{\prec},\vec{Y}]\big]. \qquad \square$$

**6.6.10 Remark** We may even allow free set variables in the defining formula of the relation $\prec$. Therefore we can extend Lemma 6.6.8 and Theorem 6.6.9 to binary relations which are $\Sigma^1_1$-definable. (Cf. [8]).

Observe that the transitivity of the order relation is crucial (cf. Exercise 6.7.6 below).

**6.6.11 Corollary** *Let $\prec$ be a transitive order relation. Then*

$$\models^{\alpha} \neg Prog(\prec,X), \underline{n} \in X$$

*implies* $\text{otyp}_{\prec}(n) \leq \alpha$.

*Proof* From

$$\models^{\alpha} \neg Prog(\prec,X), \underline{n}\, \varepsilon\, X \tag{i}$$

we obtain from the Boundedness Theorem for accessible parts

$$\mathbb{N} \models n\, \varepsilon\, \text{Acc}_{\prec}^{\alpha}. \tag{ii}$$

This implies by Lemma 6.5.2 $\text{otyp}(\prec) \leq \alpha$. □

## 6.7 The $\Pi_1^1$-Ordinal of a Theory

We defined a first-order theory as a set of first-order sentences. Liberalizing that to a set of pseudo-$\Pi_1^1$-sentences still leaves us in the realm of first-order logic and therefore allows for complete formal deduction formalisms. As pointed out in Sect. 4.6 second-order theories cannot possess a complete formal deduction formalism. However, in Sect. 4.4 we have shown that there are sound calculi for second-order theories.

To cover also the situation of second-order theories we introduce the notion of *formal systems*. A *formal rule* is a syntactically given figure of the form

(R) $\quad F_1,\ldots,F_n \vdash F.$

The formulas $F_1,\ldots,F_n$ are the *premises*, the formula $F$ is the *conclusion* of the rule (R).

A *formal system* is a decidable set $T$ of axioms together with a set $\mathscr{R}$ of formal rules. The set of formulas which are derivable in $(T,\mathscr{R})$ is inductively defined by the following clauses.

- If $F \in T$ then $(T,\mathscr{R}) \vdash F$
- If $(T,\mathscr{R}) \vdash F_1, \ldots, (T,\mathscr{R}) \vdash F_n$ and $F_1,\ldots,F_n \vdash F$ is an instantiation of a rule (R) in $\mathscr{R}$ then $(T,\mathscr{R}) \vdash F$.

Examples for formal system are the Tait calculi described in Sect. 4.3. GÖDEL's Completeness Theorem (Theorem 4.4.6) formulated in terms of general formal systems, then takes the form:

*For a decidable set of pseudo-$\Pi_1^1$-sentences $T$ there is a formal system $(T,\mathscr{R})$ such $T \models_L F$ if and only if $(T,\mathscr{R}) \vdash F$.*

As remarked above we have only the direction

$$(T,\mathscr{R}) \vdash F \;\Rightarrow\; T \models_L F$$

for second-order logic. In first-order logic the set of consequences of a theory is independent of the special choice of the formalism. The set of consequences of formal system in second-order logic, however, depends also on its set of rules.

For the rest of this section let us assume that the language of a theory or a formal system comprises the language of arithmetic, either directly or by interpretation.

Let

$$TI(\prec,X) \;:\Leftrightarrow\; Prog(\prec,X) \to (\forall x \,\varepsilon\, field(\prec))[x \,\varepsilon\, X]$$

and

$$TI(\prec) \;:\Leftrightarrow\; (\forall X) TI(\prec,X)$$

express the fact that the relation $\prec$ is well-founded. Inspired by Gentzen's work we define the *proof-theoretic ordinal* of a formal system $(T,\mathscr{R})$ to be the ordinal

$$\|(T,\mathscr{R})\| := \sup\{\mathrm{otyp}(\prec) \mid \prec \text{ is primitive recursive } \wedge (T,\mathscr{R}) \vdash TI(\prec)\}.$$

Since the set of consequences of a first-order theory is independent of the choice of the formalism, i.e., of the set of rules, provided that the formalism is complete, we define the proof-theoretic ordinal of a first-order theory by

$$\|T\| := \sup\{\mathrm{otyp}(\prec) \mid \prec \text{ is primitive recursive } \wedge T \models_L TI(\prec,X)\}. \tag{6.5}$$

Equation (6.5) therefore requires that $T$ is able to prove (pseudo-)$\Pi^1_1$-sentences.[1]

For theories $T$ and formal systems $(T,\mathscr{R})$ which prove (pseudo-)$\Pi^1_1$-sentences we define the $\Pi^1_1$-*ordinal* by

$$\|T\|_{\Pi^1_1} := \sup\{\mathrm{tc}(F)+1 \mid F \text{ is a (pseudo) } \Pi^1_1\text{-sentence } \wedge T \models_L F\}$$

and

$$\|(T,\mathscr{R})\|_{\Pi^1_1} := \sup\{\mathrm{tc}(F)+1 \mid F \text{ is a (pseudo) } \Pi^1_1\text{-sentence } \wedge (T,R) \vdash F\}.$$

A theory or a formal system is $\Pi^1_1$-*sound* if

$$T \vdash F \;\Rightarrow\; \mathbb{N} \models F \;\text{ or }\; (T,\mathscr{R}) \vdash F \;\Rightarrow\; \mathbb{N} \models F$$

for all (pseudo-) $\Pi^1_1$-sentences $F$. The next theorem is then an immediate consequence of the definition of the $\Pi^1_1$-ordinal.

**6.7.1 Theorem** *Let $T$ be a theory or a formal system. Then $T$ is $\Pi^1_1$-sound if and only if $\|T\|_{\Pi^1_1} < \omega_1^{CK}$.*

We modify the Boundedness Theorem (Theorem 6.6.9) for accessible parts of primitive recursive transitive relations.

**6.7.2 Theorem** *(Boundedness Theorem) Let $\prec$ be a primitive recursive, binary, transitive and well-founded relation. If* $\mathrm{otyp}(\prec) \in Lim$ *then* $\mathrm{otyp}(\prec) = \mathrm{tc}(TI(\prec))$.

[1] In Sect. 10 we discuss the situation for a theory, which only proves arithmetical sentences.

*Proof* Let $\alpha := \mathrm{tc}\big(TI(\prec)\big)$. For $n \in field(\prec)$ let

$$TI(\prec,n) \ :\Leftrightarrow\ \neg Prog(\prec,X) \lor \underline{n}\,\varepsilon\,X$$

and $\alpha_n := \mathrm{tc}\big(TI(\prec,n)\big)$. Then we have

$$\models^{\alpha_n} \neg Prog(\prec,X), \underline{n}\,\varepsilon\,X \tag{i}$$

for all $n \in field(\prec)$ which by Corollary 6.6.11 entails $\mathrm{otyp}_\prec(n) \le \alpha_n \le \alpha$ for all $n \in field(\prec)$. Hence $\mathrm{otyp}(\prec) \le \alpha + 1$ and thus, $\mathrm{otyp}(\prec) \le \alpha$ since $\mathrm{otyp}(\prec) \in Lim$.

For $s \in field(\prec)$ we obtain by induction on $\mathrm{otyp}_\prec(s)$

$$\models^{5\cdot(\mathrm{otyp}_\prec(s)+1)} \neg(\forall x)[(\forall y \prec x)(y\,\varepsilon\,X) \to x\,\varepsilon\,X], s\,\varepsilon\,X. \tag{6.6}$$

For $t \prec s$ we have

$$\models^{5\cdot(\mathrm{otyp}_\prec(t)+1)} \neg(\forall x)[(\forall y \prec x)(y\,\varepsilon\,X) \to x\,\varepsilon\,X], t\,\varepsilon\,X \tag{ii}$$

by the induction hypothesis. If $s \preceq t$ then $(\neg t \prec s) \in Diag(\mathbb{N})$ and we obtain

$$\models^{0} \neg(\forall x)[(\forall y \prec x)(y\,\varepsilon\,X) \to x\,\varepsilon\,X], \neg t \prec s \tag{iii}$$

by an inference $(\bigwedge)$ with empty premise. Hence

$$\models^{5\cdot(\mathrm{otyp}_\prec(t)+1)} \neg(\forall x)[(\forall y \prec x)(y\,\varepsilon\,X) \to x\,\varepsilon\,X], \neg t \prec s,\ t\,\varepsilon\,X \tag{iv}$$

for all $t$. From (iv) we get

$$\models^{5\cdot(\mathrm{otyp}_\prec(s))+3} \neg(\forall x)[(\forall y \prec x)(y\,\varepsilon\,X) \to x\,\varepsilon\,X], (\forall y)[\neg y \prec s \lor y\,\varepsilon\,X] \tag{v}$$

by two inferences $(\bigvee)$ followed by an inference according to $(\bigwedge)$. From the axiom

$$\models^{0} s\,\not\varepsilon\,X, s\,\varepsilon\,X \tag{vi}$$

and (v) we obtain by the structural rule (STR)[2] together with an inference $(\bigwedge)$

$$\models^{5\cdot(\mathrm{otyp}_\prec(s))+4} \begin{array}{l}\neg(\forall x)[(\forall y \prec x)(y\,\varepsilon\,X) \to x\,\varepsilon\,X],\\ (\forall y)[\neg y \prec s \lor y\,\varepsilon\,X] \land s\,\not\varepsilon\,X, s\,\varepsilon\,X.\end{array} \tag{vii}$$

From (vii), however, we obtain

$$\models^{5\cdot(\mathrm{otyp}_\prec(s))+5} \neg(\forall x)[(\forall y \prec x)(y\,\varepsilon\,X) \to x\,\varepsilon\,X], s\,\varepsilon\,X$$

by an inference $(\bigvee)$. This finishes the proof of (6.6). But (6.6) implies

$$\models^{5\cdot(\mathrm{otyp}_\prec(s)+1)} \neg Prog(\prec,X), s\,\not\in field(\prec), s\,\varepsilon\,X$$

for all term $s$ and since $\mathrm{otyp}(\prec) \in Lim$ we get $5\cdot(\mathrm{otyp}_\prec(n)+1)+2 < \mathrm{otyp}(\prec)$ for all $s \in field(\prec)$ and thus

---

[2] In fact we needed the structural rule also in the inferences before. Applications of the structural rule (STR) remain mostly unmentioned.

$$\models^{\mathrm{otyp}(\prec)} \neg Prog(\prec, X), (\forall x \,\varepsilon\, field(\prec))[x \,\varepsilon\, X].$$

Hence $\mathrm{tc}\big(TI(\prec)\big) \leq \mathrm{otyp}(\prec)$. □

The following theorem is an immediate consequence of the Boundedness Theorem.

**6.7.3 Theorem** *Let $T$ be a theory or a formal system. Then $\|T\| \leq \|T\|_{\Pi^1_1}$.*

The objective of an ordinal analysis for a theory is the computation of its proof-theoretic ordinal. As we will see in the remainder of this book this is commonly done in two steps. First we calculate an upper bound $\beta \geq \|T\|_{\Pi^1_1}$ and in the second step show that for all $\alpha < \beta$ there is a primitive recursive ordering $\prec$ such that $T \vdash TI(\prec)$ and $\mathrm{otyp}(\prec) = \alpha$. This implies $\beta \leq \|T\| \leq \|T\|_{\Pi^1_1} \leq \beta$ and therefore $\|T\| = \|T\|_{\Pi^1_1}$.

There is, however, also an abstract argument showing that $\|T\|_{\Pi^1_1}$ and $\|T\|$ coincide for certain theories. We formulate the theorem but give only a sketch of its proof because we will never apply the theorem.

**6.7.4 Theorem** *Let $T$ be a theory which is strong enough to prove the $\omega$-Completeness Theorem. Then $\|T\| = \|T\|_{\Pi^1_1}$.*

*Proof* (Sketch) We assume that $T$ (or a conservative extension of $T$ in which we add comprehension for elementary formulas) is strong enough[3] to prove that a pseudo-$\Pi^1_1$-sentence $F$ is equivalent to the well-foundedness of the search tree for $F$, i.e.,

$$T \vdash F(\vec{X}) \leftrightarrow TI(\prec_F) \tag{i}$$

where $\prec_F$ is the relation

$$s \prec_F t \;:\Leftrightarrow\; s \,\varepsilon\, S_{\langle F \rangle} \wedge t \subsetneq s.$$

The relation $\prec_F$ is obviously primitive recursive and for $s \in S_{\langle F \rangle}$ we have $\mathrm{otyp}_{S_{\langle F \rangle}}(s) = \mathrm{otyp}_{\prec_F}(s)$. If $\alpha < \|T\|_{\Pi^1_1}$ then there is a pseudo-$\Pi^1_1$-sentence $F$ such that $T \vdash F$ and $\alpha \leq \mathrm{tc}\big(F\big)$. But then $T \vdash TI(\prec_F)$ which shows that $\alpha \leq \mathrm{tc}\big(F\big) \leq \mathrm{otyp}(\prec_F) < \|T\|$. So we have $\|T\|_{\Pi^1_1} \leq \|T\|$ which together with Theorem 6.7.3 yields the claim. □

The next theorem is accredited to Kreisel. It limits the importance of the $\Pi^1_1$-ordinal of a theory. But as we will see in the course of the book the real importance of the $\Pi^1_1$-ordinal rest in its computation and not so much in the bare knowledge of its size. Nevertheless the proof-theoretic ordinal of a theory $T$ has become a very common measure for its proof-theoretic strength.

[3] All we need to prove the $\omega$-Completeness Theorem (Theorem 5.4.9) is Königs Lemma. According to [94] the theory $(\mathsf{ACA})_0$ of arithmetical comprehension suffices. This theory is a conservative second-order extension of PA.

**6.7.5 Theorem** *Let T be a theory or a formal system for which we have* $\|T\| = \|T\|_{\Pi^1_1}$ *and F a* $\Sigma^1_1$*-sentence which is true in the standard structure* $\mathbb{N}$*. Then* $\|T\| = \|T+F\|$.

*Proof* Assume

$$T+F \models_L TI(\prec,X) \tag{i}$$

for a primitive recursive order relation $\prec$. Then we obtain

$$T \models_L \neg F \vee TI(\prec,X) \tag{ii}$$

which implies

$$\models^{\alpha} \neg F, \neg Prog(\prec,X), (\forall x \varepsilon field(\prec))[x \varepsilon X] \tag{iii}$$

for some $\alpha < \|T\|$. By the Boundedness Theorem for accessible parts (Theorem 6.6.9) we then obtain

$$\mathbb{N} \models \neg F \vee (\forall x \in field(\prec))[x \varepsilon \mathrm{Acc}^{\alpha}_{\prec}]. \tag{iv}$$

Since $\mathbb{N} \models F$ this entails $(\forall x \in field(\prec))[x \varepsilon \mathrm{Acc}^{\alpha}_{\prec}]$, which means $\mathrm{otyp}(\prec) \leq \alpha$. Hence $\|T+F\| \leq \|T\|$. The opposite inequality holds trivially. □

**6.7.6 Exercise** (Beckmann) Let $\prec \subseteq \mathbb{N} \times \mathbb{N}$ be defined by

$$i \prec k \;:\Leftrightarrow\; k = i+1.$$

(a) Compute $\mathrm{otyp}_{\prec}(n)$ for $n \in \mathbb{N}$

(b) Show $\models^{4 \cdot n} \neg Prog(\prec,X), \underline{2^n} \varepsilon X$.

Hint: Prove

$$\models^{4 \cdot k} \neg Prog(\prec,X), \underline{n} \not\varepsilon X, \underline{n} + \underline{2^k} - \underline{1} \varepsilon X$$

for all $n$ by induction on $k$. The induction begin is trivial. For the induction step you have the induction hypotheses

$$\models^{4 \cdot k} \neg Prog(\prec,X), \underline{n} \not\varepsilon X, \underline{n} + \underline{2^k} - \underline{1} \varepsilon X$$

and

$$\models^{4 \cdot k} \neg Prog(\prec X), \underline{n} + \underline{2^k} \not\varepsilon X, \underline{n} + \underline{2^{k+1}} - \underline{1} \varepsilon X.$$

Combine these hypotheses to get the claim.

Observe that Exercise 6.7.6 shows that the hypothesis of transitiveness in the Boundedness Lemma (Lemma 6.6.8), in the Boundedness Theorem (Theorem 6.6.9) and Theorem 6.7.2 is indispensable.

**6.7.7 Exercise** Show that the boundedness property of generalized recursion theory follows from the Stage Theorem (Theorem 6.6.1), i.e.,

let $Rel(X)\ :\Leftrightarrow\ (\forall x \varepsilon X)[Seq(x) \wedge lh(x) = 2]$ denote that $X$ codes a binary relation and

$$Wf(X)\ :\Leftrightarrow\ Rel(X) \wedge$$
$$(\forall Y)\,\big[Y \subseteq field(X) \wedge (\exists x)(x \,\varepsilon\, Y) \rightarrow (\exists x)\big[x \,\varepsilon\, Y \wedge (\forall y)[\langle y,x\rangle \,\varepsilon\, X \rightarrow y \not\varepsilon\, Y]\big]\big]$$

denote that the relation coded by $X$ is well-founded.

(a) Assume that $F(X)$ is a $\Sigma^1_1$-formula such that $(\forall X)[F(X) \rightarrow Wf(X)]$. Show $\sup\{\mathrm{otyp}(X) \mid F(X)\} < \omega_1^{CK}$ using the Stage Theorem.

(b) Let $\prec$ be a $\Sigma^1_1$-definable binary relation. Show that the order-type of $\prec$ is less than $\omega_1^{CK}$ provided that $\prec$ is well-founded.

Hint: Cf. [8].

# Chapter 7
# The Ordinal Analysis for PA

This sections repeats Gentzen's ordinal analysis for arithmetic as a paradigmatic example for an ordinal analysis.

## 7.1 The Theory PA

The most familiar axiom system for the number theory is the system PA, which is commonly accredited to Guiseppe Peano [67]. It is likely that Peano was inspired by Dedekind's work on natural numbers as presented in his article "Was sind und was sollen die natürlichen Zahlen" [20]. The language $\mathscr{L}(\mathsf{PA})$ of Peano Arithmetic comprises the constants $\underline{0}$ and $\underline{1}$ for the natural numbers 0 and 1 and the binary function symbols $+$ for addition and $\cdot$ for multiplication. It is formulated in a first-order logic with identity whose nonlogical axioms are:

(NF) $(\forall x)[x+1 \neq \underline{0}]$

(INJ) $(\forall x)(\forall y)[x+\underline{1}=y+\underline{1} \;\rightarrow\; x=y]$

(PL0) $(\forall x)[x+\underline{0}=x]$

(PL1) $(\forall x)(\forall y)[x+(y+\underline{1})=(x+y)+\underline{1}]$

(MU0) $(\forall x)[x\cdot\underline{0}=\underline{0}]$

(MU1) $(\forall x)(\forall y)[x\cdot(y+\underline{1})=(x\cdot y)+x]$

(IND) $F(\underline{0}) \land (\forall x)[F(x) \rightarrow F(x+\underline{1})] \;\rightarrow\; (\forall x)F(x)$

Here we have two groups of axioms. The first group comprises the defining axioms for the non logical symbols $\underline{0}$, $\underline{1}$, $+$ and $\cdot$, the second group consists of the scheme (IND), which tries to fix the ontology of our structure $\mathbb{N}$. It follows from (IND) that every natural number is either 0 or the successor of some other natural number. To see that regard the formula

$$F(u) \;:\Leftrightarrow\; u=\underline{0} \vee (\exists y)[u=y+\underline{1}].$$

Then we have $F(\underline{0})$ and obtain $F(u+\underline{1})$ with $u$ as an witness for $y$. Therefore we obtain by (IND) $(\forall x)F(x)$ that says that every member of any model $\mathfrak{M}$ of PA is either $\underline{0}^{\mathfrak{M}}$ or $(k+\underline{1})^{\mathfrak{M}}$ for some $k \in \mathfrak{M}$.

**7.1.1 Exercise** Replace the scheme (IND) by the second-order axiom

$$(\mathrm{Ind})^2 \quad (\forall X)\big[\underline{0}\,\varepsilon\, X \wedge (\forall x)[x\,\varepsilon\, X \to x+\underline{1}\,\varepsilon\, X] \;\;\to\;\; (\forall x)[x\,\varepsilon\, X]\big]$$

and show that the so obtained axiom system $\mathsf{PA}^2$ is categorical for $\mathbb{N}$, i.e., all models of $\mathsf{PA}^2$ are isomorphic to the standard structure $\mathbb{N}$ of natural numbers.

Hint: Show that for any second-order model $\mathcal{N} \models \mathsf{PA}^2$ the mapping $n \mapsto \underline{n}^{\mathcal{N}}$ is an isomorphism.

## 7.2 The Theory NT

Instead of analyzing the axioms in PA we do that for a richer language that has constants for all primitive recursive functions.

The language $\mathcal{L}(\mathsf{NT})$ is a first-order language which contains set parameters denoted by capital Latin letters $X, Y, Z, X_1, \ldots$ and constants for 0 and all primitive recursive functions. We assume that the symbols for primitive recursive functions are built up from the symbols $S$ for the successor function, $C_k^n$ for the constant function, $P_k^n$ for the projection on the $k$-th component by the substitution operator *Sub* and the recursion operator *Rec* (cf. Chap. 2).

If $F(x_1,\ldots,x_n)$ is a formula that contains only the shown free variables we call the sentence $(\forall x_1)\ldots(\forall x_n)F(x_1,\ldots,x_n)$ the *universal closure* of the formula $F(x_1,\ldots,x_n)$.

The theory NT comprises the universal closures of the following formulas:

The successor axioms

$$(\forall x)[\neg \underline{0} = Sx]$$
$$(\forall x)(\forall y)[S(x)=S(y) \Rightarrow x=y].$$

The defining axioms for the function symbols are the universal closures of the following formulas:

$$C_k^n(x_1,\dots,x_n) = \underline{k}$$
$$P_k^n(x_1,\dots,x_n) = x_k$$
$$Sub(g,h_1,\dots,h_m)(x_1,\dots,x_n) = g(h_1(x_1,\dots,x_n)\dots h_m(x_1,\dots,x_n))$$
$$Rec(g,h)(0,x_1,\dots,x_n) = g(x_1,\dots,x_n)$$
$$Rec(g,h)(Sy,x_1,\dots,x_n) = h(y,Rec(g,h)(y,x_1,\dots,x_n),x_1,\dots,x_n).$$

The scheme of Mathematical Induction

$$F(\underline{0}) \land (\forall x)[F(x) \to F(S(x))] \to (\forall x)F(x)$$

for all $\mathcal{L}$(NT)-formulas $F(u)$.

The identity axioms IDEN

$$(\forall x)[x = x]$$
$$(\forall x)(\forall y)[x = y \to y = x]$$
$$(\forall x)(\forall y)(\forall z)[x = y \land y = z \to x = z]$$
$$(\forall \vec{x})(\forall \vec{y})[x_1 = y_1 \land \dots \land x_n = y_n \quad\to\quad f\vec{x} = f\vec{y}]$$
$$(\forall \vec{x})(\forall \vec{y})[x_1 = y_1 \land \cdots \land x_n = y_n \quad\to\quad (F_{v_1,\dots,v_n}(\vec{x}) \to F_{v_1,\dots,v_n}(\vec{y}))]$$

where $F(v_1,\dots,v_n)$ is an arbitrary formula in the language of NTand $\vec{x}$ and $\vec{y}$ stand for $x_1,\dots,x_n$ and $y_1,\dots,y_n$, respectively.

The theory NT is an extension of the theory PA by definitions. That means that for every formula $F$ in the language of NT there is a formula $F^0$ in the language of PA such that NT $\vdash F \leftrightarrow F^0$. On the other side, every formula in the language of PA which is provable in NT is already provable in PA.

We will also freely use symbols for primitive recursive predicates in the language of NT although they are not among the basic symbols. This is possible since for every primitive recursive predicate $R$ we have its characteristic function $\chi_R$ among the basic symbols and use $(Rt_1,\dots,t_n)$ as an abbreviation for $\chi_R(t_1,\dots,t_n) = 1$.

Recall the definition of NT $\vdash F$ in (4.4) on Chap. 61. The following theorem ruminates Theorem 4.4.6.

**7.2.1 Theorem** *A formula $F$ in the language of $\mathcal{L}$(NT) is provable in NT iff there are finitely many axioms $A_1,\dots,A_n$ in NT and a $m < \omega$ such that $\underset{T}{\overset{m}{\models}} \neg A_1,\dots,\neg A_n, F$.*

**7.2.2 Exercise** Let ($\Delta_0^1$–CA) be the second-order theory in the language $\mathcal{L}$(NT) whose axioms comprise all the axioms of $\mathbb{N}$ together with the axiom-scheme

$$(\Delta_0^1\text{–}CA) \quad (\exists X)(\forall x)[x \,\varepsilon\, X \leftrightarrow F(x)] \text{ for } \Delta_0^1\text{-formulas } F(x)$$

of arithmetical comprehension. Let ($\Delta_0^1$–CA)$_0$ denote the theory ($\Delta_0^1$–CA) in which the scheme (IND) of Mathematical Induction is replaced by the axiom (Ind)$^2$(cf. Exercise 7.1.1).

(a) Show that $(\Delta^1_0\text{–CA})_0$ is a conservative extension of NT.

(b) Is this also true for $(\Delta^1_0\text{–CA})$?

Hint: For (a) show that every model of NT is extendable to a model of $(\Delta^1_0\text{–CA})_0$. To answer (b) solve Exercise 7.4.8

**7.2.3 Remark** The second-order theory in the language $\mathscr{L}(\mathsf{PA})$ is based on the axioms of PA rather than those of NT but together with the scheme $(\Delta^1_0\text{–}CA)$ of arithmetical comprehension, and the scheme (IND) replaced by the single second-order axiom $(\mathrm{Ind})^2$is known as $(\mathsf{ACA})_0$. Obviously $(\Delta^1_0\text{–CA})_0$ and $(\mathsf{ACA})_0$ have the same proof-theoretical strength. $(\mathsf{ACA})_0$ plays a prominent role in the program of reverse mathematics (cf. [94]).

## 7.3 The Upper Bound

We start this section by a simple observation. We call two pseudo $\Pi^1_1$-sentences *numerically equivalent* if they only differ in number terms that yield the same values. It is plain that numerically equivalent formulas have the same rank.

**7.3.1 Lemma** *Let F and G be numerical equivalent formulas. Then* $\models^{\alpha} \Delta, F$ *implies* $\models^{\alpha} \Delta, G$

*Proof* We prove

$$s^{\mathbb{N}} = t^{\mathbb{N}} \text{ and } \models^{\alpha} \Delta(s) \;\Rightarrow\; \models^{\alpha} \Delta(t) \tag{7.1}$$

by induction on $\alpha$. The lemma is an obvious consequence of (7.1). If

$$\models^{\alpha} \Delta(s) \tag{i}$$

holds by (Ax) then $\{t_1 \,\varepsilon\, X,\ t_2 \not\varepsilon\, X\} \subseteq \Delta(s)$ such that $t_1^{\mathbb{N}} = t_2^{\mathbb{N}}$. If $s$ is different from $t_i$ for $i = 1,2$ we also have $\{t_1 \,\varepsilon\, X,\ t_2 \not\varepsilon\, X\} \subseteq \Delta(t)$. If $s$ is $t_i$ for some $i \in \{1,2\}$, say $s$ is $t_1$, then $\{t \,\varepsilon\, X, t_2 \not\varepsilon\, X\} \subseteq \Delta(t)$ and $t^{\mathbb{N}} = s^{\mathbb{N}} = t_2^{\mathbb{N}}$. Hence $\models^{\alpha} \Delta(t)$ by (Ax). If (i) holds by an instance of $(\bigvee)$ there is a formula $F(s) \in \Delta(s) \cap \bigvee\text{–type}$ and we have $\models^{\alpha_0} \Delta(s), A(s)$ for some $A(s) \in \mathrm{CS}(F(s))$. But then $a(t) \in \mathrm{CS}(F(t))$ and $\models^{\alpha_0} \Delta(t), F(t)$ by the induction hypothesis. By an inference $(\bigvee)$ we obtain the claim. If (i) holds by an instance of $(\bigwedge)$ there is a formula $F(s) \in \Delta(s) \cap \bigwedge\text{–type}$ which is critical for this inference. If $\mathrm{CS}(F(s)) = \emptyset$ then $F(s) \in Diag(\mathbb{N})$. But $t^{\mathbb{N}} = s^{\mathbb{N}}$ implies also $F(t) \in Diag(\mathbb{N})$ and we obtain $\models^{\alpha} \Delta(t)$ by an inference $(\bigwedge)$. If $\mathrm{CS}(F(s)) \neq \emptyset$ then we have $\models^{\alpha_G} \Delta(s), G(s)$ for all $G(s) \in \mathrm{CS}(F(s))$ that entails $\models^{\alpha_G} \Delta(t), G(t)$ by the induction hypothesis. But $\mathrm{CS}(F(t)) = \{G(t) \,|\, G(s) \in \mathrm{CS}(F(s))\}$ and we obtain $\models^{\alpha} \Delta(t), F(t)$ by an inference $(\bigwedge)$. This ends the proof of (7.1). □

Next we show that any derivation in the first-order logic can be translated into a derivation in the infinite verification calculus.

**7.3.2 Theorem** *Let $\Delta(\vec{u})$ be a finite set of formulas in the language of arithmetic with all free number variables shown. Then $\underset{\mathsf{T}}{\overset{m}{\vdash}} \Delta(\vec{u})$ implies $\overset{m}{\models} \Delta(\underline{\vec{n}})$ for all tuples $\underline{\vec{n}}$ of numerals.*

*Proof* We show the theorem by induction on $m$. Assume first that

$$\underset{\mathsf{T}}{\overset{m}{\vdash}} \Delta(\vec{u}) \tag{i}$$

holds by (AxL). Then there is an atomic formula $A(\vec{u})$ such that $\{A(\vec{u}), \neg A(\vec{u})\} \subseteq \Delta(\vec{u})$. If $A(\vec{u})$ does not contain a free set variable we have $A(\underline{\vec{n}}) \in Diag(\mathbb{N})$ or $\neg A(\underline{\vec{n}}) \in Diag(\mathbb{N})$ and obtain $\overset{m}{\models} \Delta(\underline{\vec{n}})$ by an inference $(\bigwedge)$ with empty premise. If $A(\vec{u})$ is a formula $(t(\vec{u})\, \varepsilon\, X)$ then $\{t(\underline{\vec{n}})\, \varepsilon\, X,\ t(\underline{\vec{n}}) \not\varepsilon\, X\} \subseteq \Delta(\underline{\vec{n}})$ and we obtain $\overset{m}{\models} \Delta(\underline{\vec{n}})$ by (Ax).

Now assume that (i) holds by an inference $(\vee)$ or $(\exists)$. Let $F(\vec{u})$ be the critical formula. Then $F$ is a formula $(A_1(\vec{u}) \vee A_2(\vec{u}))$ or $(\exists y)A(y,\vec{u})$. We have the premise

$$\underset{\mathsf{T}}{\overset{m_0}{\vdash}} \Delta(\vec{u}), A_i(\vec{u}) \tag{ii}$$

or

$$\underset{\mathsf{T}}{\overset{m_0}{\vdash}} \Delta(\vec{u}), A(t(\vec{u},\vec{v}),\vec{u}) \tag{iii}$$

for some term $t(\vec{u},\vec{v})$. Defining $G$ to be the formula $A_i(\underline{\vec{n}})$ or the formula $A(t(\underline{\vec{n}},\vec{0}),\underline{\vec{n}})$ we get

$$\overset{m_0}{\models} \Delta(\underline{\vec{n}}), G \tag{iv}$$

by the induction hypothesis. Since $G \in \mathrm{CS}((A_1(\underline{\vec{n}}) \vee A_2(\underline{\vec{n}})))$ in the first case and $G \in \mathrm{CS}((\exists y)A(y,\underline{\vec{n}}))$ in the second case, we obtain $\overset{m}{\models} \Delta(\underline{\vec{n}})$ by an inference $(\bigvee)$.

Finally, if (i) is obtained by an inference $(\wedge)$ whose critical formula is $(A_0(\vec{u}) \wedge A_1(\vec{u}))$ or by an inference $(\forall)$ with critical formula $(\forall y)A(y,\vec{u})$, we have the premise(s)

$$\underset{\mathsf{T}}{\overset{m_0}{\vdash}} \Delta(\vec{u}), A_i(\vec{u}) \ \text{ for } i \in \{0,1\} \tag{v}$$

or

$$\underset{\mathsf{T}}{\overset{m_0}{\vdash}} \Delta(\vec{u}), A(v,\vec{u}) \tag{vi}$$

for a variable $v$ that is different from all variables in the list $\vec{u}$. Let $G_i$ be the formula $A_i(\underline{\vec{n}})$ for $i = 0,1$ or the formula $A(\underline{i},\underline{\vec{n}})$ for $i \in \omega$. Abbreviating the critical formula by $F(\vec{u})$ we get $\mathrm{CS}(F(\underline{\vec{n}})) = \langle G_0, G_1\rangle$ or $\mathrm{CS}(F(\underline{\vec{n}})) = \langle G_i : i \in \omega\rangle$ and we have

$$\overset{m_0}{\models} \Delta(\underline{\vec{n}}), G_i \tag{vii}$$

for all $G_i \in \mathrm{CS}(F(\underline{\vec{n}}))$ by the induction hypothesis. We obtain the claim from (vii) by an inference $(\bigwedge)$. □

It follows from Theorems 7.2.1 and 7.3.2 that for any provable pseudo $\Pi^1_1$-sentence $F$ of NT there are finitely many axioms $A_1,\dots,A_n \in \mathsf{NT}$ such that

$$\models^{m} \neg A_1,\dots,\neg A_n, F. \tag{7.2}$$

To determine the $\Pi^1_1$-ordinal of NT we have to compute $\mathrm{tc}(F)$. Our strategy will be the following. First we compute an upper bound, say $\alpha$, for the truth complexities of all axioms in NT. This gives

$$\models^{\alpha} A_i \tag{7.3}$$

for all axioms $A_i$. Then we extend the verification calculus to an infinitary calculus with cut and use the cut rule to get rid of all the axioms. In the next step, we have to develop a procedure to eliminate cuts, which allows us to keep control over the length of the infinite derivations. The depth of the resulting cut free derivation then provides us with an upper bound for the truth complexity of $F$.

We start with the computation of the truth complexities of the axioms of NT.

All numerical instances of the defining axioms for primitive recursive functions belong to the diagram $Diag(\mathbb{N})$. Therefore we obtain their universal closure by a finite number of applications of the $\bigwedge$-rule. The same is true for all identity axioms except the last one.

So far we have seen that all mathematical and identity axioms (but the last) of NT have truth complexities below $\omega$. But we have not yet analyzed the last identity axiom and the induction scheme. Here we need a preparatory lemma.

**7.3.3 Lemma** *(Tautology Lemma) Let $F$ and $F'$ be numerically equivalent $\mathcal{L}(\mathsf{NT})$-formulas. Then* $\overset{2\cdot\mathrm{rnk}(F)}{\models\!\!=\!\!=} \Delta, \neg F', F$.

*Proof* Induction on $\mathrm{rnk}(F)$.

If $F$ is an atomic formula $s = t$ then $F'$ is a formula $s' = t'$ such that $s^{\mathbb{N}} = s'^{\mathbb{N}}$ and $t^{\mathbb{N}} = t'^{\mathbb{N}}$. We either have $(s = t) \in \bigwedge\text{–type}$, hence $(s' = t') \in \bigwedge\text{–type}$, or $(s \neq t) \in \bigwedge\text{–type}$, hence $(s' \neq t') \in \bigwedge\text{–type}$, and obtain $\models^{0} \Delta, s = t, s' \neq t'$ by an inference $(\bigwedge)$ with empty premise.

If $F$ is a formula $t \,\varepsilon\, X$ then $F'$ is a formula $t' \,\varepsilon\, X$ with $t^{\mathbb{N}} = t'^{\mathbb{N}}$. Then we obtain $\models^{0} \Delta, t \,\varepsilon\, X, t' \not\varepsilon\, X$ by (Ax).

If $F \in \bigwedge\text{–type}$ and $\mathrm{CS}(F) \neq \emptyset$ we obtain by the induction hypothesis

$$\overset{2\cdot\mathrm{rnk}(G)}{\models\!\!=\!\!=} \Delta, G, \neg G', F, \neg F' \tag{i}$$

for all $G \in \mathrm{CS}(F)$. Because of $\mathrm{CS}(\neg F') = \{\neg G' \mid G \in \mathrm{CS}(F)\}$ we obtain from (i)

$$\overset{2\cdot\mathrm{rnk}(G)+1}{\models\!\!=\!\!=} \Delta, G, F, \neg F' \tag{ii}$$

for all $G \in \mathrm{CS}(F)$ by an inference $(\bigvee)$. Since $2\cdot\mathrm{rnk}(G)+1 < 2\cdot\mathrm{rnk}(F)$ we finally obtain

$$\Big|\!\overset{2\cdot\mathrm{rnk}(F)}{=\!=\!=\!=}\; \Delta, F, \neg F'$$

from (ii) by an inference $(\bigwedge)$. The remaining cases are completely symmetrical. □

If $F(v_1,\ldots,v_n)$ is a formula of $\mathscr{L}(\mathsf{NT})$ and $\underline{k}_1,\ldots,\underline{k}_n$ and $\underline{l}_1,\ldots,\underline{l}_n$ are $n$-tuples of numerals we get

$$\Big|\!\overset{2\cdot\mathrm{rnk}(F)}{=\!=\!=\!=}\; \underline{k}_1 \neq \underline{l}_1,\ldots,\underline{k}_n \neq \underline{l}_n, \neg F_{v_1,\ldots,v_n}(\underline{k}_1,\ldots,\underline{k}_n), F_{v_1,\ldots,v_n}(\underline{l}_1,\ldots,\underline{l}_n)$$

either by an inference $(\bigwedge)$ with empty premise or by the Tautology Lemma. Abbreviating $x_1,\ldots,x_n$ by $\vec{x}$ and $y_1,\ldots,y_n$ by $\vec{y}$ we, therefore, obtain

$$\models^{k} (\forall\vec{x})(\forall\vec{y})[x_1 = y_1 \wedge \cdots \wedge x_n = y_n \rightarrow F_{v_1,\ldots,v_n}(\vec{x}) \rightarrow F_{v_1,\ldots,v_n}(\vec{y})]$$

for some $k < \omega$.

**7.3.4 Lemma** *(Induction Lemma) For any natural number $n$ and any $\mathscr{L}(\mathsf{NT})$-sentence $F(\underline{n})$ we have*

$$\Big|\!\overset{2\cdot[\mathrm{rnk}(F(\underline{n}))+n]}{=\!=\!=\!=}\; \neg F(\underline{0}), \neg(\forall x)[F(x) \rightarrow F(S(x))], F(\underline{n}).$$

*Proof* We induct on $n$. For $n = 0$ this is an instance of the Tautology Lemma. For the induction step we have

$$\Big|\!\overset{2\cdot[\mathrm{rnk}(F(\underline{n}))+n]}{=\!=\!=\!=}\; \neg F(\underline{0}), \neg(\forall x)[F(x) \rightarrow F(S(x))], F(\underline{n}) \tag{i}$$

by the induction hypothesis and obtain

$$\Big|\!\overset{2\cdot\mathrm{rnk}(F(\underline{n}))}{=\!=\!=\!=}\; \neg F(\underline{0}), \neg(\forall x)[F(x) \rightarrow F(S(x))], \neg F(S(\underline{n})), F(S(\underline{n})) \tag{ii}$$

by the Tautology Lemma. From (i) and (ii) we get by $(\bigwedge)$

$$\Big|\!\overset{2\cdot[\mathrm{rnk}(F(\underline{n}))+n]+1}{=\!=\!=\!=}\; \neg F(\underline{0}), \neg(\forall x)[F(x) \rightarrow F(S(x))], F(\underline{n}) \wedge \neg F(S(\underline{n})), F(S(\underline{n})). \tag{iii}$$

By a clause $(\bigvee)$ we get from (iii)

$$\Big|\!\overset{2\cdot[\mathrm{rnk}(F(\underline{n}))+n]+2}{=\!=\!=\!=}\; \neg F(\underline{0}), \neg(\forall x)[F(x) \rightarrow F(S(x))], F(S(\underline{n})).$$

Since $S(\underline{n})^{\mathbb{N}} = \underline{Sn}^{\mathbb{N}}$ this implies by Observation 7.3.1

$$\Big|\!\overset{2\cdot[\mathrm{rnk}(F(\underline{n}))+n]+2}{=\!=\!=\!=}\; \neg F(\underline{0}), \neg(\forall x)[F(x) \rightarrow F(S(x))], F(\underline{Sn}).$$

□

If $G$ is an instance $F(\underline{0}) \wedge (\forall x)[F(x) \rightarrow F(S(x))] \rightarrow (\forall x)F(x)$ of Mathematical Induction then $G^{\nabla}$ is the finite set $\neg F(\underline{0}), \neg(\forall x)[F(x) \rightarrow F(S(x))], (\forall x)F(x)$ and we get by Lemma 7.3.4 $\mathrm{tc}(G) \leq \omega$. Thus, together with our previous remarks, we have

$$\models^{\omega} A_i, \text{ i.e., } \mathrm{tc}(A_i) \leq \omega \tag{7.4}$$

for all identity and nonlogical axioms $A_i$ of $\mathsf{NT}$.

So far we have computed $\omega$ as an upper bound for the ordinal $\alpha$ in (7.3) on p. 110. To bridge (7.2) and (7.3) we extend the verification calculus $\models^{\alpha} \Delta$ to an infinitary calculus, $\vdash^{\alpha}_{\rho} \Delta$ which we are going to call a *semi-formal* system.[1]

**7.3.5 Definition** For a finite set $\Delta$ of pseudo $\Pi^1_1$-sentences we define the semi-formal derivability relation $\vdash^{\alpha}_{\rho} \Delta$ inductively by the following clauses:

$(Ax)$ If $s^{\mathbb{N}} = t^{\mathbb{N}}$ then $\vdash^{\alpha}_{\rho} \Delta, s \,\varepsilon\, X, t \,\not\varepsilon\, X$ for all ordinals $\alpha$ and $\rho$.

$(\bigwedge)$ If $F \in \Delta \cap \bigwedge$–type and $(\forall G \in \mathrm{CS}(F))(\exists \alpha_G < \alpha)\left[\vdash^{\alpha_G}_{\rho} \Delta, G\right]$ then $\vdash^{\alpha}_{\rho} \Delta$.

$(\bigvee)$ If $F \in \Delta \cap \bigvee$–type and $(\exists G \in \mathrm{CS}(F))(\exists \alpha_G < \alpha)\left[\vdash^{\alpha_G}_{\rho} \Delta, G\right]$ then $\vdash^{\alpha}_{\rho} \Delta$.

(cut) If $\vdash^{\alpha_0}_{\rho} \Delta, F$ , $\vdash^{\alpha_0}_{\rho} \Delta, \neg F$ and $\mathrm{rnk}(F) < \rho$ then $\vdash^{\alpha}_{\rho} \Delta$ for all $\alpha > \alpha_0$.

We call $F$ the *critical formula* of the clauses $(\bigwedge)$ and $(\bigvee)$. The critical formulas of an axiom $(Ax)$ are $s \in X$ and $t \notin X$. A cut possesses no critical formula.

Observe that we have

$$\vdash^{\alpha}_{0} \Delta \Leftrightarrow \models^{\alpha} \Delta. \tag{7.5}$$

Thus, by Theorem 7.3.2 we obtain for a finite set $\Delta$

$$\vdash^{m}_{\mathsf{T}} \Delta \;\Rightarrow\; \vdash^{m}_{0} \Delta. \tag{7.6}$$

There are two obvious properties of $\vdash^{\alpha}_{\rho} \Delta$.

**7.3.6 Lemma** *(Soundness) If $\vdash^{\alpha}_{\rho} F_1,\dots,F_n$ then $\mathbb{N} \models (F_1 \vee \cdots \vee F_n)[\Phi]$ for every assignment $\Phi$ of subsets of $\mathbb{N}$ to the set parameters in $F_1,\dots,F_n$.*

*Proof* The proof is by induction on $\alpha$. It is exactly the proof of Lemma 5.4.4 with the additional case that the last inference is a cut

$$\vdash^{\alpha_0}_{\rho} F_1,\dots,F_n,A \text{ and } \vdash^{\alpha_0}_{\rho} F_1,\dots,F_n,\neg A \;\Rightarrow\; \vdash^{\alpha}_{\rho} F_1,\dots,F_n.$$

For an assignment $\Phi$ we then obtain

$$\mathbb{N} \models (F_1 \vee \cdots \vee F_n \vee A)[\Phi]$$

as well as

$$\mathbb{N} \models (F_1 \vee \cdots \vee F_n \vee \neg A)[\Phi],$$

which implies $\mathbb{N} \models (F_1 \vee \cdots \vee F_n)[\Phi]$. □

---

[1] There is a certain inconsequence of terminology. See Note 7.3.17 for explanation.

**7.3.7 Lemma** *(Structural Lemma)* *If* $\overset{\alpha}{\underset{\rho}{\vdash}} \Delta$ *then* $\overset{\beta}{\underset{\sigma}{\vdash}} \Gamma$ *holds true for all* $\beta \geq \alpha$, $\sigma \geq \rho$ *and* $\Gamma \supseteq \Delta$.

*Proof* Induction on $\alpha$. If

$$\overset{\alpha}{\underset{\rho}{\vdash}} \Delta \tag{i}$$

holds by (Ax) then $\{t \varepsilon X, s \not\varepsilon X\} \subseteq \Delta \subseteq \Gamma$ which entails $\overset{\beta}{\underset{\rho}{\vdash}} \Gamma$ for all $\beta$ and $\sigma$.

If (i) is derived by an inference $(\bigwedge)$ then there is a formula $F \in \Delta \cap \bigwedge\text{-type} \subseteq \Gamma \cap \bigwedge\text{-type}$ such that

$$\overset{\alpha_G}{\underset{\rho}{\vdash}} \Delta, G \tag{ii}$$

with $\alpha_G < \alpha \leq \beta$ for all $G \in \mathrm{CS}(F)$. Since $\Delta, G \subseteq \Gamma, G$ we obtain

$$\overset{\alpha_G}{\underset{\sigma}{\vdash}} \Gamma, G \tag{iii}$$

for all $G \in \mathrm{CS}(F)$ from the induction hypothesis and obtain $\overset{\beta}{\underset{\sigma}{\vdash}} \Gamma$ from (iii) by an inference $(\bigwedge)$.

The remaining case that (i) holds by an inference $(\bigvee)$ is completely dual. □

As a consequence of Theorem 7.3.2, (7.4) and (7.5) we obtain an embedding theorem for NT.

**7.3.8 Theorem** *(Embedding theorem for* NT*) For every pseudo-*$\Pi^1_1$*-sentence* $F$ *which is provable in* NT *there are finite ordinals m and r such that* $\overset{\omega+m}{\underset{r}{\vdash}} F$.

*Proof* If $\mathsf{NT} \vdash F$ there are finitely many axioms $A_1, \dots, A_n$ of NT and a finite ordinal $m_0$ such that $\overset{m_0}{\underset{\mathsf{T}}{\vdash}} \neg A_1, \dots, \neg A_n, F$. By Theorem 7.3.2 and (7.5) this implies $\overset{m_0}{\underset{0}{\vdash}} \neg A_1, \dots, \neg A_n, F$. By (7.4) and (7.5) we have $\overset{\omega+4}{\underset{0}{\vdash}} A_i$ for $i = 1, \dots, n$. Using the Structural Lemma and $n$ cuts we obtain $\overset{\omega+m}{\underset{r}{\vdash}} F$ for $r > \max\{\mathrm{rnk}(A_1), \dots, \mathrm{rnk}(A_n)\}$ and $m \geq 4 + n$. □

We prepare a cut-elimination theorem for the semi-formal system by a few observations.

**7.3.9 Lemma** *(Inversion Lemma)* *If* $F \in \bigwedge\text{-type}$, $\mathrm{CS}(F) \neq \emptyset$ *and* $\overset{\alpha}{\underset{\rho}{\vdash}} \Delta, F$ *then* $\overset{\alpha}{\underset{\rho}{\vdash}} \Delta, G$ *for all* $G \in \mathrm{CS}(F)$.

*Proof* We induct on $\alpha$. If the critical formula of the last inference is different from $F$, we have an inference of the form

$$\overset{\alpha_\iota}{\underset{\rho}{\vdash}} \Delta_\iota, F \text{ for } \iota \in I \;\Rightarrow\; \overset{\alpha}{\underset{\rho}{\vdash}} \Delta, F \tag{i}$$

and obtain $\overset{\alpha_\iota}{\underset{\rho}{\vdash}} \Delta_\iota, G$ for all $\iota \in I$ from the induction hypothesis. With the same inference we then get $\overset{\alpha}{\underset{\rho}{\vdash}} \Delta, G$.

Now assume that $F$ is the critical formula of the last inference. Since $\mathrm{CS}(F) \neq \emptyset$ we have the premises

$$\vdash_{\rho}^{\alpha_G} \Delta, F, G \tag{ii}$$

for all $G \in \mathrm{CS}(F)$. From (ii) we obtain

$$\vdash_{\rho}^{\alpha_G} \Delta, G \tag{iii}$$

by the induction hypothesis and finally $\vdash_{\rho}^{\alpha} \Delta, G$ from (iii) by the Structural Lemma. □

**7.3.10 Lemma** *($\vee$-Exportation) If* $\vdash_{\rho}^{\alpha} \Delta, F_1 \vee \cdots \vee F_n$ *then* $\vdash_{\rho}^{\alpha} \Delta, F_1, \ldots, F_n$.

*Proof* We show the lemma by induction on $\alpha$. If the critical formula of the last inference is different from $F_1 \vee \cdots \vee F_n$ we obtain the claim directly from the induction hypothesis. Therefore assume that $F \equiv (F_1 \vee \cdots \vee F_n) :\equiv F_1 \vee (F_2 \vee \cdots \vee F_n)$ is the critical formula of the last inference. Then we have the premise

$$\vdash_{\rho}^{\alpha_0} \Delta, F, F_1 \quad \text{or} \quad \vdash_{\rho}^{\alpha_0} \Delta, F, (F_2 \vee \cdots \vee F_n)$$

and obtain by the induction hypothesis

$$\vdash_{\rho}^{\alpha_0} \Delta, F_1, \ldots, F_n. \tag{i}$$

The claim then follows from (i) by the Structural Lemma. □

**7.3.11 Lemma** *If $F \in Diag(\mathbb{N})$ and* $\vdash_{\rho}^{\alpha} \Delta, \neg F$ *then* $\vdash_{\rho}^{\alpha} \Delta$.

*Proof* Induction on $\alpha$. Since $\neg F$ contains no logical symbols and is not in $\bigwedge$–type it cannot be the critical formula of the last inference

$$\vdash_{\rho}^{\alpha_\iota} \Delta_\iota, \neg F \text{ for } \iota \in I \;\Rightarrow\; \vdash_{\rho}^{\alpha} \Delta, \neg F.$$

By the induction hypothesis we, therefore, get $\vdash_{\rho}^{\alpha_\iota} \Delta_\iota$ for all $\iota \in I$ and then by the same inference $\vdash_{\rho}^{\alpha} \Delta$. □

**7.3.12 Lemma** *(Reduction Lemma) Let $F \notin \bigvee$–type and $\rho = \mathrm{rnk}(F)$. If* $\vdash_{\rho}^{\alpha} \Delta, F$ *and* $\vdash_{\rho}^{\beta} \Gamma, \neg F$ *then* $\vdash_{\rho}^{\alpha+\beta} \Delta, \Gamma$.

*Proof* The proof is by induction on $\beta$. First assume that $\neg F$ is not the critical formula of the last inference

$$\vdash_{\rho}^{\beta_\iota} \Gamma_\iota, \neg F \text{ for } \iota \in I \;\Rightarrow\; \vdash_{\rho}^{\beta} \Gamma, \neg F. \tag{i}$$

If $I = \emptyset$ then either $\Gamma \cap Diag(\mathbb{N}) \neq \emptyset$, which entails $\Delta, \Gamma \cap Diag(\mathbb{N}) \neq \emptyset$, and we obtain $\frac{\alpha+\beta}{\rho}\ \Delta, \Gamma$ by an inference $(\bigwedge)$ with empty premise, or there are terms $s$ and $t$ such that $\{s \,\varepsilon\, X, t \,\not\varepsilon\, X\} \subseteq \Gamma$ and $s^{\mathbb{N}} = t^N$ which immediately implies $\frac{\alpha+\beta}{\rho}\ \Delta, \Gamma$ by $(Ax)$. Otherwise we get

$$\left|\frac{\alpha+\beta_\iota}{\rho}\ \Delta, \Gamma_\iota \right. \tag{ii}$$

for all $\iota \in I$ by the induction hypothesis and obtain $\left|\frac{\alpha+\beta}{\rho}\ \Delta, \Gamma\right.$ from (ii) by the same inference.

Now assume that $\neg F$ is the critical formula. If $\rho = 0$ then $\neg F$ is atomic. If $F \in \bigwedge$–type we have $F \in Diag(\mathbb{N})$ and obtain $\left|\frac{\alpha+\beta}{\rho}\ \Delta, \Gamma\right.$ by Lemmas 7.3.11 and 7.3.7. If $F \equiv (s \,\varepsilon\, X)$ we show

$$\left|\frac{\alpha}{\rho}\ \Delta, \Gamma\right. \tag{iii}$$

by a side induction on $\alpha$. First we observe that there is a formula $t \,\varepsilon\, X$ with $t^{\mathbb{N}} = s^{\mathbb{N}}$ in $\Gamma$ since $\left|\frac{\beta}{\rho}\ \Gamma, \neg F\right.$ holds by $(Ax)$. If $F$ is not the critical formula of $\left|\frac{\alpha}{\rho}\ \Delta, F\right.$ then we have the premises $\left|\frac{\alpha_\iota}{\rho}\ \Delta_\iota, F\right.$ for $\iota \in I$. If $I = \emptyset$ we get $\left|\frac{\alpha}{\rho}\ \Delta, \Gamma\right.$ directly and for $I \neq \emptyset$ from the induction hypothesis by the same inference. If $F$ is the critical formula we are in the case of $(Ax)$ which entails that there is a formula $r \,\not\varepsilon\, X$ in $\Delta$ with $r^{\mathbb{N}} = s^{\mathbb{N}} = t^{\mathbb{N}}$. But then we obtain $\left|\frac{\alpha}{\rho}\ \Delta, \Gamma\right.$ by $(Ax)$. The case $F \equiv (s \,\not\varepsilon\, X)$ is symmetrical. From (iii) we get $\left|\frac{\alpha+\beta}{\rho}\ \Delta, \Gamma\right.$ by the Structural Lemma.

Now assume $\rho > 0$. Then $\neg F \in \bigvee$–type and we have the premise

$$\left|\frac{\beta_0}{\rho}\ \Gamma, \neg F, \neg G\right. \tag{iv}$$

for some $G \in \mathrm{CS}(F)$. Then we obtain

$$\left|\frac{\alpha+\beta_0}{\rho}\ \Delta, \Gamma, \neg G\right. \tag{v}$$

by induction hypothesis. From $\left|\frac{\alpha}{\rho}\ \Delta, F\right.$ we get

$$\left|\frac{\alpha+\beta_0}{\rho}\ \Delta, \Gamma, G\right. \tag{vi}$$

by the Inversion Lemma and the Structural Lemma. Since $\mathrm{rnk}(G) < \mathrm{rnk}(F) = \rho$ we obtain the claim from (v) and (vi) by (cut). □

**7.3.13 Lemma** *(Basic Elimination Lemma) If* $\left|\frac{\alpha}{\rho+1}\ \Delta\right.$ *then* $\left|\frac{2^\alpha}{\rho}\ \Delta\right.$.

*Proof* Induction on $\alpha$. If the last inference is not a cut of complexity $\rho$ we obtain the claim immediately from the induction hypothesis and the fact that $\lambda\xi\,.\,2^\xi$ is order preserving. The critical case is a cut

$$\left|\frac{\alpha_0}{\rho+1}\right. \Delta, F \text{ and } \left|\frac{\alpha_0}{\rho+1}\right. \Delta, \neg F \Rightarrow \left|\frac{\alpha}{\rho+1}\right. \Delta$$

with $\text{rnk}(F) = \rho$. By the induction hypothesis and the Reduction Lemma we obtain $\left|\frac{2^{\alpha_0}+2^{\alpha_0}}{\rho}\right. \Delta$ and we have $2^{\alpha_0} + 2^{\alpha_0} = 2^{\alpha_0+1} \leq 2^{\alpha}$. □

Observe that our language so far only comprises formulas of finite rank. But we have designed the semi-formal calculus in such a way that it will also work for languages with formulas of complexities $\geq \omega$. The following results masters this situation, too.

**7.3.14 Lemma** *(Predicative Elimination Lemma) If* $\left|\frac{\alpha}{\beta+\omega^{\rho}}\right. \Delta$ *then* $\left|\frac{\varphi_{\rho}(\alpha)}{\beta}\right. \Delta$.

*Proof* Induction on $\rho$ with side induction on $\alpha$. For $\rho = 0$ we obtain $\left|\frac{2^{\alpha}}{\beta}\right. \Delta$ by the Basic Elimination Lemma. Since $2^{\alpha} \leq \omega^{\alpha} = \varphi_0(\alpha)$ this entails the claim. Now assume $\rho > 0$. If the last clause was not a cut of rank $\geq \beta$ we obtain the claim from the induction hypotheses and the fact that the function $\varphi_{\rho}$ is order preserving. Therefore assume that the last inference is

$$\left|\frac{\alpha_0}{\beta+\omega^{\rho}}\right. \Delta, F \text{ and } \left|\frac{\alpha_0}{\beta+\omega^{\rho}}\right. \Delta, \neg F \Rightarrow \left|\frac{\alpha}{\beta+\omega^{\rho}}\right. \Delta$$

such that $\beta \leq \text{rnk}(F) < \beta + \omega^{\rho}$. But then there is an ordinal $\phi$ such that $\text{rnk}(F) = \beta + \phi$, which, writing $\phi$ in Cantor normal form, means $\text{rnk}(F) = \beta + \omega^{\sigma_1} + \ldots + \omega^{\sigma_n} < \beta + \omega^{\rho}$. Hence $\sigma_1 < \rho$ and we get $\text{rnk}(F) < \beta + \omega^{\sigma_1} \cdot (n+1)$. By the side induction hypothesis we have $\left|\frac{\varphi_{\rho}(\alpha_0)}{\beta}\right. \Delta, F$ and $\left|\frac{\varphi_{\rho}(\alpha_0)}{\beta}\right. \Delta, \neg F$. By a cut it follows $\left|\frac{\varphi_{\rho}(\alpha_0)+1}{\beta+\omega^{\sigma_1}\cdot(n+1)}\right. \Delta$. If we define $\varphi^0_{\sigma_1}(\xi) := \xi$ and $\varphi^{n+1}_{\sigma_1}(\xi) := \varphi_{\sigma_1}(\varphi^n_{\sigma_1}(\xi))$ we obtain from $\sigma_1 < \rho$ by $n+1$-fold application of the main induction hypothesis $\left|\frac{\varphi^{n+1}_{\sigma_1}(\varphi_{\rho}(\alpha_0)+1)}{\beta}\right. \Delta$. Finally we show $\varphi^n_{\sigma_1}(\varphi_{\rho}(\alpha_0)+1) < \varphi_{\rho}(\alpha)$ by induction on $n$. For $n = 0$ we have $\varphi^0_{\sigma_1}(\varphi_{\rho}(\alpha_0)+1) = \varphi_{\rho}(\alpha_0)+1 < \varphi_{\rho}(\alpha)$ since $\alpha_0 < \alpha$ and $\varphi_{\rho}(\alpha) \in Cr(0)$. For the induction step we have $\varphi^{n+1}_{\sigma_1}(\varphi_{\rho}(\alpha_0)+1) = \varphi_{\sigma_1}(\varphi^n_{\sigma_1}(\varphi_{\rho}(\alpha_0)+1)) < \varphi_{\rho}(\alpha)$, since $\sigma_1 < \rho$ and $\varphi^n_{\sigma_1}(\varphi_{\rho}(\alpha_0)+1) < \varphi_{\rho}(\alpha)$ by the induction hypothesis. Hence $\left|\frac{\varphi_{\rho}\alpha}{\beta}\right. \Delta$. □

By iterated application of the Predicative Elimination Lemma we obtain

**7.3.15 Theorem** *(Elimination Theorem) Let* $\left|\frac{\alpha}{\rho}\right. \Delta$ *and assume that* $\rho =_{NF} \omega^{\rho_1} + \ldots + \omega^{\rho_n}$. *Then* $\left|\frac{\varphi_{\rho_1}(\varphi_{\rho_2}(\cdots\varphi_{\rho_n}(\alpha)\cdots))}{0}\right. \Delta$.

**7.3.16 Theorem** *(The upper bound for* NT*) Let $F$ be a pseudo-$\Pi^1_1$-sentence. If* NT $\vdash F$ *then* $\text{tc}(F) < \varepsilon_0$. *Hence*

$$\|\mathsf{NT}\| \leq \|\mathsf{NT}\|_{\Pi^1_1} \leq \varepsilon_0.$$

*Proof* If NT $\vdash F$ we get by the Embedding Theorem (Theorem 7.3.8)

$$\vdash^{\omega+\omega}_{r} F \tag{i}$$

for some finite $r$. By the Elimination Theorem (or just by the iterated application of the Basis Elimination Lemma) this entails

$$\vdash^{\varphi_0^r(\omega+\omega)}_{0} F. \tag{ii}$$

Hence $\models^{\varphi_0^r(\omega+\omega)} F$ and we get $\mathrm{tc}(F) < \varepsilon_0$, since $\varphi_0^r(\omega+\omega) < \varepsilon_0$ holds true for all finite ordinals $r$. □

**7.3.17 Note** Calculi whose rules may have infinitely many premises have already been proposed by Hilbert (cf. [44], [9]). Their systematic use in proof theory is due to Schütte who also used the term "semi-formal system" to describe such calculi. In his sense, a semi-formal system consists of a decidable[2] set $T$ of axioms together with a collection $\mathscr{R}$ of rules. In contrast to a formal system, the rules $R \in \mathscr{R}$ of a semi-formal system may have infinitely many premises. In a more strict sense, however, one requires that the premises of a rule of a semi-formal systems are primitive recursively enumerated, i.e., that there is a primitive recursive coding of the formulas and a primitive recursive function $f$ such that for an infinitary rule $\langle \Delta_i \mid i \in \mathbb{N} \rangle \vdash \Delta$ we obtain $\ulcorner \Delta_i \urcorner = f(\ulcorner \Delta \urcorner, i)$. Semi-formal systems in this strict sense have subtle applications in proof theory (cf. e.g. [90]). It is not too difficult to see that the semi-formal system of Definition 7.3.5 can be replaced by a semi-formal system in the strict sense. Also cut-elimination holds true for strict semi-formal systems. In this book, however, we will not need the properties of semi-formal systems in the strict sense. Therefore we only introduced the liberated notion, omitting the requirement of primitive recursive enumerability of the premises of an infinite inference rule. In that sense also the verification calculus $\models^{\alpha}$ can be viewed as a semi-formal system. This becomes especially clear by property (7.5) on pp. 112. But observe that some of the properties of the verification calculus may be lost by turning it into a semi-formal system in the strict sense. An example of such a property is in fact (7.5) where the direction from right to left becomes false for semi-formal systems in the strict sense.

Since there is, in principle, a qualitative difference between the verification calculus $\models^{\alpha}$, whose root is the truth definition for sentences, and the inference system $\vdash^{\alpha}_{\rho}$, which is a generalization of the finitary calculus for predicate logic, we made the difference between the verification calculus and the semi-formal calculus. This difference will become even more evident when we later introduce operator controlled derivations. In general, the verification calculus cannot be controlled by an operator.

---

[2] In some settings he also allowed axioms of the form $(\forall x)F_u(x)$ (or $F(u)$) provided that $F_u(\underline{z})$ is a true atomic formula for every instantiation $\underline{z}$ for $u$.

**7.3.18 Exercise** (a) Show that $\frac{\alpha}{1} \neg Cl_F(X), t_1 \notin X, \ldots, t_n \notin X, \Delta[X,\vec{Y}]$ for a set $\Delta[X,\vec{Y}]$ of $X$-positive pseudo-$\Pi^1_1$-sentences implies $\mathbb{N} \models \bigvee \Delta[I^{<2^\alpha}_{F[t_1,\ldots,t_n]}, \vec{S}]$ for all assignments $\vec{S}$ to the set variables $\vec{Y}$ in $\Delta[X,\vec{Y}]$.

(b) Prove the Boundedness Lemma (Lemma 6.6.8) with $\models^{\alpha}$ replaced by $\vdash^{\alpha}_{1}$.

(c) Conclude that $\vdash^{\alpha}_{1} TI(\prec, X)$ implies $\operatorname{otyp}(\prec) \leq \alpha$ for transitive relations $\prec$.

Hint: Just add the additional case of an atomic cut to the proof of Lemmas 6.6.3 and 6.6.8.

**7.3.19 Exercise** Let $\mathsf{PA}^-$ be the theory PA without the scheme of Mathematical Induction (IND). For each $n \in \mathbb{N}$ let $\mathsf{PA}_n$ be the theory $\mathsf{PA}^- + (\mathrm{IND})_n$, where $(\mathrm{IND})_n$ is the restriction of (IND) to $\Sigma^0_n$-formulas. These theories are better known under the name $I\Sigma_n$. We define the quantifier rank $\mathrm{rnk}^*(F)$ for $\mathcal{L}(\mathsf{PA})$-formulas $F$ by the following clauses:

$$\mathrm{rnk}^*(F) := \begin{cases} 0 & \text{if } F \text{ contains no quantifiers} \\ \max\{\mathrm{rnk}^*(F_1), \mathrm{rnk}^*(F_2)\}+1 & \text{if } F \equiv F_1 \circ F_2 \text{ with } \circ \in \{\wedge, \vee\} \\ & \text{and } F \text{ contains quantifiers} \\ \mathrm{rnk}^*(F_0)+1 & \text{if } F \equiv (\forall x)F_0(x) \text{ or } F \equiv (\exists x)F_0(x) \end{cases}$$

Let $\mathsf{IR}_n$ denote the rules of the Tait-calculus enlarged by the cut rule

(cut) *If* $\vdash^{m_0}_{r} \Delta, F$, $\vdash^{m_0}_{r} \Delta, \neg F$ *and* $\mathrm{rnk}^*(F) < r$ *then* $\vdash^{m}_{r} \Delta$ *for all* $m > m_0$

and the rule of $\Sigma^0_n$-induction

$(\mathrm{IR})_n$ *If* $\vdash^{m_0}_{r} \Delta, F(0)$, $\vdash^{m_0}_{r} \Delta, \neg F(u), F(Su)$ *for a* $\Sigma^0_n$*-formula* $F(u)$ *and* $u \notin FV(\Delta)$ *then* $\vdash^{m}_{r} \Delta, F(t)$ *for all* $m > m_0$ *and all* $\mathcal{L}(\mathsf{NT})$*-terms* $t$.

Let $\mathsf{PA}\text{–}\mathsf{IR}_n$ denote the formal system $(\mathsf{PA}^-, \mathsf{IR}_n)$. We use $\mathsf{PA}\text{–}\mathsf{IR}_n \vdash \Delta$ as abbreviation for the fact that there is an $m$ and an $r$ such that $\mathsf{PA}\text{–}\mathsf{IR}_n \vdash^{m}_{r} \Delta$.
We finally define $* \vdash^{\alpha}_{\rho} \Delta$ like $\vdash^{\alpha}_{\rho} \Delta$ where $\mathrm{rnk}(F)$ is replaced by $\mathrm{rnk}^*(F)$.

(a) Show $\mathsf{PA}\text{–}\mathsf{IR}_n \vdash F \Leftrightarrow \mathsf{PA}_n \vdash F$ for every $\mathcal{L}(\mathsf{PA})$-formula $F$.

For the following assume that $\Delta$ is a finite set of $\mathcal{L}(\mathsf{PA})$-formulas.

(b) Show $\mathsf{PA}\text{–}\mathsf{IR}_n \vdash^{m}_{n+1+r} \Delta \Rightarrow \mathsf{PA}\text{–}\mathsf{IR}_n \vdash^{2_r(m)}_{n+1} \Delta$ and

$\mathsf{PA}\text{–}\mathsf{IR}_n \vdash^{m}_{n+1} \Delta(\vec{x}) \Rightarrow * \vdash^{\omega \cdot (m+1)}_{n+1} \Delta(\vec{k})$ for all $\vec{k} \in \mathbb{N}$,
where $\vec{x}$ are all the first-order variables that are free in $\Delta$.

(c) Show $* \vdash^{\alpha}_{1+m+1} \Delta \Rightarrow * \vdash^{2^\alpha}_{1+m} \Delta$ for all $m \in \mathbb{N}$.

(d) Show $* \vdash^{\alpha}_{1} \Delta \Rightarrow \vdash^{\omega \cdot \alpha}_{1} \Delta$.

(e) Use (d) to compute an upper bound for $\|\mathsf{PA}_n\|$ for each $n \in \mathbb{N}$.

(f) Can you improve (d) in case that we work in the language $\mathscr{L}(\mathsf{NT})$? Will this influence $\|\mathsf{NT}_n\|$ where $\mathsf{NT}_n$ is defined analogously to $\mathsf{PA}_n$?

## 7.4 The Lower Bound

We want to show that the bound given in Theorem 7.3.16 is the best possible one. By Theorem 6.7.3 it suffices to prove Theorem 7.4.1 below because then we obtain $\varepsilon_0 \leq \|\mathsf{NT}\| \leq \|\mathsf{NT}\|_{\Pi^1_1} \leq \varepsilon_0$.

**7.4.1 Theorem** *For every ordinal $\alpha < \varepsilon_0$ there is a primitive recursive well-ordering $\prec$ on the natural numbers of order type $\alpha$ such that* $\mathsf{NT} \vdash TI(\prec, X)$.

In Sect. 3.3.1, we developed a notation system for the ordinals below $\varepsilon_0$. There we defined a primitive recursive set $OT \subseteq \mathbb{N}$ and a relation $\prec$ that corresponds to the ordinals below $\varepsilon_0$, as summarized in Theorem 3.3.17. Thus we can talk about ordinals $< \varepsilon_0$ in $\mathscr{L}(\mathsf{NT})$. To increase readability, we will not distinguish between ordinals and their representations in $\mathscr{L}(\mathsf{NT})$ and regard formulas $(\forall\alpha)[\ldots]$ as abbreviations for $(\forall x)[x \,\varepsilon\, OT \rightarrow \ldots]$ and formulas $(\exists\alpha)[\ldots]$ as abbreviation for $(\exists x)[x \,\varepsilon\, OT \wedge \ldots]$. We also write $\alpha < \beta$ instead of $\alpha \prec \beta$. We introduce the following formulas:

- $\alpha \subseteq X \;:\Leftrightarrow\; (\forall\xi)[\xi < \alpha \rightarrow \xi \,\varepsilon\, X]$
- $\mathsf{Prog}(X) \;:\Leftrightarrow\; (\forall\alpha)[\alpha \subseteq X \rightarrow \alpha \,\varepsilon\, X]$
- $\mathsf{TI}(\alpha, X) \;:\Leftrightarrow\; \mathsf{Prog}(X) \rightarrow \alpha \subseteq X$

Our aim is to show $\mathsf{TI}(\alpha, X)$ for all $\alpha < \varepsilon_0$. Since $\varepsilon_0 = \sup\{\varphi_0^n(0) \mid n \in \omega\} = \sup\{exp^n(\omega, 0) \mid n \,\varepsilon\, \omega\}$ and $\mathsf{TI}(0, X)$ holds trivially, we are done as soon as we succeed in proving

$$\mathsf{NT} \vdash \mathsf{TI}(\alpha, X) \;\Rightarrow\; \mathsf{NT} \vdash \mathsf{TI}(\omega^\alpha, X) \tag{7.7}$$

because $\mathsf{NT} \vdash \mathsf{TI}(\alpha, X)$ and $\beta < \alpha$ obviously entail $\mathsf{NT} \vdash \mathsf{TI}(\beta, X)$. As a preparation, we need a simple lemma about substitutions. Recall that if $F(X)$ is a formula containing a set variable $X$ and $G(v)$ is a formula with a number variable $v$ we denote by $F(\{x \mid G_v(x)\})$ the formula that is obtained from $F(X)$ by replacing all occurrences of $t \,\varepsilon\, X$ by $G_v(t)$ and those of $t \not\varepsilon\, X$ by $\neg G_v(t)$.

**7.4.2 Lemma** *Let $\Delta(X)$ be a finite set of formulas (in an arbitrary first-order language) containing a free set variable and $G(v)$ a formula containing an object variable $v$. If $\frac{m}{\mathsf{T}} \Delta(X)$ then there is a $k$ such that $\frac{k}{\mathsf{T}} \Delta(\{x \mid G_v(x)\})$.*

*Proof* The proof is a simple induction on $m$. There are only two cases, which need some attention. First the case of an axiom. If $\Delta(X) = \Delta_0(X), t \,\varepsilon\, X, t \not\varepsilon X$ we obtain $\Delta(\{x \mid G_v(x)\}) = \Delta_0(\{x \mid G_v(x)\}), G_v(t), \neg G_v(t)$. Here we have to satisfy ourselves that for an arbitrary formula $A$ (and not only for an atomic formula $A$) there is a $k$ such that $\underset{T}{\overset{k}{\models}} \Delta, A, \neg A$. But this is shown as in the proof of the Tautology Lemma (Lemma 7.3.3, which also shows that $k = 2{\cdot}\mathrm{rnk}(A)$). The second critical case is that of an application of the rule $(\forall)$. Here it might happen that the variable condition is violated. To avoid this, we have to show that $\underset{\mathsf{T}}{\overset{m}{\models}} \Delta(v)$ entails $\underset{\mathsf{T}}{\overset{m}{\models}} \Delta_v(u)$ which is obvious by induction on $m$. In the case of an inference according to $(\forall)$ we then have the premise $\underset{\mathsf{T}}{\overset{m_0}{\models}} \Delta_0(X), A(X,u)$ and choose a new free object variable $w$ that does not occur in $G_v(t)$. Then we obtain $\underset{\mathsf{T}}{\overset{m_0}{\models}} \Delta_0(X), A(X,u)_u(w)$ and, using the induction hypothesis, $\underset{\mathsf{T}}{\overset{k_0}{\models}} \Delta_0(\{x \mid G_v(x)\}), A(\{x \mid G_v(x)\}, u)_u(w)$ for some $k_0$ and may now apply an inference $(\forall)$ to obtain $\underset{\mathsf{T}}{\overset{k}{\models}} \Delta_0(\{x \mid G_v(x)\}), (\forall x)A(\{x \mid G_v(x)\}, u)_u(x)$. The remaining cases are all simple. □

The next observation is a consequence of the previous lemma, which, nevertheless, still needs some care.

**7.4.3 Lemma** *(Substitution for* NT*) Let $F(X)$ and $G(v)$ be a formula in the language of number theory. Then*

$$\mathsf{NT} \vdash F(X) \;\Rightarrow\; \mathsf{NT} \vdash F(\{x \mid G(x)\}). \tag{7.8}$$

*Proof* To prove (7.8) assume

$$\mathsf{NT} \vdash F(X). \tag{i}$$

Then there are finitely many axioms of $A_1, \dots, A_n \subseteq \mathsf{NT}$ and an $m$ such that

$$\underset{\mathsf{T}}{\overset{m}{\models}} \neg A_1(X), \dots, \neg A_n(X), F(X). \tag{ii}$$

By Lemma 7.4.2, we then obtain

$$\underset{\mathsf{T}}{\overset{k}{\models}} \neg A_1(\{x \mid G(x)\}), \dots, \neg A_n(\{x \mid G(x)\}), F(\{x \mid G(x)\}). \tag{iii}$$

To conclude

$$\mathsf{NT} \vdash F(\{x \mid G(x)\}),$$

we have to ensure that $A_i(\{x \mid G(x)\}) \in \mathsf{NT}$ holds for $i = 1, \dots, n$. But this is obvious since we formulated the only critical axioms, the last identity axiom and Mathematical Induction, as schemes.[3] □

---

[3] It is possible to replace the last identity scheme by the single axiom $(\forall x)(\forall y)[x = y \to (x \,\varepsilon\, X \to y \,\varepsilon\, X)]$. Formulating Mathematical Induction as a scheme is, however, inevitable.

Let

$$\mathcal{J}(X) := \{\alpha \mid (\forall\xi)[\xi \subseteq X \to \xi + \omega^\alpha \subseteq X]\}$$

denote the *jump* of $X$. Then, if we assume

$$\mathsf{NT} \vdash \mathsf{Prog}(X) \to \mathsf{Prog}(\mathcal{J}(X)), \tag{i}$$

we obtain

$$\mathsf{NT} \vdash \mathsf{TI}(\alpha, \mathcal{J}(X)) \to \mathsf{TI}(\omega^\alpha, X). \tag{ii}$$

To prove (ii) assume (working informally in NT) $\mathsf{TI}(\alpha, \mathcal{J}(X))$, i.e.,

$$\mathsf{Prog}(\mathcal{J}(X)) \to \alpha \subseteq \mathcal{J}(X) \tag{iii}$$

which entails

$$\mathsf{Prog}(\mathcal{J}(X)) \to \alpha \;\varepsilon\; \mathcal{J}(X). \tag{iv}$$

Choosing $\xi = 0$ in the definition of the jump turns (iv) into

$$\mathsf{Prog}(\mathcal{J}(X)) \to \omega^\alpha \subseteq X, \tag{v}$$

that together with (i), gives

$$\mathsf{Prog}(X) \to \omega^\alpha \subseteq X, \tag{vi}$$

i.e., $\mathsf{TI}(\omega^\alpha, X)$. Once we have (ii) we also get (7.7) because $\mathsf{NT} \vdash \mathsf{TI}(\alpha, X)$ implies $\mathsf{NT} \vdash \mathsf{TI}(\alpha, \mathcal{J}(X))$ by (7.8).

It remains to prove (i). Again we work informally in NT. Assume

$$\mathsf{Prog}(X). \tag{vii}$$

We want to prove $\mathsf{Prog}(\mathcal{J}(X))$, i.e., $(\forall\alpha)[\alpha \subseteq \mathcal{J}(X) \to \alpha \;\varepsilon\; \mathcal{J}(X)]$. Thus, assuming also

$$\alpha \subseteq \mathcal{J}(X), \tag{viii}$$

we have to show $\alpha \;\varepsilon\; \mathcal{J}(X)$, i.e., $(\forall\xi)[\xi \subseteq X \to \xi + \omega^\alpha \subseteq X]$. That means that for

$$\eta < \xi + \omega^\alpha \tag{ix}$$

we have to prove $\eta \;\varepsilon\; X$ under the additional hypothesis

$$\xi \subseteq X. \tag{x}$$

If $\eta < \xi$, we obtain $\eta \;\varepsilon\; X$ by (x). Let $\xi \leq \eta < \xi + \omega^\alpha$. If $\alpha = 0$ then $\eta = \xi$ and we obtain $\eta \;\varepsilon\; X$ by (x) and (vii). If $\alpha > 0$ then there is a $\sigma < \alpha$ and a natural number, (i.e., a numeral in NT), such that $\eta < \xi + \underbrace{\omega^\sigma + \ldots + \omega^\sigma}_{n\text{-fold}} =: \omega^\sigma \cdot n$.[4] We show

$$\sigma < \alpha \to \xi + \omega^\upsilon \cdot n \subseteq X \tag{xi}$$

[4] C.f. the proof of the Predicative Elimination Lemma (Lemma 7.3.14).

by induction on $n$. For $n = 0$ this is (x). For $n := m + 1$ we have

$$\xi + \omega^\sigma \cdot m \subseteq X \tag{xii}$$

by the induction hypothesis. From $\sigma < \alpha$ we obtain $\sigma \,\varepsilon\, \mathscr{J}(X)$ by (viii). This together with (xii) entails $\xi + \omega^\sigma \cdot n = \xi + \omega^\sigma \cdot m + \omega^\sigma \subseteq X$. This finishes the proof of (i), hence also that of (7.7) that in turn implies Theorem 7.4.1. □

Summing up we have shown

**7.4.4 Theorem** *(Ordinal Analysis of* NT*)* $\|\mathsf{NT}\| = \|\mathsf{NT}\|_{\Pi^1_1} = \varepsilon_0$.

The next theorem is a consequence of Theorem 7.3.16 and (the proof of) Theorem 7.4.1.

**7.4.5 Theorem** *There is a $\Pi^1_1$-sentence $(\forall X)(\forall x)F(X,x)$ that is true in the standard structure $\mathbb{N}$ such that* $\mathsf{NT} \vdash F(X,\underline{n})$ *for all $n \in \mathbb{N}$ but* $\mathsf{NT} \nvdash (\forall x)F(X,x)$.

To prove the theorem choose $F(X,x) \;:\Leftrightarrow\; \mathsf{Prog}(X) \to x \,\varepsilon\, OT \to x \,\varepsilon\, X$. □

**7.4.6 Remark** Theorem 7.4.5 can be regarded as a weakened form of Gödel's first incompleteness theorem. Gödel's theorem states that there is already a true $\Pi^0_1$-sentence $(\forall x)G(x)$ that is unprovable in NT while all its instantiations $G(\underline{n})$ are provable in NT. Theorem 7.4.5 can be read as there is a $\Pi^0_3$-formula $(\forall x)F(x,X)$ with this property, which is a true pseudo-$\Pi^1_1$-sentence. Observe that the presence of the free set variable in the $\Pi^0_3$-formula is crucial for the proof given here, which is completely different from GÖDEL's proof. Theorem 7.4.5 is stronger in the aspect that provability of all instances $F(\underline{n},X)$ is not obvious, while the provability of $G(\underline{n})$ follows from the fact that all true $\Sigma^0_1$-sentences are provable in NT. Getting rid of free set variables could, however, be of importance for the separation of theories in bounded arithmetic (cf. [7]). But even the application of highly sophisticated methods of impredicative proof theory to NT will only yield independence of $\Pi^0_2$-sentences (cf. Corollary 10.5.11 below).

**7.4.7 Exercise** Show that $\varphi^n_0(\omega) \leq \|\mathsf{NT}_n\|$ holds true for all $n \in \omega$. Confer the result with Exercise 7.3.19 to obtain exact bounds for $n \geq 1$.

Hint: Use mathematical induction to show $\mathsf{NT}_0 \vdash \mathsf{TI}(\omega,X)$ and define $\mathscr{J}(X)$ more parsimoniously.

**7.4.8 Exercise** Prove $\varphi_1(\varepsilon_0) \leq \|(\Delta^1_0\text{–CA})\|$.

Hint: Show

$$(\Delta^1_0\text{–CA}) \vdash (\forall X)\mathsf{TI}(\alpha,X) \to (\forall X)\mathsf{TI}(\varphi^n_0(\alpha),X)$$

using the scheme of Mathematical Induction. Observe that $\xi < \varphi_1(\alpha)$ and $\alpha \neq 0$ imply that there is a natural number $n$ and an ordinal $\gamma < \alpha$ (whose code is primitive recursively computable from the codes of $\xi$ and $\alpha$) such that $\xi < \varphi^n_0(\varphi_1(\gamma+1))$. Conclude that

$$(\Delta^1_0\text{–CA}) \vdash (\forall \xi < \alpha)(\forall X)\mathsf{TI}(\varphi_1(\xi),X) \to (\forall X)\mathsf{TI}(\varphi_1(\alpha),X)$$

.

## 7.5 The Use of Gentzen's Consistency Proof for Hilbert's Programme

In this section, we want to discuss the results just obtained in the light of the original aims of proof theory as posed in the Hilbert's programme. We start with some general remarks on the consistency of formal systems.

### 7.5.1 On the Consistency of Formal and Semi-Formal Systems

**7.5.1 Definition** A (semi) formal system $T$ is *semantically consistent* if there is no formula $A$ such that $T \vdash A$ and $T \vdash \neg A$.

If $T$ is a theory that as a model $\mathfrak{M}$ then $\mathfrak{M} \models A$ for all $A$ such that $T \vdash A$. Since $\mathfrak{M} \models A$ and $\mathfrak{M} \models \neg A$ is impossible it follows that $T$ is semantically consistent. Why is it not possible to satisfy Hilbert's programme by just showing that $T$ has a model? For a theory $T$ that contains some kind of infinity axiom it is impossible to prove by finitist means that $T$ possesses a model. An infinite model cannot be constructed by finitist means. It is difficult to give an exact description of "finitist means". Let $(\Sigma_1^0\text{–Ind})$ be the theory NT in which the scheme (IND) of mathematical induction is restricted to $\Sigma_1^0$-formulas, i.e., to formulas of the shape $(\exists x)A$ where $A$ is quantifier free. For our purposes, it is sufficient to regard everything that can be formalized within the system $(\Sigma_1^0\text{–Ind})$ as finitist.[5] It is easy to see that the common definition of "$\mathfrak{M} \models \mathsf{NT}$" cannot be carried out in NT let alone within $(\Sigma_1^0\text{–Ind})$.[6]

Trying to construct full models is apparently the wrong way to come to finitist consistency proofs. Still following the ideas of Hilbert's programme we define the notion of *syntactical consistency*.

**7.5.2 Definition** A (semi-) formal system $T$ is *syntactically consistent* iff there is a formula $A$ such that $T \not\vdash A$.

For reasonable formal systems we can show that semantical and syntactical consistency coincide. To explain which formal systems are reasonable we need some preparations. Let us first recall some basic notions. A propositional atom is either an atomic formula or a formula whose outermost logical symbol is a quantifier. A propositional assignment is a map that assigns a truth value to every propositional atom. Here we require that atoms that are dual in the Tait-language (i.e., atoms $A$ and $B$ such that $({\sim}A) = B$ and $({\sim}B) = A$) obtain opposite truth values.

---

[5] This system proves the same $\Pi_2^0$-sentences as the system in which Mathematical Induction is restricted to quantifier free formulas. There are good reasons why this system is regarded as finitist that have been widely discussed in the literature. We do not want to add further arguments to that discussion.

[6] It follows by GÖDEL's second incompleteness theorem that this is impossible in principle.

The propositional truth value of a formula under a propositional assignment is calculated according to the truth tables for the logical connectives $\land$ and $\lor$. A formula is *propositionally valid* if it becomes true under all propositional assignments.

**7.5.3 Definition** (a) Let $(T,\mathscr{R})$ be a (semi-) formal system. We say that a rule

$$A_1,\dots,A_n \vdash F$$

is permissible in $(T,\mathscr{R})$ if

$$(T,\mathscr{R}) \vdash A_1,\cdots,(T,\mathscr{R}) \vdash A_n \;\Rightarrow\; (T,\mathscr{R}) \vdash F.$$

(b) A rule

$$A_1,\dots,A_n \vdash F$$

is a *propositional rule* if the formula $\neg A_1 \lor \cdots \lor \neg A_n \lor F$ is propositionally valid.

(c) A (semi-) formal system $(T,\mathscr{R})$ is *propositionally closed* if every propositional rule is a permissible rule of $(T,\mathscr{R})$.

**7.5.4 Theorem** *A propositionally closed (semi-) formal system is semantically consistent if and only if it is syntactically consistent.*

*Proof* If $(T,\mathscr{R})$ is semantically consistent we have $(T,\mathscr{R}) \not\vdash (A \land \neg A)$. So $(T,\mathscr{R})$ is syntactically consistent. If $(T,\mathscr{R})$ is semantically inconsistent there is a formula $A$ such that $(T,\mathscr{R}) \vdash A \land \neg A$. Since $(A \land \neg A) \vdash F$ is a propositional rule for all formulas $F$ we obtain $(T,\mathscr{R}) \vdash F$ for all formulas $F$. So $(T,\mathscr{R})$ is syntactically inconsistent. □

### 7.5.2 The Consistency of NT

A theory is of course propositionally closed. To obtain the semantical consistency of NT it therefore suffices to show that there is a formula $A$ that is not derivable from the axioms in NT. We will prove that via the syntactical consistency of the semi-formal system introduced in Definition 7.3.5.

**7.5.5 Lemma** *If* $\vdash^{\alpha}_{0} A$ *for an atomic sentence* $A$ *then* $A \in Diag(\mathbb{N})$.

*Proof* We show the claim by induction on $\alpha$. Since $\mathrm{CS}(A) = \emptyset$ and the last inference cannot be a cut, the only possibility is an application of $(\bigwedge)$ with empty premises and critical formula $A$. But then $A \in \bigwedge$–type , i.e., $A \in Diag(\mathbb{N})$. □

**7.5.6 Corollary** *The semi-formal system given in Definition 7.3.5 is syntactically consistent.*

*Proof* If $\neg A \in Diag(\mathbb{N})$ and we assume $\frac{\alpha}{\rho} A$ then we obtain by the Elimination Theorem $\frac{\beta}{0} A$ for some ordinal $\beta$. This, however, contradicts Lemma 7.5.5.

**7.5.7 Lemma** *The semi-formal system given in Definition 7.3.5 is propositionally closed.*

*Proof* Assume that $\neg A_1 \vee \cdots \vee \neg A_n \vee F$ is a propositionally valid formula and

$$\left|\frac{\alpha_i}{\rho_i}\right. A_i \qquad \text{(i)}$$

for $i = 1,\ldots,n$. Let

$$\rho = \max(\{\rho_i \,|\, i = 1,\ldots,n\} \cup \{\mathrm{rnk}(A_i) \,|\, i = 1,\ldots,n\}) + 1$$

and $\alpha := \max\{\alpha_1,\ldots,\alpha_n\}$. Since $\neg A_1 \vee \cdots \vee \neg A_n \vee F$ is propositionally valid it is also logically valid and by the Completeness Theorem for first-order logic (Theorem 4.4.1) and Theorem 7.3.2 there is a natural number $m$ such that [7]

$$\left|\frac{m}{0}\right. \neg A_1 \vee \cdots \vee \neg A_n \vee F. \qquad \text{(ii)}$$

From (i) and (ii) we obtain $\left|\frac{\beta}{\rho}\right. F$ for some $\beta < \alpha + \omega$. □

**7.5.8 Theorem** *The semi-formal system given in Definition 7.3.5 is semantically consistent.*

*Proof* This follows from Corollary 7.5.6, Lemma 7.5.7 together with Theorem 7.5.4. □

**7.5.9 Theorem** *The theory* NT *is semantically consistent.*

*Proof* We already remarked that theories are always propositionally closed. So it remains to show that NT is syntactically consistent. Assuming that NT is syntactically inconsistent we obtain $\mathsf{NT} \vdash A$ for all sentences $A$. By Theorem 7.3.8 this implies $\left|\frac{\omega+\omega}{r}\right. A$ contradicting Corollary 7.5.6. □

Having seen this consistency proof we have to answer the question: "What makes this proof more 'constructive' in comparison to the proof via the construction of a model for NT?" Since this leads us away from the main topic of this book, we will be quite sketchy. Nevertheless, we want to at least touch this issue. First we observe that the infinitary proof tree constructed in the proof of Theorem 7.3.2 is primitive recursive, hence definable in $\mathsf{NT}_0$, a theory in the language $\mathcal{L}(\mathsf{NT})$ in which the scheme of Mathematical Induction is restricted to quantifier free formulas.[8]

[7] The proof via the completeness theorem works only for first-order formulas. We will later extend the semi-formal system to a language with infinitely long formulas. To show that these systems are also propositionally closed we will then have to reprove that all propositionally valid formulas are derivable.

[8] This system is also known as Primitive Recursive Arithmetic PRA.

Secondly, we have to secure that the cut-elimination operations, as given by the Reduction Lemma (Lemma 7.3.12) and Lemma 7.3.13, are primitive recursive, hence executable in $\mathsf{NT}_0$. By a slight modification of the cut-elimination procedure it is even possible to define the reduction steps for non well-founded trees. This method is known as mints' continuous cut-elimination.[9] But since we have not introduced continuous cut-elimination, we need the well-foundedness of the infinitary proof tree to show that the cut-elimination procedure terminates. If we assume an arithmetization of the formulas we can define a predicate

$$e \overset{\alpha}{\underset{\rho}{\vdash}} \ulcorner\Delta\urcorner$$

that formalizes the fact that $e$ is an index of a recursive proof tree whose nodes are tagged with notations for ordinals $\leq \alpha$ and finite sets of codes for formulas in $\mathcal{L}(\mathsf{NT})$ such that its root is tagged with (codes for) $\alpha$ and $\Delta$. Then there is a recursive function $f$ such that

$$\mathsf{NT} + (TI(exp^n(\omega,\alpha),X)) \vdash \left( e \overset{\alpha}{\underset{n}{\vdash}} \ulcorner\Delta\urcorner \to f(e) \overset{exp^n(\omega,\alpha)}{\underset{0}{\vdash}} \ulcorner\Delta\urcorner \right). \tag{i}$$

To formalize the embedding procedure let $T$ be a theory in the language of arithmetic. We assume that there is an elementary coding for the language of arithmetic and that there is a predicate

$$Prf_T(i,v) \;:\Leftrightarrow\; \text{“}i \text{ codes a proof from } T \text{ of the formula coded by } v\text{”}.$$

By the above embedding procedure we then obtain

$$\mathsf{NT} \vdash \left( (\exists x)[Prf_{\mathsf{NT}}(x, \ulcorner F\urcorner)] \to (\exists x)[(x)_0 \overset{\omega+\omega}{\underset{(x)_1}{\vdash}} \ulcorner F\urcorner] \right). \tag{ii}$$

From (ii) and (i) it follows

$$\mathsf{NT} + (TI(\varepsilon_0,X)) \vdash (\exists x)[Prf_{\mathsf{NT}}(x, \ulcorner F\urcorner)] \to (\exists x)[(x)_0 \overset{(x)_1}{\underset{0}{\vdash}} \ulcorner F\urcorner]. \tag{iii}$$

Formalizing the fact that no false atomic sentence is derivable in the semi-formal system yields

$$\mathsf{NT} + (TI(\varepsilon_0,X)) \vdash \neg(\exists x)[(x)_0 \overset{(x)_1}{\underset{0}{\vdash}} \ulcorner A\urcorner] \tag{iv}$$

for all formulas $A$ such that $\neg A \in Diag(\mathbb{N})$. But from (iv) and (iii) we obtain

$$\mathsf{NT} + (TI(\varepsilon_0,X)) \vdash \neg(\exists x)[Prf_{\mathsf{NT}}(x, \ulcorner A\urcorner)]$$

for all false atomic sentences.

So we have shown by transfinite induction along the notation system developed in Sect. 3.3.1 that NT is syntactically consistent and therefore also semantically consistent. On the other hand we have seen in Sect. 7.4 that NT proves the well-foundedness of all initial segments of the notation system. In this sense the result is optimal.

---

9 The details of a similar construction are in [12].

Using MINTS' continuous cut-elimination the result can be sharpened in so far that (iii) becomes formalizable in PRA. Since all paths in the cut-free derivation are primitive recursive it even suffices to know that every primitive recursively definable $\prec$ descending sequence is finite to get (iv). Thus, PRA together with the principle $PRWO(\prec)$, which says that every primitive recursively definable $\prec$-descending sequence is finite, suffices to prove the consistency of NT.

### *7.5.3 Kreisel's Counterexample*

Having in mind the fact that $\mathsf{PRA} + PRWO(\prec)$ for a primitive recursive well-ordering $\prec$ proves the consistency of the theory NT while NT proves the well-foundedness of all initial segments of $\prec$, it is tempting to define the proof-theoretic ordinal of a theory $T$ as the order type of the shortest primitive recursive well-ordering that is needed to show the consistency of $T$ over the basis theory PRA. That this is malicious has been pointed out by Kreisel who presented the following counterexample.

To sketch it let us work in PRA. A formula is $\Delta_0$ iff it only contains bounded quantifiers $(\forall x < a)$ or $(\exists x < a)$. For a theory $T$ we define the *provability predicate*

$$\Box_T x \;:\Leftrightarrow\; (\exists y) Prf_T(y,x).$$

Let $\bot$ be a false atomic sentence, i.e., let $(\sim\bot) \in Diag(\mathbb{N})$, and define

$$Con(T) \;:\Leftrightarrow\; \neg\Box_T \ulcorner\bot\urcorner.$$

**7.5.10 Theorem** *(Kreisel) For any consistent theory T there is a primitive recursive well-ordering $\prec_T$ of order type $\omega$ such that*

$$\mathsf{PRA} + PRWO(\prec_T) \vdash Con(T).$$

*Proof* Define

$$x \prec_T y \;:\Leftrightarrow \begin{cases} x < y & \text{if } (\forall i < x)[\neg Prf_T(i, \ulcorner\bot\urcorner)] \\ y < x & \text{otherwise} \end{cases} \tag{i}$$

and let

$$F(x) \;:\Leftrightarrow\; (\forall i \le x)[\neg Prf_T(i, \ulcorner\bot\urcorner)]. \tag{ii}$$

Now we obtain

$$\mathsf{PRA} \vdash (\forall x \prec_T y) F(x) \to F(y) \tag{iii}$$

since, if we assume $\neg F(y)$, we have $(\exists i \le y)[Prf_T(i, \ulcorner\bot\urcorner)]$ and get $y+1 \prec_T y$ and thus together with the premise of (iii) also $F(y+1)$. But this implies $F(y)$, a contradiction.

Since $\prec_T$ is primitive recursive we obtain from (iii)

$$\mathsf{PRA} + PRWO(\prec_T) \vdash (\forall x) F(x) \tag{iv}$$

and thus

$$\mathsf{PRA} + PRWO(\prec_T) \vdash Con(T). \tag{v}$$

Since $Con(T)$ is true we have $\prec_T = <$ and thus $\mathrm{otyp}(\prec_T) = \omega$. $\square$

It follows from Kreisel's counterexample that the naive definition of the proof-theoretic ordinal of $T$ will always yield proof-theoretic ordinals $\leq \omega$. It is therefore not possible to define the proof-theoretic ordinal of a theory or a formal system as the order type of the "shortest primitive recursive well-ordering" by which the consistency of the theory can be established in a "finitist" way. The well-ordering in the counterexample given here is admittedly extremely artifical. But whatever definition we choose, it is nearly impossible to avoid artifical counter examples. The true reason for that is the fact that we have to refer to presentations of ordinals by primitive recursive well-orderings (or equivalently to ordinal notations) instead of ordinals themselves. Therefore these definitions become extremely sensitive to the way in which ordinals are presented. In all cases in that we have a *"canonical"* representation or a canonical notation system these abnormalities will not occur. But until today all attempts to give a mathematical definition of a *"canonical well-ordering"* failed.

### 7.5.4 Gentzen's Consistency Proof in the Light of Hilbert's Programme

In the former sections, we did not do much more than repeating Gentzen's consistency proof for pure number theory in a different language. Our motivation, however, was mainly ordinal analysis and not primarily consistency. When Gentzen first presented his proof it was viewed as a salvation of HILBERT's programme. As cited in the introduction, Bernays regarded Gentzen's proof as a proof in the guise of Hilbert's programme. Bernays believed that only a tiny extension of Hilbert's finitist standpoint was needed to justify Gentzen's proof.

We want to discuss how tiny the extension really is. For this purpose we imagine an opponent who seriously doubts the consistency of the theory NT (or an equivalent theory) and whom we try to convince by Gentzen's proof. Of course we assume that our opponent is able to follow mathematical reasoning in an unbiased way. We present the proof as constructive as possible (which we did not do here) and he or she will probably accept all steps in the proof but finally utter his or her uneasiness with the induction up to $\varepsilon_0$ that we needed in establishing the syntactical consistency. This (s)he says is a bit beyond his or her finitist horizon. We, therefore, try to substantiate this induction as "finitistly" as possible. We avoid talking about ordinals and just want to convince him or her that the order relation we used in showing the syntactical consistency is well-founded. Therefore we introduce the order relation as in Sect. 3.3.1 (without mentioning its "ordinal origin") and, in showing its well-foundedness, will more or less inevitably end up with an argumentation that is essentially that of Sect. 7.4.

Given an $\alpha < \varepsilon_0$ we find an $n$ such that $\alpha < exp^n(\omega,0)$ and it suffices to secure the well-foundedness up to $exp^n(\omega,0)$, i.e., to prove $TI(exp^n(\omega,0),X)$. This is obtained by iterated application of (7.7). Analyzing the proof of (7.7) more closely we see that the crucial argument there is an iterated application of

$$\mathsf{NT} \vdash \mathsf{TI}(\alpha,\mathscr{J}(X)) \;\rightarrow\; \mathsf{TI}(\omega^\alpha,X). \tag{i}$$

Defining $\mathscr{J}_0(X) = X$ and $\mathscr{J}_{k+1}(X) = \mathscr{J}(\mathscr{J}_k(X))$, we have to start with the trivial statement $TI(0,\mathscr{J}_n(X))$ and then decrease the number of jumps until we reach $TI(exp^n(\omega,0),X)$. The main point in proving (i), however, was to show

$$\mathsf{NT} \vdash \mathsf{Prog}(X) \;\rightarrow\; \mathsf{Prog}(\mathscr{J}(X)) \tag{ii}$$

which in turn needed the proof of

$$\sigma < \alpha \;\rightarrow\; \xi + \omega^\sigma \cdot k \subseteq X \tag{iii}$$

by Mathematical Induction on $k$. So to show

$$\mathsf{NT} \vdash \mathsf{Prog}(\mathscr{J}_l(X)) \;\rightarrow\; \mathsf{Prog}(\mathscr{J}_{l+1}(X)) \tag{iv}$$

we need Mathematical Induction for the formula

$$\sigma < \alpha \;\rightarrow\; \xi + \omega^\sigma \cdot k \subseteq \mathscr{J}_l(X). \tag{v}$$

This means that we need the induction scheme for formulas of the complexity of $\mathscr{J}_l(X)$. We did not pay attention to define $\mathscr{J}(X)$ parsimoniously. But even under the most careful definition of $\mathscr{J}(X)$ we will have at least $l$ quantifiers in the defining formulas for $\mathscr{J}_l(X)$ which means that $\mathscr{J}_l(X)$ belongs at least to the $l$th level in the arithmetical hierarchy. To reach $\varepsilon_0$ we, therefore, must not restrict the complexities of the formulas in the scheme of Mathematical Induction. At this point our opponent will argue that in doing so we exhaust full first-order number theory and even a bit more.[10] But (s)he doubts full number theory. Therefore (s)he cannot accept the proof. We hardly can advance a mathematical argument against that.

This situation is in complete accordance with Gödel's second incompleteness theorem. If Gödel's second incompleteness theorem is more than a mere formal triviality but has a genuine content one cannot expect to bypass it by a tiny extension of the finitist standpoint. Therefore one also cannot expect to obtain proof-theoretical results in the spirit of Hilbert's programme in the narrow sense as he had put it in [41].

But Hilbert's programme has also a more general aspect. At some occasions, e.g. in his talk *"Über das Unendliche"* [42], he talks about the elimination of ideal objects. In his opinion infinite sets (and similar mathematical objects) only serve as ideal elements that are needed to obtain information about concrete objects. This part of Hilbert's programme is at least partially realized by Friedman's and Simpson's programme of reverse mathematics. Surprisingly much of mathematics

---

[10] The "bit more" is that we obtain $\mathsf{NT} \vdash TI(exp^n(\omega,0))$ **for all** $n$ only outside of NT. The quantifier *"for all n"* cannot be formalized in NT.

can already be obtained in a theory $(\mathsf{RCA})_0$ whose proof-theoretical strength is that of $(\Sigma^0_1\text{–Ind})$ (hence incorporates what we wanted to call finitist).[11]

Although Gentzen's result is of little help in the spirit of Hilbert's finitist programme it sheds some light on the consistency of number theory, which is more in the spirit of BROUWER's approach. Looking more carefully at the consistency proof (as we have sketched it in Sect. 7.5.2) we see that the consistency proof for NT can be formalized within a formal system that is based on a constructive logic, (i.e., a logic that avoids proofs by contradiction and, therefore, excludes the possibility to prove the existence of mathematical objects without having constructed them) and whose only nonfinitist feature is an induction over quantifier free formulas along an elementary definable ordering that possesses no infinite primitive recursively definable paths. Since such an ordering (which is essentially the ordering we introduced in Sect. 3.3.1) is easy to visualize it is intuitively plain that this system should be consistent (although its proof-theoretical ordinal is above $\varepsilon_0$). This form of *reductive proof theory* is in full coherence with Gödel's second incompleteness theorem.

---

[11] One should, however, be aware that just combinatorial statements, e.g. Ramsey's theorem in the strengthened form by Paris and Harrington, are the statements that turn out to be unprovable in NT and are thus not finitistly provable although they talk only about finite objects. This thwarts Hilbert's idea of justifying the infinite by the finite.

# Chapter 8
# Autonomous Ordinals and the Limits of Predicativity

The ordinal analysis of NT is a paradigmatic example for predicative proof theory. We discuss the notion of predicativity in Sect. 8.3 below. The rest of the chapter is dedicated to the presentation of the famous result of Solomon Feferman and Kurt Schütte about the limits of predicativity.[1]

## 8.1 The Language $\mathcal{L}_\kappa$

**8.1.1 Definition** Let $\kappa$ be an ordinal. We introduce the language $\mathcal{L}_\kappa$. The logical symbols of $\mathcal{L}_\kappa$ are

- Countably many free set variables, denoted by $X, Y, Z, X_1, \ldots$
- The binary relation symbols $=$, $\neq$, $\varepsilon$, $\not\varepsilon$.
- The logical connectives $\bigwedge$ and $\bigvee$.

The nonlogical symbols of $\mathcal{L}_\kappa$ are

- A constant $\underline{0}$ for the natural number 0.
- Constants for all primitive recursive functions.

*Terms* are defined inductively by the clauses

- The constant $\underline{0}$ is a term.
- If $t_1, \ldots, t_n$ are terms and $\underline{f}$ is a symbol for an $n$-ary primitive recursive function then $(\underline{f}t_1, \ldots, t_n)$ is a term.

Observe that the value $t^{\mathbb{N}}$ of a term $t$ is effectively, i.e., recursively, computable.

[1] With the exception of the definition of the language $\mathcal{L}_\kappa$ this chapter will not be used in the later chapters.

W. Pohlers, *Proof Theory: The First Step into Impredicativity*, Universitext,

We define *formulas* and their types inductively by the following clauses

- Let $s$ and $t$ be terms. If $s^{\mathbb{N}} = t^{\mathbb{N}}$ then $(s = t) \in \bigwedge\text{-type}$ and $(s \neq t) \in \bigvee\text{-type}$. If $s^{\mathbb{N}} \neq t^{\mathbb{N}}$ then $(s = t) \in \bigvee\text{-type}$ and $(s \neq t) \in \bigwedge\text{-type}$. It is $FV(s = t) = FV(s \neq t) = \emptyset$.

- If $s$ is a term and $X$ a set variable then $(s\, \varepsilon\, X)$ and $(s \not\varepsilon X)$ are atomic formulas. These atomic formulas have no type. It is $FV(s\, \varepsilon\, X) = FV(s \not\varepsilon X) = \{X\}$.

- If $\lambda < \kappa$ and $\langle F_\xi \mid \xi < \lambda \rangle$ is a sequence of formulas such that $\bigcup_{\xi<\lambda} FV(F_\xi)$ is finite then $(\bigwedge_{\xi<\lambda} F_\xi) \in \bigwedge\text{-type}$ and $(\bigvee_{\xi<\lambda} F_\xi) \in \bigvee\text{-type}$ and $FV(\bigwedge_{\xi<\lambda} F_\xi) = FV(\bigvee_{\xi<\lambda} F_\xi) = \bigcup_{\xi<\lambda} FV(F_\xi)$.

- Atomic formulas and all members of $\bigwedge\text{-type} \cup \bigvee\text{-type}$ are formulas.
- We define the *characteristic sequence* of a formula in $\bigvee\text{-type} \cup \bigwedge\text{-type}$ by
$$\mathrm{CS}(F) = \begin{cases} \emptyset & \text{if } F \text{ is of the shape } s = t \text{ or } s \neq t \\ \langle F_\xi \mid \xi < \lambda \rangle & \text{if } F = \bigcirc_{\xi<\lambda} F_\xi \text{ for } \bigcirc \in \{\bigwedge, \bigvee\}. \end{cases}$$

The formal negation $\sim\! F$ of a formula $F$ is defined by

- $\sim\!(s = t)$ is $(s \neq t)$ and $\sim\!(s \neq t)$ is $(s = t)$.

- $\sim\!(s\, \varepsilon\, X)$ is $(s \not\varepsilon X)$ and $\sim\!(s \not\varepsilon X)$ is $(s\, \varepsilon\, X)$.

- $\sim\!(\bigwedge_{\xi<\eta} F_\xi)$ is $(\bigvee_{\xi<\eta} \sim\! F_\xi)$ and $\sim\!(\bigvee_{\xi<\eta} F_\xi)$ is $\bigwedge_{\xi<\eta} \sim\! F_\xi$.

## 8.2 Semantics for $\mathscr{L}_\kappa$

Since we are only interested in the meaning of formulas of $\mathscr{L}_\kappa$ in the standard model $\mathbb{N}$ we will only define $\mathbb{N} \models F[\Phi]$ for an assignment of subsets of $\mathbb{N}$ to the set variables in $FV(F)$.

**8.2.1 Definition** Let $\Phi\colon FV(F) \longrightarrow \mathrm{Pow}(\mathbb{N})$ be an assignment. We define

$$\mathbb{N} \models F[\Phi] \;:\Leftrightarrow \begin{cases} F \text{ is a formula } (s\, \varepsilon\, X) & \text{and } s^{\mathbb{N}} \in \Phi(X) \\ F \text{ is a formula } (s \not\varepsilon X) & \text{and } s^{\mathbb{N}} \notin \Phi(X) \\ F \in \bigwedge\text{-type} & \text{and } \mathbb{N} \models G[\Phi] \text{ for all } G \in \mathrm{CS}(F) \\ F \in \bigvee\text{-type} & \text{and } \mathbb{N} \models G[\Phi] \text{ for some } G \in \mathrm{CS}(F). \end{cases}$$

Again we get directly

$$\mathbb{N} \models {\sim}F[\Phi] \;\Leftrightarrow\; \mathbb{N} \not\models F[\Phi]$$

for any assignment $\Phi$. Therefore we regularly use $\neg F$ as synonym for ${\sim}F$.

Observe that the TAIT-language of arithmetic may be regarded as a sublanguage of $\mathscr{L}_{\omega_1}$. We leave atomic formulas untouched and put $(A \wedge B)^* := \bigwedge\langle A, B\rangle$, $(A \vee B)^* := \bigvee\langle A, B\rangle$, $(\forall x)F_u(x)^* := \bigwedge_{n\in\omega} F_u(\underline{n})$ and $(\exists x)F_u(x)^* := \bigvee_{n\in\omega} F_u(\underline{n})$.

Then we get

$$\{F^* \,|\, F \in \mathscr{L}(\mathsf{NT})\} \subseteq \mathscr{L}_{\omega_1} \text{ and } \mathbb{N} \models F[\Phi] \;\Leftrightarrow\; \mathbb{N} \models F^*[\Phi]$$

for all assignments $\Phi$.

It follows from Definition 8.2.1 that we can transfer Definition 5.4.3 of the verification calculus for $\mathscr{L}(\mathsf{NT})$ to the language $\mathscr{L}_\kappa$. Therefore we define

$$\models^{\alpha} \Delta \tag{8.1}$$

for finite sets of $\mathscr{L}_\kappa$-formulas literally as in Definition 5.4.3. The difference is that in the new definition we may have characteristic sequences which are longer than $\omega$. Then

$$\models^{\alpha} \Delta \;\Rightarrow\; \mathbb{N} \models \bigvee \Delta[\Phi] \quad \text{for every assignment } \Phi, \tag{8.2}$$

follows directly from Definition 8.2.1. The opposite direction of (8.2), however, does not hold in general. It fails for $\omega_1 < \kappa$. For $\kappa \le \omega_1$, however, we can adapt the proof of Theorem 5.4.9. First we have to modify Definition 5.4.5. Since all ordinals are countable we assume that for every ordinal $\alpha < \omega_1$ there is an enumeration $\langle \alpha_i \mid i \in \omega\rangle$ of the ordinals less than $\alpha$. We modify clauses $(S_\bigwedge)$ and $(S_\bigvee)$ in Definition 5.4.5 as follows:

($\bigwedge$) If the redex of $\delta(s)$ is $\bigwedge_{\xi<\alpha} A_\xi$ then $s^\frown\langle i\rangle \in S_\Delta$ for all $i \in \omega$ and $\delta(s^\frown\langle i\rangle) := \langle\delta(s)\rangle^{r\frown}\langle A_{\alpha_i}\rangle$ where $\langle \alpha_i \mid i \in \omega\rangle$ enumerates the ordinals less than $\alpha$

and

($\bigvee$) If the redex of $\delta(s)$ is $\bigvee_{\xi<\alpha} A_\xi$ then $s^\frown\langle 0\rangle \in S_\Delta$ and we define $\delta(s^\frown\langle 0\rangle) := \langle\delta(s)\rangle^{r\frown}\langle A_{\alpha_i}, \bigvee_{\xi<\alpha} A_\xi\rangle$ for the first number $i$ such that $A_{\alpha_i} \notin \bigcup_{r\subseteq s}\delta(r)$, where $\langle \alpha_i \mid i \in \omega\rangle$ enumerates the ordinals less than $\alpha$, or $\langle\delta(s)\rangle^r$ if such an $i$ does not exist.

It is obvious that this definition cannot be extended to languages $\mathscr{L}_\kappa$ with $\omega_1 < \kappa$.

The Syntactical Main Lemma carries over literally and so does the Semantical Main Lemma. What is changed is the order-type of the search trees. All search trees are still countably branching but in general not recursive. Since $\omega_1$ is a regular cardinal every countably branching well-founded tree has an order-type below $\omega_1$. The $\omega$-completeness theorem (Theorem 5.4.9) is thus changed in the following way.

**8.2.2 Theorem** *Let $F$ be an $\mathscr{L}_{\omega_1}$-formula. Then*

$$(\forall\Phi)[\mathbb{N} \models F[\Phi]] \;\Leftrightarrow\; (\exists\alpha < \omega_1)[\models^{\alpha} F].$$

**8.2.3 Exercise** Show that the Completeness Theorem fails for $\mathcal{L}_\kappa$ with $\kappa > \omega_1$.
Hint: For $M \subseteq \mathbb{N}$ define

$$F_M(\underline{n}) := \begin{cases} \underline{n} \in X & \text{if } n \in M \\ \underline{n} \notin X & \text{if } n \notin M. \end{cases}$$

and

$$F_M^\wedge := \bigwedge \{F_M(\underline{n}) \mid n \in \mathbb{N}\}.$$

Let

$$F := \bigvee \{F_M^\wedge \mid M \subseteq \mathbb{N}\}.$$

First show $\mathbb{N} \models F[\Phi]$ for all assignments $\Phi$. Then prove

- *If* $\models^{\alpha} \Delta$ *then* $\Delta$ *cannot have the form* $F, F_{M_{i_1}}^\wedge, \dots, F_{M_{i_m}}^\wedge, F_{M_{j_1}}(\underline{k_1}), \dots, F_{M_{j_n}}(\underline{k_n})$ *for natural numbers* $i_1, \dots, i_m$ *and* $j_1 \dots j_n$ *and pairwise different natural numbers* $k_1, \dots, k_n$

by induction on the definition of $\models^{\alpha} \Delta$.

## 8.3 Autonomous Ordinals

The basis for the following definition is the notion of predicativity. The notion of predicativity is still controversial. Therefore we define and discuss here predicativity in a pure mathematical – and thus perhaps oversimplified – setting. We consider a notion to be impredicative if it is defined under recourse to an entity to which it belongs itself. The standard (an evil) example of an impredicative definition is Russell's paradoxical set $R := \{x \mid x \notin x\}$. The set $R$ is defined by recourse to all sets and is supposed to be a set itself. Therefore the paradoxical question $R \in R$ is permitted. But observe that our definition of the fixed-point of a monotone operator $\Gamma$ as the least $\Gamma$-closed set is also impredicative in this sense. How can such impredicative notions be avoided? The safe way is to introduce ramifications. Every set gets a stage, say an ordinal $\alpha$, and all members of a set of stage $\alpha$ are supposed to have lower stages. This is essentially the way proposed by Russell. Such an approach, however, does not reflect mathematical praxis. In everyday mathematics we do not talk about ramified objects, e.g. ramified real numbers. Therefore Russell introduced the axiom of reducibility to get rid again of ramifications. This axiom, however, has been regarded as unsatisfactory in different ways. A possible way to introduce ramified sets is given by the constructible hierarchy of sets introduced by Gödel. The constructible hierarchy has therefore attained a predominant role in (impredicative) proof theory.

We postpone, however, the introduction of constructible sets until Sect. 11 and discuss here predicativity in the framework of the infinitary language $\mathcal{L}_\kappa$.

Assume informally that we want to construct a mathematical universe of subsets of $\mathbb{N}$ from below. We start from scratch and accept only those sets which can be obtained by elementary methods. These, in our sense, are exactly the finite sets which can be defined by formulas in $\mathcal{L}_\omega$. It is then at least plausible that this cannot lead us beyond $\omega$. To get further we therefore have to accept $\omega$, i.e., the axiom of infinity. But then we can also accept all objects which we can reach from $\omega$ in finitely many steps. Therefore we accept infinitary formulas of $\mathscr{L}_{\omega_1}$ of length $\omega+n$ for arbitrary finite $n$ and also infinite proof trees of lengths $\omega+n$. That means that we may work in the semi-formal system of Definition 7.3.5 with formulas and proof-trees of lengths below $\omega\cdot 2$. So we accept everything which we can prove within this segment of the semi-formal system. The obvious way to come to stronger infinities is then to define an ordering $\prec$ on $\omega$ and to prove its well-foundedness within the restricted semi-formal system. Whenever we succeed in showing $\left|\frac{\omega+n}{\omega+k}\right. TI(\prec)$ for such an ordering $\prec$ we accept $\alpha := \mathrm{otyp}(\prec)$ as a new infinity and also everything which can be reached from $\alpha$ in finitely many steps. That means that we can extend the semi-formal systems to formulas and proof trees of lengths below $\alpha+\omega$. Again we use the so extended semi-formal system to obtain stronger infinities and iterate this process as far as possible. This approach is apparently perfectly predicative. To reach new infinities we only use the means which we have so far secured in a very strict sense. In a very strict sense because we do not only require that all notions in the **definition** of a new object are previously secured but also that a **proof** of the infinity, i.e., the well-foundedness, of the new objects only needs previously secured means.[2] We therefore talk about predicativity in the narrow sense.

The obvious question is whether this procedure will eventually come to a standstill and, if so, how large a segment of the ordinals can be secured by this method. These questions will be answered in this section. To make these questions mathematically precise we introduce some some notions.

**8.3.1 Definition** For a formula $F \in \mathscr{L}_{\omega_1}$ we define

$$\mathrm{rnk}(F) := \begin{cases} 0, & \text{if } F \text{ is atomic or a formula } s=t \text{ or } s \neq t \\ \sup\{\mathrm{rnk}(F_\xi)+1 \mid \xi < \lambda\} & \text{if } F = \bigcirc_{\xi<\lambda} F_\xi \text{ for } \bigcirc \in \{\bigwedge, \bigvee\}. \end{cases}$$

We extend Definition 7.3.5 to formulas of $\mathscr{L}_{\omega_1}$ and call that semi-formal system $\mathsf{NT}_{\omega_1}$. In Sect. 7.3 we have formulated and proved all theorems in such a way that they also hold for the system $\mathsf{NT}_{\omega_1}$. To grasp the informal description of the construction of an universe from below we define the *autonomous closure* $\mathrm{Aut}(\alpha)$ of an ordinal $\alpha$. First let $\alpha^* := \min\{\lambda \mid \lambda \in Lim \wedge \alpha < \lambda\}$ denote the first limit ordinal above $\alpha$. Since all ordinals below $\alpha^*$ can be accessed from $\alpha$ by finitely many steps we anticipate that all ordinals below $\alpha^*$ are predicatively accessible from $\alpha$.

---

[2] This is the difference to the familiar definition of the constructible hierarchy which is based on iterating definability and whose definition is therefore only locally predicative (cf. Sect. 11). A similar remark applies to the definition of the stages of an arithmetically definable inductive definition. Here too, every single step is predicatively defined. But to iterate the stages until the fixed-point is reached we simultaneously have to secure the well-foundedness of the next iteration steps only using the means of the so far obtained iterations. In that sense also inductive definition are only locally predicative. This fact is used in the ordinal analysis of arithmetically definable inductive definitions (cf. Sect. 9).

**8.3.2 Definition** The autonomous closure $\mathrm{Aut}(\alpha)$ of an ordinal $\alpha$ is inductively defined by the following clauses.

- $\alpha \in \mathrm{Aut}(\alpha)$
- If $\beta \in \mathrm{Aut}(\alpha)$ then $\beta^* \subseteq \mathrm{Aut}(\alpha)$
- If $\lambda \in \mathrm{Aut}(\alpha)$ and $\prec$ is an ordering of $\omega$ which is definable by a formula of rank less than $\lambda$ and there are ordinals $\xi$ and $\rho$ less than $\lambda$ such that $\mathsf{NT}_{\omega_1} \,\frac{\xi}{\rho}\, \mathsf{TI}(\prec, X)$ then $\mathrm{otyp}(\prec)$ belongs to $\mathrm{Aut}(\alpha)$.

The subsets of $\mathbb{N}$ which are definable in $\mathscr{L}_{\mathrm{Aut}(\omega)}$ are apparently the members of the predicative universe constructed on $\omega$.

## 8.4 The Upper Bound for Autonomous Ordinals

To obtain an upper bound for $\mathrm{Aut}(\alpha)$ we modify the definition of $\mathrm{Aut}(\alpha)$ slightly and define

$$\Delta_{\omega_1}(\alpha) := \{\mathrm{tc}(F) \mid F \in \mathscr{L}_{\omega_1} \wedge \mathrm{rnk}(F) < \alpha \wedge (\exists \xi < \alpha)(\exists \rho < \alpha)[\mathsf{NT}_{\omega_1} \,\frac{\xi}{\rho}\, F]\} \quad (8.3)$$

and put

- $\Delta^*_{\omega_1}(\alpha)$ is the least set which contains $\alpha$ as element and is closed under ordinal successor and the function $\Delta_{\omega_1}$.
- We say that an ordinal is $\Delta_{\omega_1}$-closed if $\Delta_{\omega_1}(\xi) \subseteq \alpha$ holds for all $\xi < \alpha$.

**8.4.1 Lemma** *We have* $\mathrm{Aut}(\alpha) \subseteq \Delta^*_{\omega_1}(\alpha)$ *for all countable ordinals* $\alpha$.

*Proof* We show that $\beta \in \mathrm{Aut}(\alpha)$ implies $\beta \in \Delta^*_{\omega_1}(\alpha)$ by induction on the definition of $\beta \in \mathrm{Aut}(\alpha)$. This is obvious for $\beta = \alpha$ and follows from the closure of $\Delta^*_{\omega_1}(\alpha)$ under ordinal successor if $\beta < \lambda^*$ for some $\lambda \in \mathrm{Aut}(\alpha)$. Now let $\beta = \mathrm{otyp}(\prec)$ and $\mathsf{NT}_{\omega_1} \,\frac{\xi}{\rho}\, \mathsf{TI}(\prec, X)$ for some $\xi$ and $\rho$ less than $\lambda \in \mathrm{Aut}(\alpha)$. Without loss of generality we can assume that $\beta$ is a limit ordinal and obtain $\mathrm{tc}(\mathsf{TI}(\prec, X)) = \beta$ by the Boundedness Theorem (Theorem 6.7.2). Hence $\beta \in \Delta^*_{\omega_1}(\alpha)$. □

**8.4.2 Theorem** *The ordinals in* $\{\omega\} \cup SC$ *are* $\Delta_{\omega_1}$*-closed.*

*Proof* Assuming $\alpha \in \{\omega\} \cup SC$, $m, n < \omega$ and $\mathsf{NT}_{\omega_1} \,\frac{m}{n}\, F$, we obtain $\mathsf{NT}_{\omega_1} \,\frac{exp^n(2,m)}{0}\, F$ by Lemma 7.3.13 which entails $\mathrm{tc}(F) < \omega$ by the Boundedness Theorem. So $\omega$ is $\Delta_{\omega_1}$-closed. If $\alpha \in SC$, $\xi, \eta < \alpha$ and $\mathsf{NT}_{\omega_1} \,\frac{\xi}{\eta}\, F$ we obtain $\mathrm{tc}(F) \le \varphi_\xi(\eta) < \alpha$ by Lemma 7.3.14 and the Boundedness Theorem. So $\alpha$ is $\Delta_{\omega_1}$-closed. □

Since all ordinals in $\{\omega\} \cup SC$ are limit ordinals both next corollaries are immediate.

**8.4.3 Corollary** *All ordinals in* $\{\omega\} \cup SC$ *are* $\Delta^*_{\omega_1}$ *closed.*

**8.4.4 Corollary** *We have* $\mathrm{Aut}(0) = \omega$ *and* $\mathrm{Aut}(\omega) \subseteq \Gamma_0$.

**8.4.5 Exercise** We use class terms as abbreviations as introduced in Sect. 8.5 and extend the the translation of the first-order TAIT-language of arithmetic to the second-order language by defining

$$(\forall X)F_U(X)^* := \bigwedge\{F_U(\{x \mid A(x)\}) \mid \mathrm{rnk}(A) < \omega\}$$

and

$$(\exists X)F_U(X)^* := \bigvee\{F_U(\{x \mid A(x)\}) \mid \mathrm{rnk}(A) < \omega\}.$$

(a) Prove that for every $\mathcal{L}(\mathsf{NT})$-formula $F(X_1,\dots,X_k,x_1,\dots,x_l)$ there is a finite ordinal $j$ such that $\mathrm{rnk}(F(A_1,\dots,A_k,\underline{n}_1,\dots,\underline{n}_l)^*) \leq \omega + j$ holds true for all tuples of arithmetical formulas $A_1,\dots,A_k$ and all tuples $n_1,\dots,n_l$ of natural numbers.

(b) Let $\Delta$ be a finite set of $\mathcal{L}_{\omega_1}$-formulas. Use $\Delta_U(A)$ as an abbreviation for $\{F_U(\{x \mid A(x)\}) \mid F \in \Delta\}$. Assume $\mathsf{NT}_{\omega_1} \mathrel{\substack{\alpha \\ \vdash \\ 0}} \Delta$ and show that this implies $\mathsf{NT}_{\omega_1} \mathrel{\substack{2\cdot\mathrm{rnk}(A)+\alpha \\ \vdash \\ 0}} \Delta_U(A)$. Prove moreover that $\mathsf{NT}_{\omega_1} \mathrel{\substack{\alpha \\ \vdash \\ \rho}} \Delta$ for $\rho \neq 0$ implies $\mathsf{NT}_{\omega_1} \mathrel{\substack{2\cdot\mathrm{rnk}(A)+\alpha \\ \vdash \\ \mathrm{rnk}(A)+\rho}} \Delta_U(A)$.

(c) Let $\mathrel{\substack{n \\ \vdash \\ \mathsf{T}^2}} \Delta(X_1,\dots,X_k,x_1,\dots,x_l)$ be a derivation in the sense of second-order logic (cf. Sect. 4.6). Show that $\mathsf{NT}_{\omega_1} \mathrel{\substack{\omega+n \\ \vdash \\ 0}} \Delta(A_1,\dots,A_k,\underline{n}_1,\dots,\underline{n}_l)^*$ holds true for all tuples $A_1,\dots,A_k$ of arithmetical formulas and all tuples $n_1,\dots,n_l$ of natural numbers.

(d) Show that $\mathsf{NT}_{\omega_1} \mathrel{\substack{\omega+\omega+k \\ \vdash \\ 0}} A^*$ holds true for all axioms of $(\Delta^1_0\text{–}\mathsf{CA})$.

(e) Use the previous results to prove $\|(\Delta^1_0\text{–}\mathsf{CA})\| \leq \varphi_1(\varepsilon_0)$ and confer the result with Exercise 7.4.8.

**8.4.6 Exercise** Modify part (d) of Exercise 8.4.5 to obtain $\|(\Delta^1_0\text{–}\mathsf{CA})_0\| \leq \varepsilon_0$.

**8.4.7 Exercise** Let $(\Delta^1_0\text{–}\mathsf{CA}) + (\mathsf{BR})$ be the formal system which comprises all axioms of $(\Delta^1_0\text{–}\mathsf{CA})$, all rules of weak second-order logic $\mathrel{\substack{m \\ \vdash \\ \mathsf{T}^2}}$ together with a cut-rule and the bar-rule

(BR) $\underset{r}{\overset{m}{\models}} \mathsf{TI}(\prec, X) \Rightarrow \underset{r}{\overset{n}{\models}} \mathsf{TI}(\prec, F)$ for all $n > m$ and all formulas $F$ in the second-order language $\mathscr{L}(\mathsf{NT})$ if $\prec$ is a primitive recursive ordering.

(a) Show that there is an embedding theorem of the form

$$(\Delta_0^1\text{–}\mathsf{CA}) + (\mathsf{BR}) \vdash F \;\Rightarrow\; (\exists k < \omega)(\exists l < \omega)\left[\mathsf{NT}_{\omega_1} \underset{\omega+k}{\overset{\varphi_1^l(0)}{\models}} F^*\right].$$

Hint: Use cut-elimination, boundedness and a modification of equation (6.6) to obtain a translation of the bar-rule.

(b) Use (a) to prove $\|(\Delta_0^1\text{–}\mathsf{CA}) + (\mathsf{BR})\| \leq \varphi_2(0)$. Is this bound exact?

## 8.5 The Lower Bound for Autonomous Ordinals

It follows from Corollary 8.4.4 that $\Gamma_0$ is an upper bound for predicativity in the narrow sense delineated above. This has first been observed by Schütte [85], [87] and [88] and independently by Feferman [23]. Both authors could, however, also show that

$$\Gamma_0 \subseteq \mathrm{Aut}(\omega). \tag{8.4}$$

So, once we have accepted $\omega$, the ordinal $\Gamma_0$ is the exact bound of predicativity in the narrow sense. The proof of (8.4) is the aim of the present section.

In Sect. 3.4.3 we introduced a notation system for the ordinals below $\Gamma_0$. Therefore we can talk about ordinals less or equal than $\Gamma_0$ in the language of arithmetic. Again we will identify ordinals and their notations. Lower case Greek letters will vary over ordinals. We use all the abbreviations introduced in Sect. 7.4. Recall, especially the formulas $\alpha \subseteq \beta$, $\mathsf{Prog}(X)$ and $\mathsf{TI}(\alpha, X)$ defined there.

Since we have constants for all characteristic functions of primitive recursive relations, all primitive recursive relations are available in $\mathscr{L}_{\omega_1}$. To improve readability we will, however, always write $R(t_1, \ldots, t_n)$ instead of $(\underline{\chi}_R t_1, \ldots, t_n) = 1$.

We regard $\mathscr{L}(\mathsf{NT})$ as a sublanguage of $\mathscr{L}_{\omega_1}$ and use the familiar notations. So $A \to B$ stands for $\bigvee\{\neg A, B\}$, $(\forall x)F(x)$ for $\bigwedge\{F(\underline{n}) \mid n \in \omega\}$ etc.

We will also freely use class terms of the form $\{x \mid F(x)\}$ although they do not belong to the language. The formula $t \;\varepsilon\; \{x \mid F(x)\}$ is to be read as an abbreviation for the formula $F(t)$. The rank of a class term $\{x \mid F(x)\}$ is the rank of the formula $F(t)$. By

$$\mathfrak{K}_\sigma := \{T \mid T \text{ is a class term and } \operatorname{rnk}(T) < \sigma\}$$

we denote the collection of all class terms of ranks less than $\sigma$. Let

$$\mathsf{TI}_\sigma(\alpha) \;:\Leftrightarrow\; \bigwedge\{\mathsf{TI}(\alpha, S) \mid S \in \mathfrak{K}_\sigma\}.$$

Since ordinal bounds for derivations are crucial in the following proofs we have to prove everything step by step. To facilitate proving (and reading) we collect some properties which are often used. By $\frac{\alpha}{\rho}\,\Delta$ we always understand in this section $\mathsf{NT}_{\omega_1} \overset{\alpha}{\underset{\rho}{\vdash}} \Delta$. In the TAIT-calculus a claim $F$ derived from a finite set $\Delta$ of hypotheses appears in the form $\overset{\alpha}{\underset{\rho}{\vdash}} \neg\Delta, F$, where $\neg\Delta := \{\neg G \mid G \in \Delta\}$.

First we prove a more or less obvious property of $\mathscr{L}_\kappa$-sentences.

**8.5.1 Lemma** *Let $F$ be a true $\mathscr{L}_\kappa$-sentence with* $\operatorname{rnk}(F) = \alpha$. *Then* $\overset{\alpha}{\underset{0}{\vdash}} \Delta, F$ *for any finite set of $\mathscr{L}_\kappa$-formulas $\Delta$.*

*Proof* We induct on $\operatorname{rnk}(F)$. If $\operatorname{rnk}(F) = 0$ then $F \in \bigwedge$–type and we obtain $\overset{0}{\underset{0}{\vdash}} \Delta, F$ by an inference $(\bigwedge)$ with empty premise.

If $\operatorname{rnk}(F) = \alpha > 0$ then we obtain $\operatorname{rnk}(G) < \alpha$ for all $G \in \operatorname{CS}(F)$. If $F \in \bigvee$–type there is some $G \in \operatorname{CS}(F)$ such that $\mathbb{N} \models G$. Hence $\overset{\alpha_G}{\underset{0}{\vdash}} \Delta, G$ for $a_G := \operatorname{rnk}(G) < \alpha$. By an inference $(\bigvee)$ we then obtain $\overset{\alpha}{\underset{0}{\vdash}} \Delta, F$.

If $F \in \bigwedge$–type we have $\mathbb{N} \models G$ for all $G \in \operatorname{CS}(F)$ and thus $\overset{\alpha_G}{\underset{0}{\vdash}} \Delta, G$ for all $G \in \operatorname{CS}(F)$ and $\alpha_G := \operatorname{rnk}(G) < \alpha$. By an inference $(\bigwedge)$ we then obtain $\overset{\alpha}{\underset{0}{\vdash}} \Delta, F$. □

**8.5.2 Lemma** *From* $\overset{\alpha}{\underset{\rho}{\vdash}} \Delta, F(t)$ *for* $2\cdot\operatorname{rnk}(F(t)) \le \alpha$ *and* $\operatorname{rnk}(F) < \rho$ *we obtain* $\overset{\alpha+1}{\underset{\rho}{\vdash}} \Delta, s \ne t, F(s)$.

*Proof* If $s^{\mathbb{N}} = t^{\mathbb{N}}$ we obtain

$$\overset{2\cdot\operatorname{rnk}(F(t))}{\underset{0}{\vdash}} \Delta, s \ne t, \neg F(t), F(s) \tag{i}$$

by the Tautology Lemma (Lemma 7.3.3). From (i) and the hypothesis $\overset{\alpha}{\underset{\rho}{\vdash}} \Delta, F(t)$ we obtain the claim by cut. If $s^{\mathbb{N}} \ne t^{\mathbb{N}}$ then $(t \ne s) \in \bigwedge$–type and we obtain $\overset{2\cdot\operatorname{rnk}(F(t))}{\underset{0}{\vdash}} \Delta, s \ne t, F(s)$ by an inference $(\bigwedge)$ with empty premises. □

**8.5.3 Lemma** *(Conjunction Lemma) Assume* $\overset{\alpha_0}{\underset{\rho_0}{\vdash}} \Delta, A$ *and* $\overset{\alpha_1}{\underset{\rho_1}{\vdash}} \Gamma, B$. *Then we obtain* $\overset{\alpha+1}{\underset{\rho}{\vdash}} \Delta, \Gamma, A \land B$ *for all* $\alpha \ge \max\{\alpha_0, \alpha_1\}$ *and all* $\rho \ge \max\{\rho_0, \rho_1\}$.

*Proof* We obtain $\overset{\alpha}{\underset{\rho}{\vdash}} \Delta, \Gamma, A, A \land B$ and $\overset{\alpha}{\underset{\rho}{\vdash}} \Delta, \Gamma, B, A \land B$ by the structural rule (Lemma 7.3.7). The claim follows now by an inference $(\bigwedge)$. □

**8.5.4 Lemma** *If* $\overset{\alpha}{\underset{\rho_0}{\vdash}} \Delta, F$ *and* $\overset{\beta}{\underset{\rho_1}{\vdash}} \Gamma, \neg F$ *then* $\overset{\gamma}{\underset{\rho+1}{\vdash}} \Delta, \Gamma$ *holds true for all ordinals* $\gamma > \max\{\alpha, \beta\}$ *and* $\rho \ge \max\{\rho_0, \rho_1, \operatorname{rnk}(F) + 1\}$.

*Proof* The proof is straight forward by the Structural Lemma and cut. □

Similar applications of the Structural Lemma will not longer be explicitely mentioned. Combining Lemmas 8.5.4 and 8.5.1, we obtain the Detachment Rule

**8.5.5 Lemma** *If* $\vdash^{\alpha}_{\rho} \Delta, F$ *and* $F$ *is a false sentence such that* $\mathrm{rnk}(F) < \min\{\alpha+1, \rho\}$. *Then* $\vdash^{\alpha+1}_{\rho} \Delta$.

For the rest of this section $\lambda$ will always denote a limit ordinal.

A rule which will often be used (mostly without mentioning it) is stated in the following lemma.

**8.5.6 Lemma** *(* $\bigvee$*-Importation) Assume* $\vdash^{\alpha}_{\rho} \Delta, F_1, \ldots, F_n$ *for* $n > 1$. *Then we obtain* $\vdash^{\alpha+n}_{\rho} \Delta, F_1 \vee \cdots \vee F_n$

*Proof* This is obvious by the structural rule and repeated applications of ($\bigvee$)-inferences. □

Lower case Greek letters in formulas are supposed to vary over ordinal notations. To improve the readability of the following proofs we agree upon the following convention. Whenever we write

$$\vdash^{\alpha}_{\rho} \Delta(\xi_1, \ldots, \xi_n)$$

we have to read it as

$$\vdash^{\alpha}_{\rho} \xi_1 \not\varepsilon OT, \ldots, \xi_n \not\varepsilon OT, \Delta(\xi_1, \ldots, \xi_n).$$

It would only be confusing to make all the "hypotheses $\xi_i \;\varepsilon\; OT$" explicit. According to this agreement it suffices to have $\vdash^{\alpha_\xi}_{\rho} \Delta, F(\xi)$ with $\alpha_\xi + 2 < \alpha$ for all ordinals $\xi$ to derive $\vdash^{\alpha}_{\rho} \Delta, (\forall \xi) F(\xi)$. This holds true because $\vdash^{\alpha_\xi}_{\rho} \Delta, F(\xi)$ stands for $\vdash^{\alpha_\xi}_{\rho} \Delta, \xi \not\varepsilon OT,\ F(\xi)$. If $\xi$ is not an ordinal notation we have $\vdash^{0}_{0} \Delta, \xi \not\varepsilon OT,\ F(\xi)$ by Lemma 8.5.1. So we have $\vdash^{\alpha_\xi}_{\rho} \Delta, \xi \not\varepsilon OT,\ F(\xi)$ for all $\xi$ which by $\bigvee$-Importation implies $\vdash^{\alpha_\xi+2}_{\rho} \Delta, \xi \not\varepsilon OT \vee F(\xi)$. By an inference ($\bigwedge$) we finally get $\vdash^{\alpha}_{\rho} \Delta, (\forall \xi) F(\xi)$.

Because of $\vdash^{0}_{0} \xi \not\varepsilon OT,\ \xi \;\varepsilon\; OT$ we can derive $\vdash^{\alpha+1}_{\rho} \Delta(\xi),\ \xi \;\varepsilon\; On \wedge F$ from the hypothesis $\vdash^{\alpha}_{\rho} \Delta(\xi), F$. We are going to use these facts tacitly.

Put

$$Jp(S) := \{\eta \mid \eta \;\varepsilon\; OT \wedge (\forall \xi)[\xi \subseteq S \rightarrow \xi + \eta \subseteq S]\}. \tag{8.5}$$

We call $Jp(S)$ the *jump* of the class $S$. Observe that $S \in \mathfrak{K}_\lambda$ entails also $Jp(S) \in \mathfrak{K}_\lambda$.

**8.5.7 Lemma** *For* $S \in \mathfrak{K}_\lambda$ *there are ordinals* $\alpha$ *and* $\rho$ *less than* $\lambda$ *such that* $\vdash^{\alpha}_{\rho} \eta \not\varepsilon Jp(S), \eta \subseteq S$.

*Proof* By Tautology (Lemma 7.3.3) we obtain

$$\vdash_0^{2\cdot\beta} \eta \not\varepsilon Jp(S), \eta \varepsilon Jp(S) \tag{i}$$

for $\beta = \mathrm{rnk}(Jp(S)) < \lambda$. Hence

$$\vdash_0^{2\cdot\beta} \eta \not\varepsilon Jp(S), \neg 0 \subseteq S, \eta \subseteq S \tag{ii}$$

by $\bigwedge$-Inversion and $\bigvee$-Exportation (Lemma 7.3.10). But

$$\vdash_0^3 \neg\xi \varepsilon OT \vee \neg\xi < 0 \vee \xi \varepsilon S$$

holds by Lemma 8.5.1 for all ordinals $\xi$. This implies

$$\vdash_0^4 0 \subseteq S \tag{iii}$$

by an $(\bigwedge)$-inference. From (ii) and (iii) the claim follows by a cut. □

**8.5.8 Lemma** *For $S \in \mathfrak{K}_\lambda$ there are ordinals $\alpha$ and $\rho$ less than $\lambda$ such that*

$$\vdash_\rho^\alpha \neg\mathsf{Prog}(Jp(S)),\ \neg(\eta \subseteq Jp(S)),\ \eta \subseteq S$$

*for all ordinals $\eta$.*

*Proof* We use Tautology to prove

$$\vdash_0^{\alpha_0} \neg\mathsf{Prog}(Jp(S)),\ \mathsf{Prog}(Jp(S)) \tag{i}$$

for $\alpha_0 = 2{\cdot}\mathrm{rnk}(\mathsf{Prog}(Jp(S))) < \lambda$. By $\bigwedge$-Inversion and $\bigvee$-Exportation we obtain from (i)

$$\vdash_0^{\alpha_0} \neg\mathsf{Prog}(Jp(S)),\ \neg\eta \subseteq Jp(S),\ \eta \varepsilon Jp(S). \tag{ii}$$

From (ii) and Lemma 8.5.7 we obtain by cut

$$\vdash_\rho^\alpha \neg\mathsf{Prog}(Jp(S)),\ \neg\eta \subseteq Jp(S),\ \eta \subseteq S$$

for $\alpha$ and $\rho$ less than $\lambda$. □

**8.5.9 Lemma** *For $S \in \mathfrak{K}_\lambda$ there is an ordinal $\alpha$ less than $\lambda$ such that*

$$\vdash_0^\alpha \eta \not\varepsilon Jp(S), \neg(\xi \subseteq S),\ \xi + \eta \varepsilon S$$

*for all ordinals $\xi$ and $\eta$.*

*Proof* Use Tautology to derive $\vdash_0^{\alpha_0} \eta \not\varepsilon Jp(S),\ \eta \varepsilon Jp(S)$ and then $\bigwedge$-Inversion and $\bigvee$-Exportation to obtain the claim. □

**8.5.10 Lemma** *For $S \in \mathfrak{K}_\lambda$ there are ordinals $\alpha$ and $\rho$ less than $\lambda$ such that*

$$\vdash_\rho^\alpha \neg\mathsf{Prog}(S),\ \mathsf{Prog}(Jp(S)).$$

*Proof* By Tautology, $\bigwedge$-Inversion and $\bigvee$-Exportation we obtain ordinals $\alpha_0$, $\alpha_1$ and $\alpha_2$ less than $\lambda$ such that

$$\vdash^{\alpha_0}_{0} \neg(\xi \subseteq S),\ \neg(\zeta < \xi),\ \zeta\, \varepsilon\, S, \tag{i}$$

$$\vdash^{\alpha_1}_{0} \neg(\eta \subseteq Jp(S)),\ \neg(\eta_0 < \eta),\ \eta_0\, \varepsilon\, Jp(S) \tag{ii}$$

and

$$\vdash^{\alpha_2}_{0} \eta_0 \not\varepsilon\, Jp(S),\ \neg(\xi \subseteq S),\ \xi + \eta_0 \subseteq S \tag{iii}$$

for all ordinals $\eta_0$. Cutting (iii) and (ii) we therefore obtain

$$\vdash^{\alpha_3}_{\rho_0} \neg(\eta \subseteq Jp(S)),\ \neg(\xi \subseteq S),\ \neg(\eta_0 < \eta),\ \xi + \eta_0 \subseteq S \tag{iv}$$

with $\alpha_3 < \lambda$ and $\rho_0 < \lambda$. Again by Tautology, $\bigwedge$-Inversion and $\bigvee$-Exportation we get an ordinal $\alpha_4 < \lambda$ such that

$$\vdash^{\alpha_4}_{0} \neg\mathsf{Prog}(S),\ \neg(\xi + \eta_0 \subseteq S),\ \xi + \eta_0\, \varepsilon\, S. \tag{v}$$

From (iv) and (v) we obtain cut

$$\vdash^{\alpha_5}_{\rho_1} \neg\mathsf{Prog}(S),\ \neg(\eta \subseteq Jp(S)),\ \neg(\xi \subseteq S),\ \neg(\eta_0 < \eta),\ \xi + \eta_0\, \varepsilon\, S \tag{vi}$$

for $\alpha_5$ and $\rho_1$ less than $\lambda$. From (vi) we obtain by Lemma 8.5.2

$$\vdash^{\alpha_5}_{\rho_1} \neg\mathsf{Prog}(S),\ \neg(\eta \subseteq Jp(S)),\ \neg(\xi \subseteq S),\ \neg(\eta_0 < \eta),\ \neg(\zeta = \xi + \eta_0),\ \zeta\, \varepsilon\, S \tag{vii}$$

and thus by inferences $(\bigvee)$ and $(\bigwedge)$

$$\vdash^{\alpha_6}_{\rho_1} \neg\mathsf{Prog}(S),\ \neg(\eta \subseteq Jp(S)),\ \neg(\xi \subseteq S),\ \neg(\exists\eta_0)[\eta_0 < \eta \wedge \zeta = \xi + \eta_0],\ \zeta\, \varepsilon\, S \tag{viii}$$

for $\alpha_6$ still less than $\lambda$. By Lemma 8.5.1 we obtain

$$\vdash^{n}_{0} \neg(\zeta < \xi + \eta),\ \zeta < \xi,\ (\exists\eta_0)[\eta_0 < \eta \wedge \zeta = \xi + \eta_0] \tag{ix}$$

for some $n < \omega \leq \lambda$. Cutting (viii) and (ix) yields

$$\vdash^{\alpha_7}_{\rho} \neg\mathsf{Prog}(S),\ \neg(\eta \subseteq Jp(S)),\ \neg(\xi \subseteq S),\ \neg(\zeta < \xi + \eta),\ \zeta < \xi,\ \zeta\, \varepsilon\, S \tag{x}$$

with $\alpha_7 < \lambda$ and $\rho < \lambda$ for all ordinals $\zeta$. By (i) and(x) we obtain with an inference $(\bigwedge)$

$$\vdash^{\alpha_8}_{\rho} \neg\mathsf{Prog}(S),\ \neg(\eta \subseteq Jp(S)),\ \neg(\xi \subseteq S),\ \neg(\zeta < \xi + \eta),\ \zeta < \xi \wedge \neg(\zeta < \xi),\ \zeta\, \varepsilon\, S \tag{xi}$$

which yields an ordinal $\alpha_9$ less than $\lambda$ such that

$$\vdash^{\alpha_9}_{\rho} \neg\mathsf{Prog}(S),\ \neg(\eta \subseteq Jp(S)),\ \neg(\xi \subseteq S),\ \neg(\zeta < \xi + \eta),\ \ \zeta\, \varepsilon\, S \tag{xii}$$

for all ordinals $\zeta$ by the Detachment Rule (Lemma 8.5.5). By $(\bigvee)$-Importation and an inference $(\bigwedge)$ we obtain

$$\vdash_{\rho}^{\alpha_{10}} \neg\mathsf{Prog}(S),\ \neg(\eta \subseteq Jp(S)),\ \neg(\xi \subseteq S),\ \xi+\eta \subseteq S \tag{xiii}$$

for all $\xi$ from (xii). Hence

$$\vdash_{\rho}^{\alpha_{11}} \neg\mathsf{Prog}(S),\ \neg(\eta \subseteq Jp(S)),\ \eta\ \varepsilon\ Jp(S) \tag{xiv}$$

by $(\bigvee)$-Importation and $(\bigwedge)$. Finally we obtain again by $(\bigvee)$-Importation and an inference $(\bigwedge)$ ordinals $\alpha < \lambda$ and $\rho$ less than $\lambda$ such that

$$\vdash_{\rho}^{\alpha} \neg\mathsf{Prog}(S),\ \mathsf{Prog}(Jp(S)).$$

□

**8.5.11 Lemma** *If $\lambda \leq \sigma$ there is for every $S \in \mathfrak{K}_\lambda$ an ordinal $\alpha < \lambda$ such that*

$$\vdash_{0}^{\alpha} \neg\mathsf{TI}_\sigma(\xi),\ \neg\mathsf{Prog}(S),\ \xi \subseteq S.$$

*Proof* Let $S \in \mathfrak{K}_\lambda$. By Tautology and $\bigvee$-Exportation we obtain

$$\vdash_{0}^{\alpha_0} \neg(\mathsf{Prog}(S) \to \xi \subseteq S),\ \neg\mathsf{Prog}(S),\ \xi \subseteq S \tag{i}$$

for $\alpha_0 = 2{\cdot}\mathrm{rnk}(\mathsf{Prog}(S) \to \xi \subseteq S) < \lambda$. Hence

$$\vdash_{0}^{\alpha_1} \neg(\forall\xi)[\mathsf{Prog}(S) \to \xi \subseteq S],\ \neg\mathsf{Prog}(S),\ \xi \subseteq S \tag{ii}$$

by an inference $(\bigvee)$. Since $(\forall\xi)[\mathsf{Prog}(S) \to \xi \subseteq S]$ is the formula $\mathsf{TI}(\xi,S)$ and $S \in \mathfrak{K}_\lambda \subseteq \mathfrak{K}_\sigma$ we finally conclude

$$\vdash_{0}^{\alpha} \neg\bigwedge\{\mathsf{TI}(\xi,S)\,|\,S \in \mathfrak{K}_\sigma\}, \neg\mathsf{Prog}(S), \xi \subseteq S$$

by an inference $(\bigvee)$. □

**8.5.12 Corollary** *If $\gamma \leq \sigma$ and $\gamma$ is a limit ordinal then $\vdash_{0}^{\gamma} \neg\mathsf{TI}_\sigma(\eta), \mathsf{TI}_\gamma(\eta)$ holds true for all $\eta$.*

*Proof* This is an immediate consequence of Lemma 8.5.11. □

**8.5.13 Lemma** *For all ordinals $\xi$ and $\eta$ we obtain*

$$\vdash_{\lambda}^{\lambda} \neg\mathsf{TI}_\lambda(\xi),\ \neg\mathsf{TI}_\lambda(\eta),\ \mathsf{TI}_\lambda(\xi+\eta).$$

*Proof* Let $S \in \mathfrak{K}_\lambda$. Then also $Jp(S) \in \mathfrak{K}_\lambda$ and by Lemma 8.5.11 we get ordinals $\alpha_0, \alpha_1 < \lambda$ such that

$$\vdash_{0}^{\alpha_0} \neg\mathsf{TI}_\lambda(\xi),\ \neg\mathsf{Prog}(S),\ \xi \subseteq S \tag{i}$$

and

$$\vdash_0^{\alpha_1} \neg \mathsf{TI}_\lambda(\eta), \neg \mathsf{Prog}(Jp(S)), \eta \subseteq Jp(S). \tag{ii}$$

By Tautology, $\bigwedge$-Inversion and $\bigvee$-Exportation we get

$$\vdash_0^{\alpha_2} \neg \mathsf{Prog}(Jp(S)), \neg \eta \subseteq Jp(S), \eta \,\varepsilon\, Jp(S) \tag{iii}$$

for some $\alpha_2 < \lambda$. But (ii) and (iii) imply

$$\vdash_\rho^{\alpha_3} \neg \mathsf{TI}_\lambda(\eta), \neg \mathsf{Prog}(Jp(S)), \eta \,\varepsilon\, Jp(S). \tag{iv}$$

From (iv) and Lemma 8.5.10 we obtain by cut

$$\vdash_\rho^{\alpha_5} \neg \mathsf{TI}_\lambda(\eta), \neg \mathsf{Prog}(S), \eta \,\varepsilon\, Jp(S) \tag{v}$$

for $\alpha_5 < \lambda$ and $\rho < \lambda$ sufficiently large. Statements (i) and (v) imply

$$\vdash_\rho^{\alpha_6} \neg \mathsf{TI}_\lambda(\xi), \neg \mathsf{TI}_\lambda(\eta), \neg \mathsf{Prog}(S), \xi \subseteq S \wedge \eta \,\varepsilon\, Jp(S) \tag{vi}$$

for some $\alpha_6 < \lambda$ by an inference $(\bigwedge)$. Lemma 8.5.9 together with (vi) yield

$$\vdash_\rho^{\alpha_7} \neg \mathsf{TI}_\lambda(\xi), \neg \mathsf{TI}_\lambda(\eta), \neg \mathsf{Prog}(S), \xi + \eta \subseteq S. \tag{vii}$$

Since (vii) holds for all $S \in \mathfrak{K}_\lambda$ we obtain

$$\vdash_\rho^{\alpha} \neg \mathsf{TI}_\lambda(\xi), \neg \mathsf{TI}_\lambda(\eta), \mathsf{TI}_\lambda(\xi + \eta)$$

for some $\alpha < \lambda$ by $\bigvee$-Importation and an inference $(\bigwedge)$. □

**8.5.14 Lemma** $\vdash_0^{\lambda} \neg \mathsf{TI}_\lambda(\eta), \neg(\xi < \eta), \mathsf{TI}_\lambda(\xi)$.

*Proof* By Lemma 8.5.11, we obtain for $S \in \mathfrak{K}_\lambda$ an $\alpha_1 < \lambda$ such that

$$\vdash_0^{\alpha_1} \neg \mathsf{TI}_\lambda(\eta), \neg \mathsf{Prog}(S), \eta \subseteq S \tag{i}$$

which by $\bigwedge$-Inversion and $\bigvee$-Exportation yields

$$\vdash_0^{\alpha_1} \neg \mathsf{TI}_\lambda(\eta), \neg \mathsf{Prog}(S), \neg(\zeta < \eta), \zeta \,\varepsilon\, S. \tag{ii}$$

By Lemma 8.5.1 and $\bigvee$-Exportation we have

$$\vdash_0^{n} \neg(\xi < \eta), \neg(\zeta < \xi), \zeta < \eta \tag{iii}$$

for some finite ordinal $n$. From (ii) and (iii) we get

$$\vdash_0^{\alpha_2} \neg \mathsf{TI}_\lambda(\eta), \neg \mathsf{Prog}(S), \neg(\xi < \eta), \neg(\zeta < \xi), \zeta \,\varepsilon\, S \tag{iv}$$

for $\alpha_2 = \alpha + n$ by the Reduction Lemma (Lemma 7.3.12). From (iv) we get by $\bigvee$-Importation and the $(\bigwedge)$-rule an $\alpha_3 < \lambda$ with

$$\vdash_0^{\alpha_3} \neg \mathsf{TI}_\lambda(\eta), \neg(\xi < \eta), \mathsf{Prog}(S) \rightarrow \xi \subseteq S.$$

Since this holds for all $S \in \mathfrak{K}_\lambda$ we obtain by an application of the $(\bigwedge)$-rule

$$\big|\frac{\lambda}{0}\ \neg \mathsf{TI}_{\lambda}(\eta), \neg(\xi < \eta), \mathsf{TI}_{\lambda}(\xi).$$ □

**8.5.15 Definition** For $\alpha =_{NF} \alpha_1 + \cdots + \alpha_n$ put $h(\alpha) := \alpha_1$.

**8.5.16 Theorem** *For a limit ordinal $\lambda$ we obtain $\big|\frac{\alpha}{\rho}\ \neg \mathsf{TI}_{\lambda}(h(\eta)),\ \mathsf{TI}_{\lambda}(\eta)$ for $\alpha$ and $\rho$ less than $\lambda^*$.*

*Proof* Let $\eta =_{NF} \eta_1 + \cdots + \eta_n$. Then $\eta_i \leq h(\eta)$ holds for $i = 1, \ldots, n$. By Lemma 8.5.14 and the Detachment Rule we obtain

$$\big|\frac{\lambda+1}{1}\ \neg \mathsf{TI}_{\lambda}(h(\eta)),\ \mathsf{TI}_{\lambda}(\eta_i) \tag{i}$$

for $i = 1, \ldots, n$. By Lemma 8.5.13 we therefore get

$$\big|\frac{\lambda+3}{\lambda+1}\ \neg \mathsf{TI}_{\lambda}(h(\eta)),\ \mathsf{TI}_{\lambda}(\eta_1 + \eta_2). \tag{ii}$$

Iterating this $n$-fold yields the claim. □

Having shown that $TI_{\lambda}$ is closed under ordinal addition the next aim is to study the closure behavior of $TI_{\lambda}$ under the VEBLEN function $\varphi_{\sigma}$. We introduce the VEBLEN-$\sigma$-jump as a class term

$$VJp_{\lambda}(\sigma) := \{\eta \mid \eta\ \varepsilon\ OT \wedge \mathsf{TI}_{\lambda}(\eta) \ \rightarrow\ \mathsf{TI}_{\lambda}(\varphi_{\sigma}(\eta))\}.$$

Observe that $VJp_{\lambda}(\sigma) \in \mathfrak{K}_{\lambda^*}$.

**8.5.17 Lemma** *There is finite ordinal $n$ such that $\big|\frac{n}{0}\ \mathsf{TI}_{\lambda}(0)$.*

*Proof* For $S \in \mathfrak{K}_{\lambda}$ we get $\big|\frac{0}{0}\ \neg \mathsf{Prog}(S),\ \neg(\xi < 0),\ \xi\ \varepsilon\ S$ for all $\xi$ by Lemma 8.5.1. Hence $\big|\frac{4}{0}\ \neg \mathsf{Prog}(S), 0 \subseteq S$ by $\bigvee$-Importation and the $(\bigwedge)$-rule. The claim now follows by another $\bigvee$-Importation and an application of an $(\bigwedge)$-inference. □

The next (quite technical) lemma provides the induction step in the proof of Lemma 8.5.19 which will be proved by induction on the complexity of ordinal notations. We therefore define the *norm* of an ordinal $\alpha < \Gamma_0$ by

$$N(\alpha) := \begin{cases} 0 & \text{if } \alpha = 0 \\ \sum_{i=1}^{n} (N(\alpha_i) + N(\beta_i) + 1) & \text{if } \alpha =_{NF} \varphi_{\alpha_1}(\beta_1) + \cdots + \varphi_{\alpha_n}(\beta_n). \end{cases} \tag{8.6}$$

**8.5.18 Lemma** *For every limit ordinal $\lambda$ there are ordinals $\rho$ and $\alpha$ less than $\lambda^*$ such that*

$$\big|\frac{\alpha}{\rho}\ \begin{array}{l} \neg(\forall \xi < \sigma)(\forall \eta)[\eta\ \varepsilon\ VJp_{\lambda}(\xi)],\ \neg(\forall \eta < \tau)[\mathsf{TI}_{\lambda}(\varphi_{\sigma}(\eta))], \\ \neg(\forall \nu)[N(\nu) < N(\mu) \wedge \nu < \varphi_{\sigma}(\tau) \rightarrow \mathsf{TI}_{\lambda}(\nu)],\ \neg \mathsf{TI}_{\lambda}(\tau), \\ \neg(\mu < \varphi_{\sigma}(\tau)),\ \mathsf{TI}_{\lambda}(\mu) \end{array}$$

*holds true for all ordinals $\mu$, $\sigma$ and $\tau$.*

*Proof* Let us abbreviate the premises of the lemma by

$$A_\lambda(\sigma) :\Leftrightarrow (\forall\xi < \sigma)(\forall\eta)[\eta \varepsilon VJp_\lambda(\xi)] \Leftrightarrow (\forall\xi <)\sigma(\forall\eta)[\mathsf{TI}_\lambda(\eta) \to \mathsf{TI}_\lambda(\varphi_\xi(\eta))]$$

$$B_\lambda(\sigma,\tau) \;:\Leftrightarrow\; (\forall\eta < \tau)[\mathsf{TI}_\lambda(\varphi_\sigma(\eta))]$$

and

$$C_\lambda(\mu,\sigma,\tau) \;:\Leftrightarrow\; (\forall\nu)[N(\nu) < N(\mu) \wedge \nu < \varphi_\sigma(\tau) \to \mathsf{TI}_\lambda(\nu)].$$

If $\varphi_\sigma(\tau) \leq \mu$ we obtain

$$\vdash^{0}_{0} \neg A_\lambda(\sigma),\ \neg B_\lambda(\sigma,\tau),\ \neg C_\lambda(\mu,\sigma,\tau),\ \neg\mathsf{TI}_\lambda(\tau), \neg(\mu < \varphi_\sigma(\tau)),\ \mathsf{TI}_\lambda(\mu) \quad \text{(i)}$$

by Lemma 8.5.1. Therefore assume $\mu < \varphi_\sigma(\tau)$. If $\mu = 0$ we obtain from Lemma 8.5.17

$$\vdash^{\omega}_{0} \neg A_\lambda(\sigma),\ \neg B_\lambda(\sigma,\tau),\ \neg C_\lambda(\mu,\sigma,\tau),\ \neg\mathsf{TI}_\lambda(\tau), \neg(\mu < \varphi_\sigma(\tau)),\ \mathsf{TI}_\lambda(\mu). \quad \text{(ii)}$$

So assume $\mu > 0$. According to Theorem 3.4.9 we distinguish the following cases.

1. It is $\mu = \varphi_{\mu_1}(\mu_2)$ such that $\mu_1 < \sigma$ and $\mu_2 < \varphi_\sigma(\tau)$. By Tautology, $\bigwedge$-Inversion and $\bigvee$-Exportation we have

$$\vdash^{\alpha_1}_{0} \neg A_\lambda(\sigma),\ \neg(\mu_1 < \sigma),\ \neg\mathsf{TI}_\lambda(\mu_2), \mathsf{TI}_\lambda(\varphi_{\mu_1}(\mu_2)) \quad \text{(iii)}$$

for $\alpha_1 = 2\cdot\mathrm{rnk}(A_\lambda(\sigma)) < \lambda^*$. By the Detachment Rule and Lemma 8.5.2 we derive form (iii)

$$\vdash^{\alpha_2}_{\rho} \neg A_\lambda(\sigma),\ \neg\mathsf{TI}_\lambda(\mu_2),\ \mathsf{TI}_\lambda(\mu) \quad \text{(iv)}$$

for $\rho = \mathrm{rnk}(\mathsf{TI}_\lambda(\mu)) + 1 < \lambda^*$. By Tautology, $\bigwedge$-Inversion and $\bigvee$-Exportation we also have

$$\vdash^{\alpha_3}_{0} \neg C_\lambda(\mu,\sigma,\tau),\ \neg(N(\mu_2) < N(\mu)),\ \neg(\mu_2 < \varphi_\sigma(\tau)),\ \mathsf{TI}_\lambda(\mu_2) \quad \text{(v)}$$

for $\alpha_3 = 2\cdot\mathrm{rnk}(C_\lambda(\mu,\sigma,\tau)) < \lambda^*$. By the Detachment Rule we derive from (v)

$$\vdash^{\alpha_4}_{\rho} \neg C_\lambda(\mu,\sigma,\tau),\ \mathsf{TI}_\lambda(\mu_2) \quad \text{(vi)}$$

for $\alpha_4 < \lambda^*$. Cutting (iv) and (vi) we conclude

$$\vdash^{\alpha_5}_{\rho} \neg A_\lambda(\sigma),\ \neg C_\lambda(\mu,\sigma,\tau),\ \mathsf{TI}_\lambda(\mu) \quad \text{(vii)}$$

for $\alpha_5$ less than $\lambda^*$.

2. It is $\mu = \varphi_{\mu_1}(\mu_2)$ such that $\mu_1 = \sigma$ and $\mu_2 < \tau$. By Tautology, $\bigwedge$-Inversion and $\bigvee$-Exportation we get

$$\vdash^{\alpha_6}_{0} \neg B_\lambda(\sigma,\tau),\ \neg(\mu_2 < \tau),\ \mathsf{TI}_\lambda(\varphi_\sigma(\mu_2)) \quad \text{(viii)}$$

for $\alpha_6 = 2\cdot\mathrm{rnk}(B_\lambda(\sigma,\tau)) < \lambda^*$. Using the Detachment Rule and Lemma 8.5.2 we infer from (viii)

$$\vdash^{\alpha_7}_{\rho} \neg B_\lambda(\sigma,\tau),\ \mathsf{TI}_\lambda(\mu) \quad \text{(ix)}$$

for $\rho = \text{rnk}(\mathsf{TI}_\lambda(\mu)) + 1 < \lambda^*$ and $\alpha_7 < \lambda^*$.

3. If $\mu < \tau$ we obtain by Lemma 8.5.14 and the Detachment Rule

$$\Big|\frac{\lambda+1}{1}\ \neg\mathsf{TI}_\lambda(\tau),\ \mathsf{TI}_\lambda(\mu). \tag{x}$$

4. It is $\mu =_{NF} \mu_1 + \cdots + \mu_n$ for $n > 1$. Then $h(\mu) = \mu_1 < \mu$ and $N(\mu_1) < N(\mu)$. By Tautology, $\bigwedge$-Inversion, $\bigvee$-Exportation and Lemma 8.5.2 we obtain

$$\Big|\frac{\alpha_8}{\rho}\ \neg C_\lambda(\mu,\sigma,\tau),\ \neg(N(h(\mu)) < N(\mu)),\ \neg(h(\mu) < \mu),\ \mathsf{TI}_\lambda(h(\mu)) \tag{xi}$$

for $\alpha_8 = 2{\cdot}\text{rnk}(C_\lambda(\mu,\sigma,\tau)) < \lambda^*$ and $\rho = \text{rnk}(\mathsf{TI}_\lambda(h(\mu))) + 1 < \lambda^*$. By the Detachment Rule it follows

$$\Big|\frac{\alpha_9}{\rho}\ \neg C_\lambda(\mu,\sigma,\tau),\ \mathsf{TI}_\lambda(h(\mu)) \tag{xii}$$

and by Theorem 8.5.16 we obtain

$$\Big|\frac{\alpha_{10}}{\rho}\ \neg C_\lambda(\mu,\sigma,\tau),\ \mathsf{TI}_\lambda(\mu) \tag{xiii}$$

for $\alpha_{10} < \lambda^*$.

Collecting (i), (ii), (vii), (x) and (xiii) yields the claim. □

We are now going to discard step by step the superfluous premises in Lemma 8.5.18.

**8.5.19 Lemma** *For every limit ordinal $\lambda$ there is an ordinal $\rho < \lambda^*$ such that for all ordinals $\mu$ there is an ordinal $\alpha_\mu < \lambda^*$ with*

$$\Big|\frac{\alpha_\mu}{\rho}\ \neg(\forall\xi < \sigma)(\forall\eta)[\eta\ \varepsilon\ VJp_\lambda(\xi)],\ \neg(\forall\eta < \tau)[\mathsf{TI}_\lambda(\varphi_\sigma(\eta))],\ \neg\mathsf{TI}_\lambda(\tau),\ \neg(\mu < \varphi_\sigma(\tau)),\ \mathsf{TI}_\lambda(\mu)$$

*for all ordinals $\sigma$ and $\tau$.*

*Proof* We prove the lemma by meta-induction on $N(\mu)$ and use the abbreviations of the proof of Lemma 8.5.18. If $N(\mu) = 0$ then $\mu = 0$ and we obtain the claim from Lemma 8.5.17. So assume $N(\mu) \neq 0$. Then

$$\Big|\frac{\alpha_{\mu_0}}{\rho_1}\ \neg A_\lambda(\sigma),\ \neg B_\lambda(\sigma,\tau),\ \neg\mathsf{TI}_\lambda(\tau),\ \neg(N(\mu_0) < N(\mu)),\ \neg(\mu_0 < \varphi_\sigma(\tau)),\ \mathsf{TI}_\lambda(\mu_0) \tag{i}$$

because either $N(\mu) \leq N(\mu_0)$ and (i) holds by Lemma 8.5.1 or $N(\mu_0) < N(\mu)$ and (i) holds by induction hypothesis. From (i) we get

$$\Big|\frac{\alpha_1}{\rho_1}\ \neg A_\lambda(\sigma),\ \neg B_\lambda(\sigma,\tau),\ \neg\mathsf{TI}_\lambda(\tau),\ C_\lambda(\mu,\sigma,\tau) \tag{ii}$$

by $\bigvee$-Importation and an $(\bigwedge)$-inference for $\alpha_1 = \sup\{\alpha_{\mu_0} + 6 \mid N(\mu_0) < N(\mu)\}$. Since there are only finitely many such $\mu_0$ we have $\alpha_1 < \lambda^*$. From (ii) and Lemma 8.5.18 we obtain by cut

$$\Big|\frac{\alpha_\mu}{\rho}\ \neg A_\lambda(\sigma),\ \neg B_\lambda(\sigma,\tau),\ \neg\mathsf{TI}_\lambda(\tau),\ \neg(\mu < \varphi_\sigma(\tau)),\ \mathsf{TI}_\lambda(\mu)$$

for $\alpha_\mu = \max\{\alpha,\alpha_1\} + 1 < \lambda^*$ and $\rho = \max\{\rho_1,\rho_2,\text{rnk}(C_\lambda(\mu,\sigma,\tau)) + 1\} < \lambda^*$ where $\alpha$ and $\rho_2$ are the ordinals stemming from Lemma 8.5.18. □

**8.5.20 Lemma** *There are ordinals $\alpha$ and $\rho$ less than $\lambda^*$ such that*

$$\vdash^{\alpha}_{\rho} \neg(\tau \subseteq VJp_\lambda(\sigma)), \neg\mathsf{TI}_\lambda(\tau), (\forall\eta < \tau)\mathsf{TI}_\lambda(\varphi_\sigma(\eta))$$

*for all ordinals $\sigma$ and $\tau$.*

*Proof* By Tautology and $\bigvee$-Exportation we obtain

$$\vdash^{\alpha_0}_{0} \eta \not\varepsilon VJp_\lambda(\sigma), \neg\mathsf{TI}_\lambda(\eta), \mathsf{TI}_\lambda(\varphi_\sigma(\eta)) \tag{i}$$

for $\alpha_0 = 2{\cdot}\mathrm{rnk}(VJp_\lambda(\sigma)) < \lambda^*$. By Lemma 8.5.14 we have

$$\vdash^{\lambda}_{0} \neg\mathsf{TI}_\lambda(\tau), \neg(\eta < \tau), \mathsf{TI}_\lambda(\eta). \tag{ii}$$

Cutting (i) and (ii) yields

$$\vdash^{\alpha_1}_{\rho} \eta \not\varepsilon VJp_\lambda(\sigma), \neg(\eta < \tau), \neg\mathsf{TI}_\lambda(\tau), \mathsf{TI}_\lambda(\varphi_\sigma(\eta)) \tag{iii}$$

with $\alpha_1$ and $\rho$ less than $\lambda^*$. By Lemma 8.5.1 we have

$$\vdash^{0}_{0} \eta < \tau, \neg\mathsf{TI}_\lambda(\tau), \neg(\eta < \tau), \mathsf{TI}_\lambda(\varphi_\sigma(\eta)) \tag{iv}$$

and obtain from (iii) and (iv) by an inference $(\bigwedge)$

$$\vdash^{\alpha_1+1}_{\rho} \eta < \tau \wedge \eta \not\varepsilon VJp_\lambda(\sigma), \neg\mathsf{TI}_\lambda(\tau), \neg(\eta < \tau), \mathsf{TI}_\lambda(\varphi_\sigma(\eta)). \tag{v}$$

From (v) we obtain by an inference $(\bigvee)$

$$\vdash^{\alpha_1+4}_{\rho} \neg(\tau \subseteq VJp_\lambda(\sigma)), \neg\mathsf{TI}_\lambda(\tau), \neg(\eta < \tau), \mathsf{TI}_\lambda(\varphi_\sigma(\eta)) \tag{vi}$$

and from (vi) by $\bigvee$-Importation and an inference $(\bigwedge)$

$$\vdash^{\alpha}_{\rho} \neg(\tau \subseteq VJp_\lambda(\sigma)), \neg\mathsf{TI}_\lambda(\tau), (\forall\eta < \tau)\mathsf{TI}_\lambda(\varphi_\sigma(\eta))$$

for ordinals $\alpha$ and $\rho$ less than $\lambda^*$. □

**8.5.21 Lemma** $\vdash^{\lambda}_{\lambda} \neg(\forall\mu < \sigma)\mathsf{TI}_\lambda(\mu), \mathsf{TI}_\lambda(\sigma)$.

*Proof* For every class term $S \in \mathfrak{K}_\lambda$ there is by Lemma 8.5.11 an ordinal $\alpha_0 < \lambda$ such that

$$\vdash^{\alpha_0}_{0} \neg\mathsf{TI}_\lambda(\mu), \neg\mathsf{Prog}(S), \mu \subseteq S \tag{i}$$

and an ordinal $\alpha_1 < \lambda$ such that

$$\vdash^{\alpha_1}_{0} \neg\mathsf{Prog}(S), \neg(\mu \subseteq S), \mu \,\varepsilon\, S. \tag{ii}$$

Since we also have

$$\vdash^{0}_{0} \neg(\mu < \sigma), \mu < \sigma \tag{iii}$$

by Lemma 8.5.1, we obtain from (i), (ii) and (iii) by cut and an inference $(\bigwedge)$

$$\Big|\frac{\alpha_2}{\lambda}\ \mu < \sigma \land \neg\mathsf{TI}_\lambda(\mu),\ \neg\mathsf{Prog}(S),\ \neg(\mu < \sigma),\ \mu\ \varepsilon\ S \tag{iv}$$

with $\alpha_2 < \lambda$. From (iv) we get

$$\Big|\frac{\alpha_2+3}{\lambda}\ \neg(\forall\mu < \sigma)\mathsf{TI}_\lambda(\mu),\ \neg\mathsf{Prog}(S),\ \neg(\mu < \sigma),\ \mu\ \varepsilon\ S \tag{v}$$

by an inference $(\bigvee)$ and from (v) by $\bigvee$-Importation and an inference $(\bigwedge)$

$$\Big|\frac{\alpha_3}{\lambda}\ \neg(\forall\mu < \sigma)\mathsf{TI}_\lambda(\mu),\ \neg\mathsf{Prog}(S),\ \sigma \subseteq S. \tag{vi}$$

So we obtain $\alpha_4 < \lambda$ such that

$$\Big|\frac{\alpha_4}{\lambda}\ \neg(\forall\mu < \sigma)\mathsf{TI}_\lambda(\mu),\ \mathsf{Prog}(S) \to \sigma \subseteq S \tag{vii}$$

for all $S \in \mathfrak{K}_\lambda$ from (vi) by $\bigvee$-Importation. The claim follows from (vii) by an inference $(\bigwedge)$. □

**8.5.22 Lemma** *There are ordinals $\alpha < \lambda^{**}$ and $\rho < \lambda^*$ such that*

$$\Big|\frac{\alpha}{\rho}\ \neg(\forall\xi < \sigma)(\forall\eta)[\eta\ \varepsilon\ VJp_\lambda(\xi)],\ \neg\mathsf{TI}_{\lambda^*}(\tau),\ \mathsf{TI}_\lambda(\varphi_\sigma(\tau))$$

*for all ordinals $\sigma$ and $\tau$.*

*Proof* By Lemma 8.5.19, there is an ordinal $\rho_1 < \lambda^*$ and for all $\mu$ an ordinal $\alpha_\mu$ less than $\lambda^*$ such that

$$\Big|\frac{\alpha_\mu}{\rho_1}\ \neg(\forall\xi < \sigma)(\forall\eta)[\eta\ \varepsilon\ VJp_\lambda(\xi)],\ \neg(\forall\eta < \tau)\mathsf{TI}_\lambda(\varphi_\sigma(\eta)),\ \neg\mathsf{TI}_\lambda(\tau),\ \neg(\mu < \varphi_\sigma(\tau)),\ \mathsf{TI}_\lambda(\mu). \tag{i}$$

From (i) we obtain by $\bigvee$-Importation and an $(\bigwedge)$-inference

$$\Big|\frac{\lambda^*}{\rho_1}\ \neg(\forall\xi < \sigma)(\forall\eta)[\eta\ \varepsilon\ VJp_\lambda(\xi)],\ \neg(\forall\eta < \tau)\mathsf{TI}_\lambda(\varphi_\sigma(\eta)),\ \neg\mathsf{TI}_\lambda(\tau),\ (\forall\mu < \varphi_\sigma(\tau))\mathsf{TI}_\lambda(\mu). \tag{ii}$$

By (ii) and Lemma 8.5.21 we obtain

$$\Big|\frac{\lambda^*+1}{\rho_1}\ \neg(\forall\xi < \sigma)(\forall\eta)[\eta\ \varepsilon\ VJp_\lambda(\xi)],\ \neg(\forall\eta < \tau)\mathsf{TI}_\lambda(\varphi_\sigma\eta),\ \neg\mathsf{TI}_\lambda(\tau),\ \mathsf{TI}_\lambda(\varphi_\sigma(\tau).) \tag{iii}$$

By Lemma 8.5.20, we have

$$\Big|\frac{\alpha_1}{\rho_1}\ \neg(\tau \subseteq VJp_\lambda(\sigma)),\ \neg\mathsf{TI}_\lambda(\tau),\ (\forall\eta < \tau)\mathsf{TI}_\lambda(\varphi_\sigma(\eta)) \tag{iv}$$

for $\alpha_1 < \lambda^*$. Cutting (iv) and (iii) yields

$$\Big|\frac{\alpha_2}{\rho}\ \neg(\forall\xi < \sigma)(\forall\eta)[\eta\ \varepsilon\ VJp_\lambda(\xi)],\ \neg(\tau \subseteq VJp_\lambda(\sigma)),\ \neg\mathsf{TI}_\lambda(\tau),\ \mathsf{TI}_\lambda(\varphi_\sigma(\tau)) \tag{v}$$

with $\alpha_2 < \lambda^{**}$ and $\rho < \lambda^*$ large enough to majorize also all cuts to come. From (v) we obtain by $\bigvee$-Importation

$$\Big|\frac{\alpha_3}{\rho_1}\ \neg(\forall\xi < \sigma)(\forall\eta)[\eta\ \varepsilon\ VJp_\lambda(\xi)],\ \neg(\tau \subseteq VJp_\lambda(\sigma)),\ \tau\ \varepsilon\ VJp_\lambda(\sigma) \tag{vi}$$

for all ordinals $\tau$. Again by $\bigvee$-Importation and an $(\bigwedge)$-inference we obtain from (vi)

$$\mathrel{\substack{\alpha_4\\ \vdash\\ \rho}} \neg(\forall\xi < \sigma)(\forall\eta)[\eta\ \varepsilon\ VJp_\lambda(\xi)],\ \mathsf{Prog}(VJp_\lambda(\sigma)) \tag{vii}$$

with $\alpha_4$ still less than $\lambda^{**}$. By Tautology, we have

$$\mathrel{\substack{\beta\\ \vdash\\ 0}} \neg(\tau \subseteq VJp_\lambda(\sigma)),\ \tau \subseteq VJp_\lambda(\sigma) \tag{viii}$$

for $\beta = 2{\cdot}\mathrm{rnk}(\tau \subseteq VJp_\lambda(\sigma)) < \lambda^{**}$ and by the Conjunction Lemma (Lemma 8.5.3) we obtain from (viii) and (vii)

$$\mathrel{\substack{\alpha_5\\ \vdash\\ \rho}} \neg(\forall\xi < \sigma)(\forall\eta)[\eta\ \varepsilon\ VJp_\lambda(\xi)],\ \mathsf{Prog}(VJp_\lambda(\sigma)) \land \neg(\tau \subseteq VJp_\lambda(\sigma)),\ \tau \subseteq VJp_\lambda(\sigma),$$

i.e.,

$$\mathrel{\substack{\alpha_5\\ \vdash\\ \rho}} \neg(\forall\xi < \sigma)(\forall\eta)[\eta\ \varepsilon\ VJp_\lambda(\xi)],\ \neg(\mathsf{Prog}(VJp_\lambda(\sigma)) \to (\tau \subseteq VJp_\lambda(\sigma))),\ \tau \subseteq VJp_\lambda(\sigma). \tag{ix}$$

Since $VJp_\lambda(\sigma) \in \mathfrak{K}_{\lambda^*}$ we obtain from (ix) by an inference $(\bigvee)$

$$\mathrel{\substack{\alpha_5+1\\ \vdash\\ \rho}} \neg(\forall\xi < \sigma)(\forall\eta)[\eta\ \varepsilon\ VJp_\lambda(\xi)],\ \neg\mathsf{TI}_{\lambda^*}(\tau),\ \tau \subseteq VJp_\lambda(\sigma). \tag{x}$$

By Tautology, $\bigwedge$-Inversion and $\bigvee$-Exportation we get

$$\mathrel{\substack{\beta_1\\ \vdash\\ 0}} \neg\mathsf{Prog}(VJp_\lambda(\sigma)),\ \neg(\tau \subseteq VJp_\lambda(\sigma)),\ \tau\ \varepsilon\ VJp_\lambda(\sigma) \tag{xi}$$

for $\beta_1 = 2{\cdot}\mathrm{rnk}(\mathsf{Prog}(VJp_\lambda(\sigma))) < \lambda^{**}$. From (vii) and (xi) we get by cut

$$\mathrel{\substack{\alpha_6\\ \vdash\\ \rho}} \neg(\forall\xi < \sigma)(\forall\eta)[\eta \in VJp_\lambda(\xi)],\ \neg(\tau \subseteq VJp_\lambda(\sigma)),\ \tau\ \varepsilon\ VJp_\lambda(\sigma) \tag{xii}$$

and from (x) and (xii) by cut

$$\mathrel{\substack{\alpha_7\\ \vdash\\ \rho}} \neg(\forall\xi < \sigma)(\forall\eta)[\eta\ \varepsilon\ VJp_\lambda(\xi)],\ \neg\mathsf{TI}_{\lambda^*}(\tau),\ \tau\ \varepsilon\ VJp_\lambda(\sigma). \tag{xiii}$$

By $\bigvee$-Exportation we get from (xiii)

$$\mathrel{\substack{\alpha_7\\ \vdash\\ \rho}} \neg(\forall\xi < \sigma)(\forall\eta)[\eta\ \varepsilon\ VJp_\lambda(\xi)],\ \neg\mathsf{TI}_{\lambda^*}(\tau),\ \neg\mathsf{TI}_\lambda(\tau),\ \mathsf{TI}_\lambda(\varphi_\sigma(\tau)). \tag{xiv}$$

From Corollary 8.5.12 and (xiv) we obtain by cut

$$\mathrel{\substack{\alpha\\ \vdash\\ \rho}} \neg(\forall\xi < \sigma)(\forall\eta)[\eta\ \varepsilon\ VJp_\lambda(\xi)], \neg\mathsf{TI}_{\lambda^*}(\tau),\ \mathsf{TI}_\lambda(\varphi_\sigma(\tau))$$

for $\alpha < \lambda^{**}$. Checking all the cuts we see that we can keep $\rho$ below $\lambda^*$. □

Observe that Lemma 8.5.22 is the first which uses essentially the infinitary language. Passing from $\mathsf{TI}_\lambda(\tau)$ to $\mathsf{TI}_\lambda(\varphi_\sigma(\tau))$ is the essential step in the well-ordering proof. To render this step possible we have, however, still to get rid of the hypothesis $(\forall\xi < \sigma)(\forall\eta)[\eta\ \varepsilon\ VJp_\lambda(\xi)]$ and to equalize the asymmetry between $\lambda^*$ and $\lambda$. This is prepared by the following lemmas.

**8.5.23 Lemma** *Let $\lambda$ be a limit ordinal and $\{\mu_\sigma \mid \sigma < \eta\}$ a set which is unbounded in $\lambda$. Then $\mathrel{\substack{\alpha_\sigma\\ \vdash\\ \rho_\sigma}} \Delta, \mathsf{TI}_{\mu_\sigma}(\tau)$ with $\alpha_\sigma < \alpha$ and $\rho_\sigma \leq \rho$ for all $\sigma < \eta$ implies $\mathrel{\substack{\alpha\\ \vdash\\ \rho}} \Delta, \mathsf{TI}_\lambda(\tau)$.*

*Proof* Since $\{\mu_\sigma \mid \sigma < \eta\}$ is unbounded in $\lambda$ we have for every class term $S \in \mathfrak{K}_\lambda$ a $\mu_\sigma$ with $\sigma < \eta$ such that $S \in \mathfrak{K}_{\mu_\sigma}$. From the hypothesis $\vdash^{\alpha_\sigma}_{\rho_\sigma} \Delta, \mathsf{TI}_{\mu_\sigma}(\tau)$ we obtain $\vdash^{\alpha_\sigma}_{\rho_\sigma} \Delta, \mathsf{Prog}(S) \to \tau \subseteq S$ by $\bigwedge$-Inversion. Since $\alpha_\sigma < \alpha$ and $\rho_\sigma \le \rho$ we obtain $\vdash^{\alpha}_{\rho} \Delta, \mathsf{TI}_\lambda(\tau)$ by an inference $(\bigwedge)$. □

**8.5.24 Lemma** *For a limit ordinal $\eta$ we have*

$$\vdash^{\omega\cdot\eta}_{\omega\cdot\eta} \neg\mathsf{TI}_{\omega\cdot\eta}(\tau),\ \mathsf{TI}_{\omega\cdot\eta}(\varphi_0(\tau))$$

*for all ordinals $\tau$.*

*Proof* By Lemma 8.5.1 we have

$$\vdash^{\alpha_0}_{0} (\forall\xi < 0)(\forall\eta)[\eta\ \varepsilon\ VJp_\lambda(\xi)] \tag{i}$$

for $\alpha_0 < \lambda^{**}$. For limit ordinals $\lambda$ we therefore obtain from (i) and Lemma 8.5.22 by cut

$$\vdash^{\alpha_1}_{\rho} \neg\mathsf{TI}_{\lambda^*}(\tau),\ \mathsf{TI}_\lambda(\varphi_0(\tau)) \tag{ii}$$

for $\alpha_1 < \lambda^{**}$ and $\rho < \lambda^*$. By Corollary 8.5.12 we have

$$\vdash^{\lambda^*}_{\lambda^*} \neg\mathsf{TI}_{\omega\cdot\eta}(\tau),\ \mathsf{TI}_{\lambda^*}(\tau) \tag{iii}$$

for all limit ordinals $\lambda^* \le \omega\cdot\eta$. For $\lambda = \omega\cdot\xi < \omega\cdot\eta$ we obtain from (ii) and (iii)

$$\vdash^{\alpha_\xi}_{\omega\cdot(\xi+1)} \neg\mathsf{TI}_{\omega\cdot\eta}(\tau),\ \mathsf{TI}_{\omega\cdot\xi}(\varphi_0(\tau)) \tag{iv}$$

with $\alpha_\xi < \omega\cdot(\xi+2) < \omega\cdot\eta$, since $\eta \in Lim$. The set $\{\omega\cdot\xi \mid \xi < \eta\}$ is unbounded in $\omega\cdot\eta$ and we obtain by Lemma 8.5.23 and (iv)

$$\vdash^{\omega\cdot\eta}_{\omega\cdot\eta} \neg\mathsf{TI}_{\omega\cdot\eta}(\tau),\ \mathsf{TI}_{\omega\cdot\eta}(\varphi_0(\tau)).$$ □

**8.5.25 Lemma** *(Euklidian division for ordinals) For ordinals $\sigma$ and $\tau \neq 0$ there are ordinals $\rho < \tau$ and $\eta$ such that $\sigma = \tau\cdot\eta + \rho$.*

*Proof* Let $\eta_0 := \min\{\xi \mid \sigma < \tau\cdot\xi\}$. Then $\eta_0 \neq 0$. By minimality $\eta_0$ cannot be a limit ordinal. Hence $\eta_0 = \eta + 1$ and we obtain $\tau\cdot\eta \le \sigma$. But then there is a $\rho$ such that $\tau\cdot\eta + \rho = \sigma < \tau\cdot\eta + \tau$ which entails $\rho < \tau$. □

**8.5.26 Lemma** *Let $\nu$ and $\eta \neq 0$ be ordinals. Then*

$$\vdash^{\omega^{1+\nu+1}\cdot\eta}_{\omega^{1+\nu+1}\cdot\eta} \neg\mathsf{TI}_{\omega^{1+\nu+1}\cdot\eta}(\tau),\ \mathsf{TI}_{\omega^{1+\nu+1}\cdot\eta}(\varphi_\nu(\tau))$$

*for all ordinals $\tau$.*

*Proof* We induct on $\nu$.

Since $\omega{\cdot}\eta \in Lim$ we get for $\nu = 0$ by Lemma 8.5.24

$$\Big|\frac{\omega^2\cdot\eta}{\omega^2\cdot\eta} \neg\mathsf{TI}_{\omega^2\cdot\eta}(\tau),\ \mathsf{TI}_{\omega^2\cdot\eta}(\varphi_\nu(\tau)).$$

So assume $\nu \neq 0$. For $\mu < \nu$ we obtain a $\delta$ such that $\nu = \mu + 1 + \delta$. For $\xi \neq 0$ we thus have $\omega^{1+\nu}{\cdot}\xi = \omega^{1+\mu+1}{\cdot}\omega^\delta{\cdot}\xi =: \omega^{1+\mu+1}{\cdot}\zeta$ for some $\zeta \neq 0$. Therefore we obtain by the induction hypothesis

$$\Big|\frac{\omega^{1+\nu}\cdot\xi}{\omega^{1+\nu}\cdot\xi} \neg\mathsf{TI}_{\omega^{1+\nu}\cdot\xi}(\tau),\ \mathsf{TI}_{\omega^{1+\nu}\cdot\xi}(\varphi_\mu(\tau)) \tag{i}$$

which by two inferences $(\bigvee)$ entails

$$\Big|\frac{\omega^{1+\nu}\cdot\xi+2}{\omega^{1+\nu}\cdot\xi} \tau\,\varepsilon\, VJp_{\omega^{1+\nu}\cdot\xi}(\mu) \tag{ii}$$

for all ordinals $\tau$. By an inference $(\bigwedge)$ this implies

$$\Big|\frac{\omega^{1+\nu}\cdot\xi+5}{\omega^{1+\nu}\cdot\xi} (\forall\tau)[\tau\,\varepsilon\, VJp_{\omega^{1+\nu}\cdot\xi}(\mu)] \tag{iii}$$

for all $\mu < \nu$. By an inference $(\bigvee)$ together with Lemma 8.5.1 (in case that $\nu \leq \mu$) we then obtain

$$\Big|\frac{\omega^{1+\nu}\cdot\xi+6}{\omega^{1+\nu}\cdot\xi} \neg(\mu < \nu) \vee (\forall\tau)[\tau\,\varepsilon\, VJp_{\omega^{1+\nu}\cdot\xi}(\mu)] \tag{iv}$$

for all ordinals $\mu$. From (iv) we get

$$\Big|\frac{\omega^{1+\nu}\cdot\xi+9}{\omega^{1+\nu}\cdot\xi} (\forall\mu < \nu)(\forall\tau)[\tau\,\varepsilon\, VJp_{\omega^{1+\nu}\cdot\xi}(\mu)] \tag{v}$$

with an inference $(\bigwedge)$. By Lemma 8.5.22 there is an ordinal $\alpha < \omega^{1+\nu}{\cdot}\xi + \omega{\cdot}2$ and an ordinal $\rho < \omega^{1+\nu}{\cdot}\xi + \omega$ such that

$$\Big|\frac{\alpha}{\rho} \neg(\forall\mu < \nu)(\forall\tau)[\tau\,\varepsilon\, VJp_{\omega^{1+\nu}\cdot\xi}(\mu)],\ \neg\mathsf{TI}_{\omega^{1+\nu}\cdot\xi+\omega}(\tau),\ \mathsf{TI}_{\omega^{1+\nu}\cdot\xi}(\varphi_\nu(\tau)) \tag{vi}$$

for all ordinals $\tau$. From (vi) and (v) we obtain by cut

$$\Big|\frac{\alpha_1}{\omega^{1+\nu}\cdot\xi+\omega} \neg\mathsf{TI}_{\omega^{1+\nu}\cdot\xi+\omega}(\tau),\ \mathsf{TI}_{\omega^{1+\nu}\cdot\xi}(\varphi_\nu(\tau)) \tag{vii}$$

for an ordinal $\alpha_1 < \omega^{1+\nu}{\cdot}\xi + \omega{\cdot}2$. To apply Lemma 8.5.23 we show that the set

$$M := \{\zeta \,|\, (\exists\xi)[\xi \neq 0 \wedge \zeta = \omega^{1+\nu}{\cdot}\xi \wedge (\forall n < \omega)[\omega^{1+\nu}{\cdot}\xi + \omega{\cdot}n < \omega^{1+\nu+1}{\cdot}\eta]]\}$$

is unbounded in $\omega^{1+\nu+1}{\cdot}\eta$. Let $\sigma < \omega^{1+\nu+1}{\cdot}\eta$. We have to find an ordinal $\xi \neq 0$ such that $\sigma < \omega^{1+\nu}{\cdot}\xi$ and $\omega^{1+\nu}{\cdot}\xi + \omega{\cdot}n < \omega^{1+\nu+1}{\cdot}\eta$ for all $n \in \omega$. By Euklidian division there are ordinals $\rho < \omega^{1+\nu}$ and $\xi_0$ such that $\sigma = \omega^{1+\nu}{\cdot}\xi_0 + \rho < \omega^{1+\nu}{\cdot}(\xi_0 + 1)$. Since $\omega^{1+\nu}{\cdot}\xi_0 \leq \sigma < \omega^{1+\nu+1}{\cdot}\eta$ we obtain $\xi_0 < \omega{\cdot}\eta$ and therefore also $\xi_0 + n < \omega{\cdot}\eta$ for all $n < \omega$. Hence $\omega^{1+\nu}{\cdot}(\xi_0 + n) < \omega^{1+\nu}{\cdot}\omega{\cdot}\eta = \omega^{1+\nu+1}{\cdot}\eta$. Choosing $\xi := \xi_0 + 1$ we get $\sigma < \omega^{1+\nu}{\cdot}\xi$ and $\omega^{1+\nu}{\cdot}\xi + \omega{\cdot}n \leq \omega^{1+\nu}{\cdot}\xi_0 + \omega^{1+\nu} + \omega{\cdot}n \leq \omega^{1+\nu}{\cdot}(\xi_0 + n + 1) < \omega^{1+\nu+1}{\cdot}\eta$.

Now let $\zeta \in M$. By (vii) and Corollary 8.5.12 we obtain

$$\Big|\frac{\zeta+\omega\cdot 2}{\omega^{1+\nu+1}\cdot\eta} \neg \mathsf{TI}_{\omega^{1+\nu+1}\cdot\eta}(\tau),\ \mathsf{TI}_{\zeta}(\varphi_\nu(\tau)). \tag{viii}$$

Since $\zeta + \omega\cdot 2 < \omega^{1+\nu+1}\cdot\eta$ for all $\zeta \in M$ and $M$ is unbounded in $\omega^{1+\nu+1}\cdot\eta$ we obtain

$$\Big|\frac{\omega^{1+\nu+1}\cdot\eta}{w^{1+\nu+1}\cdot\eta} \neg \mathsf{TI}_{\omega^{1+\nu+1}\cdot\eta}(\tau),\ \mathsf{TI}_{\omega^{1+\nu+1}\cdot\eta}(\varphi_\nu(\tau)) \tag{ix}$$

for all ordinals $\tau$ by Lemma 8.5.23. □

The aim of this section is to prove (8.4) in p. 138. Therefore put

$$\zeta_0 := \varphi_1(0) = \varepsilon_0 \text{ and } \zeta_{n+1} := \varphi_{\zeta_n}(0).$$

We show

$$\Gamma_0 = \sup\{\zeta_n \,|\, n \in \omega\}. \tag{8.7}$$

First we prove $\zeta_n < \Gamma_0$ for all $n$ by induction on $n$. Since $\Gamma_0 \in \mathit{Lim}$ we get by Corollary 3.4.10 and Lemma 3.4.12 (2) $\zeta_0 = \varphi_1(0) < \varphi_{\Gamma_0}(0) = \Gamma_0$. In the induction step we get from the induction hypothesis $\zeta_n < \Gamma_0$ by Corollary 3.4.10 and Lemma 3.4.12 (2) $\zeta_{n+1} = \varphi_{\zeta_n}(0) < \varphi_{\Gamma_0}(0) = \Gamma_0$. Next we show that $\{\zeta_n \,|\, n \in \omega\}$ is unbounded in $\Gamma_0$. For $\sigma < \Gamma_0$ we prove by induction on $N(\sigma)$ that there is an $n < \omega$ such that $\sigma < \zeta_n$. This is obvious for $\sigma = 0$. If $\sigma =_{NF} \sigma_1 + \cdots + \sigma_m$ then there is by induction hypothesis an $n$ such that $\sigma_1 < \zeta_n$. Since $\zeta_n \in \mathbb{H}$ we obtain $\sigma < \zeta_n$. If $\sigma =_{NF} \varphi_{\sigma_1}(\sigma_2)$ then there is by induction hypothesis an $n$ such that $\sigma_i < \zeta_n$ for $i = 1, 2$. But then $\sigma = \varphi_{\sigma_1}(\sigma_2) < \varphi_{\zeta_n}(\zeta_n) < \varphi_{\zeta_{n+1}}(0) = \zeta_{n+2}$. This finishes the proof of (8.7).

**8.5.27 Lemma** $\Big|\frac{\zeta_n\cdot\omega+1}{\zeta_n\cdot\omega+1} \mathsf{TI}_{\zeta_n\cdot\omega+1}(\zeta_{n+1})$.

*Proof* Since $\omega^{1+\zeta_n+1} = \zeta_n\cdot\omega$ for all $n \in \omega$ we obtain from Lemma 8.5.26

$$\Big|\frac{\zeta_n\cdot\omega}{\zeta_n\cdot\omega} \neg \mathsf{TI}_{\zeta_n\cdot\omega}(0),\ \mathsf{TI}_{\zeta_n\cdot\omega}(\varphi_{\zeta_n}(0)).$$

Cutting that with Lemma 8.5.17 yields the claim. □

**8.5.28 Lemma** $\Big|\frac{\omega^2+\omega}{\omega^2+\omega} \neg \mathsf{TI}_{\omega^2}(\zeta_n),\ \mathsf{TI}_{\omega^2}(\zeta_n\cdot\omega+1)$

*Proof* By Theorem 8.5.16, we have ordinals $\alpha_0$ and $\rho$ less than $\omega^2 + \omega$ such that

$$\Big|\frac{\alpha_0}{\rho} \neg \mathsf{TI}_{\omega^2}(\zeta_n),\ \mathsf{TI}_{\omega^2}(\zeta_n + 1) \tag{i}$$

and from Lemma 8.5.24 we obtain by $\bigvee$-Exportation

$$\Big|\frac{\omega^2}{\omega^2} \neg \mathsf{TI}_{\omega^2}(\zeta_n + 1),\ \mathsf{TI}_{\omega^2}(\omega^{\zeta_n+1}). \tag{ii}$$

Again by Theorem 8.5.16, we obtain ordinals $\alpha_1$ and $\rho$ less than $\omega^2+\omega$ such that

$$\frac{\alpha_1}{\rho} \neg\mathsf{TI}_{\omega^2}(\omega^{\zeta_n+1}),\ \mathsf{TI}_{\omega^2}(\zeta_n\cdot\omega+1) \tag{iii}$$

because $h(\zeta_n\cdot\omega+1)=\omega^{\zeta_n+1}$. Cutting (i), (ii) and (iii) yields the claim. □

**8.5.29 Lemma** $\left|\frac{7\cdot(\lambda+\alpha+1)}{0}\right.\ \mathsf{TI}_\lambda(\alpha)$.

*Proof* Let $\rho := \mathrm{rnk}(S)$. We show

$$\left|\frac{7\cdot(\rho+\alpha)}{0}\right.\ \neg\mathsf{Prog}(S),\ \alpha\subseteq S \tag{i}$$

by induction on $\alpha$. By induction hypothesis and Lemma 8.5.1 we have

$$\left|\frac{7\cdot(\rho+\beta)}{0}\right.\ \neg\mathsf{Prog}(S),\ \neg(\beta<\alpha),\ \beta\subseteq S \tag{ii}$$

for all ordinals $\beta$ and by Tautology also

$$\left|\frac{2\cdot\rho}{0}\right.\ \beta\not\varepsilon S,\ \beta\,\varepsilon\, S. \tag{iii}$$

From (ii) and (iii) we then obtain

$$\left|\frac{7\cdot(\rho+\beta)+1}{0}\right.\ \neg\mathsf{Prog}(S),\ \beta\subseteq S\wedge\beta\not\varepsilon S,\ \neg(\beta<\alpha),\ \beta\,\varepsilon\, S \tag{iv}$$

for all $\beta$ by the Conjunction Lemma (Lemma 8.5.3). By an inference $\bigvee$ we obtain from (iv)

$$\left|\frac{7\cdot(\rho+\beta)+2}{0}\right.\ \neg\mathsf{Prog}(S),\neg(\beta<\alpha),\ \beta\,\varepsilon\, S \tag{v}$$

and finally

$$\left|\frac{7\cdot(\rho+\alpha)}{0}\right.\ \neg\mathsf{Prog}(S),\ \alpha\subseteq S$$

from (v) by $\bigvee$-Importation and an $(\bigwedge)$-inference. This proves (i). From (i) we obtain the claim by $\bigvee$-Importation and an inference $(\bigwedge)$. □

**8.5.30 Lemma** $\Gamma_0\subseteq\mathrm{Aut}(\omega)$.

*Proof* We show

$$\zeta_n\in\mathrm{Aut}(\omega) \tag{i}$$

by induction on $n$. Since $\mathrm{Aut}(\omega)$ is transitive by Lemma 8.5.14, we then obtain $\Gamma_0\subseteq\mathrm{Aut}(\omega)$ by (8.7).

In the proof of Theorem 7.4.1, we have shown $\mathsf{NT}\vdash\mathsf{TI}(\alpha,X)$ for every ordinal $\alpha$ less than $\varepsilon_0$, i.e., less than $\zeta_0$. By Lemma 7.4.3 this implies $\mathsf{NT}\vdash\mathsf{TI}(\alpha,S)$ for every $\alpha<\varepsilon_0$ and every $S\in\mathfrak{K}_\omega$. By the Embedding Theorem for $\mathsf{NT}$ (Theorem 7.3.8) and Lemma 8.5.1 there are finite ordinals $n_S$ such that

$$\Big|\frac{\omega+n_S}{\omega}\ \neg(\alpha<\varepsilon_0),\ \mathsf{TI}(\alpha,S) \tag{ii}$$

for all ordinals $\alpha$ and all $S \in \mathfrak{K}_\omega$. Hence $\varepsilon_0 \subseteq \mathrm{Aut}(\omega)$ and

$$\Big|\frac{\omega\cdot 2+3}{\omega}\ (\forall\xi<\varepsilon_0)\mathsf{TI}_\omega(\xi). \tag{iii}$$

Together with Lemma 8.5.21 this entails

$$\Big|\frac{\omega\cdot 2+4}{\omega\cdot 2}\ \mathsf{TI}_\omega(\varepsilon_0) \tag{iv}$$

which implies $\zeta_0 = \varepsilon_0 \in \mathrm{Aut}(\omega)$.

Let $\zeta_n \in \mathrm{Aut}(\omega)$ by induction hypothesis. From Lemma 8.5.29 we obtain $\Big|\frac{\zeta_n+7}{0}\ \mathsf{TI}_{\omega^2}(\zeta_n)$ which by Lemma 8.5.28 implies $\Big|\frac{\zeta_n+8}{\omega^2+\omega}\ \mathsf{TI}_{\omega^2}(\zeta_n\cdot\omega+1)$. Hence $\zeta_n\cdot\omega+1 \in \mathrm{Aut}(\omega)$ by $\bigwedge$-Inversion and we obtain with Lemma 8.5.27 $\zeta_{n+1} \in \mathrm{Aut}(\omega)$. □

**8.5.31 Theorem** *(S. Feferman, K. Schütte)* $\mathrm{Aut}(\omega) = \Gamma_0$.

*Proof* This is immediate by Corollary 8.4.4 and Lemma 8.5.30. □

It follows from Theorem 8.5.31 that the Schütte-Feferman-ordinal $\Gamma_0$ is the limit ordinal for predicativity in the strict sense.

**8.5.32 Exercise*** Let $A(X,x)$ be an arithmetical formula and $\Phi_A$ the operator induced by $A(X,x)$ (which needs not to be monotone). Let $\prec$ be a well-ordering. We define the iterations of the arithmetical operator $\Phi_A$ in the following way. For $x \in \mathit{field}(\prec)$ we put $\Phi_A^x := \{n \in \mathbb{N} \mid \mathbb{N} \models A[\Phi_A^{\prec x}, n]\}$ where $\Phi_A^{\prec x}$ stands for $\{x \in N \mid (\exists z \prec x)[x \in \Phi_A^z]\}$. These iterations can be described in the second-order language of $\mathsf{NT}$. Let $LO(X)$ express that $X$ is a set of pairs such that the relation $x \prec_X y :\Leftrightarrow \langle x,y\rangle \in X$ is a linear ordering and $WO(X) :\Leftrightarrow LO(X) \land Wf(X)$ express that $X$ represents a well-ordering. For a set $X$ we denote by $X_a := \{x \mid \langle x,a\rangle\ \varepsilon\ X\}$ the $a$-slice of $X$ and define $X^{\prec a} := \{x \mid (\exists z \prec a)[x \in X_z]\}$. The formula

$$\mathit{Hier}_A(X,Y)\ :\Leftrightarrow\ (\forall x)[x \in Y \leftrightarrow A(Y^{\prec_X (x)_1}, (x)_0)]$$

expresses that $Y$ represents the iterations of the operator $\Phi_A$ along the ordering $\prec_X$. The scheme of arithmetical transfinite recursion (cf. [94]) is then expressed by

$$(ATR)\qquad (\forall X)[WO(X)\ \rightarrow\ (\exists Y)\mathit{Hier}_A(X,Y)].$$

Let (ATR) denote the second-order theory which comprises all axioms of $\mathsf{NT}$ together with the scheme $(\Delta_0^1\text{–}CA)$ for arithmetical comprehension and the scheme *(ATR)* of arithmetical transfinite recursion. By $(\mathsf{ATR})_0$ we denote the theory where the scheme (IND) of Mathematical Induction is replaced by the second-order axiom $(\mathrm{Ind})^2$.

(a) Show that for every $\alpha < \Gamma_0$ there is a primitive recursive order relation $\prec$ such that $\mathrm{otyp}(\prec) = \alpha$ and $(\mathsf{ATR})_0 \vdash TI(\prec)$.

(b) Show that for every $\alpha < \Gamma_{\varepsilon_0} = \varphi_{1,0}(\varepsilon_0)$ there is a primitive recursive order relation $\prec$ such that $\mathrm{otyp}(\prec) = \alpha$ and $(\mathsf{ATR}) \vdash TI(\prec)$.

Hint: This is by far not simple. One possibility is to show that $\mathsf{NT}_{\omega_1}$ is formalizable within $(\mathsf{ATR})_0$ and then refer to Sect. 8.5. Although tedious this is quite straight forward. You have to represent the infinitary proof trees by arithmetically definable trees tagged with ordinal notations. Of course there are direct proofs in $(\mathsf{ATR})_0$ which are much more satisfying. The basic ideas of these proofs are essentially the same as in Sect. 8.5 but become much clearer in $(\mathsf{ATR})_0$ where they are not obscured by the tedious computations of derivation lengths. A well-ordering proof for ordinals below $\Gamma_0$ in a different system $\mathsf{IR}$ is in [23] and another for the system $(\mathsf{ATR})_0$ in [27].

# Chapter 9
# Ordinal Analysis of the Theory for Inductive Definitions

Having exhausted the limits of predicativity, we are ready for the first step into impredicativity. This book will solely be restricted to this first step. The further steps are more complicated and will be treated in a forthcoming volume. As mentioned in the discussion on predicativity at the beginning of Sect. 8.3 the definition of the least fixed-point of a monotone inductive definition $\Gamma$ as the intersection of all $\Gamma$-closed sets carries the features of an impredicative definition. An axiom system for a theory, which allows the formation of such fixed-points will therefore be the first (and simplest) example for an impredicative theory.

## 9.1 The Theory $\mathsf{ID}_1$

We want to axiomatize the theory for positively definable inductive definitions over the natural numbers. According to Corollary 6.5.6, we can express $\Pi^1_1$-relations by inductively defined relations. Therefore we can dispense with set parameters in the theory, to save some case distinctions and also to give examples for some of the phenomena which are characteristic for impredicative proof theory. To define the language of the theory for inductive definitions, let us assume that we have $n$-ary relation variables (instead of only unary ones) in the language $\mathscr{L}(\mathsf{NT})$. Since there are primitive recursive coding functions in $\mathscr{L}(\mathsf{NT})$ this is not really necessary but it facilitates the introduction of new constants for fixed-points. We will, however, continue to talk about "set"-variables and "set"-parameters.

**9.1.1 Definition** The language $\mathscr{L}(\mathsf{ID})$ comprises the first-order language of $\mathsf{NT}$. For every $X$-positive formula $F(X,x_1,\dots,x_n)$ of $\mathscr{L}(\mathsf{NT})$, where $X$ is supposed to be an $n$-ary predicate variable, we introduce a new $n$-ary set constant $I_F$.

The theory $\mathsf{ID}_1$ comprises $\mathsf{NT}$ including the scheme for Mathematical Induction for the extended language (without set variables) together with the defining axioms for the set constants

$(ID^1_1)$ $(\forall\vec{x})[F(I_F,\vec{x}) \;\rightarrow\; \vec{x}\,\varepsilon\, I_F]$

W. Pohlers, *Proof Theory: The First Step into Impredicativity*, Universitext, 

expressing that $I_F$ is closed under the operator defined by the $X$-positive formula $F(X,\vec{x})$ and

$(ID_1^2)$ $(\forall\vec{x})[F(G,\vec{x}) \to G(\vec{x})] \quad \to \quad (\forall\vec{x})[\vec{x}\,\varepsilon\, I_F \to G(\vec{x})]$,

expressing that it is the least such set. Here the notion $F(G,\vec{x})$ stands for the formula obtained from $F(X,\vec{x})$ by replacing all occurrences of $\vec{t}\,\varepsilon\, X$ by $G(\vec{t})$ and $\vec{t}\,\not\varepsilon\, X$ by $\neg G(\vec{t})$. We frequently use the abbreviation

$$Cl_F(G) \;:\Leftrightarrow\; (\forall\vec{x})[F(G,\vec{x}) \to G(\vec{x})] \tag{9.1}$$

to express that the "class" $\{\vec{x}\,|\,G(\vec{x})\}$ is closed under the operator $\Gamma_F$ induced by $F(X,\vec{x})$.

The standard interpretation for $I_F$ is the least fixed point $I_F$ of the operator $\Phi_F$ as introduced in Definition 6.4.1. The following two properties are left as exercises.

$$\mathbb{N} \models \vec{n}\,\varepsilon\, I_F \;\Leftrightarrow\; \mathbb{N} \models (\forall X)[Cl_F(X) \to \vec{n}\,\varepsilon\, X] \tag{9.2}$$

$$\mathsf{ID}_1 \vdash (\forall\vec{x})[F(I_F,\vec{x}) \leftrightarrow \vec{x}\,\varepsilon\, I_F]. \tag{9.3}$$

In the following, we distinguish the theory $\mathsf{ID}_1(\vec{X})$, which allows set parameters in its language, from the theory $\mathsf{ID}_1$ without set parameters. Since there are no defining axioms for the set parameters, every model for $\mathsf{ID}_1$ can easily be expanded to a model of $\mathsf{ID}_1(\vec{X})$. Therefore $\mathsf{ID}_1(\vec{X})$ is a conservative extension of $\mathsf{ID}_1$.

The definition of the proof-theoretic ordinal as well as the definition of the $\Pi^1_1$-ordinal of a theory require, (pseudo)-$\Pi^1_1$-sentences in the language of the theory. Therefore only $\|\mathsf{ID}_1(\vec{X})\|$ and $\|\mathsf{ID}_1(\vec{X})\|_{\Pi^1_1}$ are meaningful. In the first step we want to show that for impredicative theories comprising $\mathsf{ID}_1$ set parameters become dispensable.

**9.1.2 Definition** We define the ordinal

$$\kappa^{\mathsf{ID}_1} := \sup\{|n|_F + 1 \,|\, F(X,x) \text{ is } X\text{-positive } \wedge \mathsf{ID}_1 \vdash \underline{n}\,\varepsilon\, I_F\}.$$

The set $\{|n|_F + 1 \,|\, F(X,x)$ is $X$-positive $\wedge\, \mathsf{ID}_1 \vdash \underline{n}\,\varepsilon\, I_F\}$ is a $\Sigma$-definable set of ordinals less than $\omega_1^{CK}$. This implies $\kappa^{\mathsf{ID}_1} < \omega_1^{CK}$.[1] We are going to show that computation of $\kappa^{\mathsf{ID}_1}$ yields an ordinal analysis for $\mathsf{ID}_1(\vec{X})$. First we obtain

$$\mathsf{ID}_1(\vec{X}) \vdash F(X) \;\Rightarrow\; \mathsf{ID}_1(\vec{X}) \vdash F(G) \tag{9.4}$$

in the same way as in the case of NT in Lemma 7.4.2. Then we claim

$$\mathsf{ID}_1(\vec{X}) \vdash TI(\prec,X) \;\Leftrightarrow\; \mathsf{ID}_1 \vdash (\forall x \varepsilon \mathit{field}(\prec))\,[x\,\varepsilon\, \mathrm{Acc}_\prec] \tag{9.5}$$

[1] For details consult [4]. Cf. also Sect. 11.

where $\mathrm{Acc}_{\prec}$ has been defined in Definition 6.5.1 as the fixed-point of the formula

$$F_{\prec} \;:\Leftrightarrow\; \mathrm{Acc}(X,x,\prec) \;\Leftrightarrow\; x\,\varepsilon\,\mathit{field}(\prec) \wedge (\forall y \prec x)[y\,\varepsilon\,X].$$

To check (9.5) observe that $TI(\prec,X)$ means $Cl_{F_{\prec}}(X) \to (\forall x\,\varepsilon\,\mathit{field}(\prec))[x\,\varepsilon\,X]$. The assumption $\mathsf{ID}_1(\vec{X}) \vdash TI(\prec,X)$ therefore entails

$$\mathsf{ID}_1(\vec{X}) \vdash Cl_{F_{\prec}}(\mathrm{Acc}_{\prec}) \to (\forall x\,\varepsilon\,\mathit{field}(\prec))[x\,\varepsilon\,\mathrm{Acc}_{\prec}]$$

by (9.4). But $Cl_{F_{\prec}}(\mathrm{Acc}_{\prec})$ is an axiom $(ID_1^1)$ of $\mathsf{ID}_1(\vec{X})$. Hence

$$\mathsf{ID}_1(\vec{X}) \vdash (\forall x\,\varepsilon\,\mathit{field}(\prec))[x\,\varepsilon\,\mathrm{Acc}_{\prec}].$$

For the opposite direction we notice that

$$Cl_{F_{\prec}}(X) \to (\forall x)[x\,\varepsilon\,\mathrm{Acc}_{\prec} \to x\,\varepsilon\,X]$$

is an instance of axiom $(ID_2^1)$ of $\mathsf{ID}_1(\vec{X})$. From

$$\mathsf{ID}_1(\vec{X}) \vdash (\forall x\,\varepsilon\,\mathit{field}(\prec))[x\,\varepsilon\,\mathrm{Acc}_{\prec}]$$

we therefore immediately get

$$\mathsf{ID}_1 \vdash Cl_{F_{\prec}}(X) \to (\forall x)[x\,\varepsilon\,\mathit{field}(\prec) \to x\,\varepsilon\,X],$$

i.e., $\mathsf{ID}_1(\vec{X}) \vdash TI(\prec,X)$.

If $\prec$ is an arithmetically definable transitive relation such that $\mathsf{ID}_1(\vec{X}) \vdash TI(\prec,X)$ we get by (9.5) and Lemma 6.5.2

$$\begin{aligned}\mathrm{otyp}(\prec) &= \sup\{\mathrm{otyp}_{\prec}(x)+1 \mid x\,\varepsilon\,\mathit{field}(\prec)\}\\ &\le \sup\{|x|_{F_{\prec}}+1 \mid \mathsf{ID}_1 \vdash x\,\varepsilon\,\mathrm{Acc}_{\prec}\} \le \kappa^{\mathsf{ID}_1},\end{aligned} \tag{9.6}$$

hence $\|\mathsf{ID}_1(\vec{X})\| \le \kappa^{\mathsf{ID}_1}$.

But we also have

$$\mathsf{ID}_1 \vdash \vec{t}\,\varepsilon\,I_F \;\Leftrightarrow\; \mathsf{ID}_1(\vec{X}) \vdash Cl_F(X) \to \vec{t}\,\varepsilon\,X. \tag{9.7}$$

The direction from left to right is true since $Cl_F(X) \to \vec{t}\,\varepsilon\,I_F \to \vec{t}\,\varepsilon\,X$ is an instance of the axiom $(ID_1^2)$ and the opposite direction follows because $\mathsf{ID}_1 \vdash \vec{t}\,\varepsilon\,I_F$ is obvious from the instantiation $Cl_F(I_F) \to \vec{t}\,\varepsilon\,I_F$ and axiom $(ID_1^1)$ which says $Cl_F(I_F)$.

By the Stage Theorem (Theorem 6.6.1) we have $\kappa^{\mathsf{ID}_1} \le 2^{\|\mathsf{ID}_1(\vec{X})\|_{\Pi_1^1}}$. Together with (9.6) and (9.7) this implies

$$\|\mathsf{ID}_1(\vec{X})\| \le \kappa^{\mathsf{ID}_1} \le 2^{\|\mathsf{ID}_1(\vec{X})\|_{\Pi_1^1}} = 2^{\|\mathsf{ID}_1(\vec{X})\|} = \|\mathsf{ID}_1(\vec{X})\| \tag{9.8}$$

where we anticipated that $\|\mathsf{ID}_1(\vec{X})\|$ is an $\varepsilon$-number. This fact will follow only from the well-ordering proof in Sect. 9.6.2. However, ruminating the well-ordering proof in Sect. 7.4, we see even now that the proof-theoretic ordinal of any theory which comprises the scheme of Mathematical Induction is closed under $\omega$-powers, hence an $\varepsilon$-number.

Equation (9.8) confirms our decision to not include set parameters in the language of $\mathsf{ID}_1$. Our aim is therefore to compute $\kappa^{\mathsf{ID}_1}$.[2]

## 9.2 The Language $\mathscr{L}_\infty(\mathsf{NT})$

The aim of the next section is to provide a fragment of the language $\mathscr{L}_{\omega_1+1}$, which on the one side is large enough to allow an embedding of the theory $\mathsf{ID}_1$ and on the other side is simple enough to be handled proof-theoretically. The basic idea is to represent the stages of arithmetically definable monotone operators by infinitely long formulas.

**9.2.1 Definition** (The language $\mathscr{L}_\infty(\mathsf{NT})$) We define the language $\mathscr{L}_\infty(\mathsf{NT})$ as a fragment of the language $\mathscr{L}_{\omega_1+1}$ defined in Definition 8.1.1. We dispense with set parameters and restrict the formation of disjunctions and conjunctions to finite sequences, sequences of the form $\langle F_u(\underline{n}) \mid n \in \omega \rangle$ and to sequences of the form

$$\langle F(I_F^{<\xi}, \vec{t}\,) \mid \xi < \lambda \leq \omega_1 \rangle$$

and

$$\langle \sim F(I_F^{<\xi}, \vec{t}\,) \mid \xi < \lambda \leq \omega_1 \rangle$$

for an $X$-positive $\mathscr{L}(\mathsf{NT})$-formula $F(X, \vec{x})$ where $\vec{t} \,\varepsilon\, I_F^{<\xi}$ is recursively defined by

$$\vec{t} \,\varepsilon\, I_F^{<\xi} \;:\Leftrightarrow\; \bigvee_{\eta<\xi} F(I_F^{<\eta}, \vec{t}\,). \tag{9.9}$$

Since $\mathscr{L}_\infty(\mathsf{NT})$ is a sublanguage of $\mathscr{L}_{\omega_1+1}$, the infinite verification calculus $\models^{\alpha} \Delta$ of Sect. 8 is also a correct verification relation for $\mathscr{L}_\infty(\mathsf{NT})$. We mentioned in Sect. 8.1 that there is no completeness theorem for $\mathscr{L}_{\omega_1+1}$. For the fragment $\mathscr{L}_\infty(\mathsf{NT})$, however, we get completeness nearly for free. The reason is the absence of set variables in this fragment. In $\mathscr{L}_\infty(\mathsf{NT})$ there are only sentences which belong either to $\bigwedge$–type or to $\bigvee$–type. Therefore we can dispense with the rule (Ax) in the definition of $\models^{\alpha} \Delta$. All we need are the following two clauses (cf. Definitions 5.4.3 and 8.2.1).

($\bigwedge$) If $F \in (\Delta \cap \bigwedge\text{–type})$ and $(\forall G \in \mathrm{CS}(F))(\exists \alpha_G < \alpha)\big[\models^{\alpha_G} \Delta, G\big]$ then $\models^{\alpha} \Delta$.

---

[2] As a word of warning we want to emphasize that the identicalness of $\kappa^{\mathsf{ID}_1}$ and $\|\mathsf{ID}_1(\vec{X})\|$ depends on the presence of axiom $(ID_1^2)$. There are meta-predicative theories in which (among others) $(ID_1^1)$ is replaced by $(\forall \vec{x})[F(I_F, \vec{x}) \leftrightarrow \vec{x} \,\varepsilon\, I_F]$ and $(ID_1^2)$ is omitted. In such theories $T$ the ordinals $\kappa^T$ and $\|T\|$ may differ. On the side of subsystems of set theories the ordinal $\kappa^T$ corresponds to the ordinal $\|T\|_{\Sigma_1^{\omega_1^{CK}}}$ (cf. Remark 11.7.3). A similar effect of incoherence of $\|T\|$ and $\|T\|_{\Sigma_1^{\omega_1^{CK}}}$ for subsystems of set theory with restricted foundation occurs in Theorem 12.6.14.

($\bigvee$) If $F \in (\Delta \cap \bigvee\text{–type})$ and $(\exists G \in \mathrm{CS}(F))(\exists \alpha_G < \alpha)\left[\,\overset{\alpha_G}{\models} \Delta, G\right]$ then $\overset{\alpha}{\models} \Delta$.

The completeness of the verification relation is stated in the next lemma.

**9.2.2 Lemma** *For $F \in \mathcal{L}_\infty(\mathsf{NT})$ we have*

$$\mathbb{N} \models F \;\Rightarrow\; \overset{\mathrm{rnk}(F)}{\models\!\!=} F.$$

*Proof* We induct on $\mathrm{rnk}(F)$. If $\mathrm{CS}(F) = \emptyset$ then $F \in \bigwedge\text{–type}$, which immediately implies $\overset{0}{\models} F$.

If $F \in \bigwedge\text{–type}$ and $G \in \mathrm{CS}(F)$ then we have $\mathbb{N} \models G$ and thus by induction hypothesis

$$\overset{\mathrm{rnk}(G)}{\models\!\!=} G \tag{i}$$

for all $G \in \mathrm{CS}(F)$. From (i) we obtain

$$\overset{\mathrm{rnk}(F)}{\models\!\!=} F$$

by an inference $\bigwedge$.

If $F \in \bigvee\text{–type}$ then $\mathrm{CS}(F) \neq \emptyset$ and there is a sentence $G \in \mathrm{CS}(F)$ such that $\mathbb{N} \models G$. Thus

$$\overset{\mathrm{rnk}(G)}{\models\!\!=} G \tag{ii}$$

by induction hypothesis, which implies

$$\overset{\mathrm{rnk}(F)}{\models\!\!=} F$$

by an inference $\bigvee$. □

There is an obvious embedding of the language $\mathcal{L}(\mathsf{NT})$ into the fragment of $\mathcal{L}_\infty(\mathsf{NT})$.

**9.2.3 Definition** We define an embedding $^*\colon \mathcal{L}(\mathsf{NT}) \longrightarrow \mathcal{L}_\infty(\mathsf{NT})$ by

- $(s = t)^* := (s = t)$ and $(s \neq t)^* := (s \neq t)$
- $(A \wedge B)^* := \bigwedge\langle A^*, B^*\rangle$
- $(A \vee B)^* := \bigvee\langle A^*, B^*\rangle$
- $((\forall x)F_u(x))^* := \bigwedge\limits_{n<\omega} F_u(\underline{n})^*$
- $((\exists x)F_u(x))^* := \bigvee\limits_{n<\omega} F_u(\underline{n})^*$

The language $\mathscr{L}(\mathsf{NT})$ can therefore be regarded as a sublanguage of $\mathscr{L}_\infty(\mathsf{NT})$. However, by (9.9) also the stages of an inductive definition over $\mathbb{N}$ can be easily expressed in $\mathscr{L}_\infty(\mathsf{NT})$.

**9.2.4 Lemma** *Let $F(X,\vec{x})$ be a formula in $\mathscr{L}(\mathsf{NT})$. Recall the definition of $(\vec{t}\;\varepsilon\;\mathsf{I}_F^{<\xi})$ in (9.9). As a shorthand we define*

$$\vec{t}\;\varepsilon\;\mathsf{I}_F^{\alpha} \;:\Leftrightarrow\; F(\mathsf{I}_F^{<\alpha},\vec{t}\,)$$

*Then we obtain*

$$\mathbb{N}\models\vec{t}\;\varepsilon\;\mathsf{I}_F^{\alpha} \;\Leftrightarrow\; \vec{t}^{\,\mathbb{N}}\in I_F^{\alpha} \tag{9.10}$$

*where $I_F^\alpha$ denote the stages of the inductive definition induced by $F$ in the sense of Definition 6.4.5 and Definition 6.3.1.*

*Proof* We prove the lemma by induction on $\alpha$. We have

$$\mathbb{N}\models\vec{t}\;\varepsilon\;\mathsf{I}_F^{<\alpha} \;\Leftrightarrow\; \vec{t}^{\,\mathbb{N}}\in I_F^{<\alpha} \tag{i}$$

by the induction hypothesis. But from (i) we immediately obtain

$$\mathbb{N}\models F(\mathsf{I}_F^{<\alpha},\vec{t}\,) \;\Leftrightarrow\; \mathbb{N}\models F(I_F^{<\alpha},\vec{t}\,) \;\Leftrightarrow\; \vec{t}^{\,\mathbb{N}}\in\Phi_F(I_F^{<\alpha}) \;\Leftrightarrow\; \vec{t}^{\,\mathbb{N}}\in I_F^{\alpha}. \qquad\square$$

In short, we also use

$$\vec{t}\;\not\varepsilon\;\mathsf{I}_F^{\alpha} \;:\Leftrightarrow\; \neg F(\mathsf{I}_F^{<\alpha},\vec{t}\,).$$

If $F(X,\vec{x})$ is an $X$-positive $\mathscr{L}(\mathsf{NT})$-formula, we get $|F|\le\omega_1^{CK}$ by Theorem 6.6.4. Hence $I_F = I_F^{<\omega_1^{CK}} = I_F^{<\omega_1}$. Let us use $\Omega$ as a symbol which can either be interpreted by $\omega_1^{CK}$ or $\omega_1$.[3]

We may now extend the embedding $^*\colon\mathscr{L}(\mathsf{NT})\longrightarrow\mathscr{L}_\infty(\mathsf{NT})$ to an embedding

$$^*\colon\mathscr{L}(\mathsf{ID})\longrightarrow\mathscr{L}_\infty(\mathsf{NT}) \tag{9.11}$$

by adding the clause

- $(\vec{t}\;\varepsilon\;\mathsf{I}_F)^* := (\vec{t}\;\varepsilon\;\mathsf{I}_F^{<\Omega})$ and $(\vec{t}\;\not\varepsilon\;\mathsf{I}_F)^* := (\vec{t}\;\not\varepsilon\;\mathsf{I}_F^{<\Omega})$

to Definition 9.2.3.

**9.2.5 Lemma** *For any $\mathscr{L}(\mathsf{ID})$-sentence $G$ we get*

$$\mathbb{N}\models G \;\Leftrightarrow\; \mathbb{N}\models G^*.$$

*Proof* The proof is a straightforward induction on $\mathrm{rnk}(G)$. The only remarkable case is that $G$ is a formula $\vec{t}\;\varepsilon\;\mathsf{I}_F^{<\Omega}$ or $\vec{t}\;\not\varepsilon\;\mathsf{I}_F^{<\Omega}$. By Lemma 9.2.4 we obtain $\{\vec{n}\mid\mathbb{N}\models\underline{\vec{n}}\;\varepsilon\;\mathsf{I}_F^{<\Omega}\}=I_F$ irrespective of the interpretation of $\Omega$. This settles also this case. $\square$

[3] Possible additional alternative interpretations for $\Omega$ are discussed in Sect. 9.7.

For an $\mathscr{L}(\mathsf{ID})$-sentence $F$ we define its truth complexity by

$$\mathrm{tc}(F) := \mathrm{tc}(F^*) = \begin{cases} \min\{\alpha \mid \models^{\alpha} F^*\} & \text{if this exists} \\ \infty & \text{otherwise.} \end{cases}$$

The following observation (9.12) shows that the "truth complexity" of the provable sentences of $\mathsf{ID}_1$ also provides an upper bound for $\kappa^{\mathsf{ID}_1}$.

$$\models^{\alpha} n \,\varepsilon\, I_F^{<\Omega} \text{ implies } \models^{\alpha} n \,\varepsilon\, I_F^{<\alpha} \text{ and thus also } |n|_F < \alpha. \tag{9.12}$$

It is not difficult to prove (9.12) directly (cf. Exercise 9.3.18). However, we omit the proof because we need it in the modified form of Corollary 9.3.17 which will be proved there.

## 9.3 The Semi-Formal System for $\mathscr{L}_\infty(\mathsf{NT})$

In the previous chapters, we have seen that cut elimination is the main tool in the ordinal analysis of predicative theories. The aim of this introductory section is to study if this is also true for impredicative theories.

### *9.3.1 Semantical Cut-Elimination*

**9.3.1 Definition** We adopt Definition 7.3.5 of $\models^{\alpha}_{\rho} \Delta$ to finite sets $\Delta$ of sentences of the language $\mathscr{L}_\infty(\mathsf{NT})$. Since there are no free set variables, we can dispense with clause *(Ax)* in the definition of $\mathscr{L}_\infty(\mathsf{NT}) \models^{\alpha}_{\rho} \Delta$. So only the following clauses are left.

($\bigwedge$) If $F \in \Delta \cap \bigwedge$–type and $(\forall G \in \mathrm{CS}(F))(\exists \alpha_G < \alpha)\left[\mathscr{L}_\infty(\mathsf{NT}) \models^{\alpha_G}_{\rho} \Delta, G\right]$ then $\mathscr{L}_\infty(\mathsf{NT}) \models^{\alpha}_{\rho} \Delta$.

($\bigvee$) If $F \in \Delta \cap \bigvee$–type and $(\exists G \in \mathrm{CS}(F))(\exists \alpha_G < \alpha)\left[\mathscr{L}_\infty(\mathsf{NT}) \models^{\alpha_G}_{\rho} \Delta, G\right]$ then $\mathscr{L}_\infty(\mathsf{NT}) \models^{\alpha}_{\rho} \Delta$.

(cut) If $\mathscr{L}_\infty(\mathsf{NT}) \models^{\alpha_0}_{\rho} \Delta, F$, $\mathscr{L}_\infty(\mathsf{NT}) \models^{\alpha_0}_{\rho} \Delta, \neg F$ for some $\mathscr{L}_\infty(\mathsf{NT})$ formula $F$ such that $\mathrm{rnk}(F) < \rho$ then $\mathscr{L}_\infty(\mathsf{NT}) \models^{\alpha}_{\rho} \Delta$ for all $\alpha > \alpha_0$.

The equivalence (7.5) on p. 112 remains true for $\mathscr{L}_\infty(\mathsf{NT}) \models^{\alpha}_{0} \Delta$. Instead of $\mathscr{L}_\infty(\mathsf{NT}) \models^{\alpha}_{\rho} \Delta$, we will from now on briefly write $\models^{\alpha}_{\rho} \Delta$.

Our first observation is that, due to the fact that there are only sentences in $\mathscr{L}_\infty(\mathsf{NT})$, cut elimination for this infinitary system comes nearly for free.

**9.3.2 Theorem** *(Semantical cut elimination) Let $\Gamma$ be a finite set of $\mathcal{L}_\infty(\mathsf{NT})$-sentences. Then $\frac{\alpha}{\rho}\!\!\vdash \Gamma$ already implies $\models^{\alpha} \Gamma$.*

*Proof* We prove

$$\vdash^{\alpha}_{\rho} \Gamma, \Delta \text{ and } \mathbb{N} \not\models F \text{ for all } F \in \Delta \;\Rightarrow\; \models^{\alpha} \Gamma \tag{i}$$

by induction on $\alpha$. The claim then follows from (i) taking $\Delta$ to be empty.

In the case of a cut we have the premises

$$\vdash^{\alpha_0}_{\rho} \Gamma, \Delta, F \quad \text{and} \quad \vdash^{\alpha_0}_{\rho} \Gamma, \Delta, \neg F. \tag{ii}$$

Then either $\mathbb{N} \not\models F$ or $\mathbb{N} \not\models \neg F$. Using the induction hypothesis on the corresponding premise we get the claim.

In case of an inference according to $(\bigwedge)$ there is a formula $F \in \bigwedge\text{–type} \cap (\Gamma \cup \Delta)$ and we have the premises

$$\vdash^{\alpha_G}_{\rho} \Gamma, \Delta, G \tag{iii}$$

for all $G \in \mathrm{CS}(F)$. If $F \in \Delta$ there is a $G \in \mathrm{CS}(F)$ such that $\mathbb{N} \not\models G$ and we obtain

$$\models^{\alpha} \Gamma$$

from (iii) by the induction hypothesis and the structural rule. If $F \in \Gamma$ we obtain

$$\models^{\alpha_G} \Gamma, G \tag{iv}$$

for all $G \in \mathrm{CS}(F)$ by the induction hypothesis and $\models^{\alpha} \Gamma$ from (iv) by an inference $(\bigwedge)$.

In case of an inference according to $(\bigvee)$ there is a formula $F \in \bigvee\text{–type} \cap (\Delta \cup \Gamma)$ and we have the premise

$$\vdash^{\alpha_0}_{\rho} \Gamma, \Delta, G_0 \tag{v}$$

for some $G_0 \in \mathrm{CS}(F)$. If $F \in \Delta$ we have $\mathbb{N} \not\models G$ for all $G \in \mathrm{CS}(F)$ and obtain the claim from (v) by the induction hypothesis and the structural rule. If $F \in \Gamma$ we get $\models^{\alpha_0} \Gamma, G_0$ from (v) by the induction hypothesis and from that $\models^{\alpha} \Gamma$ by an inference $(\bigvee)$. □

**9.3.3 Remark** There is also a semantical proof of the Cut-Elimination Theorem for $\mathsf{NT}_{\omega_1}$. From $\mathsf{NT}_{\omega_1} \vdash^{\alpha}_{\rho} F$ we obtain by Lemma 7.3.6 $\mathbb{N} \models F[\Phi]$ for all assignments $\Phi$. By the Completeness Theorem (Theorem 8.2.2) we then get an ordinal $\delta < \omega_1$ such that $\models^{\delta} F$. On the one hand, no further information of the size of $\delta$ is provided by this proof (which makes it useless for ordinal analysis). On the other hand, this proof is much deeper than that of Theorem 9.3.2 since it uses the Completeness Theorem (which we regard as a deeper theorem)[4] although Theorem 9.3.2 provides a

[4] The Completeness Theorem for $\mathcal{L}_\infty(\mathsf{NT})$ is Lemma 9.2.2 which is also close to trivial. A completeness theorem for $\mathcal{L}_\infty(\mathsf{NT}, \vec{X})$ with free set variables fails (cf. Exercise 8.2.3). But this can

bound for $\delta$, namely $\alpha$ itself. This illuminates that cut elimination alone cannot play the same exclusive role in the ordinal analysis for $\mathsf{ID}_1$ as it does in the ordinal analysis of predicative theories. We have seen that the language $\mathscr{L}_\infty(\mathsf{NT})$ is sufficiently expressive to allow an embedding of the language of $\mathsf{ID}_1$. Theorem 9.3.2 explains us that the ordinal obtained by the embedding procedure is already an upper bound for the truth complexities of the provable sentences of $\mathsf{ID}_1$. No real cut elimination procedure is needed. This raises the suspicion that the ordinal bound obtained by embedding is too coarse. We are going to examine this phenomenon closer.

In view of Remark 9.3.3 we have to check the embedding strategy carefully. First we notice that already the translation of pure logic needs some extra care. We cannot transfer Theorem 7.3.2 literally. The sentence $\vec{t}\,\varepsilon\, I_F$ is an atomic sentence of $\mathscr{L}(\mathsf{ID})$, while its translation $(\vec{t}\,\varepsilon\, I_F)^*$ is not longer an atomic sentence of $\mathscr{L}_\infty(\mathsf{NT})$. Because of $\mathrm{rnk}(\vec{t}\,\varepsilon\, I_F^{<\Omega}) = \Omega$ we obtain by the Tautology Lemma (Lemma 7.3.3)

$$\models^{\Omega} \Delta, \vec{t} \notin I_F^{<\Omega}, \vec{t}\,\varepsilon\, I_F^{<\Omega}. \tag{9.13}$$

Thus Theorem 7.3.2 can be modified to

**9.3.4 Theorem** *If* $\underset{\mathsf{T}}{\models^{m}} \Delta(\vec{x})$ *holds for a finite set of* $\mathscr{L}(\mathsf{ID})$*-formulas then* $\models^{\Omega+m} \Delta(\underline{\vec{n}})^*$ *for all tuples* $\underline{\vec{n}}$ *of numerals.*

The truth complexities of the defining axioms for primitive recursive functions are not altered. More caution is again needed for the last identity axiom

$$(\forall\vec{x})(\forall\vec{y})[\vec{x} = \vec{y} \;\rightarrow\; F(\vec{x}) \rightarrow F(\vec{y})].$$

But here we get

$$\models^{\Omega+n} (\forall\vec{x})(\forall\vec{y})[\vec{x} = \vec{y} \;\rightarrow\; F(\vec{x}) \rightarrow F(\vec{y})]$$

for some $n < \omega$ again by the Tautology Lemma and the fact that all translations of $\mathscr{L}(\mathsf{ID})$-formulas have ranks $\Omega + m$ for some finite ordinal $m$.

By the same fact and the Induction Lemma (Lemma 7.3.4) we also obtain

$$\models^{\Omega+\omega+4} G^*$$

for all instances $G$ of the scheme of Mathematical Induction in $\mathsf{ID}_1$. It remains to check the truth complexities for the axioms $(ID_1^1)$ and $(ID_1^2)$. The Induction Lemma can be generalized to

$$\models^{2\cdot\mathrm{rnk}(G)+\omega\cdot(\alpha+1)} \neg(\forall\vec{x})[F(G,\vec{x}) \rightarrow G(\vec{x})],\, \vec{t} \notin I_F^{\alpha},\, G(\vec{t}) \tag{9.14}$$

which yields $\Omega\cdot 2+\omega$ as an upper bound for the truth complexities for translations of axiom $ID_1^2$. We are not going to prove (9.14) here. It will follow from Lemma 9.5.3 given below.

be rectify by replacing $\omega_1$ by $\omega_1^{CK}$. This is sufficient to embed $\mathsf{ID}_1$ and makes $\mathscr{L}_\infty(\mathsf{NT},\vec{X})$ to a fragment of $\mathscr{L}_{\omega_1}$ for which Theorem 8.2.2 holds.

Up to now we never used the peculiarities of the language $\mathscr{L}_\infty(\mathsf{NT})$. All the mentioned results hold for any language $\mathscr{L}_\kappa$ even with free set variables. Of another quality is the translation $(ID_1^1)^*$ of axiom $(ID_1^1)$. The sentence $(ID_1^1)^*$ can be regarded as the defining axiom for the ordinal symbol $\Omega$ since it postulates that $\Omega$ represents an ordinal which is an upper bound for the closure ordinal for all arithmetically definable inductive definitions. We cannot expect to import the properties of $\Omega$ from outside into the semi-formal proof relation in a similar way as we did it in the case of the Induction Lemma and its generalization (9.14). To obtain a verification for $(ID_1^1)$ we have to fall back on Lemma 9.2.2 by which we obtain

$$\left|\overline{\underline{\phantom{xx}}}\right.^{\Omega+n} Cl_F(I_F^{<\Omega})$$

since $\mathrm{rnk}(Cl_F(I_F^{<\Omega})) = \Omega + n$ for some $n < \omega$.

Combining that with Theorem 9.3.4 we obtain infinite derivations $\left|\frac{\Omega\cdot 2+\omega}{\Omega+m}\right. F^*$ for sentences $F$ which are provable in $\mathsf{ID}_1$. It follows from Theorem 9.3.2 that we can also get a cut-free derivation $\left|\overline{\underline{\phantom{xx}}}\right.^{\Omega\cdot 2+\omega} F^*$ but this bound is much too big for an ordinal analysis. Since obviously $\kappa^{\mathsf{ID}_1} \leq \kappa^{\mathbb{N}} = \omega_1^{CK} < \omega_1$ a better bound is already obtained by Theorem 6.6.4.

These observations show that the ordinal analysis for $\mathsf{ID}_1$ needs something new. We need a collapsing procedure that allows us to collapse the length of a derivation for the formula $\vec{n} \in I_F^{<\Omega}$ obtained by the canonical embedding into a derivation with length below $\Omega$. *The hallmark for impredicative proof theory is thus not longer cut-elimination but the necessity for a collapsing procedure.*[5] But ordinals, as hereditarily transitive sets, are in general not collapsible. A possible remedy is to use an ordinal notation system with gaps and to assign only ordinals from the notation system to the nodes of the derivation trees of the semi-formal system. This was the way originally used in [71] and [73]. Since this method uses heavily the fact that the stages of an inductive definition are locally predicatively defined we talk about the *method of local predicativity.* Here we are going to use a simplification of the method of local predicativity due to Buchholz [13]. We do not have to start with a notation system but assign ordinals which are controlled by operators. This again will leave enough gaps for a collapsing procedure. We will, however, not copy Buchholz' proof directly but introduce a slight variant which even more sharply pinpoints the role of collapsing.

### 9.3.2 Operator Controlled Derivations

Recall the notion of a strongly critical ordinal as defined in (3.10) on p. 39. Lemma 3.4.17 together with the Cantor normal-form for ordinals shows that every ordinal has a normal form

[5] We will, however, see that cut-elimination again plays the important role in the collapsing procedure.

$$\alpha =_{NF} \varphi_{\alpha_1}(\beta_1) + \cdots + \varphi_{\alpha_n}(\beta_n) \tag{9.15}$$

such that $\beta_i < \alpha$ for $i = 1, \ldots, n$ and $\varphi_{\alpha_1}(\beta_1) \geq \cdots \geq \varphi_{\alpha_n}(\beta_n)$.

**9.3.5 Definition** We define the set $SC(\alpha)$ of *strongly critical components* of an ordinal $\alpha$ as follows.

$$SC(\alpha) := \begin{cases} \emptyset & \text{if } \alpha = 0, \\ \{\alpha\} & \text{if } \alpha \in SC, \\ SC(\alpha_1) \cup SC(\alpha_2) & \text{if } \alpha = \varphi_{\alpha_1}(\alpha_2) \text{ and } \alpha_1, \alpha_2 < \alpha, \\ \bigcup_{i=1}^{n} SC(\alpha_i) & \text{if } \alpha =_{NF} \alpha_1 + \cdots + \alpha_n \text{ and } n > 1. \end{cases}$$

**9.3.6 Definition** A *Skolem-hull operator* is a function $\mathscr{H}$, which maps sets of ordinals to sets of ordinals satisfying the conditions

- For all $X \subseteq On$ it is $X \subseteq \mathscr{H}(X)$
- If $Y \subseteq \mathscr{H}(X)$ then $\mathscr{H}(Y) \subseteq \mathscr{H}(X)$
- $\overline{\overline{\mathscr{H}(X)}} = \max\{\overline{\overline{X}}, \aleph_0\}$.

A Skolem–hull operator $\mathscr{H}$ is *Cantorian closed* if it satisfies

- $\alpha \in \mathscr{H}(X) \;\Leftrightarrow\; SC(\alpha) \subseteq \mathscr{H}(X)$ for any set $X$ of ordinals.

We call $\mathscr{H}$ *transitive* if

- $\mathscr{H}(\emptyset) \cap \Omega$ is transitive.

The "least" Cantorian closed operator $\mathscr{B}_0$ is obtained by inductively defining

$$SC(X) \subseteq \mathscr{B}_0(X)$$

and

$$SC(\alpha) \subseteq \mathscr{B}_0(X) \;\Rightarrow\; \alpha \in \mathscr{B}_0(X)$$

where $SC(X)$ stands for $\bigcup_{\xi \in X} SC(\xi)$. Then $\mathscr{B}_0(X)$ satisfies

$$X \subseteq \mathscr{B}_0(X) \text{ and } \alpha \in \mathscr{B}_0(X) \;\Leftrightarrow\; SC(\alpha) \subseteq \mathscr{B}_0(X). \tag{9.16}$$

It follows from Lemma 3.4.17 that $\mathscr{B}_0(\emptyset) = \Gamma_0$ which shows that this operator is also transitive. It shows moreover that for any Cantorian closed operator $\mathscr{H}$ and any set $X$ the image $\mathscr{H}(X)$ contains all ordinals below $\Gamma_0$. To obtain "stronger" operators we have to equip them with stronger closure properties. For the theories treated in this book it suffices to assume that all operators satisfy

- $\Omega \in \mathscr{H}(\emptyset)$.

For a set $X \subseteq On$ and an operator $\mathscr{H}$ let

- $\mathscr{H}[X] := \lambda \Xi\,.\,\mathscr{H}(X \cup \Xi)$.

**9.3.7 Definition** For a sentence $G$ in the fragment of $\mathcal{L}_\infty(\mathsf{NT})$ we define

$$\mathrm{par}(G) := \{\alpha \mid I_F^{<\alpha} \text{ occurs in } G\}.^6$$

For a finite set $\Delta$ of sentences of the fragment of $\mathcal{L}_\infty(\mathsf{NT})$ we define

$$\mathrm{par}(\Delta) := \bigcup_{F\in\Delta} \mathrm{par}(F).$$

**9.3.8 Definition** Let $\mathcal{H}$ be a Skolem-hull operator. We define the relation $\mathcal{H} \frac{|\alpha}{\rho} \Delta$ by the clauses $(\bigwedge)$, $(\bigvee)$ and (cut) of Definition 7.3.5 with the additional conditions

- $\alpha \in \mathcal{H}(\mathrm{par}(\Delta))$

and for an inference

$$\mathcal{H} \frac{|\alpha_\iota}{\rho} \Delta_\iota \text{ for } \iota \in I \;\Rightarrow\; \mathcal{H} \frac{|\alpha}{\rho} \Delta$$

with finite $I$ also

- $\mathrm{par}(\Delta_\iota) \subseteq \mathcal{H}(\mathrm{par}(\Delta))$ for all $\iota \in I$.

We introduce the following abbreviation

$$\mathcal{H}_1 \subseteq \mathcal{H}_2 \;:\Leftrightarrow\; (\forall X \subseteq On)\big[\mathcal{H}_1(X) \subseteq \mathcal{H}_2(X)\big].$$

When writing $\mathcal{H} \frac{|\alpha}{\rho} \Delta$ we tacitly assume that $\mathcal{H}$ is a Cantorian closed Skolem–hull operator. The Structural Lemma of Sect. 7.3 extents to

**9.3.9 Lemma** *If $\mathcal{H}_1 \subseteq \mathcal{H}_2$, $\alpha \le \beta$, $\rho \le \sigma$, $\Delta \subseteq \Gamma$, $\beta \in \mathcal{H}_2(\mathrm{par}(\Gamma))$ and $\mathcal{H}_1 \frac{|\alpha}{\rho} \Delta$ then $\mathcal{H}_2 \frac{|\beta}{\sigma} \Gamma$.*

*Proof* We show the lemma by induction on $\alpha$. Let

$$\mathcal{H}_1 \frac{|\alpha_\iota}{\rho} \Delta_\iota \text{ for } \iota \in I \;\Rightarrow\; \mathcal{H}_1 \frac{|\alpha}{\rho} \Delta$$

be the last inference. By the induction hypothesis we then obtain

$$\mathcal{H}_2 \frac{|\alpha_\iota}{\sigma} \Gamma, \Delta_\iota \text{ for } \iota \in I. \tag{i}$$

For finite $I$ we have $\mathrm{par}(\Delta_\iota) \subseteq \mathcal{H}_1(\mathrm{par}(\Delta))$ and thus $\mathrm{par}(\Gamma, \Delta_\iota) \subseteq \mathcal{H}_1(\mathrm{par}(\Gamma)) \subseteq \mathcal{H}_2(\mathrm{par}(\Gamma))$. Because of $\rho \le \sigma$ and $\alpha \le \beta \in \mathcal{H}_2(\mathrm{par}(\Gamma))$ we obtain $\mathcal{H}_2 \frac{|\beta}{\sigma} \Gamma$ from (i) by the same inference. □

---

[6] The symbol $I_F^{<\alpha}$ is in fact an abbreviation for $\bigwedge_{\xi<\alpha} F(I_F^{<\xi}, ..)$ or $\bigvee_{\xi<\alpha} F(I_F^{<\xi}, ..)$, respectively. But only $\alpha$ is counted as parameter. The $\xi$'s are treated as "bounded".

**9.3.10 Lemma** *Let $X \subseteq \mathscr{H}(\mathrm{par}(\Delta))$ and $\mathscr{H}[X] \left|\frac{\alpha}{\rho}\right. \Delta$. Then $\mathscr{H} \left|\frac{\alpha}{\rho}\right. \Delta$.*

*Proof* Induction on $\alpha$. Assume that

$$\mathscr{H}[X] \left|\frac{\alpha_\iota}{\rho}\right. \Delta_\iota \text{ for all } \iota \in I \;\Rightarrow\; \mathscr{H}[X] \left|\frac{\alpha}{\rho}\right. \Delta$$

is the last inference. All inference rules have the property $\mathrm{par}(\Delta) \subseteq \mathrm{par}(\Delta_\iota)$. By the induction hypothesis we therefore obtain $\mathscr{H} \left|\frac{\alpha_\iota}{\rho}\right. \Delta_\iota$. But $X \subseteq \mathscr{H}(\mathrm{par}(\Delta))$ implies $\mathscr{H}[X](\mathrm{par}(\Delta)) = \mathscr{H}(\mathrm{par}(\Delta))$ and we obtain $\mathscr{H} \left|\frac{\alpha}{\rho}\right. \Delta$ by the same inference. □

**9.3.11 Lemma** *(Inversion Lemma) Let $F \in \bigwedge$–type, $\mathrm{CS}(F) \neq \emptyset$ and assume $\mathscr{H} \left|\frac{\alpha}{\rho}\right. \Delta, F$. Then $\mathscr{H}[\mathrm{par}(F)] \left|\frac{\alpha}{\rho}\right. \Delta, G$ holds true for all $G \in \mathrm{CS}(F)$.*

*Proof* The proof parallels that of Lemma 7.3.9 but needs some extra care on the parameters. We induct on $\alpha$. If $F$ is not the critical formula of the last inference

$$\mathscr{H} \left|\frac{\alpha_\iota}{\rho}\right. \Delta_\iota, F \text{ for } \iota \in I \;\Rightarrow\; \mathscr{H} \left|\frac{\alpha}{\rho}\right. \Delta, F \tag{i}$$

we obtain

$$\mathscr{H}[\mathrm{par}(F)] \left|\frac{\alpha_\iota}{\rho}\right. \Delta_\iota, G \tag{ii}$$

by the induction hypothesis. In case that $I$ is finite, we have to check the parameter condition. But then we have $\mathrm{par}(\Delta_\iota, F) \subseteq \mathscr{H}(\mathrm{par}(\Delta, F))$ which entails $\mathrm{par}(\Delta_\iota, G, F) \subseteq \mathscr{H}(\mathrm{par}(\Delta, G, F))$. Since $\alpha \in \mathscr{H}(\mathrm{par}(\Delta, F)) \subseteq \mathscr{H}(\mathrm{par}(\Delta, F, G))$ we obtain the claim by the same inference.

Now assume that $F$ is the critical formula of the last inference. Then this was an inference according to $(\bigwedge)$ and we have the premises

$$\mathscr{H} \left|\frac{\alpha_G}{\rho}\right. \Delta, F, G$$

for all $G \in \mathrm{CS}(F)$ and obtain by the induction hypothesis

$$\mathscr{H}[\mathrm{par}(F)] \left|\frac{\alpha_G}{\rho}\right. \Delta, G. \tag{iii}$$

Since $\alpha_G < \alpha \in \mathscr{H}(\mathrm{par}(\Delta, F)) \subseteq \mathscr{H}(\mathrm{par}(\Delta, G, F))$ we get the claim from (iii) by Lemma 9.3.9. □

**9.3.12 Lemma** *($\vee$-Exportation) Assume $\mathscr{H} \left|\frac{\alpha}{\rho}\right. \Delta, F_1 \vee \cdots \vee F_n$. Then we also have $\mathscr{H} \left|\frac{\alpha}{\rho}\right. \Delta, F_1, \ldots, F_n$.*

*Proof* The claim follows directly by induction on $\alpha$. □

**9.3.13 Lemma** *If $F \in Diag(\mathbb{N})$ and $\mathscr{H} \left|\frac{\alpha}{\rho}\right. \Delta, \neg F$ then $\mathscr{H} \left|\frac{\alpha}{\rho}\right. \Delta$.*

*Proof* Induction on $\alpha$. Since $F \in Diag(\mathbb{N})$ we have $\mathrm{par}(F) = \emptyset$ and $\neg F \in \bigvee$–type. Therefore $\neg F$ cannot be the critical formula of the last inference

$$\mathscr{H} \,\big|\!\frac{\alpha_\iota}{\rho}\, \Delta_\iota, \neg F \text{ for all } \iota \in I \;\Rightarrow\; \mathscr{H} \,\big|\!\frac{\alpha}{\rho}\, \Delta, \neg F. \tag{i}$$

From the induction hypothesis we get

$$\mathscr{H} \,\big|\!\frac{\alpha_\iota}{\rho}\, \Delta_\iota \tag{ii}$$

which entails $\mathscr{H} \,\big|\!\frac{\alpha}{\rho}\, \Delta$ by the same inference because $\alpha_\iota < \alpha \in \mathscr{H}(\mathrm{par}(\Delta, \neg F)) = \mathscr{H}(\mathrm{par}(\Delta))$ and, in case that $I$ is finite, $\mathrm{par}(\Delta_\iota) \subseteq \mathrm{par}(\Delta_\iota, \neg F) \subseteq \mathscr{H}(\mathrm{par}(\Delta, F)) = \mathscr{H}(\mathrm{par}(\Delta))$. □

**9.3.14 Lemma** *(Reduction Lemma) Let $F \in \bigwedge$-type, $\rho = \mathrm{rnk}(F)$ and $\mathrm{par}(F) \subseteq \mathscr{H}(\mathrm{par}(\Delta))$. If $\mathscr{H} \,\big|\!\frac{\alpha}{\rho}\, \Delta, F$ and $\mathscr{H} \,\big|\!\frac{\beta}{\rho}\, \Gamma, \neg F$ then $\mathscr{H} \,\big|\!\frac{\alpha+\beta}{\rho}\, \Delta, \Gamma$.*

*Proof* The proof is very similar to that of Lemma 7.3.12 but we need extra care on the controlling operator. If $\rho = 0$, we obtain $\mathscr{H} \,\big|\!\frac{\beta}{0}\, \Gamma$ by Lemma 9.3.13, which entails $\mathscr{H} \,\big|\!\frac{\alpha+\beta}{0}\, \Gamma, \Delta$ by Lemma 9.3.9. Therefore assume $\rho > 0$ which means $\mathrm{CS}(F) \neq \emptyset$. We induct on $\beta$. Let us first assume that $\neg F$ is not the critical formula of the last inference

$$\text{(J)} \qquad \mathscr{H} \,\big|\!\frac{\beta_\iota}{\rho}\, \Gamma_\iota, \neg F \text{ for } \iota \in I \;\Rightarrow\; \mathscr{H} \,\big|\!\frac{\beta}{\rho}\, \Gamma, \neg F.$$

If (J) is an inference according to $(\bigwedge)$ with empty premises then we have also

$$\mathscr{H} \,\big|\!\frac{\alpha+\beta}{0}\, \Delta, \Gamma.$$

Therefore assume that $I \neq \emptyset$. We still have $\mathrm{par}(F) \subseteq \mathscr{H}(\mathrm{par}(\Delta))$ and obtain $\mathscr{H} \,\big|\!\frac{\alpha+\beta_\iota}{\rho}\, \Delta, \Gamma_\iota$ by the induction hypothesis. Since $\alpha + \beta_\iota < \alpha + \beta$, and in the case of finite $I$ also $\mathrm{par}(\Delta, \Gamma_\iota) \subseteq \mathscr{H}(\mathrm{par}(\Delta, \Gamma))$, we obtain $\mathscr{H} \,\big|\!\frac{\alpha+\beta}{\rho}\, \Delta, \Gamma$ by an inference (J).

Now assume that $\neg F$ is the critical formula of the last inference in $\mathscr{H} \,\big|\!\frac{\beta}{\rho}\, \Gamma, \neg F$. Then we have the premise $\mathscr{H} \,\big|\!\frac{\beta_0}{\rho}\, \Gamma, \neg F, \neg G$ for some $G \in \mathrm{CS}(F)$ with

$$\mathrm{par}(\Gamma, F, G) \subseteq \mathscr{H}(\mathrm{par}(\Gamma, F)) \tag{i}$$

and obtain $\mathscr{H} \,\big|\!\frac{\alpha+\beta_0}{\rho}\, \Delta, \Gamma, \neg G$ by the induction hypothesis. By inversion, the Structural Lemma, the hypothesis $\mathrm{par}(F) \subseteq \mathscr{H}(\mathrm{par}(\Delta)) \subseteq \mathscr{H}(\mathrm{par}(\Delta, \Gamma))$ and Lemma 9.3.10 we obtain from the first hypothesis also $\mathscr{H} \,\big|\!\frac{\alpha+\beta_0}{\rho}\, \Delta, \Gamma, G$. It is $\alpha + \beta_0 < \alpha + \beta$ and $\mathrm{rnk}(G) < \rho$. To apply a cut we still have to check

$$\mathrm{par}(\Delta, \Gamma, G) \subseteq \mathscr{H}(\mathrm{par}(\Delta, \Gamma)). \tag{ii}$$

But this is secured by (i) and the hypothesis $\mathrm{par}(F) \subseteq \mathscr{H}(\mathrm{par}(\Delta)) \subseteq \mathscr{H}(\mathrm{par}(\Delta, \Gamma))$. □

**9.3.15 Theorem** *(Cut elimination for controlled derivations) Let $\mathscr{H}$ be a Cantorian closed Skolem–hull operator. Then*

$$\text{(i)} \quad \mathscr{H} \overset{\alpha}{\underset{\rho+1}{\vdash}} \Delta \Rightarrow \mathscr{H} \overset{2^\alpha}{\underset{\rho}{\vdash}} \Delta$$

*and*

$$\text{(ii)} \quad \mathscr{H} \overset{\alpha}{\underset{\beta+\omega^\rho}{\vdash}} \Delta \text{ and } \rho \in \mathscr{H}(\mathrm{par}(\Delta)) \Rightarrow \mathscr{H} \overset{\varphi_\rho(\alpha)}{\underset{\beta}{\vdash}} \Delta.$$

*Proof* We show (i) by induction on $\alpha$. If the last inference

$$\mathscr{H} \overset{\alpha_\iota}{\underset{\rho+1}{\vdash}} \Delta_\iota \text{ for } \iota \in I \Rightarrow \mathscr{H} \overset{\alpha}{\underset{\rho+1}{\vdash}} \Delta$$

is not a cut of rank $\rho$, we have $\mathscr{H} \overset{2^{\alpha_\iota}}{\underset{\rho}{\vdash}} \Delta_\iota$ by induction hypothesis and $\mathrm{par}(\Delta_\iota) \subseteq \mathscr{H}(\mathrm{par}(\Delta))$ in the case of finite $I$. So we get $\mathscr{H} \overset{2^\alpha}{\underset{\rho}{\vdash}} \Delta$ by the same inference.

In case that the last inference is a cut

$$\mathscr{H} \overset{\alpha_0}{\underset{\rho+1}{\vdash}} \Delta, F \quad \mathscr{H} \overset{\alpha_0}{\underset{\rho+1}{\vdash}} \Delta, \neg F \Rightarrow \mathscr{H} \overset{\alpha}{\underset{\rho+1}{\vdash}} \Delta$$

of rank $\rho$ we obtain $\mathscr{H} \overset{2^{\alpha_0}}{\underset{\rho}{\vdash}} \Delta, F$ and $\mathscr{H} \overset{2^{\alpha_0}}{\underset{\rho}{\vdash}} \Delta, \neg F$ by the induction hypothesis. But either $F \in \bigwedge\text{–type}$ or $\neg F \in \bigwedge\text{–type}$ and $\mathrm{par}(F) = \mathrm{par}(\neg F) \subseteq \mathscr{H}(\mathrm{par}(\Delta))$. Therefore we may apply the Reduction Lemma (Lemma 9.3.14) and the fact that $2^{\alpha_0} + 2^{\alpha_0} \leq 2^\alpha$ to obtain $\mathscr{H} \overset{2^\alpha}{\underset{\rho}{\vdash}} \Delta$.

Now we prove (ii) by induction on $\rho$ with side induction on $\alpha$. For $\rho = 0$ as proved by (i) and Lemma 9.3.9. Thus, let $\rho > 0$ and

$$\mathscr{H} \overset{\alpha_\iota}{\underset{\beta+\omega^\rho}{\vdash}} \Delta_\iota \text{ for all } \iota \in I \Rightarrow \mathscr{H} \overset{\alpha}{\underset{\beta+\omega^\rho}{\vdash}} \Delta \tag{i}$$

be the last inference. Then we have by the side induction hypothesis

$$\mathscr{H} \overset{\varphi_\rho(\alpha_\iota)}{\underset{\beta}{\vdash}} \Delta_\iota \text{ for all } \iota \in I. \tag{ii}$$

From $\alpha \in \mathscr{H}(\mathrm{par}(\Delta))$ and $\rho \in \mathscr{H}(\mathrm{par}(\Delta))$ we obtain $\varphi_\rho(\alpha) \in \mathscr{H}(\mathrm{par}(\Delta))$. If the last inference was not a cut of rank $\geq \beta$ we get from (ii), and in the case of a finite index set $I$ together with $\mathrm{par}(\Delta_\iota) \subseteq \mathscr{H}(\mathrm{par}(\Delta))$, the claim

$$\mathscr{H} \overset{\varphi_\rho(\alpha)}{\underset{\beta}{\vdash}} \Delta$$

by the same inference. Now assume that the last inference is a cut

$$\mathscr{H} \overset{\alpha_0}{\underset{\beta+\omega^\rho}{\vdash}} \Delta, F \quad \mathscr{H} \overset{\alpha_0}{\underset{\beta+\omega^\rho}{\vdash}} \Delta, \neg F \Rightarrow \mathscr{H} \overset{\alpha}{\underset{\beta+\omega^\rho}{\vdash}} \Delta \tag{iii}$$

such that $\beta \leq \mathrm{rnk}(F) =: \sigma < \beta + \omega^\rho$. Then $\mathrm{par}(F) \subseteq \mathscr{H}(\mathrm{par}(\Delta))$ which entails $\sigma \in \mathscr{H}(\mathrm{par}(\Delta))$ and for $\sigma =_{NF} \beta + \omega^{\sigma_0} + \cdots + \omega^{\sigma_n} < \beta + \omega^\rho$ we also get $\beta, \sigma_i \in \mathscr{H}(\mathrm{par}(\Delta))$. Then $\sigma < \beta + \omega^{\sigma_0} \cdot (n+1)$ and $\sigma_0 < \rho$ and we obtain by cut

$$\frac{|\varphi_\rho(\alpha_0)+1}{\beta+\omega^{\sigma_0}\cdot(n+1)}\ \Delta \tag{iv}$$

with $\sigma_0 \in \mathscr{H}(\mathrm{par}(\Delta))$. From (iv) we get the claim by iterated application of the main induction hypothesis together with the fact that $\varphi_{\sigma_0}(\cdots(\varphi_\rho(\alpha_0)+1)\cdots) < \varphi_\rho(\alpha)$. □

We close this section by proving (9.12) on p. 163 for operator controlled derivations. Lemma 9.3.16 given below is one of the key properties of local predicativity. It plays an important role in the elimination of the impredicative axiom $(ID_1^1)^*$.

For a finite set $\Delta$ of formulas we denote by $\Delta(F)$ that the formula $F$ occurs as sub-formula in some of the formulas in $\Delta$.

**9.3.16 Lemma** *(Boundedness) If* $\mathscr{H} \,|\frac{\alpha}{\rho}\, \Delta(\vec{t}\ \varepsilon\ I_F^{<\beta})$ *then* $\mathscr{H}[\{\beta\}] \,|\frac{\alpha}{\rho}\, \Delta(\vec{t}\ \varepsilon\ I_F^{<\gamma})$ *holds true for all* $\gamma$ *such that* $\alpha \le \gamma \le \beta$.

*Proof* We induct on $\alpha$. In the cases that $\vec{t}\ \varepsilon\ I_F^{<\beta}$ is not the critical formula of the last inference

$$\mathscr{H} \,|\frac{\alpha_\iota}{\rho}\, \Delta_\iota(\vec{t}\ \varepsilon\ I_F^{<\beta}) \text{ for } \iota \in I \;\Rightarrow\; \mathscr{H} \,|\frac{\alpha}{\rho}\, \Delta(\vec{t}\ \varepsilon\ I_F^{<\beta}) \tag{i}$$

we get

$$\mathscr{H}[\{\beta\}] \,|\frac{\alpha_\iota}{\rho}\, \Delta_\iota(\vec{t}\ \varepsilon\ I_F^{<\gamma}) \tag{ii}$$

by induction hypothesis. For a finite index set $I$ we have the additional hypothesis $\mathrm{par}(\Delta_\iota(\vec{t}\ \varepsilon\ I_F^{<\beta})) \subseteq \mathscr{H}(\mathrm{par}(\Delta(\vec{t}\ \varepsilon\ I_F^{<\beta})))$ which entails $\mathrm{par}(\Delta_\iota(\vec{t}\ \varepsilon\ I_F^{<\gamma})) \subseteq \mathscr{H}[\{\beta\}](\mathrm{par}(\Delta(\vec{t}\ \varepsilon\ I_F^{<\gamma})))$ and we obtain

$$\mathscr{H}[\{\beta\}] \,|\frac{\alpha}{\rho}\, \Delta(\vec{t}\ \varepsilon\ I_F^{<\gamma}) \tag{iii}$$

from (ii) by the same inference.

If $\vec{t}\ \varepsilon\ I_F^{<\beta}$ is the critical formula, we are in the case of an $(\bigvee)$ inference with premise

$$\mathscr{H} \,|\frac{\alpha_0}{\rho}\, \Delta_0, \vec{t}\ \varepsilon\ I_F^{<\beta}, \vec{t}\ \varepsilon\ I_F^{\xi} \tag{iv}$$

for some $\xi < \beta$. Applying the induction hypothesis twice, we obtain

$$\mathscr{H}[\{\beta,\xi\}] \,|\frac{\alpha_0}{\rho}\, \Delta_0, \vec{t}\ \varepsilon\ I_F^{<\gamma}, \vec{t}\ \varepsilon\ I_F^{\alpha_0}. \tag{v}$$

From $\alpha_0 \in \mathscr{H}(\mathrm{par}(\Delta_0, \vec{t}\ \varepsilon\ I_F^{<\beta}, \vec{t}\ \varepsilon\ I_F^{\xi}))$ and $\xi \in \mathscr{H}(\mathrm{par}(\Delta_0, \vec{t}\ \varepsilon\ I_F^{<\beta}))$ we obtain $\alpha_0 \in \mathscr{H}(\mathrm{par}(\Delta_0, I_F^{<\beta})) \subseteq \mathscr{H}[\{\beta\}](\mathrm{par}(\Delta_0, I_F^{<\gamma}))$ and $\mathscr{H}[\{\beta,\xi\}](\mathrm{par}(\Delta_0, I_F^{<\gamma})) = \mathscr{H}[\{\beta\}](\mathrm{par}(\Delta_0, I_F^{<\gamma}))$. Since $\alpha_0 < \alpha \le \gamma$ we can apply Lemma 9.3.10 and an inference $(\bigvee)$ to obtain

$$\mathscr{H}[\{\beta\}] \,|\frac{\alpha}{\rho}\, \Delta_0, \vec{t}\ \varepsilon\ I_F^{<\gamma}.$$ □

**9.3.17 Corollary** *If* $\mathcal{H} \,\substack{\alpha\\ \vdash\\ \rho}\, \underline{n} \,\varepsilon\, I_F^{<\beta}$ *for* $\alpha \leq \beta \leq \Omega$ *then* $|n|_F < \alpha$.

*Proof* Forgetting the controlling operator we get by the Boundedness Lemma $\,\substack{\alpha\\ \vdash\\ \rho}\, \underline{n} \,\varepsilon\, I_F^{<\alpha}$, which by Theorem 9.3.2 implies $\overset{\alpha}{\models} \underline{n} \,\varepsilon\, I_F^{<\alpha}$. Hence $\mathbb{N} \models n \,\varepsilon\, I_F^{<\alpha}$. □

**9.3.18 Exercise** Prove claim (9.12) in p. 163.
Hint. Adapt the proof of the Boundedness Lemma.

**9.3.19 Exercise** Prove claim (9.14) in p. 165.

## 9.4 The Collapsing Theorem for $ID_1$

Let $\mathcal{H}$ be a Cantorian closed operator. We define its iterations $\mathcal{H}_\alpha$.

**9.4.1 Definition** For $X \subseteq On$ let $\mathcal{H}_\alpha(X)$ be the least set of ordinals containing $X \cup \{0, \Omega\}$ which is closed under $\mathcal{H}$ and the collapsing function $\psi_{\mathcal{H}} \restriction \alpha$ where

$$\psi_{\mathcal{H}}(\alpha) := \min\{\xi \mid \xi \notin \mathcal{H}_\alpha(\emptyset)\}.$$

We need a few facts about the operators $\mathcal{H}_\alpha$. Here it is comfortable to interprete $\Omega$ as the first uncountable cardinal. Interpreting $\Omega$ as $\omega_1^{CK}$ makes the following considerations much harder. We return to alternative interpretations for $\Omega$ in Sect. 9.7.

First we observe

$$|\mathcal{H}_\alpha(X)| = \max\{|X|, \omega\}, \tag{9.17}$$

which implies

$$\psi_{\mathcal{H}}(\alpha) < \Omega \tag{9.18}$$

showing that $\psi_{\mathcal{H}}$ is in fact collapsing. Clearly the operators $\mathcal{H}_\alpha$ are Cantorian closed Skolem-hull operators which are cumulative, i.e.,

$$\alpha \leq \beta \;\Rightarrow\; \mathcal{H}_\alpha \subseteq \mathcal{H}_\beta \text{ and } \psi_{\mathcal{H}}(\alpha) \leq \psi_{\mathcal{H}}(\beta). \tag{9.19}$$

Since for $\alpha \in \mathcal{H}_\beta(\emptyset) \cap \beta$ implies $\psi_{\mathcal{H}}(\alpha) \in \mathcal{H}_\beta(\emptyset)$ we have

$$\alpha \in \mathcal{H}_\beta(\emptyset) \cap \beta \;\Rightarrow\; \psi_{\mathcal{H}}(\alpha) < \psi_{\mathcal{H}}(\beta). \tag{9.20}$$

From (9.20) we get

$$\mathcal{H}_\alpha(\emptyset) \cap \Omega = \psi_{\mathcal{H}}(\alpha). \tag{9.21}$$

The "$\supseteq$"-direction follows from the definition of $\psi_{\mathcal{H}}(\alpha)$ and (9.18). For the opposite inclusion, observe that $\psi_{\mathcal{H}}(\alpha)$ is closed under all operations in $\mathcal{H}$ – which means especially $\psi_{\mathcal{H}}(\alpha)$ is strongly critical – and show

$$\xi \in \mathscr{H}_\alpha(\emptyset) \cap \Omega \;\Rightarrow\; \xi < \psi_{\mathscr{H}}(\alpha)$$

by induction on the definition of $\xi \in \mathscr{H}_\alpha(\emptyset)$. If $\xi$ is obtained by operations in $\mathscr{H}$ we get $\xi < \psi_{\mathscr{H}}(\alpha)$ immediately. In case that $\xi = \psi_{\mathscr{H}}(\eta)$ we have $\eta \in \mathscr{H}_\alpha(\emptyset) \cap \alpha$ which by (9.20) implies $\xi = \psi_{\mathscr{H}}(\eta) < \psi_{\mathscr{H}}(\alpha)$.

From (9.21) we see that all the iterations $\mathscr{H}_\alpha$ are again transitive operators.

Another immediate property of the iterated operators is

$$\lambda \in Lim \;\Rightarrow\; \mathscr{H}_\lambda(X) = \bigcup_{\xi<\lambda} \mathscr{H}_\xi(X).$$

This can even be focused to

$$\lambda \in Lim \;\Rightarrow\; \mathscr{H}_\lambda(X) = \bigcup\{\mathscr{H}_\xi(X) \mid \xi \in \mathscr{H}_\xi(X) \cap \lambda\}. \tag{9.22}$$

The inclusion from right to left in (9.22) follows from the equation above. For the opposite inclusion, let $\xi \in \mathscr{H}_\lambda(X)$ and $\eta_0 := \min\{\eta \mid \xi \in \mathscr{H}_\eta(X)\}$. Then $\eta_0 < \lambda$ and $\eta_0 \notin Lim$. Let $\eta_0 = \eta + 1$. Then $\eta \in \mathscr{H}_\eta(X)$, hence $\eta_0 \in \mathscr{H}_{\eta_0}(X)$, because otherwise we had $\xi \in \mathscr{H}_{\eta+1}(X) = \mathscr{H}_\eta(X)$ contradicting the minimality of $\eta_0$. This ends the proof of (9.22).

By definition we have

$$\psi_{\mathscr{H}_\gamma}(0) = \min\{\xi \mid \xi \notin \mathscr{H}_\gamma(\emptyset)\} = \psi_{\mathscr{H}}(\gamma).$$

This implies that

$$\gamma \in \mathscr{H}_\gamma(X) \;\Rightarrow\; (\mathscr{H}_\gamma)_1(X) = \mathscr{H}_{\gamma+1}(X)^7 \tag{9.23}$$

holds true for all sets $X$ of ordinals.We show

$$\xi \in \mathscr{H}_{\gamma+1}(X) \Rightarrow \xi \in (\mathscr{H}_\gamma)_1(X) \tag{i}$$

by induction on the definition of $\xi \in \mathscr{H}_{\gamma+1}(X)$. If $\xi \in X$ or $\xi$ is obtained by operations in $\mathscr{H}$ then (i) holds immediately. So assume $\xi = \psi_{\mathscr{H}}(\eta)$ for some $\eta \in \mathscr{H}_{\gamma+1}(X) \cap \gamma+1$. Then $\eta \in (\mathscr{H}_\gamma)_1(X)$ by induction hypothesis and for $\eta < \gamma$ we get $\xi \in \mathscr{H}_\gamma(X) \subseteq (\mathscr{H}_\gamma)_1(X)$. If $\eta = \gamma$ we directly obtain $\xi = \psi_{\mathscr{H}}(\gamma) = \psi_{\mathscr{H}_\gamma}(0) \in (\mathscr{H}_\gamma)_1(X)$.

For the opposite direction

$$(\mathscr{H}_\gamma)_1(X) \subseteq \mathscr{H}_{\gamma+1}(X) \tag{ii}$$

we observe that $(\mathscr{H}_\gamma)_1(X)$ is the closure of $\mathscr{H}_\gamma(X) \cup \{\psi_{\mathscr{H}}(\gamma)\}$ under $\mathscr{H}$. But $\mathscr{H}_{\gamma+1}(X)$ is closed under $\mathscr{H}$, too. Moreover, we have $\mathscr{H}_\gamma(X) \subseteq \mathscr{H}_{\gamma+1}(X)$ and get $\psi_{\mathscr{H}}(\gamma) \in \mathscr{H}_{\gamma+1}(X)$ from the hypothesis $\gamma \in \mathscr{H}_\gamma(X)$. Hence (ii) and thus also (9.23).

The next lemma shows that property (9.23) can be extended.

**9.4.2 Lemma** *Let $\mathscr{H}$ be a Cantorian closed operator. Then $\alpha + \beta \in \mathscr{H}_{\alpha+\beta}(X)$ implies $(\mathscr{H}_\alpha)_\beta(X) = \mathscr{H}_{\alpha+\beta}(X)$ for all $X$.*

*Proof* We prove the lemma by induction on $\beta$. The claim holds trivially for $\beta = 0$. If $\beta = \gamma + 1$ we get $\alpha + \gamma \in \mathscr{H}_{\alpha+\gamma}(X)$ since the assumption $\alpha + \gamma \notin \mathscr{H}_{\alpha+\gamma}(X)$

[7] The "normal form condition" $\gamma \in \mathscr{H}_\gamma(\emptyset)$ will play an important role in Sect. 9.6.

implies $\mathscr{H}_{\alpha+\gamma+1}(X) = \mathscr{H}_{\alpha+\gamma}(X)$, hence $\alpha+\gamma \notin \mathscr{H}_{\alpha+\gamma+1}(X)$ in contradiction to $\alpha+\gamma+1 \in \mathscr{H}_{\alpha+\gamma+1}(X)$. So we have

$$(\mathscr{H}_\alpha)_\gamma(X) = \mathscr{H}_{\alpha+\gamma}(X) \tag{i}$$

by induction hypothesis which implies $\gamma \in (\mathscr{H}_\alpha)_\gamma(X)$. By (9.23) and (i) we therefore obtain

$$(\mathscr{H}_\alpha)_{\gamma+1}(X) = ((\mathscr{H}_\alpha)_\gamma)_1(X) = (\mathscr{H}_{\alpha+\gamma})_1(X) = \mathscr{H}_{\alpha+\gamma+1}(X).$$

If $\beta \in \mathit{Lim}$ we get by (9.22) and the induction hypothesis

$$\begin{aligned}\mathscr{H}_{\alpha+\beta}(X) &= \bigcup\{\mathscr{H}_{\alpha+\xi}(X) \mid \alpha+\xi \in \mathscr{H}_{\alpha+\xi}(X) \cap \alpha+\beta\}\\ &= \bigcup\{(\mathscr{H}_\alpha)_\xi(X) \mid \xi \in (\mathscr{H}_\alpha)_\xi(X) \cap \beta\} = (\mathscr{H}_\alpha)_\beta(X).\end{aligned}$$ □

The next claim is a direct corollary of Lemma 9.4.2.

**9.4.3 Corollary** *If* $\alpha+\beta \in \mathscr{H}_{\alpha+\beta}(\emptyset)$ *then* $\psi_{\mathscr{H}}(\alpha+\beta) = \psi_{\mathscr{H}_\alpha}(\beta)$.

The following collapsing property is crucial for the ordinal analysis of $\mathsf{ID}_1$. Already in Sect. 9.3, we mentioned that axiom $ID_1^1$ is the critical axiom of $\mathsf{ID}_1$. Its translation $Cl_F(I_F^{<\Omega}) \;:\Leftrightarrow\; (\forall\vec{x})[F(I_F^{<\Omega},\vec{x}) \to \vec{x}\,\varepsilon\, I_F^{<\Omega}]$ looks already formally impredicative. The complexity of its premise $F(I_F^{<\Omega},\vec{x})$ is bigger than that of its conclusion $\vec{x}\,\varepsilon\, I_F^{<\Omega}$. As pointed out in Sect. 9.3.1, $Cl_F(I_F^{<\Omega})$ can be viewed as the defining axiom for $\Omega$ as an ordinal satisfying $Cl_F(I_F^{<\Omega})$. Since $\Omega$ looses this property when there are gaps below $\Omega$, we can hardly expect to obtain an operator controlled derivation of $Cl_F(I_F^{<\Omega})$. However, the truth complexity of true $\mathscr{L}_\infty(\mathsf{NT})$-sentences which do not contain conjunctions of length $\geq \Omega$ is certainly below $\Omega$ which means that they possess infinitary derivations of length less than $\Omega$. Let us characterize this class of formulas by the following definition.

**9.4.4 Definition** We say that a sentence in the fragment $\mathscr{L}_\infty(\mathsf{NT})$ is in $\bigvee^\Omega$–type if it does not contain sub-formulas of the shape $\vec{t} \not\varepsilon I_F^{<\Omega}$, i.e., if it does not contain conjunctions of length $\geq \Omega$.

Since infinitary proofs stemming from translations of formal proofs in $\mathsf{ID}_1$ show a big uniformity, there is some hope that sentences in $\bigvee^\Omega$–type that are translations of $\mathsf{ID}_1$-theorems possess an operator controlled derivation which does not use axiom $(ID_1^1)^*$ and whose length can be computed to be less than $\Omega$. That this hope can be substantiated is prepared by the following collapsing lemma. It shows that an operator controlled derivation of an $\bigvee^\Omega$–type-sentence depending on instances of $(ID_1^1)^*$ can be transformed into an operator controlled derivation of the same sentence which no longer uses $(ID_1^1)^*$ and has a computable length less than $\Omega$. The cost is that we have to iterate the controlling operator. A study of the proof of the collapsing theorem shows that the instances of $(ID_1^1)^*$ are replaced by a series of cuts. So there is a faint resemblance to Gentzen's method of resolving applications of instances of the scheme of mathematical induction by a series of cuts.

**9.4.5 Lemma** *(Collapsing Lemma)* *Let* $\Delta \subseteq \bigvee^{\Omega}$*–type such that* $\mathrm{par}(\Delta) \subseteq \mathscr{H}(\emptyset)$ *and* $\mathscr{H} \overset{\beta}{\underset{\Omega}{\vdash}} \neg Cl_{F_1}(I_{F_1}^{<\Omega}),\ldots,\neg Cl_{F_k}(I_{F_k}^{<\Omega}),\Delta$*. Then* $\mathscr{H}_{\omega^\beta+1} \overset{\psi_{\mathscr{H}}(\omega^\beta)}{\underset{\Omega}{\vdash}} \Delta$.

*Proof* The proof is by induction on $\beta$. The key property is

$$\beta \in \mathscr{H}(\emptyset) \text{ and } \omega^\beta < \gamma \Rightarrow \psi_{\mathscr{H}}(\omega^\beta) < \psi_{\mathscr{H}}(\gamma) \tag{i}$$

which is obvious by (9.20) since we have $\omega^\beta \in \mathscr{H}(\emptyset) \cap \gamma \subseteq \mathscr{H}_\gamma(\emptyset) \cap \gamma$. Other observations are

$$\mathscr{H}(\mathrm{par}(\Delta)) = \mathscr{H}(\emptyset) \tag{ii}$$

because $\mathrm{par}(\Delta) \subseteq \mathscr{H}(\emptyset)$ and

$$\beta \in \mathscr{H}(\emptyset) \Rightarrow \omega^\beta \in \mathscr{H}_{\omega^\beta+1}(\emptyset) \text{ and } \psi_{\mathscr{H}}(\omega^\beta) \in \mathscr{H}_{\omega^\beta+1}(\emptyset) \tag{iii}$$

which is clear by (9.19) and the closure properties of $\mathscr{H}_{\omega^\beta+1}(\emptyset)$.

Let

$$\Theta_k :\Leftrightarrow \neg Cl_{F_1}(I_{F_1}^{<\Omega}),\ldots,\neg Cl_{F_k}(I_{F_k}^{<\Omega})$$

and assume that the critical part of the last inference

$$\mathscr{H} \overset{\beta_\iota}{\underset{\Omega}{\vdash}} \Theta_k,\Delta_\iota \text{ for } \iota \in I \Rightarrow \mathscr{H} \overset{\beta}{\underset{\Omega}{\vdash}} \Theta_k,\Delta \tag{iv}$$

belongs to a sentence in $\Delta$. Observe that $\mathrm{par}(\Theta_k) = \{\Omega\}$. So we have to bother only about the parameters of $\Delta$. We claim

$$\mathrm{par}(\Delta_\iota) \subseteq \mathscr{H}(\emptyset). \tag{v}$$

If $I$ is finite then we have $\mathrm{par}(\Delta_\iota) \subseteq \mathscr{H}(\mathrm{par}(\Delta)) = \mathscr{H}(\emptyset)$ because $\mathrm{par}(\Delta) \subseteq \mathscr{H}(\emptyset)$.

If $I$ is infinite and the critical formula has the shape $((\forall x)F_u(x))^*$, then $I = \omega$ and $\Delta_n = \Delta, F_u(\underline{n})$. But then $\mathrm{par}(\Delta_n) = \mathrm{par}(\Delta)$ and (v) holds. If the critical formula of the inference is $\vec{t} \notin I_F^{<\xi}$ then $\xi < \Omega$ because $\Delta \subseteq \bigvee^{\Omega}$–type. Then $\Delta_\iota = \Delta, G$ for some $G \in \mathrm{CS}(\vec{t} \notin I_F^{<\xi})$, which means that $\mathrm{par}(\Delta_\iota) \subseteq \mathrm{par}(\Delta) \cup \{\eta\}$ for some $\eta < \xi$. But $\xi \in \mathscr{H}(\emptyset) \cap \Omega$ entails $\xi \subseteq \mathscr{H}(\emptyset)$ by the transitivity of $\mathscr{H}$. Hence $\mathrm{par}(\Delta_\iota) \subseteq \mathscr{H}(\emptyset)$ for all $\iota \in I$ and the proof of (v) is completed.

Next we claim

$$\Delta_\iota \subseteq \bigvee^{\Omega}\text{–type}. \tag{vi}$$

This follows from $\Delta \subseteq \bigvee^{\Omega}$–type for inferences that are no cuts. In case that the inference in (iv) is a cut, its cut-sentence is of rank $< \Omega$ which ensures that it belongs to $\bigvee^{\Omega}$–type, too. Because of (v) and (vi) the induction hypothesis applies to the premises of (iv) and we obtain

$$\mathscr{H}_{\omega^{\beta_\iota}+1} \overset{\psi_{\mathscr{H}}(\omega^{\beta_\iota})}{\underset{\Omega}{\vdash}} \Delta_\iota. \tag{vii}$$

From $\beta_\iota \in \mathscr{H}(\emptyset) \cap \beta$ we obtain $\psi_{\mathscr{H}}(\omega^{\beta_\iota}) < \psi_{\mathscr{H}}(\omega^\beta)$ by (i) and from $\beta \in \mathscr{H}(\emptyset)$ also $\psi_{\mathscr{H}}(\omega^\beta) \in \mathscr{H}_{\omega^\beta+1}(\emptyset)$. Since also $\mathrm{par}(\Delta_\iota) \subseteq \mathscr{H}(\emptyset) \subseteq \mathscr{H}_{\omega^\beta+1}(\emptyset)$ we get

$$\mathscr{H}_{\omega^\beta+1} \left|\frac{\psi_{\mathscr{H}}(\omega^\beta)}{\Omega}\right. \Delta \tag{viii}$$

from (vii) by the same inference.

Now assume that the critical formula of the last inference is

$$\neg Cl_{F_i}(I_{F_i}^{<\Omega}), \text{ i.e., the formula } (\exists x)[F_i(I_{F_i}^{<\Omega},x) \land x \not\varepsilon I_{F_i}^{<\Omega}]. \tag{ix}$$

Then we have the premise

$$\mathscr{H} \left|\frac{\beta_0}{\Omega}\right. \neg Cl_{F_1}(I_{F_1}^{<\Omega}),\dots,\neg Cl_{F_k}(I_{F_k}^{<\Omega}), F_i(I_{F_i}^{<\Omega},t) \land t \not\varepsilon I_{F_i}^{<\Omega}, \Delta \tag{x}$$

with $\beta_0 \in \mathscr{H}(\mathrm{par}(\Delta) \cup \{\Omega\}) = \mathscr{H}(\emptyset)$. By inversion we obtain from (x)

$$\mathscr{H} \left|\frac{\beta_0}{\Omega}\right. \neg Cl_{F_1}(I_{F_1}^{<\Omega}),\dots,\neg Cl_{F_k}(I_{F_k}^{<\Omega}), F_i(I_{F_i}^{<\Omega},t), \Delta \tag{xi}$$

and

$$\mathscr{H} \left|\frac{\beta_0}{\Omega}\right. \neg Cl_{F_1}(I_{F_1}^{<\Omega}),\dots,\neg Cl_{F_k}(I_{F_k}^{<\Omega}), t \not\varepsilon I_{F_i}^{<\Omega}, \Delta. \tag{xii}$$

Applying the induction hypothesis to (xi) and then using boundedness, gives

$$\mathscr{H}_{\omega^{\beta_0}+1} \left|\frac{\psi_{\mathscr{H}}(\omega^{\beta_0})}{\Omega}\right. F_i(I_{F_i}^{<\psi_{\mathscr{H}}(\omega^{\beta_0})},t), \Delta, \tag{xiii}$$

i.e.,

$$\mathscr{H}_{\omega^{\beta_0}+1} \left|\frac{\psi_{\mathscr{H}}(\omega^{\beta_0})}{\Omega}\right. t\,\varepsilon\, I_{F_i}^{\psi_{\mathscr{H}}(\omega^{\beta_0})}, \Delta. \tag{xiv}$$

From (xii) we obtain by inversion

$$\mathscr{H} \left|\frac{\beta_0}{\Omega}\right. \neg Cl_{F_1}(I_{F_1}^{<\Omega}),\dots,\neg Cl_{F_k}(I_{F_k}^{<\Omega}), t \not\varepsilon I_{F_i}^{\psi_{\mathscr{H}}(\omega^{\beta_0})}, \Delta \tag{xv}$$

which entails

$$\mathscr{H}_{\omega^{\beta_0}+1} \left|\frac{\beta_0}{\Omega}\right. \neg Cl_{F_1}(I_{F_1}^{<\Omega}),\dots,\neg Cl_{F_k}(I_{F_k}^{<\Omega}), t \not\varepsilon I_{F_i}^{\psi_{\mathscr{H}}(\omega^{\beta_0})}, \Delta. \tag{xvi}$$

Since $\psi_{\mathscr{H}}(\omega^{\beta_0}) \in \mathscr{H}_{\omega^{\beta_0}+1}(\emptyset)$ the induction hypothesis applies to (xvi) and we obtain

$$(\mathscr{H}_{\omega^{\beta_0}+1})_{\omega^{\beta_0}+1} \left|\frac{\psi_{\mathscr{H}_{\omega^{\beta_0}+1}}(\omega^{\beta_0})}{\Omega}\right. t \not\varepsilon I_{F_i}^{\psi_{\mathscr{H}}(\omega^{\beta_0})}, \Delta. \tag{xvii}$$

By Lemma 9.4.2 and Corollary 9.4.3, this entails

$$\mathscr{H}_{\omega^{\beta_0}+1+\omega^{\beta_0}+1} \left|\frac{\psi_{\mathscr{H}}(\omega^{\beta_0}+1+\omega^{\beta_0})}{\Omega}\right. t \not\varepsilon I_{F_i}^{\psi_{\mathscr{H}}(\omega^{\beta_0})}, \Delta \tag{xviii}$$

and we obtain

$$\mathscr{H}_{\omega^\beta+1} \left|\frac{\psi_{\mathscr{H}}(\omega^\beta)}{\Omega}\right. \Delta$$

from (xiv) and (xvii) by the Structural Lemma and (cut). □

**9.4.6 Remark** Although we will not need it for the ordinal analysis of $\mathsf{ID}_1$ we want to remark that the Collapsing Lemma may be strengthened to

$$\mathscr{H} \Big|\frac{\beta}{\Omega+1} \neg Cl_{F_1}(I_{F_1}^{<\Omega}), \ldots, \neg Cl_{F_k}(I_{F_k}^{<\Omega}), \Delta \Rightarrow \mathscr{H}_{\omega^\beta+1} \Big|\frac{\psi_{\mathscr{H}}(\omega^\beta)}{\psi_{\mathscr{H}}(\omega^\beta)} \Delta.$$

For $k = 0$ it can be modified to

$$\mathscr{H} \Big|\frac{\beta}{\Omega} \Delta \Rightarrow \mathscr{H}_{\beta+1} \Big|\frac{\psi_{\mathscr{H}}(\beta)}{\psi_{\mathscr{H}}(\beta)} \Delta.$$

*Proof* We have to do three things. First we observe that in the case of a cut of rank $< \Omega$, we have $\mathrm{par}(F) \subseteq \mathscr{H}(\emptyset) \cap \Omega \subseteq \mathscr{H}_{\omega^\beta} \cap \Omega = \psi_{\mathscr{H}}(\omega^\beta)$. Since $\mathrm{rnk}(F) < \max \mathrm{par}(F) + \omega$, we obtain $\mathrm{rnk}(F) < \psi_{\mathscr{H}}(\omega^\beta)$. If the cut rank is $\Omega + 1$ we have the additional case of a cut of rank $\Omega$. Then the cut sentence is $t \,\varepsilon\, I_F^{<\Omega}$ and we have the premises

$$\mathscr{H} \Big|\frac{\beta_0}{\Omega+1} \neg Cl_{F_1}(I_{F_1}^{<\Omega}), \ldots, \neg Cl_{F_k}(I_{F_k}^{<\Omega}), \Delta, t \,\varepsilon\, I_F^{<\Omega} \tag{i}$$

and

$$\mathscr{H} \Big|\frac{\beta_0}{\Omega+1} \neg Cl_{F_1}(I_{F_1}^{<\Omega}), \ldots, \neg Cl_{F_k}(I_{F_k}^{<\Omega}), \Delta, t \,\not\varepsilon\, I_F^{<\Omega}. \tag{ii}$$

But we may apply the induction hypothesis to (i) and then proceed as in the last case in the proof of the Collapsing Lemma. The resulting cut sentence is $t \,\varepsilon\, I_F^{\psi_{\mathscr{H}}(\omega^{\beta_0})}$ which shows that the cut sentence has rank $< \psi_{\mathscr{H}}(\omega^\beta)$.

Finally we observe that only in the case that the critical formula of the last inference is one of the formulas $Cl_{F_i}(I_{F_i}^{<\Omega})$ we needed the fact that $\omega^\beta$ is additively indecomposable. If $k = 0$ we may therefore replace $\omega^\beta$ by $\beta$. □

## 9.5 The Upper Bound

To get an upper bound for $\kappa^{\mathsf{ID}_1}$ we have to strengthen Theorem 9.3.4 in the following way.

**9.5.1 Theorem** *If* $\Big|\frac{m}{\mathsf{T}} \Delta(\vec{x})$ *holds true for a finite set of* $\mathscr{L}(\mathsf{ID})$*-formulas then we get* $\mathscr{H} \Big|\frac{\Omega+m}{0} \Delta(\underline{\vec{n}})$ *for all tuples* $\underline{\vec{n}}$ *of numerals and all Cantorian closed Skolem–hull operators* $\mathscr{H}$.

Recall that two formulas are *numerically equivalent*, if they only differ in terms $s_1, \ldots, s_n$ and $t_1, \ldots, t_n$ such that $t_i^{\mathbb{N}} = s_i^{\mathbb{N}}$ for $i = 1, \ldots, n$. The key in the proof of Theorem 9.5.1 is a refinement of the Tautology Lemma.

**9.5.2 Lemma** *(Controlled Tautology) We have* $\mathscr{H} \,\frac{|^{2\cdot\mathrm{rnk}(F)}}{|_0} \, \Delta, \neg F', F$ *for numerically equivalent* $\mathscr{L}_\infty(\mathsf{NT})$*-sentences* $F$ *and* $F'$ *and all Cantorian closed Skolem–hull operators* $\mathscr{H}$.

The proof by induction on $\mathrm{rnk}(F)$ is easy. First observe that $2\cdot\mathrm{rnk}(F) \in \mathscr{H}(\mathrm{par}(F))$ for every Cantorian closed Skolem–hull operator because $\mathrm{rnk}(F) = \max\mathrm{par}(F) + n$ for some $n < \omega$. Assume without loss of generality that $F \in \bigwedge\text{–type}$ . By induction hypothesis we have

$$\mathscr{H} \,\frac{|^{2\cdot\mathrm{rnk}(G)}}{|_0}\, \Delta, \neg F', F, G, \neg G' \tag{i}$$

for all $G \in \mathrm{CS}(F)$ where $G'$ corresponds to $G$ in the same way as $F'$ to $F$. Since $\mathrm{par}(\Delta, \neg F', F, G, \neg G') \subseteq \mathscr{H}(\mathrm{par}(\Delta, \neg F', F, G))$ we obtain from (i)

$$\mathscr{H} \,\frac{|^{2\cdot\mathrm{rnk}(G)+1}}{|_0}\, \Delta, \neg F', F, G, \tag{ii}$$

for all $G \in \mathrm{CS}(F)$ by an inference $(\bigvee)$. From (ii) and $2\cdot\mathrm{rnk}(G)+1 < 2\cdot(\mathrm{rnk}(G)+1) \leq 2\cdot\mathrm{rnk}(F)$, however, we immediately get

$$\mathscr{H} \,\frac{|^{2\cdot\mathrm{rnk}(F)}}{|_0}\, \Delta, \neg F', F$$

by an inference $(\bigwedge)$. ☐

Now we prove Theorem 9.5.1 by induction on $m$. If $\frac{|^{m}}{|_{\mathsf{T}}} \Delta, \neg F(\vec{x}), F(\vec{x})$ holds by (AxL) we get $\mathscr{H} \,\frac{|^{\Omega+m}}{|_0}\, \Delta, \neg F(\underline{\vec{n}}), F(\underline{\vec{n}})$ by Lemma 9.5.2 and possibly Lemma 9.3.9.

In case of an inference $(\wedge)$ we have the induction hypotheses $\mathscr{H} \,\frac{|^{\Omega+m_i}}{|_0}\, \Delta(\underline{\vec{n}}), A_i(\underline{\vec{n}})$ for $i \in \{1,2\}$. By Lemma 9.3.9 we obtain $\mathscr{H} \,\frac{|^{\Omega+m_i}}{|_0}\, \Delta(\underline{\vec{n}}), A_i(\vec{n}), (A_1 \wedge A_2)(\underline{\vec{n}})$ and by an inference according to $(\bigwedge)$ finally $\mathscr{H} \,\frac{|^{\Omega+m}}{|_0}\, \Delta(\underline{\vec{n}}), (A_1 \wedge A_2)(\underline{\vec{n}})$.

In the case of an inference according to $(\forall)$ we have the premise $\frac{|^{m_0}}{|_{\mathsf{T}}} \Delta(\vec{x}), A(u)$ where $u$ does not occur in $\Delta(\vec{x})$ and obtain $\mathscr{H} \,\frac{|^{\Omega+m_0}}{|_0}\, \Delta(\underline{\vec{n}}), A_u(\underline{k})$ for all numerals $\underline{k}$. By Lemma 9.3.9 and an inference $(\bigwedge)$ we obtain the claim.

The cases of inferences according to $(\vee)$ and $(\exists)$ are treated similarly. ☐

It is obvious that all defining axioms and also all identity axioms, but the last, are controlled derivable with a derivation depth below $\omega$. With controlled tautology we also immediately get cut free controlled derivations of depths below $\Omega + \omega$ for all translations of the last identity axiom. Ruminating the proof of the Induction Lemma (Lemma 7.3.4) shows that this proof is controlled by any Cantorian closed Skolem–hull operator. Summing up we get

$$\mathscr{H} \,\frac{|^{\Omega+\omega+4}}{|_0}\, G^* \tag{9.24}$$

for every axiom $G$ of $\mathsf{NT}$ in the language $\mathscr{L}(\mathsf{ID})$ where $\mathscr{H}$ may be an arbitrary Cantorian closed Skolem–hull operator.

So it remains to check the schemes $(ID_1^1)$ and $(ID_1^2)$. By the Collapsing Lemma (Lemma 9.4.5) we have only to deal with $(ID_1^2)$.

**9.5.3 Lemma** *(Generalized Induction) Let $F(X,\vec{x})$ be an X-positive* NT *formula. Then*

$$\mathscr{H} \Big|\frac{2\cdot \mathrm{rnk}(G)+\omega\cdot(\alpha+1)}{0} \neg Cl_F(G),\ \vec{\underline{n}} \notin I_F^{\alpha},\ G(\vec{\underline{n}})$$

*holds true for any sentence $G(\vec{n})$ in the fragment $\mathscr{L}_\infty(\mathsf{NT})$ and for any Cantorian closed Skolem–hull operator $\mathscr{H}$.*

From the Generalized Induction Lemma we obtain

$$\mathscr{H} \Big|\frac{\Omega\cdot 2+3}{0} \neg Cl_F(G), (\forall \vec{x})[\vec{x} \,\varepsilon\, I_F^{<\Omega} \to G(\vec{x})] \tag{9.25}$$

which is the translation of the scheme $(ID_1^2)$.

The proof of Lemma 9.5.3 still needs a preparing lemma.

**9.5.4 Lemma** *(Monotonicity Lemma) Assume $\mathscr{H} \Big|\frac{\alpha}{\rho} \Delta, \neg G(\vec{n}), H(\vec{n})$ for all tuples $\vec{n}$ of numerals and let $F(X,\vec{x})$ be an X-positive $\mathscr{L}(\mathsf{NT})$-formula. Then we obtain $\mathscr{H} \Big|\frac{\alpha+2\cdot\mathrm{rnk}(F)}{\rho} \Delta, \neg F(G,\vec{n}), F(H,\vec{n})$ for all tuples $\vec{n}$.*

*Proof* Induction on $\mathrm{rnk}(F)$. In the case that $F$ is the formula $(\vec{x} \in X)$, we obtain the claim from the hypothesis $\mathscr{H} \Big|\frac{\alpha}{\rho} \Delta, \neg G(\vec{n}), H(\vec{n})$. The remaining cases are as in the proof of the Controlled Tautology Lemma. □

*Proof* of the Generalized Induction Lemma. If $\alpha = 0$ then

$$\mathscr{H} \Big|\frac{2\cdot\mathrm{rnk}(G)+\omega\cdot\alpha+1}{0} \neg Cl_F(G),\ \vec{\underline{n}} \notin I_F^{<\alpha}, G(\vec{\underline{n}}) \tag{i}$$

for all $\vec{\underline{n}}$ by an inference $(\bigwedge)$ with empty premise. If $\alpha > 0$ we get (i) by induction hypothesis. From (i) we obtain

$$\mathscr{H} \Big|\frac{2\cdot\mathrm{rnk}(G)+\omega\cdot\alpha+2\cdot\mathrm{rnk}(F)+1}{0} \neg Cl_F(G),\ \vec{\underline{n}} \notin I_F^{\alpha}, F(G,\vec{\underline{n}}) \tag{ii}$$

for all $\vec{\underline{n}}$ by the Monotonicity Lemma. By controlled tautology we have

$$\mathscr{H} \Big|\frac{2\cdot\mathrm{rnk}(G)}{0} \neg Cl_F(G),\ \vec{\underline{n}} \notin I_F^{\alpha}, \neg G(\vec{\underline{n}}), G(\vec{\underline{n}}). \tag{iii}$$

From (ii) and (iii) we get

$$\mathscr{H} \Big|\frac{2\cdot\mathrm{rnk}(G)+\omega\cdot\alpha+(2\cdot\mathrm{rnk}(F))+2}{0} \neg Cl_F(G),\ \vec{\underline{n}} \notin I_F^{\alpha}, F(G,\vec{\underline{n}}) \land \neg G(\vec{\underline{n}}), G(\vec{\underline{n}}) \tag{iv}$$

by an inference $(\bigwedge)$. From (iv) we finally obtain

$$\mathscr{H} \Big|\frac{2\cdot\mathrm{rnk}(G)+\omega\cdot\alpha+2\cdot\mathrm{rnk}(F)+3}{0} \neg Cl_F(G),\ \vec{\underline{n}} \notin I_F^{\alpha}, G(\vec{\underline{n}})$$

by an inference $(\bigvee)$. Since $2\cdot\mathrm{rnk}(G)+\omega\cdot\alpha+2\cdot\mathrm{rnk}(F)+3 < 2\cdot\mathrm{rnk}(G)+\omega\cdot(\alpha+1)$ we are done. □

**9.5.5 Theorem** *Assume* $\mathsf{ID}_1 \vdash F(\vec{x})$ *for a formula* $F(\vec{x})$ *whose free variables occur all in the list* $\vec{x}$. *Then there are finitely many instances* $Cl_{F_1}(I_{F_1}^{<\Omega}),\dots,Cl_{F_k}(I_{F_k}^{<\Omega})$ *of translations of axiom* $(ID_1^1)$ *and an* $n<\omega$ *such that*

$$\mathscr{H} \,\Big|\!\frac{\Omega\cdot 2+\omega}{\Omega+n}\; \neg Cl_{F_1}(I_{F_1}^{<\Omega}),\dots,\neg Cl_{F_k}(I_{F_k}^{<\Omega}),F^*(\vec{m})$$

*holds true for any tuple* $\vec{m}$ *of the length of* $\vec{x}$ *and for any Cantorian closed Skolem–hull operator.*

*Proof* If $\mathsf{ID}_1 \vdash F(\vec{x})$, then there are finitely many axioms $A_1,\dots,A_r$ and a natural number $p$ such that $\Big|\!\frac{p}{\mathsf{T}}\; \neg A_1,\dots,\neg A_r,F(\vec{x})$. By Theorem 9.5.1 this implies

$$\mathscr{H} \,\Big|\!\frac{\Omega+p}{0}\; \neg A_1^*,\dots,\neg A_r^*,F^*(\vec{m}) \tag{i}$$

for any Cantorian closed Skolem–hull operator $\mathscr{H}$. From (i), (9.24) and (9.25) we obtain the claim by some cuts. □

Let $\mathscr{B}(X)$ be the least set $Y\supseteq X$ such that $\{0,\Omega\}\subseteq Y$ and $\alpha\in Y \;\Leftrightarrow\; SC(\alpha)\subseteq Y$. Then $\mathscr{B}(0)$ is a Cantorian closed operator such that $\mathscr{B}(\emptyset)\cap\Omega=\Gamma_0$, which shows that $\mathscr{B}(0)$ is also transitive. It is the smallest possible extension of the minimal operator $\mathscr{B}_0$ which contains $\Omega$ (cf.(9.16) in p. 167). We construct the hierarchy $\mathscr{B}_\alpha$ of Cantorian closed operators based on $\mathscr{B}$ and put $\psi(\alpha):=\psi_{\mathscr{B}}(\alpha)$. The ordinal $\psi(\varepsilon_{\Omega+1})$ is known as the Bachmann–Howard ordinal.

**9.5.6 Theorem** *(The Upper Bound for* $\mathsf{ID}_1$*)* *It is* $\kappa^{\mathsf{ID}_1}\le\psi(\varepsilon_{\Omega+1})$.

*Proof* If $\mathsf{ID}_1 \vdash \underline{m}\,\varepsilon\, I_F$ we obtain by Theorem 9.5.5

$$\mathscr{B} \,\Big|\!\frac{\Omega\cdot 2+\omega}{\Omega+n}\; \neg Cl_{F_1}(I_{F_1}^{<\Omega}),\dots,\neg Cl_{F_k}(I_{F_k}^{<\Omega}),\underline{m}\,\varepsilon\, I_F^{<\Omega}. \tag{i}$$

By Theorem 9.3.15, we obtain an $\alpha\in\mathscr{B}_{\varepsilon_{\Omega+1}}(\emptyset)\cap\varepsilon_{\Omega+1}$ such that

$$\mathscr{B} \,\Big|\!\frac{\alpha}{\Omega}\; \neg Cl_{F_1}(I_{F_1}^{<\Omega}),\dots,\neg Cl_{F_k}(I_{F_k}^{<\Omega}),\underline{m}\,\varepsilon\, I_F^{<\Omega}. \tag{ii}$$

From (ii) and the Collapsing Lemma (Lemma 9.4.5) it follows

$$\mathscr{B}_{\omega^\alpha+1} \,\Big|\!\frac{\psi(\omega^\alpha)}{\Omega}\; \underline{m}\,\varepsilon\, I_F^{<\Omega}$$

which by Corollary 9.3.17 implies $|m|_F<\psi(\omega^\alpha)<\psi(\varepsilon_{\Omega+1})$. □

## 9.6 The Lower Bound

### *9.6.1 Coding Ordinals in* $\mathscr{L}(\mathsf{NT})$

It follows from the previous sections that $\mathscr{B}_{\varepsilon_{\Omega+1}}(\emptyset)\cap\varepsilon_{\Omega+1}$ is the set of ordinals that is relevant in the computation of an upper bound for $\kappa^{\mathsf{ID}_1}$. To prove that $\psi(\varepsilon_{\Omega+1})$ is the exact bound it suffices to show that $\mathsf{ID}_1$ proves $\underline{n}\,\varepsilon\,\mathrm{Acc}^{\alpha}_{\prec}$ for some arithmetical definable relation $\prec$ and all $\alpha < \psi(\varepsilon_{\Omega+1})$. We are even going to show that for each $\alpha < \psi(\varepsilon_{\Omega+1})$ there is a primitive recursive relation $\prec$ such that $\mathrm{otyp}(\prec) = \alpha$ and $\mathsf{ID}_1$ proves $(\forall x\,\varepsilon\,\mathit{field}(\prec))[x \in \mathrm{Acc}_{\prec}]$. Therefore, we get by (9.5) (on p. 158) also $\mathsf{ID}_1(X) \vdash TI(\prec, X)$ which entails $\psi(\varepsilon_{\Omega+1}) \leq \|\mathsf{ID}_1(X)\|$. Hence $\|\mathsf{ID}_1(X)\| = \psi(\varepsilon_{\Omega+1})$.

Since we cannot talk about ordinals in $\mathcal{L}(\mathsf{ID})$ we need codes for the ordinals in $\mathcal{B}_{\varepsilon_{\Omega+1}}(\emptyset)$. The only parameters occurring on $\mathcal{B}_{\varepsilon_{\Omega+1}}(\emptyset)$ are 0 and $\Omega$. Therefore every ordinal in $\mathcal{B}_{\varepsilon_{\Omega+1}}(\emptyset)$ possesses a term notation which is built up from 0, $\Omega$ by the functions $+$, $\varphi$ and $\psi$. This term notation, however, is not unique. To show that the set of term notations together with the induced $<$-relation on the terms are primitive recursive, we need an unique term notation. This forces us to inspect the set $\mathcal{B}_\alpha(\emptyset)$ more closely.

We define

$$\alpha =_{NF} \psi(\beta) \;:\Leftrightarrow\; \alpha = \psi(\beta) \wedge \beta \in \mathcal{B}_\beta(\emptyset).$$

Then we obtain

$$\alpha =_{NF} \psi(\beta_1) \wedge \alpha =_{NF} \psi(\beta_2) \;\Rightarrow\; \beta_1 = \beta_2 \tag{9.26}$$

since the assumption $\beta_1 < \beta_2$ would imply $\beta_1 \in \mathcal{B}_{\beta_1}(\emptyset) \cap \beta_2 \subseteq \mathcal{B}_{\beta_2}(\emptyset) \cap \beta_2$ which by (9.20) entails $\psi(\beta_1) < \psi(\beta_2)$, contradicting $\psi(\beta_1) = \psi(\beta_2)$. The assumption $\beta_2 < \beta_1$ leads to a similar contradiction.

We define a set of ordinals $T$ by the clauses

$(T_0)$ $\{0, \Omega\} \subseteq T$

$(T_1)$ $\alpha \notin SC \wedge SC(\alpha) \subseteq T \;\Rightarrow\; \alpha \in T$

$(T_2)$ $\beta \in T \wedge \alpha =_{NF} \psi(\beta) \;\Rightarrow\; \alpha \in T$.

Let $\Gamma := en_{SC}$ be the enumerating function of the strongly critical ordinals. Commonly we write $\Gamma_\xi$ instead of $\Gamma(\xi)$. Then we obtain $\Gamma_{\Omega+1} = \min\{\alpha \in SC \mid \Omega < \alpha\}$. We want to prove

$$T = \mathcal{B}_{\Gamma_{\Omega+1}}(\emptyset). \tag{9.27}$$

The inclusion $\subseteq$ in (9.27) is obvious. Troublesome is the converse inclusion. The idea is of course to prove

$$\xi \in \mathcal{B}_{\Gamma_{\Omega+1}}(\emptyset) \;\Rightarrow\; \xi \in T \tag{9.28}$$

by induction on the definition of $\xi \in \mathcal{B}_{\Gamma_{\Omega+1}}(\emptyset)$. We will therefore redefine the sets $\mathcal{B}_\alpha(\emptyset)$ more carefully by the following clauses.

$(B_0)$ $\{0, \Omega\} \subseteq B^n_\alpha$ for all $\alpha$ and all $n < \omega$

$(B_1)$ $\xi \notin SC \wedge SC(\xi) \subseteq B^n_\alpha \;\Rightarrow\; \xi \in B^{n+1}_\alpha$

$(B_2)$ $\eta \in B_\alpha^n \cap \alpha \Rightarrow \psi(\eta) \in B_\alpha^{n+1}$

$(B_3)$ $B_\alpha := \bigcup_{n\in\omega} B_\alpha^n \wedge \psi(\alpha) := \min\{\xi \mid \xi \notin B_\alpha\}$.

We first check that

$$B_\alpha = \mathscr{B}_\alpha(\emptyset) \text{ for all } \alpha \leq \Gamma_{\Omega+1} \tag{9.29}$$

which justifies the use of the same symbol $\psi$ to denote the function $\psi_{\mathscr{B}}(\alpha) := \min\{\xi \mid \xi \notin \mathscr{B}_\alpha(\emptyset)\}$ and $\psi_B(\alpha) := \min\{\xi \mid \xi \notin B_\alpha\}$. The proof of (9.29) is by induction on $\alpha$. By definition $\mathscr{B}_\alpha(\emptyset)$ is the least set, which contains $\{0,\Omega\}$ and is closed under $+$, $\varphi$ and $\psi_{\mathscr{B}}\restriction\alpha$. The functions $\psi_{\mathscr{B}}\restriction\alpha$ and $\psi_B\restriction\alpha$ coincide by induction hypothesis and $B_\alpha$ contains $\{0,\Omega\}$ and is closed under $+$, $\varphi$ and $\psi_B\restriction\alpha$ by definition. So $\mathscr{B}_\alpha(\emptyset) \subseteq B_\alpha$. For the opposite direction we show

$$\xi \in B_\alpha^n \Rightarrow \xi \in \mathscr{B}_\alpha(\emptyset)$$

by side induction on $n$. This is obvious for the cases $(B_0)$ and $(B_1)$ and needs the induction hypothesis $\psi_{\mathscr{B}}\restriction\alpha = \psi_B\restriction\alpha$ in case $(B_2)$. □

So (9.28) can be shown by proving

$$\xi \in B_\alpha^n \Rightarrow \xi \in T \tag{9.30}$$

for all $\alpha < \Gamma_{\Omega+1}$ by induction on $n$. But still troublesome in pursuing this strategy is, case $(B_2)$. In this case, we do not know if $\psi(\eta)$ is in normal-form, i.e., if $\eta \in B_\eta$. First we show that a normal-form always exists.

**9.6.1 Lemma** *For every ordinal $\alpha < \Gamma_{\Omega+1}$ there exists an ordinal $\alpha_{\mathsf{nf}}$ such that $\psi(\alpha) =_{NF} \psi(\alpha_{\mathsf{nf}})$.*

*Proof* Put $\alpha_{\mathsf{nf}} := \min\{\xi \mid \alpha \leq \xi \in B_\alpha\}$. For any ordinal $\alpha$ we have $\varphi_\Omega^n(0) \in B_\alpha$ and $\Gamma_{\Omega+1} = \sup_{n\in\omega} \varphi_\Omega^n(0)$. Therefore $\alpha_{\mathsf{nf}}$ exists. But $[\alpha, \alpha_{\mathsf{nf}}) \cap B_\alpha = \emptyset$ holds true by definition and this implies $B_\alpha = B_{\alpha_{\mathsf{nf}}}$ and thus also $\psi(\alpha) = \psi(\alpha_{\mathsf{nf}})$. Since $\alpha_{\mathsf{nf}} \in B_\alpha = B_{\alpha_{\mathsf{nf}}}$ we have $\psi(\alpha) =_{NF} \psi(\alpha_{\mathsf{nf}})$. □

Our troubles are solved as soon as we can show

$$\eta \in B_\alpha^n \Rightarrow \eta_{\mathsf{nf}} \in B_\alpha^n. \tag{9.31}$$

Then we may argue in case $(B_2)$ that for $\eta \in B_\alpha^{n-1}$ we also have $\eta_{\mathsf{nf}} \in B_\alpha^{n-1}$ and thus $\eta_{\mathsf{nf}} \in T$ which entails $\psi(\eta) =_{NF} \psi(\eta_{\mathsf{nf}}) \in T$.

We obtain (9.31) as a special case of the following lemma whose proof is admittedly tedious. Also we cannot learn much from it. Therefore one commonly includes the normal-form condition into clause $(B_2)$, which then becomes

$(B_2)'$ $\eta \in B_\alpha^{n-1} \cap \alpha \wedge \eta \in B_\eta \Rightarrow \psi(\eta) \in B_\alpha^n$.

The proof of (9.30) then becomes trivial.[8]

[8] The importance of the normal form condition appeared already in the proof of Lemma 9.4.2. A possibility to include the normal form condition in the definition of the iterations of Skolem-hull operators is to define $\psi_{\mathscr{H}}(\alpha) := \min\{\xi \mid \xi \in \mathscr{H}_\xi(\emptyset) \wedge \xi \notin \mathscr{H}_\alpha(\emptyset)\}$.

**9.6.2 Lemma** *Let $\delta(\alpha) := \min\{\xi \mid \alpha \le \xi \in B_\delta\}$. Then $\alpha \in B^n_\beta$ implies $\delta(\alpha) \in B^n_\beta$ for all $\alpha < \Gamma_{\Omega+1}$.*

*Proof* We show the lemma by induction on $n$. First observe that by the minimality of $\delta(\alpha)$ we get

$$\alpha \in \mathbb{H} \Rightarrow \delta(\alpha) \in \mathbb{H} \text{ and } \alpha \in SC \Rightarrow \delta(\alpha) \in SC. \tag{i}$$

The lemma is trivial if $\alpha \in B_\delta$. Then $\delta(\alpha) = \alpha$. Therefore we assume

$$\alpha \notin B_\delta. \tag{ii}$$

Then $\alpha < \delta(\alpha)$ and for $\alpha < \Omega$ we get by (9.21), $\delta(\alpha) = \Omega \in B^n_\delta$ for any $n$. Therefore we may also assume

$$\Omega \le \alpha. \tag{iii}$$

We have

$$\xi \notin SC \wedge \xi \in B^n_\beta \;\Rightarrow\; SC(\xi) \subseteq B^{n-1}_\beta. \tag{iv}$$

Since $(\Omega, \Gamma_{\Omega+1}) \cap SC = \emptyset$ we obtain by induction hypothesis

$$\delta(SC(\alpha)) := \{\delta(\xi) \mid \xi \in SC(\alpha)\} \subseteq B^{n-1}_\beta \cap B_\delta. \tag{v}$$

We are done if we can prove

$$SC(\delta(\alpha)) \subseteq B^{n-1}_\beta \cap B_\delta. \tag{vi}$$

To prove (vi) we extend Definition (8.6) (on p. 145) of the norm of ordinals below $\Gamma_0$ to all ordinals by putting $N(\alpha) := 0$ for $\alpha \in SC$. We prove the more general claim

$$\delta(SC(\gamma)) \subseteq B^{n-1}_\beta \cap B_\delta \;\Rightarrow\; SC(\delta(\gamma)) \subseteq B^{n-1}_\beta \cap B_\delta \tag{vii}$$

by induction on $N(\gamma)$.

1. If $\gamma \in SC$ then $\gamma \le \Omega$ and $\delta(\gamma) \in \{\gamma, \Omega\}$. For $\delta(\gamma) = \Omega$ the claim holds trivially and for $\delta(\gamma) = \gamma$ we get $SC(\delta(\gamma)) = \{\gamma\} = \delta(SC(\gamma)) \subseteq B^{n-1}_\beta \cap B_\delta$.

2. Let $\gamma =_{NF} \gamma_1 + \cdots + \gamma_k$ and put $i := \min\{j \mid \gamma_j < \delta(\gamma_j)\}$. If $i$ is undefined we have $\delta(\gamma) =_{NF} \gamma_1 + \cdots + \gamma_k = \gamma$ and $\delta(SC(\gamma)) = \delta(SC(\gamma_1)) \cup \cdots \cup \delta(SC(\gamma_k)) \subseteq B^{n-1}_\beta \cap B_\delta$ and obtain $SC(\delta(\gamma)) = SC(\delta(\gamma_1)) \cup \cdots SC(\delta(\gamma_k)) \subseteq B^{n-1}_\beta \cap B_\delta$ by induction hypothesis. If $i$ is defined we claim

$$\delta(\gamma) = \gamma_1 + \cdots + \gamma_{i-1} + \delta(\gamma_i) = \delta(\gamma_1) + \cdots + \delta(\gamma_{i-1}) + \delta(\gamma_i). \tag{viii}$$

From (viii) we obtain (vii) by induction hypothesis. Let $\eta := \gamma_1 + \cdots + \gamma_{i-1}$. Then $\gamma < \eta + \delta(\gamma_i)$ and thus $\delta(\gamma) \le \eta + \delta(\gamma_i)$. Assuming $\delta(\gamma) < \eta + \delta(\gamma_i)$ we get $\eta + \gamma_i \le \gamma \le \delta(\gamma) = \eta + \varepsilon < \eta + \delta(\gamma_i)$ for some $\varepsilon \in B_\delta$. Hence $\gamma_i \le \varepsilon < \delta(\gamma_i)$ for $\varepsilon \in B_\delta$ which contradicts the definition of $\delta(\gamma_i)$.

3. Next assume $\gamma =_{NF} \varphi_{\gamma_1}(\gamma_2)$. Then $\gamma = \varphi_{\gamma_1}(\gamma_2) \le \delta(\gamma) = \varphi_\xi(\eta) \le \varphi_{\delta(\gamma_1)}(\delta(\gamma_2))$ for some $\xi, \eta \in B_\delta$. If $\gamma_1 = \delta(\gamma_1)$ we get $\delta(\gamma_2) = \eta$ by minimality of $\delta(\gamma_2)$ and thus $\delta(\gamma) = \varphi_{\delta(\gamma_1)}(\delta(\gamma_2))$ and (vii) follows from the induction hypothesis. If $\gamma_1 <$

$\delta(\gamma_1)$ and $\gamma \le \delta(\gamma_2)$ we obtain $\delta(\gamma) \le \delta(\gamma_2) \le \delta(\gamma)$ and (vii) follows by induction hypothesis. So assume $\gamma_1 < \delta(\gamma_1)$ and $\delta(\gamma_2) < \gamma$. Let

$$\gamma_3 := \min\{\zeta \mid \gamma \le \varphi_{\delta(\gamma_1)}(\zeta)\}. \tag{ix}$$

We claim

$$\gamma_3 \in B_\beta^{n-1} \cap B_\delta. \tag{x}$$

From (x) we get $\delta(\gamma) \le \varphi_{\delta(\gamma_1)}(\gamma_3)$. If we assume $\delta(\gamma) = \varphi_\xi \eta < \varphi_{\delta(\gamma_1)}(\gamma_3)$ we have $\gamma = \varphi_{\gamma_1}(\gamma_2) < \varphi_{\delta(\gamma_1)}(\gamma_3)$. The assumption $\xi = \delta(\gamma_1)$ yields $\gamma \le \delta(\gamma) = \varphi_{\delta(\gamma_1)}(\eta) < \varphi_{\delta(\gamma_1)}(\gamma_3)$ and thus $\eta < \gamma_3$, contradicting the minimality of $\gamma_3$. Assuming $\delta(\gamma_1) < \xi$ yields $\delta(\gamma) < \gamma_3$ and $\gamma \le \delta(\gamma) = \varphi_{\delta(\gamma_1)}(\delta(\gamma))$, again contradicting the minimality of $\gamma_3$. So it remains $\xi < \delta(\gamma_1)$. But, since $\xi \in B_\delta$, this implies $\xi < \gamma_1$ which in turn entails $\gamma \le \eta \in B_\delta \cap \delta(\gamma)$ contradicting the definition of $\delta(\gamma)$. Therefore we have

$$\delta(\gamma) = \varphi_{\delta(\gamma_1)}(\gamma_3) \tag{xi}$$

and obtain (vi) from (xi) by induction hypothesis and (x).

It remains to prove (x). We are done if $\gamma_3 = 0$. If we assume $\gamma_3 \in Lim$ we get $\gamma =_{NF} \varphi_{\gamma_1}(\gamma_2) = \varphi_{\delta(\gamma_1)}(\gamma_3)$ by the continuity of $\varphi_{\delta(\gamma_1)}$. Since, $\gamma_1 < \delta(\gamma_1)$ we obtain $\gamma_2 = \gamma$ contradicting $\gamma =_{NF} \varphi_{\gamma_1}(\gamma_2)$. It remains in the case that $\gamma_3 = \mu + 1$. Then $\varphi_{\delta(\gamma_1)}(\mu) < \gamma =_{NF} \varphi_{\gamma_1}(\gamma_2) \le \varphi_{\delta(\gamma_1)}(\mu+1)$. Because of $\gamma_1 < \delta(\gamma_1)$ this implies $\varphi_{\delta(\gamma_1)}(\mu) < \gamma_2 \le \delta(\gamma_2) < \gamma = \varphi_{\delta(\gamma_1)}(\mu+1)$. Since $\delta(\gamma_2) \in B_\beta^{n-1} \cap B_\delta$ we have shown

$$B_\delta \cap B_\beta^{n-1} \cap (\varphi_{\delta(\gamma_1)}(\mu), \varphi_{\delta(\gamma_1)}(\mu+1)) \neq \emptyset. \tag{xii}$$

To finish the proof we show that in general, we have

$$B_\beta^n \cap (\varphi_\zeta(\nu), \varphi_\zeta(\nu+1)) \neq \emptyset \;\Rightarrow\; \nu+1 \in B_\beta^n. \tag{9.32}$$

From (xii) and (9.32) we then obtain $\gamma_3 \in B_\delta \cap B_\beta^{n-1}$, i.e., (x).

To prove (9.32) we first show

$$\rho \in [\varphi_\zeta(\nu), \varphi_\zeta(\nu+1)) \;\Rightarrow\; SC(\nu) \subseteq SC(\rho) \tag{xiii}$$

by induction on $N(\rho)$.

If $\rho =_{NF} \rho_1 + \cdots + \rho_k$ we have $\rho_1 \in [\varphi_\zeta(\nu), \varphi_\zeta(\nu+1))$ and obtain $SC(\nu) \subseteq SC(\rho_1) \subseteq SC(\rho)$. Let $\rho =_{NF} \varphi_{\rho_1}(\rho_2)$. If $\zeta < \rho_1$ then $\nu \le \rho < \nu+1$ and thus $\rho = \nu$. If $\zeta = \rho_1$ then $\nu = \rho_2$ and $SC(\nu) = SC(\rho_2) \subseteq SC(\rho)$. If $\rho_1 < \zeta$ then $\varphi_\zeta(\nu) \le \rho_2 < \rho < \varphi_\zeta(\nu+1)$ and we obtain $SC(\nu) \subseteq SC(\rho_2) \subseteq SC(\rho)$ by induction hypothesis. If finally $\rho \in SC$ then $\varphi_\rho(0) = \rho = \varphi_\zeta(\nu)$ and thus $\rho = \nu$ or $\rho = \zeta$ and $\nu = 0$ and the claim in both cases is obvious.

We prove (9.32) by induction on $n$. Let $\sigma \in B_\beta^n \cap (\varphi_\xi(\eta), \varphi_\xi(\eta+1))$. Then $\sigma \notin SC$ and we have $SC(\sigma) \subseteq B_\beta^{n-1}$. By (xiii) we get $SC(\nu) \subseteq SC(\sigma) \subseteq B_\beta^{n-1}$. Since $0 \in B_\beta^{n-1}$ we also have $SC(\nu+1) \subseteq B_\beta^{n-1}$ and thus obtain $\nu+1 \in B_\beta^n$. □

Now we have all the material to prove (9.27).

**9.6.3 Theorem** *It is* $\mathcal{B}_{\Gamma_{\Omega+1}}(\emptyset) = B_{\Gamma_{\Omega+1}} = T$.

*Proof* We easily obtain $\alpha \in T \Rightarrow \alpha \in \mathcal{B}_{\Gamma_{\Omega+1}}(\emptyset)$ by induction on the definition of $\alpha \in T$. For the opposite inclusion we use (9.29) (in p. 183) and prove

$$\xi \in B^n_\alpha \;\Rightarrow\; \xi \in T,$$

i.e., (9.30), by induction on $n$. If $\xi \in B^n_\alpha$ by $(B_0)$ we obtain $\xi \in T$ by $(T_0)$. If $\xi \in B^{n+1}_\alpha$ by $(B_1)$ we obtain $\xi \in T$ by the induction hypothesis and $(T_1)$. So assume that $\xi \in B^{n+1}_\alpha$ by $(B_2)$. Then $\xi = \psi(\eta)$ such that $\eta \in B^n_\alpha \cap \alpha$. By Lemma 9.6.1 we obtain $\psi(\eta) =_{NF} \psi(\eta(\eta))$ and by Lemma 9.6.2 $\eta(\eta) \in B^n_\alpha$. So $\eta(\eta) \in T$ by induction hypothesis and it follows $\xi = \psi(\eta) =_{NF} \psi(\eta(\eta)) \in T$ by $(T_2)$. □

Having shown Theorem 9.6.3 we want to develop a primitive recursive notation system for the ordinals in $T$. But still annoying is the normal-form condition in clause $(T_2)$. To define a set On of notions for ordinals in $T$ together with a $<$-relation on On by simultaneous course-of-values recursion, we should try to replace the condition $\beta \in B_\beta$ in $\alpha =_{NF} \psi(\beta)$ by a condition that refers only to proper subterms of $\beta$. To do that we have

$$\begin{aligned} \xi \in B_\beta \;\Leftrightarrow\; & \xi = 0 \vee \xi = \Omega \vee \\ & (\xi \notin SC \wedge SC(\xi) \subseteq B_\beta) \vee \\ & (\xi = \psi(\eta) \wedge \eta \in B_\beta \cap \beta). \end{aligned} \tag{9.33}$$

From (9.33) we read off the following definition.

**9.6.4 Definition** Let

$$K(\xi) := \begin{cases} \emptyset & \text{if } \xi = 0 \text{ or } \xi = \Omega \\ \bigcup \{K(\eta) \mid \eta \in SC(\xi)\} & \text{if } \xi \notin SC \\ \{\eta\} \cup K(\eta) & \text{if } \xi = \psi(\eta). \end{cases}$$

From (9.33) and Definition 9.6.4 we immediately get

**9.6.5 Lemma** *It is* $\xi \in B_\beta$ *iff* $K(\xi) \subseteq \beta$.

**9.6.6 Corollary** *We have* $\alpha =_{NF} \psi(\beta)$ *iff* $\alpha = \psi(\beta)$ *and* $K(\beta) \subseteq \beta$.

Another obstacle for a notation system that assigns an uniquely determined term notation to the ordinals in $T$ are the fixed-points of the Veblen functions $\varphi$. This was tolerable in the development of the notation system for the ordinals below $\Gamma_0$. The simultaneous definition of the relations $\prec$ and $\equiv$ by course-of-values recursion was not too difficult. In developing a notation system for the ordinals in $T$ it would, however, cause unnecessary complications. In Exercise 3.4.20 we introduced the fixed-point free versions $\bar{\varphi}$ of the Veblen-functions. For these functions we always have $\bar{\varphi}_{\xi_1}(\eta_1) = \bar{\varphi}_{\xi_2}(\eta_2) \;\Leftrightarrow\; \xi_1 = \xi_2 \wedge \eta_1 = \eta_2$. Therefore we are going to use $\bar{\varphi}$ instead of the functions $\varphi$ in the following sections.

**9.6.7 Definition** We use the known facts about ordinals in $T$ to define sets $\mathsf{SC} \subseteq \mathsf{H} \subseteq \mathsf{On} \subseteq \mathbb{N}$ of ordinal notations together with a finite set $\mathsf{K}(a) \subseteq \mathsf{On}$ of sub-terms of $a \in \mathsf{On}$, a relation $\prec\ \subseteq \mathsf{On} \times \mathsf{On}$ and an evaluation function $|\ |_{\mathcal{O}}\colon \mathsf{On} \longrightarrow T$ by the following clauses where we use $a \preceq b$ to denote $a \prec b \vee a = b$.

Definition of SC, H and On.

- Let $\langle 0\rangle \in \mathsf{On}$, $\langle 1\rangle \in \mathsf{SC}$, $|\langle 0\rangle|_{\mathcal{O}} := 0$ and $|\langle 1\rangle|_{\mathcal{O}} := \omega_1$
- If $n > 1$, $a_1,\dots,a_n \in \mathsf{H}$ and $a_1 \succeq \cdots \succeq a_n$ then we put $\langle 1,a_1,\dots,a_n\rangle \in \mathsf{On}$ and define $|\langle 1,a_1,\dots,a_n\rangle|_{\mathcal{O}} := |a_1|_{\mathcal{O}} + \cdots + |a_n|_{\mathcal{O}}$
- If $a,b \in \mathsf{On}$ then $\langle 2,a,b\rangle \in \mathsf{H}$ and $|\langle 2,a,b\rangle|_{\mathcal{O}} = \bar{\varphi}_{|a|_{\mathcal{O}}}(|b|_{\mathcal{O}})$
- If $a \in \mathsf{On}$ and $b \prec a$ for all $b \in \mathsf{K}(a)$ then $\langle 3,a\rangle \in \mathsf{SC}$ and $|\langle 3,a\rangle|_{\mathcal{O}} := \psi(|a|_{\mathcal{O}})$

Definition of $\mathsf{K}(a)$.

- $\mathsf{K}(\langle 0\rangle) = \mathsf{K}(\langle 1\rangle) = \emptyset$
- $\mathsf{K}(\langle 1,a_1,\dots,a_n\rangle) = \mathsf{K}(a_1) \cup \cdots \cup \mathsf{K}(a_n)$
- $\mathsf{K}(\langle 2,a,b\rangle) = \mathsf{K}(a) \cup \mathsf{K}(b)$
- $\mathsf{K}(\langle 3,a\rangle) = \{a\} \cup \mathsf{K}(a)$

Let $a,b \in \mathsf{On}$. Then $a \prec b$ iff one of the following conditions is satisfied.

- $a = \langle 0\rangle$ and $b \neq \langle 0\rangle$
- $a = \langle 1,a_1,\dots,a_m\rangle$, $b = \langle 1,b_1,\dots,b_n\rangle$ and $(\exists i < m)(\forall j \le i)[a_j = b_j \wedge a_{i+1} \prec b_{i+1}]$ or $m < n \wedge (\forall j \le m)[a_j = b_j]$
- $a = \langle 1,a_1,\dots,a_n\rangle$, $b \in \mathsf{H}$ and $a_1 \prec b$
- $a \in \mathsf{H}$, $b = \langle 1,b_1,\dots,b_n\rangle$ and $a \preceq b_1$
- $a = \langle 2,a_1,a_2\rangle$, $b = \langle 2,b_1,b_2\rangle$ and one of the following conditions is satisfied

  $a_1 \prec b_1$ and $a_2 \prec b$

  $a_1 = b_1$ and $a_2 \prec b_2$

  $b_1 \prec a_1$ and $a \preceq b_2$
- $a = \langle 2,a_1,a_2\rangle$, $b \in \mathsf{SC}$ and $a_1,a_2 \prec b$
- $a \in \mathsf{SC}$, $b = \langle 2,b_1,b_2\rangle$ and $a \preceq b_1$ or $a \preceq b_2$

- $a = \langle 3, a_1 \rangle$, $b = \langle 3, b_1 \rangle$ and $a_1 \prec b_1$
- $a = \langle 3, a_1 \rangle$ and $b = \langle 1 \rangle$

Collecting all the known facts about $T$ and observing that On, SC, H, K$(a)$ and $\prec$ are defined by simultaneous course-of-values recursion, we have the following theorem.

**9.6.8 Theorem** *The sets* On, H *and* SC *as well as the relations* K *and* $\prec$ *are primitive recursive. The map* $|\ |_{\mathcal{O}}$:On $\longrightarrow T$ *is one-one and onto such that* $a \prec b$ *iff* $|a|_{\mathcal{O}} < |b|_{\mathcal{O}}$.

*Proof* Since we defined On, H and SC as well as the relations K and $\prec$ by simultaneous course-of-values recursion, all these sets and relations are primitive recursive. The fact that $|a|_{\mathcal{O}} \in T$ follows immediately from the definition of the map $|\ |_{\mathcal{O}}$. It remains to show that $|\ |_{\mathcal{O}}$ is onto. We prove:

*"For every ordinal* $\alpha \in T$ *there is a notation* $\ulcorner\alpha\urcorner \in$ On *such that*

*(a)* $\alpha = |\ulcorner\alpha\urcorner|_{\mathcal{O}}$,

*(b)* $\alpha \in \mathbb{H} \Leftrightarrow \ulcorner\alpha\urcorner \in \mathsf{H}$,

*(c)* $\alpha \in SC \Leftrightarrow \ulcorner\alpha\urcorner \in \mathsf{SC}$,

*(d)* $\mathsf{K}(\ulcorner\alpha\urcorner) = \{\ulcorner\beta\urcorner \mid \beta \in K(\alpha)\}$,

*(e)* $\beta < \alpha \Leftrightarrow \ulcorner\beta\urcorner \prec \ulcorner\alpha\urcorner$ *and*

*(f)* $\alpha = \beta \Leftrightarrow \ulcorner\alpha\urcorner = \ulcorner\beta\urcorner$"

by induction on the definition of $\alpha \in T$.

We put $\ulcorner 0\urcorner := \langle 0 \rangle$ and $\ulcorner\Omega\urcorner := \langle 1 \rangle$. If $\alpha =_{NF} \bar{\varphi}_{\alpha_1}(\beta_1) + \cdots + \bar{\varphi}_{\alpha_n}(\beta_n)$ for $n > 1$, we put $\ulcorner\alpha\urcorner := \langle 1, \langle 2, \ulcorner\alpha_1\urcorner, \ulcorner\beta_1\urcorner \rangle, \ldots, \langle 2, \ulcorner\alpha_n\urcorner, \ulcorner\beta_n\urcorner \rangle\rangle$. Then (a) follows directly from the induction hypothesis and the definition of $|\ |_{\mathcal{O}}$. Claims (b) and (c) hold trivially and (d) follows from the induction hypothesis by comparing definitions 9.6.4 and 9.6.7. If $\alpha = \bar{\varphi}_{\alpha_1}(\beta_1)$ then $\alpha \notin SC$ and we define $\ulcorner\alpha\urcorner := \langle 2, \ulcorner\alpha_1\urcorner, \ulcorner\alpha_2\urcorner \rangle$ and obtain claim (a) again from the induction hypothesis and the definition of $|\ |_{\mathcal{O}}$. Claim (b) and (c) follow directly and claim (d) from the induction hypothesis and a comparison of definitions 9.6.4 and 9.6.7. If $\alpha =_{NF} \psi(\alpha_0)$ we have $K(\alpha_0) \subseteq \alpha_0$ by Corollary 9.6.6. Then we obtain $\mathsf{K}(\ulcorner\alpha_0\urcorner) = \{\ulcorner\beta\urcorner \mid \beta \in K(\alpha_0)\}$ and thus $\ulcorner\beta\urcorner \prec \ulcorner\alpha_0\urcorner$ for all $\ulcorner\beta\urcorner \in \mathsf{K}(\ulcorner\alpha_0\urcorner)$. Hence $\ulcorner\alpha\urcorner := \langle 3, \ulcorner\alpha_0\urcorner \rangle \in$ On and $\alpha = |\ulcorner\alpha\urcorner|_{\mathcal{O}}$ since $\alpha_0 = |\ulcorner\alpha_0\urcorner|_{\mathcal{O}}$ by the induction hypothesis. Claims (b) and (c) again follow directly and claim (d) from the induction hypothesis by comparing definitions 9.6.4 and 9.6.7. Claims (e) and (f) follow from the induction hypothesis, Observation 3.3.9, Exercise 3.4.20 and the fact that, for $\alpha =_{NF} \psi(\alpha_0)$ and $\beta =_{NF} \psi(\beta_0)$ we have $\alpha < \beta$ iff $\alpha_0 < \beta_0$. $\square$

**9.6.9 Corollary** $\psi(\Gamma_{\Omega+1}) < \omega_1^{CK}$.

*Proof* We have $\psi(\Gamma_{\Omega+1}) = \text{otyp}(\prec\restriction\Omega)$. Since $\prec$ is primitive recursive we get $\psi(\Gamma_{\Omega+1}) < \omega_1^{CK}$. □

## 9.6.2 The Well-Ordering Proof

In view of Theorem 9.6.8 we may talk about the ordinals in $\mathscr{B}_{\Gamma_{\Omega+1}}(\emptyset)$ in $\mathscr{L}(\mathsf{NT})$ and $\mathscr{L}(\mathsf{ID})$. For the sake of better readability we will, however, not use the codes but identify ordinals in $\mathscr{B}_{\Gamma_{\Omega+1}}(\emptyset)$ and their codes. We denote (codes of ) ordinals by lower case Greek letters and write $\alpha < \beta$ instead of $\alpha \prec \beta$. We use the abbreviations and conventions introduced in Sects. 7.4 and 8.5.

The aim of this section is to show that there is a primitive recursive relation $<_0$ such that for every $\alpha < \psi(\varepsilon_{\Omega+1})$ we get $\mathsf{ID}_1 \vdash n\, \varepsilon\, \mathrm{Acc}_{<_0}$ and $|n|_{\mathrm{Acc}_{<_0}} = \alpha$. The strategy of the proof will be the following.

- We first define a relation $<_1$ for which $\mathsf{TI}_1(\Omega, X)$ holds trivially. The relation $<_1$ is no longer arithmetically definable but needs a fixed point in its definition. Then we use the well-ordering proof of Sect. 7.4 to obtain $\mathsf{TI}_1(\alpha, X)$ provable in $\mathsf{ID}_1(\vec{X})$ for all $\alpha <_1 \varepsilon_{\Omega+1}$.
- By a *condensing argument* we then show that $\mathsf{TI}_1(\alpha, X)$ implies $\psi(\alpha)\, \varepsilon\, \mathrm{Acc}_{<_0}$.

**9.6.10 Definition** For ordinals $\alpha, \beta$ we define

- $\alpha <_0 \beta \;:\Leftrightarrow\; \alpha < \beta < \Omega$.

Let $F(u)$ be an $\mathscr{L}(\mathsf{ID})$ formula. By $\xi \subseteq_0 F$ we denote the formula $(\forall \eta <_0 \xi)[F_u(\eta)]$.

Let $\mathscr{A}cc$ be the fixed point of the operator induced by the formula

$(\forall \eta < \xi)[\xi < \Omega \wedge \eta\, \varepsilon\, X]$ i.e., $\mathscr{A}cc = \mathrm{Acc}_{<_0} \cap \Omega$.

Let $\mathsf{M} := \{\alpha \mid SC(\alpha) \cap \Omega \subseteq \mathscr{A}cc\}$. For $\alpha, \beta \in \mathsf{On}$ we define

- $\alpha <_1 \beta \;:\Leftrightarrow\; \alpha < \beta \wedge \alpha\, \varepsilon\, \mathsf{M} \wedge \beta\, \varepsilon\, \mathsf{M}$.

$\xi \subseteq_1 F$ stands for $(\forall \eta <_1 \xi)[F_u(\eta)]$.

Let

- $\mathsf{Prog}_i(F) \;:\Leftrightarrow\; (\forall \xi\, \varepsilon\, \mathit{field}(<_i))[(\forall \eta <_i \xi) F(\eta) \to F(\xi)]$
- $\mathsf{TI}_i(\alpha, F) \;:\Leftrightarrow\; \alpha\, \varepsilon\, \mathit{field}(<_i) \wedge \mathsf{Prog}_i(F) \to (\forall \xi <_i \alpha) F(\xi)$.

Observe that by the axioms of $\mathsf{ID}_1$ and (9.3) (on page158) we have

$$\mathsf{ID}_1 \vdash \alpha \,\varepsilon\, \mathscr{A}cc \;\leftrightarrow\; (\alpha < \Omega \wedge \alpha \subseteq \mathscr{A}cc) \tag{9.34}$$

$$\mathsf{ID}_1 \vdash \mathsf{Prog}_0(\mathscr{A}cc) \tag{9.35}$$

$$\mathsf{ID}_1 \vdash \mathsf{Prog}_0(F) \;\rightarrow\; (\forall \xi)[\xi \,\varepsilon\, \mathscr{A}cc \rightarrow F(\xi)] \tag{9.36}$$

**9.6.11 Lemma** *Let* $\mathsf{Prog}(F)$ *abbreviate the formula* $(\forall\alpha)[(\forall\xi < \alpha)F(\xi) \rightarrow F(\alpha)]$. *Then* $\mathsf{ID}_1 \vdash \mathsf{Prog}(F) \rightarrow \mathsf{Prog}_0(F)$ *and thus also* $\mathsf{ID}_1 \vdash \mathsf{Prog}(F) \rightarrow (\forall \xi \,\varepsilon\, \mathscr{A}cc)F(\xi)$.

*Proof* $(\forall\xi <_0 \alpha)F(\xi)$ implies $(\forall\xi < \alpha)F(\xi)$ for $\alpha < \Omega$. Together with $\mathsf{Prog}(F)$ we therefore get $F(\alpha)$, i.e., we have $\mathsf{Prog}_0(F)$. Combining this with (9.36) we obtain the second claim. □

**9.6.12 Lemma** $(\mathsf{ID}_1)$ *The class* $\mathscr{A}cc$ *is closed under ordinal addition.*

*Proof* Let $\mathscr{A}cc_+ := \{\xi \mid (\forall\eta \,\varepsilon\, \mathscr{A}cc)[\eta + \xi \,\varepsilon\, \mathscr{A}cc]\}$. We claim

$$\mathsf{Prog}_0(\mathscr{A}cc_+). \tag{i}$$

To prove (i) we have the hypothesis

$$\alpha < \Omega \text{ and } (\forall \xi < \alpha)[\xi \,\varepsilon\, \mathscr{A}cc_+] \tag{ii}$$

and have to show $\alpha \,\varepsilon\, \mathscr{A}cc_+$, i.e.,

$$(\forall \eta \,\varepsilon\, \mathscr{A}cc)[\eta + \alpha \,\varepsilon\, \mathscr{A}cc]. \tag{iii}$$

By (9.34) it suffices to have

$$\eta + \alpha \subseteq \mathscr{A}cc \tag{iv}$$

to get (iii). Let $\xi < \eta + \alpha$. If $\xi < \eta$ then we get $\xi \,\varepsilon\, \mathscr{A}cc$ from $\eta \,\varepsilon\, \mathscr{A}cc$ by (9.34). If $\eta \leq \xi < \eta + \alpha$, there is a $\rho < \alpha$ such that $\xi = \eta + \rho$. Then we obtain $\eta + \rho \,\varepsilon\, \mathscr{A}cc$ by (ii).

From (i) we obtain

$$(\forall \xi \,\varepsilon\, \mathscr{A}cc)[\xi \,\varepsilon\, \mathscr{A}cc_+] \tag{v}$$

by (9.36) which means

$$(\forall \xi \,\varepsilon\, \mathscr{A}cc)(\forall \eta \,\varepsilon\, \mathscr{A}cc)[\xi + \eta \,\varepsilon\, \mathscr{A}cc].$$

□

**9.6.13 Lemma** $\mathsf{ID}_1 \vdash \mathsf{Prog}_1(F) \;\rightarrow\; \mathsf{Prog}_0(F)$.

*Proof* We have the premises $\mathsf{Prog}_1(F)$, $\alpha < \Omega$ and $(\forall\xi <_0 \alpha)F(\xi)$ and have to show $F(\alpha)$. If $\xi <_1 \alpha$ we get $\xi <_0 \alpha$ by $\alpha < \Omega$ and thus $F(\xi)$ by $(\forall\xi <_0 \alpha)F(\xi)$. Hence $(\forall\xi <_1 \alpha)F(\xi)$, which entails $F(\alpha)$ by $\mathsf{Prog}_1(F)$. □

**9.6.14 Lemma** $(\mathsf{ID}_1)$ *The class $\mathscr{A}cc$ is closed under $\lambda\xi,\eta\,.\,\bar{\varphi}_\xi(\eta)$.*

Define

$$\mathscr{A}cc_\varphi := \{\alpha \mid (\forall\xi\,\varepsilon\,\mathscr{A}cc)[\bar{\varphi}_\alpha(\xi)\,\varepsilon\,\mathscr{A}cc] \lor \alpha \notin \mathsf{M} \lor \Omega \le \alpha\}. \tag{i}$$

We claim

$$\mathsf{Prog}_1(\mathscr{A}cc_\varphi). \tag{ii}$$

To prove (ii) we have the hypothesis

$$(\forall\xi <_1 \alpha)[\xi\,\varepsilon\,\mathscr{A}cc_\varphi] \tag{iii}$$

and have to show

$$\alpha\,\varepsilon\,\mathscr{A}cc_\varphi. \tag{iv}$$

For $\alpha \not\varepsilon\, \mathsf{M}$ or $\Omega \le \alpha$ (iv) is obvious. Therefore assume

$$\alpha\,\varepsilon\,\mathsf{M} \cap \Omega. \tag{v}$$

We have to show

$$(\forall\xi\,\varepsilon\,\mathscr{A}cc)[\bar{\varphi}_\alpha(\xi)\,\varepsilon\,\mathscr{A}cc]. \tag{vi}$$

According to Lemma 9.6.11 we may assume that we have

$$(\forall\eta < \xi)[\eta\,\varepsilon\,\mathscr{A}cc \to \bar{\varphi}_\alpha(\eta)\,\varepsilon\,\mathscr{A}cc] \tag{vii}$$

and have to show

$$\xi\,\varepsilon\,\mathscr{A}cc \to \bar{\varphi}_\alpha(\xi)\,\varepsilon\,\mathscr{A}cc \tag{viii}$$

for which by (9.34) it suffices to prove

$$\rho < \bar{\varphi}_\alpha(\xi) \;\to\; \rho\,\varepsilon\,\mathscr{A}cc. \tag{ix}$$

We show (ix) by Mathematical Induction on the length of the term notation of $\rho$. By (9.34) $0\,\varepsilon\,\mathscr{A}cc$ holds trivially. If $\rho =_{NF} \rho_1 + \cdots + \rho_n$ we have $\rho_i\,\varepsilon\,\mathscr{A}cc$ by induction hypothesis and obtain $\rho\,\varepsilon\,\mathscr{A}cc$ by Lemma 9.6.12. If $\rho \in \mathsf{SC}$ then we have $\rho \le \alpha$ or $\rho \le \xi$. If $\rho \le \xi$ we get $\rho\,\varepsilon\,\mathscr{A}cc$ from $\xi\,\varepsilon\,\mathscr{A}cc$. If $\rho \le \alpha$ we have $\rho \le \mu$ for some $\mu \in SC(\alpha)$. Since $\alpha\,\varepsilon\,\mathsf{M}$ we have $\mu\,\varepsilon\,\mathscr{A}cc$ and thence also $\rho\,\varepsilon\,\mathscr{A}cc$.

Now assume $\rho \in \mathbb{H} \setminus SC$. Then $\rho = \bar{\varphi}_{\rho_1}(\rho_2)$. There are the following cases.

1. $\rho_1 = \alpha$ and $\rho_2 < \xi$. Then we obtain $\bar{\varphi}_{\rho_1}(\rho_2)\,\varepsilon\,\mathscr{A}cc$ by (vii).
2. $\alpha < \rho_1$ and $\rho < \xi$. Then $\rho\,\varepsilon\,\mathscr{A}cc$ follows from $\xi\,\varepsilon\,\mathscr{A}cc$.
3. $\rho_1 < \alpha$ and $\rho_2 < \bar{\varphi}_\alpha(\xi)$. Then $SC(\rho_1) \cap \Omega$ is majorized by some $\mu \in SC(\alpha) \cap \Omega \subseteq \mathscr{A}cc$ which means $SC(\rho_1) \cap \Omega \subseteq \mathscr{A}cc$ and therefore $\rho_1 <_1 \alpha$. By

(iii) we obtain $\rho_1 \varepsilon \mathscr{A}cc_\varphi$. By induction hypothesis we have $\rho_2 \varepsilon \mathscr{A}cc$ and which entails $\bar{\varphi}_{\rho_1}(\rho_2) \varepsilon \mathscr{A}cc$. This finishes the proof of (ii).

To prove the lemma we have to show

$$\alpha, \beta \varepsilon \mathscr{A}cc \;\Rightarrow\; \bar{\varphi}_\alpha(\beta) \varepsilon \mathscr{A}cc. \tag{x}$$

From $\alpha, \beta \varepsilon \mathscr{A}cc$ we get $\alpha, \beta < \Omega$. Since $SC(\alpha) \subseteq \alpha + 1$ this implies $SC(\alpha) \cap \Omega \subseteq \mathscr{A}cc$. Hence $\alpha \varepsilon \mathsf{M} \cap \Omega$. From (ii) and Lemma 9.6.13 we obtain $\mathsf{Prog}_0(\mathscr{A}cc_\varphi)$ and thence $\mathscr{A}cc \subseteq \mathscr{A}cc_\varphi$ by (9.36). Together with $\beta \varepsilon \mathscr{A}cc$ this implies $\bar{\varphi}_\alpha(\beta) \varepsilon \mathscr{A}cc$. □

Put $\mathscr{A}cc_\Omega := \{\alpha \mid \alpha \not\varepsilon \mathsf{M} \vee (\exists \xi \varepsilon K(\alpha))[\alpha \leq \xi] \vee \psi(\alpha) \varepsilon \mathscr{A}cc\}$.

**9.6.15 Lemma** $\mathsf{ID}_1 \vdash \mathsf{Prog}_1(\mathscr{A}cc_\Omega)$.

*Proof* Assume

$$\alpha \varepsilon field(<_1) \text{ and } (\forall \eta <_1 \alpha)[\eta \varepsilon \mathscr{A}cc_\Omega]. \tag{i}$$

We have to show

$$\alpha \varepsilon \mathscr{A}cc_\Omega. \tag{ii}$$

For $\alpha \not\varepsilon \mathsf{M}$ or $(\exists \xi \varepsilon K(\alpha))[\alpha \leq \xi]$ (ii) is obvious. Therefore assume $\alpha \varepsilon \mathsf{M}$ and $K(\alpha) \subseteq \alpha$. To prove (ii) it remains to show

$$\psi(\alpha) \varepsilon \mathscr{A}cc. \tag{iii}$$

For (iii) in turn it suffices to have

$$\rho < \psi(\alpha) \;\rightarrow\; \rho \varepsilon \mathscr{A}cc. \tag{iv}$$

We prove (iv) by Mathematical Induction on the length of the term notation of $\rho$. If $\rho \notin SC$ we get $SC(\rho) \subseteq \mathscr{A}cc$ by induction hypothesis and thence $\rho \varepsilon \mathscr{A}cc$ by Lemma 9.6.12 and Lemma 9.6.14. If $\rho \in SC$ then there is a $\rho_0$ such that $\rho =_{NF} \psi(\rho_0)$ and $\rho_0 < \alpha$ which implies $K(\rho_0) \subseteq \rho_0 < \alpha$. For $\xi \in SC(\rho_0) \cap \Omega$ we have $\xi =_{NF} \psi(\eta)$ for some $\eta$. Then $\eta \varepsilon K(\xi) \subseteq K(\rho_0) \subseteq \alpha$ which implies $\xi = \psi(\eta) < \psi(\alpha)$. Hence $SC(\rho_0) \cap \Omega \subseteq \psi(\alpha)$. By induction hypothesis we therefore obtain $SC(\rho_0) \cap \Omega \subseteq \mathscr{A}cc$. Hence $\rho_0 <_1 \alpha$ and therefore $\rho_0 \varepsilon \mathscr{A}cc_\Omega$ by (i). Since $K(\rho_0) \subseteq \rho_0$ and we just showed $\rho_0 \varepsilon \mathsf{M}$, this implies $\rho = \psi(\rho_0) \varepsilon \mathscr{A}cc$. □

**9.6.16 Lemma** *(Condensation Lemma) Let $\alpha$ be an ordinal in $\mathscr{B}_{\Gamma_{\Omega+1}}$ such that $K(\alpha) \subseteq \alpha$, $\alpha \varepsilon \mathsf{M}$ and $\mathsf{ID}_1 \vdash \mathsf{TI}_1(\alpha, F)$. Then $\mathsf{ID}_1 \vdash \psi(\alpha) \varepsilon \mathscr{A}cc$.*

*Proof* We especially have

$$\mathsf{ID}_1 \vdash \mathsf{TI}_1(\alpha, \mathscr{A}cc_\Omega). \tag{i}$$

From (i) and Lemma 9.6.15 we obtain

$$(\forall \xi <_1 \alpha)[\xi \varepsilon \mathscr{A}cc_\Omega] \tag{ii}$$

and from (ii) and Lemma 9.6.15

$$\alpha \,\varepsilon\, \mathscr{A}cc_{\Omega}. \tag{iii}$$

But (iii) together with the other hypotheses yield $\psi(\alpha)\,\varepsilon\,\mathscr{A}cc$. □

**9.6.17 Lemma** $\mathsf{ID}_1 \vdash \mathsf{TI}_1(\Omega+1,F) \wedge K(\Omega+1) \subseteq \Omega+1 \wedge \Omega+1\,\varepsilon\,\mathsf{M}$.

*Proof* Since $SC(\Omega+1)\cap\Omega = \emptyset$ and $K(\Omega+1) = \emptyset$ we trivially have $K(\Omega+1) \subseteq \Omega+1 \wedge \Omega+1\,\varepsilon\,\mathsf{M}$. Assuming $\mathsf{Prog}_1(F)$ we have to show $(\forall\xi <_1 \Omega+1)[F(\xi)]$. If $\xi <_1 \Omega$ we obtain $SC(\xi) \subseteq \mathscr{A}cc$ and thus $\xi\,\varepsilon\,\mathscr{A}cc$ by Lemma 9.6.12 and Lemma 9.6.14. By Lemma 9.6.13 we get $\mathsf{Prog}_0(F)$ which then by (9.36) entails $F(\xi)$. So we have $(\forall\xi <_1 \Omega)[F(\xi)]$ which by $\mathsf{Prog}_1(F)$ also implies $F(\Omega)$. □

**9.6.18 Lemma** *Assume*

$$\mathsf{ID}_1 \vdash \mathsf{TI}_1(\alpha,F) \wedge K(\alpha) \subseteq \alpha \wedge \alpha\,\varepsilon\,\mathsf{M}.$$

*Then also*

$$\mathsf{ID}_1 \vdash \mathsf{TI}_1(\omega^{\alpha},F) \wedge K(\omega^{\alpha}) \subseteq \omega^{\alpha} \wedge \omega^{\alpha}\,\varepsilon\,\mathsf{M}$$

*holds true.*

*Proof* We show

$$\mathsf{ID}_1 \vdash \mathsf{TI}_1(\alpha,F) \;\Rightarrow\; \mathsf{ID}_1 \vdash \mathsf{TI}_1(\omega^{\alpha},F) \tag{i}$$

analogously to the proof of (7.7). Put

$$\mathscr{J}F(\alpha) \;:\Leftrightarrow\; \alpha\,\varepsilon\,\mathsf{M} \wedge (\forall\eta\,\varepsilon\,\mathsf{M})[\eta \subseteq_1 F \rightarrow \eta+\omega^{\alpha} \subseteq_1 F]. \tag{ii}$$

First we have to show

$$\mathsf{ID}_1 \vdash \mathsf{Prog}_1(F) \;\rightarrow\; \mathsf{Prog}_1(\mathscr{J}F). \tag{iii}$$

To prove (iii) we have the premises

$$\mathsf{Prog}_1(F) \tag{iv}$$

$$\alpha\,\varepsilon\,\mathsf{M} \wedge \alpha \subseteq_1 \mathscr{J}F \tag{v}$$

$$\eta\,\varepsilon\,\mathsf{M} \wedge \eta \subseteq_1 F \tag{vi}$$

$$\xi <_1 \eta+\omega^{\alpha} \tag{vii}$$

and have to show

$$F(\xi). \tag{viii}$$

If $\xi < \eta$ this follows from (vi). If $\xi = \eta$ this follows from (vi) together with (iv). So assume $\eta < \xi =_{NF} \eta + \omega^{\xi_1} + \cdots + \omega^{\xi_k} < \eta + \omega^\alpha$. From $\xi \,\varepsilon\, \mathsf{M}$ we also obtain $\xi_i \,\varepsilon\, \mathsf{M}$ and $\xi < \eta + \omega^{\xi_1}\cdot(k+1)$. Therefore it suffices to prove

$$\eta + \omega^{\xi_1}\cdot u \subseteq_1 F \tag{ix}$$

by mathematical induction on $u$. For $u = 0$ this is (vi). From $\xi_1 <_1 \alpha$ we obtain $\mathscr{J}F(\xi_1)$ by (v). Together with the induction hypothesis $\eta + \omega^{\xi_1}\cdot u \subseteq_1 F$ this yields $\eta + \omega^{\xi_1}\cdot(u+1) \subseteq_1 F$. So we have (ix) and thus also (iii). Now assume

$$\mathsf{ID}_1 \vdash \mathsf{TI}_1(\alpha, F) \tag{x}$$

for all $\mathscr{L}(\mathsf{ID})$-formulas $F$. This embodies especially

$$\mathsf{ID}_1 \vdash \mathsf{TI}_1(\alpha, \mathscr{J}F). \tag{xi}$$

which means

$$\mathsf{ID}_1 \vdash \alpha \,\varepsilon\, \mathsf{M} \wedge (\mathrm{Prog}_1(\mathscr{J}F) \to \alpha \subseteq_1 \mathscr{J}F). \tag{xii}$$

By (xii) we get $\alpha \,\varepsilon\, \mathscr{J}F$. But this implies $\omega^\alpha \subseteq_1 F$ and we obtain by (iii)

$$\mathsf{ID}_1 \vdash \omega^\alpha \,\varepsilon\, \mathsf{M} \wedge (\mathrm{Prog}_1(F) \to \omega^\alpha \subseteq_1 F),$$

i.e., $\mathsf{TI}_1(\omega^\alpha, F)$.

Because of $SC(\omega^\alpha) \cap \Omega = SC(\alpha) \cap \Omega$ and $K(\omega^\alpha) = K(\alpha)$ the remaining claims follow trivially. □

**9.6.19 Theorem** *(The lower bound for* $\mathsf{ID}_1$*)* *For every ordinal* $\alpha < \psi(\varepsilon_{\Omega+1})$ *there is a primitive recursive ordering* $\prec$ *such that* $\mathsf{ID}_1 \vdash \underline{n} \,\varepsilon\, \mathrm{Acc}_\prec$ *and* $\alpha \leq |n|_{\mathrm{Acc}_\prec}$. *Hence* $\psi(\varepsilon_{\Omega+1}) \leq \kappa^{\mathsf{ID}_1}$.

*Proof* We have outlined in Theorem 9.6.8 that $<$ is primitive recursive. Defining a sequence $\zeta_0 = \Omega + 1$ and $\zeta_{n+1} = \omega^{\zeta_n}$ we obtain by Lemma 9.6.17 and Lemma 9.6.18

$$\mathsf{ID}_1 \vdash \mathsf{TI}_1(\zeta_n, F) \wedge K(\zeta_n) \subseteq \zeta_n \wedge \zeta_n \,\varepsilon\, \mathsf{M}$$

for all $n$. By the Condensation Lemma (Lemma 9.6.16), this implies $\psi(\zeta_n) \,\varepsilon\, \mathscr{A}cc = \mathrm{Acc}_{<_0} \cap \Omega$ for all $n$, which, according to our convention, actually means $\ulcorner\psi(\zeta_n)\urcorner \,\varepsilon\, \mathscr{A}cc$ for all $n$. From Lemma 6.5.2 we obtain $|\ulcorner\psi(\zeta_n)\urcorner|_{\mathrm{Acc}_{<_0}} = \mathrm{otyp}_{<_0}(\ulcorner\psi(\zeta_n)\urcorner) = |\ulcorner\psi(\zeta_n)\urcorner|_{\mathscr{O}} = \psi(\zeta_n)$ for all $n$. Hence $|\psi(\zeta_n)|_{\mathrm{Acc}_{<_0}} = \psi(\zeta_n)$ for all $n$ and the claim follows because $\sup_n \psi(\zeta_n) = \psi(\varepsilon_{\Omega+1})$. □

**9.6.20 Corollary** *(Ordinal analysis of* $\mathsf{ID}_1$*)*

$$\kappa^{\mathsf{ID}_1} = \psi(\varepsilon_{\Omega+1})$$

*and*

$$\|\mathsf{ID}_1(\vec{X})\| = \|\mathsf{ID}_1(\vec{X})\|_{\Pi^1_1} = \kappa^{\mathsf{ID}_1(\vec{X})} = \psi(\varepsilon_{\Omega+1}).$$

**9.6.21 Exercise** For $\alpha \in On$ and a set $X$ of ordinals let $\mathcal{C}_\alpha(X)$ be the closure of $X \cup \{0, \Omega\}$ under $+$ and $\lambda\xi < \alpha . \lambda\eta . \vartheta_\xi(\eta)$ for $\vartheta_\gamma := en_{In(\gamma)}$ and $In(\gamma) := \{\xi \mid \xi \notin \mathcal{C}_\gamma(\xi)\}$. Show the following properties

a) $In(\alpha)$ is unbounded and $In(\alpha) \cap \Omega$ is unbounded in $\Omega$.
b) $\Omega$ is closed under $\vartheta_\alpha$.
c) $\mathcal{C}_\alpha(\emptyset) \cap \Omega = \vartheta_\alpha(0)$.
d) If $\alpha < \vartheta_\Omega(0)$ then $In(\alpha) \cap \Omega = Cr(\alpha) \cap \Omega$.
e) If $\alpha < \vartheta_\Omega(0)$ and $\beta < \Omega$ then $= \vartheta_\alpha(\beta) = \varphi_\alpha(\beta)$.
f) $\vartheta_\Omega(0) = \Gamma_0$.
g) $\psi(\varepsilon_{\Omega+1}) = \vartheta_{\vartheta_1(\Omega+1)}(0)$

Hint: For d) show first $\alpha \in \mathcal{C}_\Omega(0) \cap \Omega \Rightarrow \alpha \in \mathcal{C}_\alpha(0)$.

## 9.7 Alternative Interpretations for $\Omega$

In defining the iterated operators $\mathcal{B}_\alpha$, we interpreted the ordinal $\Omega$ as $\omega_1$, the first uncountable ordinal. We already mentioned that another possible interpretation for $\Omega$ could be $\omega_1^{CK}$, the first recursively regular ordinal. In view of the role of $\Omega$ in the ordinal analysis of the theory $\mathsf{ID}_1$, the interpretation of $\Omega$ as $\omega_1^{CK}$ would even be the more natural one. The aim of this section (which is not needed for the further reading) is to show, that the "variable" $\Omega$ may be interpreted in many ways without altering the transitive part of the resulting operators $\mathcal{B}_\alpha$.

We obtained the ordinal analysis of $\mathsf{ID}_1$ by iterating the operator $\mathcal{B}$ which essentially closes a set $X \cup \{0, \Omega\} \subseteq On$ under ordinal addition $+$ and the Veblen function $\varphi$ viewed as a binary function. Since $+$ and $\varphi$ have a fixed meaning they do not depend on the interpretation of the ordinal $\Omega$. The definition of the function $\psi_{\mathcal{B}}$, however, depends on the operator $\mathcal{B}$ which in turn depends on the interpretation of the "variable" $\Omega$.

By assigning a value $V(\Omega)$ to the "symbol" $\Omega$ we obtain an operator $\mathcal{B}^V$ together with a function $\psi_V := \psi_{\mathcal{B}^V}$ as defined in Definition 9.4.1. In the following, we will denote by $\mathcal{B}_\alpha$, the operators obtained by the standard interpretation $V(\Omega) := \omega_1$. Recall that $\mathcal{B}_\alpha \subseteq \mathcal{B}_{\Gamma_{\omega_1+1}}$ for all ordinals $\alpha$. This follows because all strongly critical ordinals which enter the sets $\mathcal{B}_\alpha$ are of the form $\psi(\beta) < \omega_1$. The first strongly critical ordinal above $\omega_1$, which is $\Gamma_{\omega_1+1}$, is therefore inaccessible for the operations that form the sets $\mathcal{B}_\alpha$.

In the next definition, we refer to the definition of ordinal terms On as presented in Definition 9.6.7. Here, however, we do not care that these terms are coded as natural numbers. To improve readability we will therefore write 0, $\Omega$ instead of $\langle 0 \rangle$, $\langle 1 \rangle$, $a_1 + \cdots + a_n$ instead of $\langle 1, a_1, \ldots, a_n \rangle$, $\bar{\varphi}_a(b)$ instead of $\langle 2, a, b \rangle$ and $\psi(a)$ instead of $\langle 3, a \rangle$. We will also write $<$ instead of $\prec$ etc. To distinguish ordinal terms, which may contain the variable $\Omega$, from their interpretations in the ordinals we are, however, going to denote them (for the moment) by lower case Roman letters.

**9.7.1 Definition** Let $V(\Omega) \in On$ be an assignment of an ordinal to $\Omega$. We define the interpretation $a^V \in On$ for ordinal terms $a \in \mathsf{On}$ by the following clauses.

- $0^V := 0$ and $\Omega^V := V(\Omega)$
- $(a_1 + \cdots + a_n)^V := a_1^V + \cdots + a_n^V$
- $(\bar{\varphi}_a(b))^V := \bar{\varphi}_{(a^V)}(b^V)$
- $(\psi(a))^V := \psi_V(a^V)$

Let $\mathsf{On}^V := \{a^V \mid a \in \mathsf{On}\}$.

The *standard interpretation* of $\Omega$ is the interpretation $St(\Omega) := \omega_1$. For the standard interpretation we obviously have $a^{St} = |a|_{\mathcal{O}}$ for all ordinal terms $a \in \mathsf{On}$ and

$$\mathsf{On}^{St} = T = \mathscr{B}_{\Gamma_{\omega_1+1}}(\emptyset) = B_{\Gamma_{\omega_1+1}} \tag{9.37}$$

as shown in Theorems 9.6.3 and 9.6.8.

If we try to develop the theory of iterations of the operator $\mathscr{B}$ as in the Sects. 9.4–9.6, on the basis of a nonstandard interpretation for $\Omega$ we see that we already fail in proving equation (9.18) (in p. 173). Here we used that $St(\Omega) = \omega_1$ is a regular cardinal. The obvious remedy is to require equation (9.18) axiomatically and to introduce

$(Ax_\Omega)$ $\quad (\forall \xi)[\psi_V(\xi) < V(\Omega)]$

as the defining axiom for $\Omega$. Ruminating Sect. 9.4 through Sect. 9.6, we will then notice that this (and at some places also $V(\Omega) \in SC$, but by Exercise 9.7.26 even this can be dispensed with) is all we need. In the rest of this section we are going to characterize the ordinals which satisfy $Ax_\Omega$.

**9.7.2 Definition** An interpretation $V$ is *good relative to an ordinal* $\Theta$ if

$$(\forall a)[a^{St} \in \mathscr{B}_\Theta \cap (\Theta + 1) \;\Rightarrow\; \psi_V(a^V) < V(\Omega)]. \tag{9.38}$$

We call $V$ a *good interpretation* if it is good relative to the ordinal $\Gamma_{\omega_1+1}$.

It follows from (9.18) on p. 173 that $St$ is a good interpretation. We will continue to write $\psi(a)$ instead of $\psi_{St}(a)$.

To avoid silly technicalities, we will from now onwards assume that any interpretation satisfies $V(\Omega) \in SC$. Then the following lemma is immediate.

**9.7.3 Lemma** *Let $V$ be an interpretation. Then $a \in \mathsf{H}$ implies $a^V \in \mathbb{H}$ and $a \in \mathsf{SC}$ implies $a^V \in SC$.*

Recall that $\Gamma$ denotes the enumerating function of the class $SC$ of strongly critical ordinals and $\Gamma'$ its derivative. For an ordinal $\alpha$ we define its strongly critical successor by $\alpha^{SC} := \min\{\eta \in SC \mid \alpha < \eta\}$. For an interpretation $V$, we abbreviate $\mathscr{B}_\alpha^V(\emptyset)$ by $\mathscr{B}_\alpha^V$.

**9.7.4 Lemma** *Let V be an interpretation such that* $\Gamma'(0) \leq V(\Omega)$. *Then*

$$\psi_V \restriction \Gamma'(0) = \Gamma \restriction \Gamma'(0). \tag{9.39}$$

*Proof* We show

$$\xi < \Gamma'(0) \;\Rightarrow\; \mathscr{B}^V_\xi \cap \Omega = \psi_V(\xi) = \Gamma(\xi)$$

by induction on $\xi$. By induction hypothesis we have

$$\eta < \xi \;\Rightarrow\; \mathscr{B}^V_\eta \cap \Omega = \psi_V(\eta) = \Gamma(\eta) > \eta. \tag{i}$$

Hence $\eta \in \mathscr{B}^V_\eta \subseteq \mathscr{B}^V_\xi$ and $\Gamma(\eta) \in \mathscr{B}^V_\xi$ for all $\eta < \xi$. For $\rho < \Gamma(\xi)$ there is an $\eta < \xi$ such that $\Gamma(\eta)$ majorizes $SC(\rho)$ which implies $\rho \in \mathscr{B}^V_\xi$, hence

$$\Gamma(\xi) \subseteq \mathscr{B}^V_\xi \cap \Omega \subseteq \psi_V(\xi).$$

But (i) and the hypothesis $\Gamma'(0) \leq V(\Omega)$ imply $\Gamma(\xi) \notin \mathscr{B}^V_\xi$ which shows also $\psi_V(\xi) \leq \Gamma(\xi)$. Hence $\psi_V(\xi) = \Gamma(\xi) = \mathscr{B}^V_\xi \cap \Omega$. □

**9.7.5 Corollary** *Let V be an interpretation such that* $\Gamma'(0) \leq V(\Omega)$. *Then* $\psi_V \restriction \Gamma'(0) = \Gamma \restriction \Gamma'(0) = \psi_{St} \restriction \Gamma'(0)$ *and* $a^V = a^{St}$ *for all ordinal terms* $a \in \mathsf{On}$ *such that* $a^{St} < \Gamma'(0)$.

*Proof* The first part of the claim is in Lemma 9.7.4. We prove the second part by induction on the length of the term $a$. If $a$ is not of the form $\psi_V(b)$ we obtain the claim directly from the induction hypothesis. If $a$ is a term $\psi(b)$ such that $a^{St} = \psi(b^{St}) < \Gamma'(0)$ then $b^{St} < \Gamma'(0)$ and we obtain $b^{St} = b^V$ and thus $a^{St} = \psi(b^{St}) = \psi_V(b^V) = a^V$. □

**9.7.6 Lemma** *Let V be an interpretation which is good relative to an ordinal* $\Theta$. *For ordinal terms* $a, b \in \mathsf{On}$ *we then obtain*

$$a^{St} \in \mathscr{B}_{\Theta+1} \;\Rightarrow\; [a^{St} < b^{St} \;\Leftrightarrow\; a^V < b^V] \tag{i}$$

$$b^{St} \in \mathscr{B}_\Theta \cap (\Theta+1) \wedge a^{St} \in \mathscr{B}_{b^{St}} \;\Rightarrow\; a^V \in \mathscr{B}^V_{b^V} \tag{ii}$$

*Proof* Let $lh(a)$ denote the length of the term $a \in \mathsf{On}$. We prove (i) and (ii) simultaneously by induction on $2^{lh(a)} + 2^{lh(b)}$. In the proof of (i) we follow the distinction by cases of Definition 9.6.7. Of course it suffices to show the direction from left to right. The opposite direction is then an immediate consequence.

If $a^{St} = 0$ and $b^{St} \neq 0$ we have $a^V = 0$ and $b^V \neq 0$. Hence $a^V < b^V$.

Now suppose $a = a_1 + \cdots + a_m$. From $a^{St} \in \mathscr{B}_{\Theta+1}$ we get $\{a_1^{St}, \ldots, a_m^{St}\} \subseteq \mathscr{B}_{\Theta+1}$. Since $2^{lh(a_i)} + 2^{lh(a_{i+1})} < 2^{lh(a)} + 2^{lh(b)}$ we obtain by the induction hypothesis $a_i^V \geq a_{i+1}^V$ thence $a^V =_{NF} a_1^V + \cdots + a_m^V$. If $b = b_1 + \cdots + b_n$ we obtain $b^V =_{NF} b_1^V + \cdots + b_n^V$

with the same argument. Since $a_i^{St} < b_i^{St} \Leftrightarrow a_i^V < b_i^V$ by the induction hypothesis we get $a^V < b^V$.

If $b \in \mathsf{H}$ we have $a_1^{St} < b^{St}$. By the induction hypothesis it follows that $a_1^V < b^V$, which entails $a^V < b^V$ since $b^V \in \mathbb{H}$ by Lemma 9.7.3.

Now suppose $a \in \mathsf{H}$ and $b = b_1 + \cdots + b_n$. Then $a^{St} \leq b_1^{St}$ and we obtain $\mathbb{H} \ni a^V \leq b_1^V$ and $b^V =_{NF} b_1^V + \ldots + b_n^V$ by the induction hypothesis. But this implies $a^V < b^V$.

Assume $a = \bar{\varphi}_{a_1}(a_2)$ and $b = \bar{\varphi}_{b_1}(b_2)$. If $a_1^{St} < b_1^{St}$ and $a_2^{St} < b^{St}$, we obtain $a_1^V < b_1^V$ and $a_2^V < b^V$ by the induction hypothesis, which implies $a^V < b^V$. If $a_1^{St} = b_1^{St}$ and $a_2^{St} < b_2^{St}$ then $a_1^V = b_1^V$ and $a_2^V < b_2^V$ by the induction hypothesis which implies $a^V < b^V$. Similarly we obtain from $b_1^{St} < a_1^{St}$ and $a^{St} \leq b_2^{St}$ by induction hypothesis $b_1^V < a_1^V$ and $a^V \leq b_2^V$, which in turn imply $a^V < b^V$.

If $a = \bar{\varphi}_{a_1}(a_2)$ and $b \in \mathsf{SC}$ such that $a_1^{St} < b^{St}$ and $a_2^{St} < b^{St}$ then $b^V \in SC$ and $a_1^V < b^V$ and $a_2^V < b^V$ by the induction hypothesis. By Lemma 9.7.3 we have $b^V \in SC$ which entails $a^V = \bar{\varphi}_{a_1^V}(a_2^V) < b^V$.

If $a \in \mathsf{SC}$ and $b = \bar{\varphi}_{b_1}(b_2)$ we have $a^{St} < b_1^{St}$ or $a^{St} < b_2^{St}$. By Lemma 9.7.3 we have $a^V \in SC$ and obtain $a^V < b_1^V$ or $a^V < b_2^V$ by the induction hypothesis. In both cases this implies $a^V < \bar{\varphi}_{b_1^V}(b_2^V)$.

Now let $a = \psi(a_0)$ and $b = \psi(b_0)$. Then $2^{lh(a_0)} + 2^{lh(a_0)} \leq 2^{lh(a)} < 2^{lh(a)} + 2^{lh(b)}$ which by the induction hypothesis for (i) entails $a_0^V < b_0^V$. Moreover we have $c \in \mathsf{K}(a_0) \Rightarrow c^{St} < a_0^{St}$, which entails $K(a_0^{St}) \subseteq a_0^{St}$ and that in turn implies $a_0^{St} \in \mathscr{B}_{a_0^{St}}$ by Lemma 9.6.5. Since $\mathscr{B}_{\Theta+1} \cap \omega_1 = \psi(\Theta+1)$ and $\psi(a_0^{St}) < \omega_1$ we obtain $\psi(a_0^{St}) < \psi(\Theta+1)$, which implies $a_0^{St} \leq \Theta$. Hence $a_0^{St} \in \mathscr{B}_\Theta \cap \Theta + 1$. By the induction hypothesis for (ii) we therefore obtain $a_0^V \in \mathscr{B}^V_{a_0^V} \subseteq \mathscr{B}^V_{b_0^V}$. But then $a^V = \psi_V(a_0^V) < \psi_V(b_0^V) = \psi_V(b^V)$.

If finally $a = \psi(a_0)$ and $b = \Omega$ we get $a_0^{St} \in \mathscr{B}_\Theta \cap \Theta + 1$ as before and thus $\psi_V(a_0^V) < \Omega^V = V(\Omega)$ since $V$ is good relative to $\Theta$.

We are now going to prove (ii). The claim is obvious for $a = 0$ and $a = \Omega$. If $a = a_1 + \cdots + a_n$ or $a = \bar{\varphi}_{a_0}(a_1)$ we obtain by the induction hypothesis for (i) and Lemma 9.7.3 $a^V =_{NF} a_1^V + \cdots + a_n^V$ or $a^V = \bar{\varphi}_{a_0^V}(a_1^V)$, respectively. By the induction hypothesis for (ii) we then have $SC(a^V) \subseteq \mathscr{B}^V_{b^V}$, which entails $a^V \in \mathscr{B}^V_{b^V}$.

If $a = \psi(a_0)$ then $a^{St} =_{NF} \psi(a_0^{St})$ and $a_0^{St} \in \mathscr{B}_{b^{St}} \cap b^{St} \subseteq \mathscr{B}_\Theta \cap (\Theta+1)$. By the induction hypothesis for (ii) and (i) we then obtain $a^V =_{NF} \psi_V(a_0^V)$ and $a_0^V \in \mathscr{B}^V_{b^V} \cap b^V$ and this implies $a^V =_{NF} \psi_V(a_0^V) \in \mathscr{B}^V_{b^V}$. □

We have seen in Theorem 9.6.8 that, for every ordinal $\alpha \in T$ there is an ordinal term $\ulcorner\alpha\urcorner \in \mathsf{On}$ such that $\alpha = |\ulcorner\alpha\urcorner|_{\mathscr{O}} = \ulcorner\alpha\urcorner^{St}$. The notation $\ulcorner\alpha\urcorner$ is uniquely defined because we used the fixed-point free versions of the Veblen functions.

**9.7.7 Definition** For an ordinal $\alpha \in T$ and an interpretation $V$ we define $\alpha^V := \ulcorner\alpha\urcorner^V$.

So if $V$ is an interpretation, which is good relative to an ordinal $\Theta$ we obtain from Lemma 9.7.6 for ordinals $\alpha, \beta \in T$.

$$\alpha \in \mathscr{B}_{\Theta+1} \Rightarrow [\alpha < \beta \Leftrightarrow \alpha^V < \beta^V] \wedge [\alpha = \beta \Leftrightarrow \alpha^V = \beta^V] \tag{9.40}$$

$$\beta \in \mathscr{B}_\Theta \cap (\Theta+1) \wedge \alpha \in \mathscr{B}_{\flat St} \;\Rightarrow\; \alpha^V \in \mathscr{B}^V_{\beta^V} \tag{9.41}$$

$$\beta \in \mathscr{B}_{\Theta+1} \wedge \alpha \in Cr(\beta) \;\Rightarrow\; \alpha^V \in Cr(\beta^V). \tag{9.42}$$

Property (9.42) holds true since $\alpha \in Cr(\beta)$ implies that there is a $\beta_0 \geq \beta$ such that $\alpha \in Cr(\beta_0) \setminus Cr(\beta_0+1)$. Then $\alpha = \bar{\varphi}_{\beta_0}(\eta)$ for some $\eta$ and thus $\alpha^V = \ulcorner\alpha\urcorner^V = \bar{\varphi}_{\ulcorner\beta_0\urcorner^V}(\ulcorner\eta\urcorner^V) \in Cr(\beta_0^V) \subseteq Cr(\beta^V)$ since $\beta^V \leq \beta_0^V$ by (9.40).

For $\beta \in \mathscr{B}_\Theta \cap (\Theta+1)$ the interpretation $^V$ is thus an embedding from $\mathscr{B}_\beta$ into $\mathscr{B}^V_{\beta^V}$ which preserves principality, strong criticality and $\beta$-criticality. We will now prove that this embedding is also onto. The proof will need a relativized version of Lemma 9.6.2 saying that if $\psi_V(\alpha) \in \mathscr{B}^{V,n}_\beta$ there is an $\alpha_0 \in \mathscr{B}^{V,n}_\beta$ such that $\alpha_0 \in \mathscr{B}^V_{\alpha_0}$ and $\psi_V(\alpha) = \psi_V(\alpha_0)$, where $\mathscr{B}^{V,n}_\beta$ is defined analogously to $B^n_\beta$. Since the proof of Lemma 9.6.2 only needs $\psi(\alpha) < \Omega$ it relativizes easily to interpretations which are good relative to some $\Theta$.

**9.7.8 Lemma** *Let $V$ be a good interpretation relative to $\Theta$ and $\beta \in \mathscr{B}_\Theta \cap (\Theta+1)$. Then for every $\alpha \in \mathscr{B}^V_{\beta^V}$ there is a $\gamma \in \mathscr{B}_\beta$ such that $\alpha = \gamma^V$. Moreover we have $\alpha \in SC$ iff $\gamma \in SC$ and $\alpha \in \mathbb{H}$ iff $\gamma \in \mathbb{H}$.*

*Proof* Let $\alpha \in \mathscr{B}^{V,n}_\beta$. We prove the lemma by induction on $\beta^V$ with side induction on $n$.

If $\alpha = 0$ we put $\gamma := 0$ and if $\alpha = V(\Omega)$ we put $\gamma := \omega_1$. Now assume $\alpha =_{NF} \alpha_1 + \cdots + \alpha_n$. Then $\mathbb{H} \ni \alpha_i < \alpha$ for $i = 1,\ldots,n$. By the main induction hypothesis there are ordinals $\gamma_i \in \mathscr{B}_\beta$ such that $\alpha_i = \gamma_i^V$ and $\gamma_i \in \mathbb{H}$. By equation (9.40) we obtain $\gamma_1 \geq \cdots \geq \gamma_n$ and put $\gamma := \gamma_1 \cdots + \gamma_n$. Then $\gamma =_{NF} \gamma_1 \cdots + \gamma_n$ and $\gamma^V = \gamma_1^V \cdots + \gamma_n^V$, $\gamma \in \mathscr{B}_\beta$ and $\gamma \notin \mathbb{H}$.

Next assume $\alpha = \bar{\varphi}_{\alpha_1}(\alpha_2)$. Then $\alpha \in \mathbb{H} \setminus SC$ and $\alpha_i < \alpha$. By the main induction hypothesis there are ordinals $\gamma_1, \gamma_2 \in \mathscr{B}_\beta$ such that $\gamma_i^V = \alpha_i$ for $i = 1,2$. Let $\gamma = \bar{\varphi}_{\gamma_1}(\gamma_2)$. Then $\gamma \in \mathscr{B}_\beta$ and $\gamma_i < \gamma$ for $i = 1,2$.

Let $\alpha = \psi_V(\eta)$ such that $\eta \in \mathscr{B}^{V,n-1}_{\beta^V} \cap \beta^V$. Then $\alpha \in SC$. By Lemma 9.6.2 there is an $\eta_0 \in \mathscr{B}^{V,n-1}_{\beta^V} \cap \beta^V$ such that $\eta_0 \in \mathscr{B}^V_{\eta_0}$ and $\alpha = \psi_V(\eta_0)$. By induction hypothesis there is an $\alpha_0$ such that $\eta_0 = \alpha_0^V$, hence $\alpha_0^V \in \mathscr{B}^V_{\alpha_0^V}$, which implies $\alpha_0 \in \mathscr{B}_{\alpha_0}$. So $\gamma := \psi(\alpha_0)$ implies $\gamma \in SC$ and $\gamma =_{NF} \psi(\alpha_0)$ and we obtain $\alpha = \psi_V(\eta_0) = \psi_V(\alpha_0^V) = \psi(\alpha_0)^V = \gamma^V$. □

**9.7.9 Theorem** *Let $V$ be an interpretation which is good relative to $\Theta$ and $\beta \in \mathscr{B}_\Theta \cap (\Theta+1)$. Then $(\mathscr{B}_\beta)^V = \mathscr{B}^V_{\beta^V}$.*

*Proof* From $\alpha \in \mathscr{B}_\beta$ we obtain $\alpha^V \in \mathscr{B}^V_{\beta^V}$ by (9.41). Hence $(\mathscr{B}_\beta)^V := \{\alpha^V \mid \alpha \in \mathscr{B}_\beta\} \subseteq \mathscr{B}^V_{\beta^V}$. Conversely we obtain for $\alpha \in \mathscr{B}^V_{\beta^V}$ a $\gamma \in \mathscr{B}_\beta$ such that $\alpha = \gamma^V \in (\mathscr{B}_\beta)^V$ by Lemma 9.7.8. Hence $\mathscr{B}^V_{\beta^V} \subseteq (\mathscr{B}_\beta)^V$. □

**9.7.10 Corollary** *For every good interpretation we have* $(\mathcal{B}_{\Gamma_{\omega_1+1}})^V = \mathcal{B}^V_{V(\Omega)^{SC}}$.

*Proof* Define $\Delta_0 := \omega_1 + 1$ and $\Delta_{n+1} = \bar{\varphi}_{\Delta_n}(0)$. Then $\sup_{n\in\omega} \Delta_n = \Gamma_{\omega_1+1}$, $\Delta_n \in \mathcal{B}_{\Delta_n}$ for all $n \in \omega$ and $\mathcal{B}_{\Gamma_{\omega_1+1}} = \bigcup\{\mathcal{B}_{\Delta_n} \mid n \in \omega\}$. The interpretation $V$ is good relative to all $\Delta_n$ for $n \in \omega$. Hence $(\mathcal{B}_{\Gamma_{\omega_1+1}})^V = \bigcup\{(\mathcal{B}_{\Delta_n})^V \mid n \in \omega\} = \bigcup\{\mathcal{B}^V_{\Delta_n^V} \mid n \in \omega\} = \mathcal{B}^V_{V(\Omega)^{SC}}$, since $\sup\{\Delta_n^V \mid n \in \omega\} = V(\Omega)^{SC}$. □

Summing up, we obtain the following theorem.

**9.7.11 Theorem** *If V is an interpretation which is good relative to* $\Theta$ *and* $\beta \in \mathcal{B}_\Theta \cap (\Theta + 1)$ *then* ${}^V$ *is an order isomorphism from* $\mathcal{B}_\beta$ *onto* $\mathcal{B}^V_{\beta^V}$.

*If V is a good interpretation then* ${}^V$ *is an order isomorphism from* $\mathcal{B}_{\Gamma_{\omega_1+1}}$ *onto* $\mathcal{B}^V_{V(\Omega)^{SC}}$.

The main goal of the present section is to show that the transitive part of $\mathcal{B}_{\Gamma_{\Omega+1}}$ does not depend on the interpretation of $\Omega$, i.e., we want to show that $\mathcal{B}_{\Gamma_{\omega_1+1}} \cap \omega_1 = \mathcal{B}^V_{V(\Omega)^{SC}} \cap V(\Omega)$ holds true for any good interpretation $V$.

**9.7.12 Lemma** *Let V be an interpretation which is good relative to* $\Theta$. *Then* $\mathcal{B}^V_{\alpha^V} \cap V(\Omega) = \psi_V(\alpha^V)$ *holds for all* $\alpha \in \mathcal{B}_\Theta \cap (\Theta + 1)$.

*Proof* Since $V$ is good relative to $\Theta$, we have $\psi_V(\alpha^V) < V(\Omega)$ for all $\alpha \in \mathcal{B}_\Theta \cap \Theta + 1$ by definition. Then we prove the lemma literally as we proved (9.21) on p. 173. □

**9.7.13 Theorem** *Let V be an interpretation which is good relative to* $\Theta$. *Then*

*(i)* $\mathcal{B}^V_{\alpha^V} \cap V(\Omega) = \mathcal{B}_\alpha \cap \omega_1$ *for all* $\alpha \in \mathcal{B}_\Theta \cap (\Theta + 1)$

*and*

*(ii)* $\alpha^V = \alpha$ *for all* $\alpha < \psi(\Theta)$.

*Proof* Since $V$ is a good interpretation relative to $\Theta$ we know from Theorem 9.7.11 that ${}^V$ is an order isomorphism from $\mathcal{B}_\alpha$ onto $\mathcal{B}^V_{\alpha^V}$ for all $\alpha \in \mathcal{B}_\Theta \cap (\Theta + 1)$ which maps $\omega_1$ to $V(\Omega)$. By (9.21) and Lemma 9.7.12 $\mathcal{B}_\alpha \cap \omega_1$ and $\mathcal{B}^V_{\alpha^V} \cap V(\Omega)$ are transitive sets. Therefore ${}^V$ is the identity on these sets. This proves (i).

Since $\alpha < \psi(\Theta) = \mathcal{B}_\Theta \cap \omega_1 \subseteq \mathcal{B}_{\Gamma_{\omega_1+1}} = T$ we have a term notation for $\alpha$. We prove (ii) by induction on $lh(\alpha) := lh(\ulcorner\alpha\urcorner)$. We trivially have $0^V = 0$. If $\alpha =_{NF} \alpha_1 + \cdots + \alpha_n$ or $\alpha =_{NF} \bar{\varphi}_{\alpha_1}(\alpha_2)$ we get $\alpha_i < \psi(\Theta)$ and thus $\alpha_i^V = \alpha_i$ for $i = 1, \ldots, n$ or $i = 1, 2$, respectively. Hence $\alpha^V = \alpha_1^V + \cdots + \alpha_n^V = \alpha_1 + \cdots + \alpha_n = \alpha$ or $\alpha^V =_{NF} \bar{\varphi}_{\alpha_1^V}(\alpha_2^V) = \bar{\varphi}_{\alpha_1}(\alpha_2) = \alpha$.

If $\alpha =_{NF} \psi(\alpha_0)$ then $\alpha_0 \in \mathcal{B}_\Theta \cap \Theta$. By (i) we thus obtain $\mathcal{B}^V_{\alpha_0^V} \cap V(\Omega) = \mathcal{B}_{\alpha_0} \cap \omega_1$. Hence $\psi_V(\alpha_0^V) = \mathcal{B}^V_{\alpha_0^V} \cap V(\Omega) = \mathcal{B}_{\alpha_0} \cap \omega_1 = \psi(\alpha_0)$. □

**9.7.14 Corollary** *For a good interpretation $V$ we have $\alpha^V = \alpha$ for all $\alpha \in \mathcal{B}_{\Gamma_{\omega_1+1}} \cap \omega_1 = \psi(\Gamma_{\omega_1+1})$.*

*Proof* This follows immediately from Theorem 9.7.13. □

It follows from Corollary 9.7.14 that a reinterpretation of $\omega_1$ does not move the ordinals below $\omega_1$ in $\mathcal{B}_{\Gamma_{\omega_1+1}}$ provided that this reinterpretation is good. This becomes, of course, false for ordinals above $\omega_1$.

Our next aim is to characterize good interpretations.

**9.7.15 Theorem** *Assume $\Theta \in \mathcal{B}_\Theta$. Then $\psi(\Theta) < V(\Omega)$ for every interpretation which is good relative to $\Theta$.*

*Proof* Assume $V(\Omega) \leq \psi(\Theta)$. By Theorem 9.7.13 it follows $V(\Omega) \leq \psi(\Theta) = \mathcal{B}_\Theta \cap \omega_1 = \mathcal{B}^V_{\Theta^V} \cap V(\Omega) = \psi_V(\Theta^V)$ contradicting the definition of "good relative to $\Theta$". □

**9.7.16 Corollary** *It is $\psi(\Gamma_{\omega_1+1}) \leq V(\Omega)$ for any good interpretation $V$.*

*Proof* Let $\Delta_n$ be the fundamental sequence for $\Gamma_{\omega_1+1}$ as introduced in the proof of Corollary 9.7.10. Then $V$ is good relative to all $\Delta_n$ and we have $\Delta_n \in \mathcal{B}_{\Delta_n}$ for all $n \in \omega$. By Theorem 9.7.15 it follows $\psi(\Delta_n) < V(\Omega)$ for all $n$ which in turn implies $\psi(\Gamma_{\omega_1+1}) \leq V(\Omega)$. □

**9.7.17 Lemma** *Let $V$ be an interpretation and $\Theta$ an ordinal such that $\Gamma'(0) \leq \psi(\Theta) \leq V(\Omega) \leq \omega_1$. Then*

*(i) $\beta^V \leq \beta$ holds for all $\beta \in \mathcal{B}_\Theta$*

*and*

*(ii) $V$ is a good interpretation relative to all $\beta < \Theta$.*

*Proof* We show both claims simultaneously by induction on $lh(\beta)$ with side induction on $\beta$. We start with proving (i).

For $\beta = 0$ we have $\beta^V = 0 \leq \beta$ and for $\beta = \omega_1$ by definition $\beta^V = V(\Omega) \leq \omega_1$.

For $\beta =_{NF} \beta_1 + \cdots + \beta_n$ or $\beta =_{NF} \varphi_{\beta_1}(\beta_2)$ we obtain the claim immediately form the main induction hypothesis.

Assume $\beta =_{NF} \psi(\beta_0)$. Then $\beta_0 \in \mathcal{B}_\Theta \cap \Theta$ and $V$ is good relative to $\beta_0$ by the main induction hypothesis for (ii). But we also have $\beta_0 \in \mathcal{B}_{\beta_0} \cap (\beta_0 + 1)$ by the normal form condition and obtain $\beta^V = \psi_V(\beta_0^V) = \psi\beta_0 = \beta$ by Theorem 9.7.13.

To prove (ii) we have to show

$$\xi \in \mathcal{B}_\beta \cap (\beta + 1) \;\Rightarrow\; \psi_V(\xi^V) < V(\Omega). \tag{i}$$

So assume $\xi \in \mathcal{B}_\beta \cap (\beta + 1)$. Then $\psi(\xi) \leq \psi(\beta)$. If $\beta = 0$, then $\xi = 0$ and we obtain $\psi_V(\xi^V) = \Gamma_0 < \Gamma'(0) \leq \psi(\Theta) \leq V(\Omega)$ by Lemma 9.7.4.

For $\beta \in Lim$ we distinguish the following cases.

1. $\xi = \beta$. Then $\xi \in \mathscr{B}_\xi$, which shows that $\psi(\xi)$ is in normal form. Hence $\xi = \beta \neq \psi(\beta) < \omega_1$. If $\beta \leq \psi(\beta)$ we therefore obtain $\beta < \psi(\beta) \leq \psi(\omega_1) = \Gamma'(0)$. By Corollary 9.7.5, this implies $\xi^V = \xi = \beta$ and $\psi_V(\xi^V) = \Gamma_\xi = \Gamma_\beta < \Gamma'(0) \leq \psi(\Theta) \leq V(\Omega)$. If $\psi\beta < \beta$ we obtain $\psi\beta \in \mathscr{B}_\Theta \cap \psi(\Theta)$ from $\beta \in \mathscr{B}_\Theta \cap \Theta$. By the side induction hypothesis for (ii) we then obtain $\psi_V(\xi^V) = \psi_V(\beta^V) = \psi(\beta)^V \leq \psi(\beta) < \psi(\Theta) \leq V(\Omega)$.

2. $\xi < \beta$. Since $\mathscr{B}_\beta = \bigcup_{\eta<\beta} \mathscr{B}_\eta$ we find an $\eta < \beta$ such that $\xi \in \mathscr{B}_\eta \cap (\eta+1)$. By the side induction hypothesis for (ii) we know that $V$ is good relative to $\eta$ and obtain $\psi_V(\xi^V) < V(\Omega)$.

Finally we assume $\beta = \beta_0 + 1$. If $\beta \leq \psi(\beta)$ we obtain $\xi \leq \beta < \psi(\beta) \leq \psi(\omega_1) = \Gamma'(0)$ as in case 1. Hence $\psi_V(\xi^V) = \Gamma_\xi < \Gamma'(0) \leq V(\Omega)$. Now let $\psi(\beta) < \beta$. There is an $\alpha \in \mathscr{B}_{\omega_1}$ such that $\psi(\xi) =_{NF} \psi(\alpha)$. In the terminology of Lemma 9.6.2 it is $\alpha = \xi(\xi) \geq \xi$. If $\xi = \beta$ then $\alpha = \xi$ as $\xi \in \mathscr{B}_\beta = \mathscr{B}_\xi$ and therefore $\xi(\xi) = \beta(\beta) = \min\{\eta \mid \beta \leq \eta \in \mathscr{B}_\beta\} = \beta$. From $\beta \in \mathscr{B}_\Theta \cap \Theta$ and $\psi(\beta) < \beta$ we obtain $\psi(\beta) \in \mathscr{B}_\Theta \cap \beta$. By the side induction hypothesis for (i) it follows $\bar{\varphi}_V(\xi^V) = \bar{\varphi}_V(\beta^V) \leq \psi(\beta) < \psi(\Theta) \leq V(\Omega)$. If $\xi < \beta$ then $\psi(\xi) =_{NF} \psi(\alpha) \in \mathscr{B}_\beta$ which implies $\alpha \in \mathscr{B}_\beta \cap \beta \subseteq \mathscr{B}_\Theta$. By the side induction hypothesis for (ii) we know that $V$ is good relative to $\beta_0$. By (9.40) we therefore obtain $\xi^V \leq \alpha^V$ from $\xi \leq \alpha$ and as $\psi(\alpha) < \psi(\beta) < \beta$ finally also $\psi_V(\xi^V) \leq \psi_V(\alpha^V) \leq \psi(\alpha) \leq \psi(\beta) < \psi(\Theta) \leq V(\Omega)$ by the side induction hypothesis for (i). □

**9.7.18 Lemma** *Suppose that $\Theta$ is a limit ordinal and $V$ an interpretation such that $\Gamma'(0) \leq \psi(\Theta) \leq V(\Omega)$. Then $\psi_V(\xi^V) = \psi(\xi)^V = \psi(\xi) < \psi(\Theta)$ holds true for all $\xi \in \mathscr{B}_\Theta \cap \Theta$.*

*Proof* If $\omega_1 \leq V(\Omega)$ we obtain $\psi_V(\xi) < \omega_1 \leq V(\Omega)$ for all ordinals $\xi$ and $V$is a good interpretation. If $V(\Omega) < \omega_1$ and $\xi \in \mathscr{B}_\Theta \cap \Theta$ then there is an $\eta < \Theta$ such that $\xi \in \mathscr{B}_\eta \cap \eta$. By the hypothesis $\psi(\Theta) \leq V(\Omega)$ it follows from Lemma 9.7.17 that $V$ is good relative to $\eta$ which by Lemma 9.7.13 entails $\psi_V(\xi^V) = \psi(\xi)^V = \psi(\xi)$. But $\psi(\xi) < \psi(\Theta)$ is clear from $\xi \in \mathscr{B}_\Theta \cap \Theta$. □

Now we are prepared for a characterization of good interpretations.

**9.7.19 Theorem** *An interpretation $V$ is good if and only if $\psi(\Gamma_{\omega_1+1}) \leq V(\Omega)$.*

*Proof* If $V$ is good then $\psi(\Gamma_{\omega_1+1}) \leq V(\Omega)$ by Corollary 9.7.16. For the opposite direction, assume $\psi(\Gamma_{\omega_1+1}) \leq V(\Omega)$ and $\xi \in \mathscr{B}_{\Gamma_{\omega_1+1}} \cap (\Gamma_{\omega_1+1})$. By Lemma 9.7.18 we obtain $\psi_V(\xi^V) = \psi(\xi) < \psi(\Gamma_{\omega_1+1}) \leq V(\Omega)$. So $V$ is a good interpretation. □

As a consequence of the characterization theorem for good interpretation we obtain the following observation.

**9.7.20 Theorem** *The following interpretations are good interpretations.*

*(A) The standard interpretation $V(\Omega) := \omega_1$.*

*(B) The recursive standard interpretation* $V(\Omega) := \omega_1^{CK}$

*(C) The term interpretation* $V(\Omega) := \psi(\Gamma_{\omega_1+1})$.

*Proof* Theorem 9.7.19 together with Equation (9.18) implies that the standard interpretation is a good interpretation. From Corollary 9.6.9 and Theorem 9.7.19 it follows that the recursive standard interpretation is good and (C) follows immediately from Theorem 9.7.19. □

Let us now return to $Ax_\Omega$.

**9.7.21 Definition** Let $V$ be an interpretation. We call $V$ a *global model of* $Ax_\Omega$ if

$$(\forall \xi)[\psi_V(\xi) < V(\Omega)].$$

If we only have

$$(\forall \xi \,\varepsilon\, \mathscr{B}^V_{V(\Omega)^{SC}})[\psi_V(\xi) < V(\Omega)],$$

we talk about a *local model of* $Ax_\Omega$.

We already mentioned in the beginning of the section that for global models we can develop the theory of the iterations $\mathscr{B}_\alpha$ as in the case of the standard interpretation. The ordinal $\Omega$ may therefore be viewed as a "virtual ordinal", which has only to be bigger than the transitive part of all $\mathscr{B}_\alpha$. Since we cannot know its size in advance we have opted to interpret $\Omega$ as $\omega_1$, which for cardinality reasons is safely outside of all the eventually obtained transitive parts.

**9.7.22 Theorem** *The following statements are equivalent.*

*(A) The interpretation V is a global model for* $Ax_\Omega$.

*(B) The interpretation V is good and* $\psi(\Gamma_{\omega_1+1}) < V(\Omega)$.

*(C) The interpretation V is a local model of* $Ax_\Omega$ *and* $\psi((V(\Omega))^{SC}) < V(\Omega)$.

*Proof* From (A) we obtain immediately that $V$ is a good interpretation. Let $\Delta_n$ be again the fundamental sequence for $\Gamma_{\omega_1+1}$ as defined in the proof of Corollary 9.7.10. Then $\Delta_n^V < V(\Omega)^{SC}$ and we obtain $\psi(\Gamma_{\omega_1+1}) = \sup\{\psi(\Delta_n) \mid n \in \omega\} = \sup\{\psi_V(\Delta_n^V) \mid n \in \omega\} \leq \psi_V(V(\Omega)^{SC}) < V(\Omega)$.

Assume (B) and choose $\xi \in \mathscr{B}^V_{V(\Omega)^{SC}}$. Then $\xi \in \mathscr{B}^V_{\Delta_n^V}$ for some $n \in \omega$. We have $\Delta_n \in \mathscr{B}_{\Delta_n} \cap (\Delta_n + 1)$ and $V$ is good relative to $\Delta_n$. By Lemma 9.7.8, $\eta \in \mathscr{B}_{\Gamma_{\Delta_n}} \subseteq \mathscr{B}_{\Gamma_{\omega_1+1}}$ such that $\xi = \eta^V$. Hence $\psi_V(\xi) = \psi_V(\eta^V) < V(\Omega)$. So $V$ is a local model for $Ax_\Omega$. Now regard the sequence $\Delta_n^V$. It is $\sup_{n\in\omega} \Delta_n^V = V(\Omega)^{SC}$ and we obtain $\Delta_n^V \in \mathscr{B}^V_\xi \cap V(\Omega)^{SC}$ for any ordinal $\xi$. Now we claim

$$\sup_{n\in\omega} \psi_V(\Delta_n^V) = \psi_V(V(\Omega)^{SC}). \tag{i}$$

From (i) and Corollary 9.7.14 we then obtain $\psi_V(V(\Omega)^{SC}) = \sup_{n\in\omega} \psi_V(\Delta_n^V) = \sup_{n\in\omega} \psi(\Delta_n) = \psi(\Gamma_{\omega_1+1}) < V(\Omega)$. It remains to show (i). We already used the obvious fact that $\sup_{n\in\omega} \psi_V(\Delta_n^V) \le \psi_V(V(\Omega)^{SC})$. By Lemma 9.7.18 we obtain $\sigma := \sup_{n\in\omega} \psi_V(\Delta_n^V) = \sup_{n\in\omega} \psi(\Delta_n) = \psi(\Gamma_{\omega_1+1}) < V(\Omega)$. Towards a contradiction, we assume $\sigma < \psi_V(V(\Omega)^{SC}) = \bigcup_{n\in\omega} \mathscr{B}^V_{\Delta_n} \cap V(\Omega)$. Then there is an $n \in \omega$ such that $\sigma \in \mathscr{B}^V_{\Delta_n^V} \cap V(\Omega) = \psi_V(\Delta_n^V)$. Hence $\psi(\Gamma_{\omega_1+1}) = \sigma < \psi_V(\Delta_n^V) = \psi(\Delta_n) < \psi(\Gamma_{\omega_1+1})$, a contradiction.

Let us finally assume (C). We first observe

$$\alpha \le \beta \wedge [\alpha,\beta) \cap \mathscr{B}^V_\beta = \emptyset \;\Rightarrow\; \mathscr{B}^V_\alpha = \mathscr{B}^V_\beta \wedge \psi_V(\alpha) = \psi_V(\beta) \tag{ii}$$

which is obvious from the definition of $\mathscr{B}^V_\beta$ and $\psi_V(\beta)$. We also prove

$$\xi \in \mathscr{B}^V_\beta \;\Rightarrow\; \xi \in \mathscr{B}^V_{V(\Omega)^{SC}} \cap V(\Omega)^{SC} \tag{iii}$$

for any ordinal $\beta$ easily by induction on the definition of $\xi \in \mathscr{B}^V_\beta$ using the fact that $V$ is a local model of $Ax_\Omega$. For $V(\Omega)^{SC} \le \xi$ we obtain by (ii) and (iii) $\psi_V(\xi) = \psi_V(V(\Omega)^{SC}) < V(\Omega)$ and for $\xi < V(\Omega)^{SC}$ by the weak monotonicity of $\psi_V$ also $\psi_V(\xi) \le \psi_V(V(\Omega)^{SC}) < V(\Omega)$. □

As an immediate consequence of Theorem 9.7.19 and Theorem 9.7.22 we obtain

**9.7.23 Corollary** *If $\psi(\Gamma_{\omega_1+1}) < V(\Omega)$ then $V$ is a global model of $Ax_\Omega$.*

Corollary 9.7.23 together with Corollary 9.6.9 yield the following theorem.

**9.7.24 Theorem** *The recursive standard interpretation $V(\Omega) := \omega_1^{CK}$ is a global model of $Ax_\Omega$.*

In Fig. 9.1 we have visualized the set of ordinals in $\bigcup_{\alpha\in On} \mathscr{B}_\alpha = \mathscr{B}_{\Gamma_{\omega_1+1}}$ and alternative interpretations for $\Omega$. For all interpretations $V(\Omega) \in (\psi(\Gamma_{\omega_1+1}), \omega_1)$ the transitive segment stays untouched but $\bigcup_{\alpha\in On} \mathscr{B}^V_\alpha \subseteq \omega_1$. All these interpretations generate a global model of $Ax_\Omega$. The term interpretation $V(\Omega) = \psi(\Gamma_{\omega_1+1})$, however, does not generate a global model of $Ax_\Omega$. As indicated in Fig. 9.1 moving

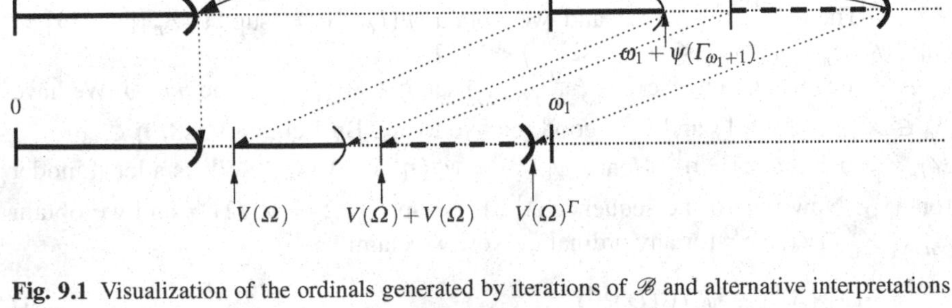


**Fig. 9.1** Visualization of the ordinals generated by iterations of $\mathscr{B}$ and alternative interpretations for $\Omega$.

$V(\Omega)$ to $\psi(\Gamma_{\omega_1+1})$ enlarges the transitive part of $\mathscr{B}^V_{V(\Omega)^{SC}}$. In fact this interpretation compresses all gaps such that $\mathscr{B}^V_{V(\Omega)^{SC}}$ becomes a transitive set whose order type is $\mathrm{otyp}(\mathscr{B}_{\Gamma_{\omega_1+1}})$.

**9.7.25 Exercise** Prove Lemma 9.7.3

**9.7.26 Exercise**

(a) Show that $\Gamma(\alpha) = \psi_V(\alpha)$ holds true for $V(\Omega) \in \Gamma'(0) \setminus SC$.

(b) Describe the behavior of $\psi_V$ for $V(\Omega) \in \Gamma'(0) \cap SC$.

For the following parts drop the hypothesis $V(\Omega) \in SC$ in the definition of an interpretation. Call $V$ good relative to an ordinal $\Theta$ if $\psi_V(a^V)V < \sup\{\lambda \in SC \mid \lambda \leq V(\Omega)\}$ holds true for all $a$ such that $a^{St} \in \mathscr{B}_\Theta \cap (\Theta+1)$.

(c) Show that Theorem 9.7.19 and Theorem 9.7.22 remain true

(d) Show that Lemma 9.7.6 (i) fails if we only require $\psi_V(a^V) < V(\Omega)$ for $a^{St} \in \mathscr{B}_\Theta \cap (\Theta+1)$ in the definition of a good interpretation, which may not be strongly critical.

Hint: For (c) extend $\mathbb{H}$ to $\mathbb{H}^V := \mathbb{H} \cup \{V(\Omega)\}$, $SC$ to $SC^V := SC \cup \{V(\Omega)\}$ and modify the normal-form conditions for ordinal terms replacing $\mathbb{H}$ and $SC$ by $\mathbb{H}^V$ and $SC^V$ accordingly. Then check that all claims of this section remain correct.

**9.7.27 Exercise** Show that $V(\Omega) =_{NF} \psi(\alpha)$ implies $\psi(\alpha) < \psi_V(\alpha^V)$.

**9.7.28 Exercise** Assume again $V(\Omega) \in SC$. Show that the following claims are equivalent.

(a) $V$ is a global model for $Ax_\Omega$

(b) $\psi_V(V(\Omega)^{SC}) = \psi(\omega_1^{SC})$ and $\psi_V(\xi^V) = \psi(\xi)$ holds true for all $\xi \in \mathscr{B}_{\omega_1^{SC}}$.

(c) $\psi_V(V(\Omega)^{SC}) = \psi(\omega_1^{SC})$ and $\psi(\xi)^V = \psi(\xi)$ holds true for all $\xi \in \mathscr{B}_{\omega_1^{SC}}$.

(d) $\psi_V(V(\Omega)^{SC}) = \mathscr{B}^V_{V(\Omega)^{SC}} \cap V(\Omega)$.

(e) $\psi_V(V(\Omega)^{SC}) < V(\Omega)$.

(f) The function $\psi^V$ is continuous and $\psi_V(\xi+1) = \psi_V(\xi)^{SC}$.

**9.7.29 Exercise** Assume again $V(\Omega) \in SC$. Show that the following claims are equivalent.

(a) $V$ is a good interpretation.

(b) $\psi_V(\xi^V) = \psi(\xi)$ holds true for all $\xi \in \mathscr{B}_{\omega_1^{SC}}$.

(c) $\psi(\xi)^V = \psi(\xi)$ holds true for all $\xi \in \mathscr{B}_{\omega_1^{SC}}$.

(d) $\mathscr{B}_\xi^V \subseteq V(\Omega)^{SC}$ holds true for all ordinals $\xi$.

(e) $\psi_V(\xi) \leq V(\Omega)^{SC}$ holds true for all ordinals $\xi$.

(f) $\mathscr{B}_{\xi^V}^V \subseteq V(\Omega)^{SC}$ holds true for all $\xi \in \mathscr{B}_{\omega_1^{SC}}$.

(g) $\psi_V(\xi^V) \leq V(\Omega)^{SC}$ holds true for all $\xi \in \mathscr{B}_{\omega_1^{SC}}$.

(h) $\psi(\xi)^V \leq V(\Omega)^{SC}$ holds true for all $\xi \in \mathscr{B}_{\omega_1^{SC}}$.

**9.7.30 Exercise** Disprove the following statements.

(a) If $\psi^V$ is continuous then $V$ is a global model for $Ax_\Omega$.

(b) If $\psi_V(\xi+1) \leq \psi_V(\xi)^{SC}$ holds true for all ordinals $\xi$ then $V$ is a global model for $Ax_\Omega$.

**9.7.31 Exercise** Let $V$ be the term interpretation. Show that $\psi_V(V(\Omega)^{SC}) = \text{otyp}(\mathscr{B}_{\omega_1^{SC}}) = \psi(\omega_1^{SC})^{SC}$.

# Chapter 10
# Provably Recursive Functions of NT

Kreisel once asked the question if it is possible to obtain upper bounds for the proof-theoretic ordinal of a theory whose language does not include free predicate variables without using Gödel's second incompleteness theorem. In Sect. 9.3 we proved in Theorem 9.3.2 that cut elimination is close to trivial for semi-formal systems which derive only sentences. Subsequently we discussed that no ordinal information can be expected from cut-elimination in such semi-formal systems. The proof of Theorem 9.3.2 fails, however, in the presence of set variables. For the standard procedure in getting an ordinal analysis of NT it is therefore crucial to include free set variables in its language. This is probably the background of Kreisel's question. On the other side, we have seen in Chap. 9 that controlling operators may allow us to obtain information also from semi-formal systems without free set variables in their languages. Weiermann observed that this is also true for "predicative" semi-formal systems. He could prove that the methods of impredicative proof theory are also applicable in predicative proof theory and lead there to better results. In particular he succeeded in (re)characterizing the provably recursive functions of NT (cf. [106] and [10]). In the following sections we present a variant of one of Weiermann's approaches.

## 10.1 Provably Recursive Functions of a Theory

Let $T$ be a theory whose language extends $\mathscr{L}(\mathsf{NT})$. In $T$ we can express all primitive recursive predicates by their characteristic functions. Again we write

$$T \vdash P(\vec{x}) \text{ instead of } T \vdash \chi_P(\vec{x}) = 1.$$

To explain what the provably recursive functions of a theory $T$ are, we recall some notions and results of elementary recursion theory.

Partial recursive function terms are defined by extending the definition of the primitive recursive function terms in Sect. 2. We replace in Definition 2.1.1, the phrase "primitive recursive function term" by "partial recursive function term" and add the clause

- If $t$ is an $n+1$-ary partial recursive function term, then $\mu x_i \,.\, t$ is for $1 \le i \le n+1$ an $n$-ary partial recursive function term.

The inductive definition of $\mathrm{ev}(f, z_1, \ldots, z_n) = z$ has to be altered to an inductive definition of $\mathrm{ev}(f, z_1, \ldots, z_n) \simeq z$ which means that $f$ is now to be interpreted by a partial function, i.e., a function whose domain $dom(f)$ is a subset of $\mathbb{N}^n$ and not necessarily the whole space $\mathbb{N}^n$. A partial function $f\colon \mathbb{N}^n \longrightarrow_p \mathbb{N}$ may be extended to a total function $\tilde{f}\colon \mathbb{N}^n \longrightarrow \mathbb{N} \cup \{\uparrow\}$ defined by

$$\tilde{f}(z_1, \ldots, z_n) := \begin{cases} f(z_1, \ldots, z_n) & \text{if } (z_1, \ldots, z_n) \in dom(f) \\ \uparrow & \text{otherwise} \end{cases}$$

where $f(z_1, \ldots, z_n) = \uparrow$ should be read as "*$f(z_1, \ldots, z_n)$ is undefined*". Then we obtain $f(z_1, \ldots, z_n) \simeq g(z_1, \ldots, z_n)$ iff $\tilde{f}(z_1, \ldots, z_n) = \tilde{g}(z_1, \ldots, z_n)$. We mostly identify $f$ and $\tilde{f}$.

We always simply write $f(z_1, \ldots, z_n) \simeq z$ instead of $\mathrm{ev}(f, z_1, \ldots, z_n) \simeq z$. The clauses of the inductive definition of $f(z_1, \ldots, z_n) \simeq z$ are those of the inductive evaluation of primitive recursive function terms as given in Definition 2.1.2 with $=$ replaced by $\simeq$ and the additional clause

$$\mu x_i \,.\, t\,(z_1, \ldots, z_n) :\simeq \begin{cases} \min\{x \mid t(z_1, \ldots, z_{i-1}, x, z_i \ldots z_n) \simeq 0\} & \text{if this exists} \\ \quad \text{and } t(z_1, \ldots, z_{i-1}, y, z_i \ldots z_n) \text{ is defined for all } y < x & \\ \uparrow & \text{otherwise.} \end{cases}$$

A function is *partial recursive* iff it is the interpretation of a partial recursive function term. The partial recursive function terms can be coded by natural numbers. The predicate "*e codes a partial recursive function term*" is definable by course of values recursion and thence primitive recursive. By $T$ we denote, Kleene's T-predicate. The meaning of $T(e, x_1, \ldots, x_n, y)$ is that $y$ codes the evaluation of the term number $e$ at the arguments $x_1, \ldots, x_n$. We cite two important theorems of elementary recursion theory. The first is Kleene's normal-form theorem. The second the Recursion Theorem.

**10.1.1 Theorem** *(Kleene's Normal-form Theorem) There are $n+2$-ary primitive recursive predicates $T^n$ and a regressive primitive recursive function $U$, i.e., a function satisfying $U(y) \le y$, such that for every $n$-ary partial recursive function $f$ there is an $e \in \omega$ such that $f(\vec{x}) \simeq U(\mu y \,.\, T^n(e, \vec{x}, y))$.*

We call $e$ an *index* for the $n$-ary partial recursive function $f$. It is a common abbreviation to put

$$\{e\}^n(\vec{x}) :\simeq U(\mu y \,.\, T^n(e, \vec{x}, y)).$$

This notation is known as Kleene-bracket. The connection between indices for functions of different arities is stated in the $S^m_n$-Theorem.

**10.1.2 Theorem** *($S^m_n$-Theorem) For each $m,n \in \mathbb{N}$ there is an $m+1$-ary primitive recursive function $S^m_n$ such that $\{e\}^{m+n}(\vec{y},\vec{x}) \simeq \{S^m_n(e,\vec{y})\}^n(\vec{x})$.*

An immediate consequence of the $S^m_n$-Theorem is the following Recursion Theorem.

**10.1.3 Theorem** *(Recursion Theorem) Let $g$ be an $n+1$-ary partial recursive function. Then there is an index $e$ such that $\{e\}^n(\vec{x}) \simeq g(e,\vec{x})$.*

*Proof* In order to emphasize that the Recursion Theorem is an immediate consequence of the purely combinatorial $S^m_n$-Theorem we give its simple proof. Define $h(y,\vec{x}) :\simeq g(S^1_n(y,y),\vec{x})$ and let $e_0$ be an index for $h$. For $e := S^1_n(e_0,e_0)$ we then compute

$$\begin{aligned}\{e\}^n(\vec{x}) &\simeq \{S^1_n(e_0,e_0)\}^n(\vec{x}) \simeq \{e_0\}^{n+1}(e_0,\vec{x})\\ &\simeq h(e_0,\vec{x}) \simeq g(S^1_n(e_0,e_0),\vec{x}) \simeq g(e,\vec{x})\end{aligned}$$

and are done. □

Since $\{e\}^n(\vec{x})$ is partial recursive in the arguments $e$ and $\vec{x}$ it follows from the Recursion Theorem that "nearly everything we can write down" is partial recursive. As an example we get an index $e$ such that $\{e\}^1(x) \simeq \{e\}^1(x)+1$. But this is not a contradiction. It only shows the function $\lambda x.\,\{e\}^1(x)$ is nowhere defined. We showed this example to stress the fact that the important part in introducing a recursive function is to show that this function is total, i.e., to show $(\forall\vec{x})(\exists y)\,T^n(e,\vec{x},y)$. This is the background of the following definition.

**10.1.4 Definition** Let $T$ be a theory which comprises the language of arithmetic (either directly or by interpretation). If $e$ is an index for a partial recursive function $f$ and $T$ proves $(\forall\vec{x})(\exists y)\,T^n(e,\vec{x},y)$ then we say that $f$ is a provably recursive function of $T$.

The aim of the following sections is to characterize the provably recursive functions of the theory $\mathsf{NT}$. The theory $\mathsf{NT}$ here serves as a paradigmatic example how an adaption of the methods of impredicative proof theories can be used to obtain better results also in predicative proof theory. This method can be extended also to stronger theories (cf. [10]).

## 10.2 Operator Controlled Derivations

In this section, we transfer the concept of operator controlled derivations to the semi-formal system defined in Definition 7.3.5. This time we only regard sentences in the language $\mathcal{L}(\mathsf{NT})$. Every sentence belongs either to $\bigwedge$–type or to $\bigvee$–type. Therefore we do not need clause *(Ax)* in the definition of $\frac{\alpha}{\rho}\ \Delta$ for a finite set $\Delta$ of $\mathcal{L}(\mathsf{NT})$-sentences. We want to refine $\frac{\alpha}{\rho}\ \Delta$ by controlling operators. In contrast to the situation in the ordinal analysis of $\mathsf{ID}_1$, we do not have ordinal parameters in the sentences of $\mathcal{L}(\mathsf{NT})$. The parameters which are of interest here are the natural numbers occurring in the sentences. Number parameters do not influence the computation of the $\Pi^1_1$-ordinal of a theory. Therefore, we could neglect them in the computation of $\|\mathsf{ID}_1\|_{\Pi^1_1}$. However, in the computation of what we will call the $\Pi^0_2$-ordinal of a theory, they play a crucial role.

For a finite set $\Delta$ of $\mathcal{L}(\mathsf{NT})$-sentences we define

$$\begin{aligned} \mathrm{par}(\Delta) := \{t^{\mathbb{N}} \in \omega \mid &\text{ The closed term } t \text{ occurs in some of the sentences in } \Delta\} \\ &\cup\{\mathrm{rnk}(F) \mid F \in \Delta\}. \end{aligned}$$

We count also parameters inside of open terms. As an example, the number 7 is among the parameters of the sentence $(\exists x)(\exists y)[\underline{7}+x=y]$.

To obtain controlling operators let

$$f\colon \mathbb{N} \longrightarrow \mathbb{N}$$

be a strictly increasing function. The function $f$ canonically induces an operator

$$\begin{aligned} &\bar{\mathsf{F}}_f\colon Pow(\mathbb{N}) \longrightarrow Pow(\mathbb{N}) \\ &\bar{\mathsf{F}}_f(X) := f[X] := \{f(x) \mid x \in X\}. \end{aligned}$$

Since we will only deal with finite sets $X \subseteq \mathbb{N}$ and are only interested in upper bounds it would be overkill to work with the induced operators. Therefore, we define for an increasing function $f\colon \mathbb{N} \longrightarrow \mathbb{N}$ and a finite set $X \subseteq \mathbb{N}$ directly

$$f(X) := \max\, \bar{\mathsf{F}}_f(X) = f(\max X). \tag{10.1}$$

To emphasize the fact that we think of operators rather than strictly increasing functions we keep on using sanserif capitals to denote strictly increasing functions and call them strictly increasing operators.

We abbreviate $(\forall x \in X)(\exists y \in Y)[x \leq y]$ by $X \leq Y$. Strictly increasing "operators" $\mathsf{F}$ are monotone in the sense that we have

$$X \leq Y \;\Rightarrow\; \mathsf{F}(X) \leq \mathsf{F}(Y).$$

As in Sect. 9.3 we use the notation

$$\mathsf{F}[X] := \lambda \Xi \,.\, \mathsf{F}(X \cup \Xi).$$

The next aim is to modify Definition 9.3.8 to operators induced by increasing functions. This is not completely obvious because $\bar{\mathsf{F}}_f(X)$ only contains natural numbers, i.e., finite ordinals, while we have to assign infinite ordinals. Here we need an observation which turns out to be crucial. In describing this observation, we restrict ourselves to ordinals below $\Gamma_0$. There we have a normal form

$$\alpha =_{NF} \varphi_{\alpha_1}(\beta_1) + \cdots + \varphi_{\alpha_n}(\beta_n)$$

as shown in Sect. 3.4.3. Recall the *norm* of $\alpha$ which has been defined by

$$N(\alpha) := \begin{cases} 0 & \text{if } \alpha = 0 \\ \sum_{i \leq n}(N(\alpha_i) + N(\beta_i) + 1) & \text{if } \alpha =_{NF} \varphi_{\alpha_1}(\beta_1) + \cdots + \varphi_{\alpha_n}(\beta_n). \end{cases}$$

Observe that $N(\alpha) < \omega$ for all ordinals $\alpha$ and $N(n) = n$ for $n < \omega$. For a finite set $X$ of ordinals let $N(X) := \{N(\alpha) \mid \alpha \in X\}$. For an operator $F$ and a finite set $X \subseteq On$ we define

$$\mathsf{O}_F(X) := \{\alpha < \Gamma_0 \mid N(\alpha) \leq F(N(X))\}.$$

Since there are only finitely many ordinals of the same norm, the set $\mathsf{O}_F(X)$ is always finite. Instead of $\mathsf{O}_F(\{\alpha\})$ we write mostly $\mathsf{O}_F(\alpha)$. There is no confusion as $\mathsf{O}_F$ takes only finite sets as arguments.

**10.2.1 Definition** Let $F$ be an operator. We define the relation $F \vdash^{\alpha}_{\rho} \Delta$ by the clauses $(\bigwedge)$, $(\bigvee)$ and (cut) of Definition 7.3.5 with the additional conditions

- $\alpha \in \mathsf{O}_F(\mathrm{par}(\Delta))$

and for an inference

$$F \vdash^{\alpha_\iota}_{\rho} \Delta_\iota \text{ for } \iota \in I \;\Rightarrow\; F \vdash^{\alpha}_{\rho} \Delta$$

with finite $I$ also

- $\mathrm{par}(\Delta_\iota) \leq F(\mathrm{par}(\Delta))$.

Of course the modified system remains sound, i.e., we still have

$$F \vdash^{\alpha}_{\rho} F_1, \ldots, F_n \;\Rightarrow\; \mathbb{N} \models F_1 \vee \cdots \vee F_n.$$

We define

$$F \leq G \;:\Leftrightarrow\; (\forall X)[F(X) \leq G(X)]$$

and say that the operator $G$ *extends* the operator $F$. The following lemma corresponds to Lemma 9.3.9.

**10.2.2 Lemma** *(Structural Lemma) Let $F$ and $G$ be strictly increasing operators which are not trivial and $\Delta$ and $\Gamma$ be finite sets of sentences. Assume $F[\mathrm{par}(\Delta)] \leq G[\mathrm{par}(\Delta)]$, $\Delta \subseteq \Gamma$, $\rho \leq \sigma$, $\alpha \leq \beta \in \mathsf{O}_G(\mathrm{par}(\Gamma))$ and $F \vdash^{\alpha}_{\rho} \Delta$. Then we also obtain $G \vdash^{\beta}_{\sigma} \Gamma$.*

The proof is a direct induction on $\alpha$. □

The operators $F$ are not Skolem-hull operators. The property $(F \circ F)(X) = F(X)$ which is characteristic for Skolem-hull operators does not hold. Therefore Lemma 9.3.10 has to be modified. This has a series of consequences.

**10.2.3 Lemma** *If $X \leq F(\mathrm{par}(\Delta))$ and $F[X] \vdash^{\alpha}_{\rho} \Delta$ then $F \circ F \vdash^{\alpha}_{\rho} \Delta$.*

*Proof* If $X \leq F(\mathrm{par}(\Delta))$ then $F[X](\mathrm{par}(\Delta)) \leq F^2(\mathrm{par}(\Delta))$. □

Next, we prove an inversion lemma for sentences in $\bigwedge$–type .

**10.2.4 Lemma** *(Inversion Lemma) Let $F \in \bigwedge$–type, $\mathrm{CS}(F) \neq \emptyset$ and assume $\mathscr{H} \overset{\alpha}{\underset{\rho}{\vdash}} \Delta, F$. Then $\mathscr{H}[\mathrm{par}(F)] \overset{\alpha}{\underset{\rho}{\vdash}} \Delta, G$ holds true for all $G \in \mathrm{CS}(F)$.*

*Proof* Induction on $\alpha$. The proof parallels completely the proof of Lemma 9.3.11. □

There is also a corresponding lemma to Lemma 9.3.13.

**10.2.5 Lemma** *(Detachment Lemma) Assume $\mathsf{F} \overset{\alpha}{\underset{\rho}{\vdash}} \Delta, F$ and $\neg F \in Diag(\mathbb{N})$. Then we already get $\mathsf{F}[\mathrm{par}(F)] \overset{\alpha}{\underset{\rho}{\vdash}} \Delta$.*

*Proof* Induction on $\alpha$. From $\neg F \in Diag(\mathbb{N})$ we conclude that $F$ is atomic and $F \in \bigvee$–type . Since $\mathrm{CS}(F) = \emptyset$, the sentence $F$ cannot be the critical formula of the last inference

(J) $\mathsf{F} \overset{\alpha_\iota}{\underset{\rho}{\vdash}} \Delta_\iota, F$ for $\iota \in J \;\Rightarrow\; \mathsf{F} \overset{\alpha}{\underset{\rho}{\vdash}} \Delta, F.$

By the induction hypothesis, we get

$$\mathsf{F}[\mathrm{par}(F)] \overset{\alpha_\iota}{\underset{\rho}{\vdash}} \Delta_\iota \text{ for } \iota \in J. \tag{i}$$

For finite $J$ we have $\mathrm{par}(\Delta_\iota, F) \leq \mathsf{F}(\mathrm{par}(\Delta, F))$, hence $\mathrm{par}(\Delta_\iota) \leq \mathsf{F}[\mathrm{par}(F)](\mathrm{par}(\Delta))$ and, because of $\alpha \in \mathsf{O}_{\mathsf{F}}(\mathrm{par}(\Delta, F)) \subseteq \mathsf{O}_{\mathsf{F}[\mathrm{par}(F)]}(\mathrm{par}(\Delta))$, we obtain $\mathsf{F}[\mathrm{par}(F)] \overset{\alpha}{\underset{\rho}{\vdash}} \Delta$ by an inference (J). □

**10.2.6 Lemma** *(Reduction Lemma) Let $\mathsf{F}$ be a controlling operator such that $2{\cdot}x \leq \mathsf{F}(x)$. Assume $\mathsf{F} \overset{\alpha}{\underset{\rho}{\vdash}} \Delta, F$ and $\mathsf{F} \overset{\beta}{\underset{\rho}{\vdash}} \Gamma, \neg F$ for a sentence $F \in \bigwedge$–type such that $\mathrm{rnk}(F) = \rho$ and $\mathrm{par}(F) \leq \mathsf{F}(\mathrm{par}(\Delta))$. Then $\mathsf{F}^3 \overset{\alpha+\beta}{\underset{\rho}{\vdash}} \Delta, \Gamma$.*

*Proof* We induct on $\beta$. Let us first check that $\alpha + \beta \in \mathsf{O}_{\mathsf{F}^3}(\mathrm{par}(\Delta, \Gamma))$. We have

$$\begin{aligned} N(\alpha+\beta) &\leq N(\alpha) + N(\beta) \leq 2{\cdot}\max\{N(\alpha), N(\beta)\} \\ &\leq 2{\cdot}\max\{\mathsf{F}(\mathrm{par}(\Delta, F)), \mathsf{F}(\mathrm{par}(\Gamma, F))\} \leq \mathsf{F}^2(\mathrm{par}(\Delta, \Gamma, F)). \end{aligned} \tag{i}$$

From $\mathrm{par}(F) \leq \mathsf{F}(\mathrm{par}(\Delta))$ we obtain $\mathsf{F}^2(\mathrm{par}(\Delta, \Gamma, F)) \leq \mathsf{F}^3(\mathrm{par}(\Delta, \Gamma))$. Hence $\alpha + \beta \in \mathsf{O}_{\mathsf{F}^3}(\mathrm{par}(\Delta, \Gamma))$.

If $\rho = 0$ then $F \in Diag(\mathbb{N})$ and we obtain $\mathsf{F}^3 \overset{\alpha+\beta}{\underset{0}{\vdash}} \Delta, \Gamma$ from the hypothesis $\mathsf{F} \overset{\beta}{\underset{0}{\vdash}} \Gamma, \neg F$ by the Detachment Lemma together with $\mathrm{par}(F) \leq \mathsf{F}(\mathrm{par}(\Delta))$, Lemma 10.2.2 and Lemma 10.2.3. Thus suppose $\rho > 0$, which entails $\mathrm{CS}(F) \neq \emptyset$. First assume that $\neg F$ is not the critical formula of the last inference

(J) $\mathsf{F} \overset{\beta_\iota}{\underset{\rho}{\vdash}} \Gamma_i, \neg F$ for $\iota \in I \;\Rightarrow\; \mathsf{F} \overset{\beta}{\underset{\rho}{\vdash}} \Gamma, \neg F.$

If (J) is an inference according to $(\bigwedge)$ with empty premises, then we have

$$\mathsf{F}^3 \overset{\alpha+\beta}{\underset{0}{\vdash}} \Delta, \Gamma.$$

Therefore assume that $I \neq \emptyset$. We have $\mathrm{par}(F) \leq \mathsf{F}(\mathrm{par}(\Delta))$ and obtain $\mathsf{F}^3 \overset{\alpha+\beta_\iota}{\underset{\rho}{\vdash}} \Delta, \Gamma_\iota$ by the induction hypothesis. Since $\alpha + \beta_\iota < \alpha + \beta$, and in the case of finite $I$ also $\mathrm{par}(\Delta, \Gamma_\iota) \subseteq \mathrm{par}(\Delta, \Gamma_\iota, F) \leq \mathsf{F}^3(\mathrm{par}(\Delta, \Gamma))$, we obtain $\mathsf{F}^3 \overset{\alpha+\beta}{\underset{\rho}{\vdash}} \Delta, \Gamma$ by an inference (J).

Now assume that $\neg F$ is the critical formula of the last inference in $\mathsf{F} \overset{\beta}{\underset{\rho}{\vdash}} \Gamma, \neg F$. Then we have the premise $\mathsf{F} \overset{\beta_0}{\underset{\rho}{\vdash}} \Gamma, \neg F, \neg G$ for some $G \in \mathrm{CS}(F)$ with

$$\mathrm{par}(\Gamma, F, G) \leq \mathsf{F}(\mathrm{par}(\Gamma, F)) \tag{ii}$$

and obtain

$$\mathsf{F}^3 \overset{\alpha+\beta_0}{\underset{\rho}{\vdash}} \Delta, \Gamma, \neg G \tag{iii}$$

by the induction hypothesis. By inversion and Lemma 10.2.3 we obtain from the hypotheses $\mathsf{F} \overset{\alpha}{\underset{\rho}{\vdash}} \Delta, F$ and $\mathrm{par}(F) \leq \mathsf{F}(\mathrm{par}(\Delta))$

$$\mathsf{F} \circ \mathsf{F} \overset{\alpha}{\underset{\rho}{\vdash}} \Delta, G. \tag{iv}$$

From (iv) we get by the Structural Lemma $\mathsf{F}^3 \overset{\alpha+\beta_0}{\underset{\rho}{\vdash}} \Delta, \Gamma, G$. It is $\alpha + \beta_0 < \alpha + \beta$ and $\mathrm{rnk}(G) < \rho$. To apply a cut we still have to check

$$\mathrm{par}(\Delta, \Gamma, G) \leq \mathsf{F}^3(\mathrm{par}(\Delta, \Gamma)). \tag{v}$$

But this is secured by (ii) and the hypothesis $\mathrm{par}(F) \leq \mathsf{F}(\mathrm{par}(\Delta))$ which implies $\mathrm{par}(\Delta, \Gamma, G) \leq \mathsf{F}(\mathrm{par}(\Delta, \Gamma, F)) \leq \mathsf{F}^3(\mathrm{par}(\Delta, \Gamma))$. □

## 10.3 Iterating Operators

The remarkable and, in comparison to Sect. 9.3, new fact is that we have to iterate the controlling operator in the Reduction Lemma. The crucial point is how to define transfinite iterations of operators. Here we will follow quite closely the pattern we used in defining the iterated Skolem-hull operators $\mathcal{H}_\alpha$.

**10.3.1 Definition** Let $\mathsf{F}$ be an operator. We define

$$\begin{aligned}\psi_{\mathsf{F}}(\alpha) &:= \min\{\eta \mid (\forall \xi \in \mathsf{O}_{\mathsf{F}}(\alpha) \cap \alpha)[\psi_{\mathsf{F}}(\xi) < \eta]\} \\ &= \sup(\{\psi_{\mathsf{F}}(\beta) + 1 \mid \beta \in \mathsf{O}_{\mathsf{F}}(\alpha) \cap \alpha\} \cup \{0\}).\end{aligned}$$

Since $\mathsf{O}_{\mathsf{F}}(\alpha) \cap \alpha$ is finite, the supremum in Definition 10.3.1 is in fact a maximum and the following theorem is immediate.

**10.3.2 Theorem** *For all ordinals $\alpha < \Gamma_0$ and all strictly increasing operators $\mathsf{F}$ we get $\psi_{\mathsf{F}}(\alpha) < \omega$.*

The finite ordinal $\psi_F(\alpha)$ is the first ordinal majorizing all the "ordinals" which are reachable by $\psi_F\restriction(O_F(\alpha)\cap\alpha)$. Observe the strong analogy to the definition of $\psi_{\mathscr{H}}(\alpha)$.

To simplify notation we introduce the abbreviation

$$\alpha \ll_F \beta \;:\Leftrightarrow\; \alpha \in O_F(\beta)\cap\beta \;\Leftrightarrow\; \alpha<\beta \wedge N(\alpha)\le F(N(\beta)).$$

Then we obtain

$$\alpha \ll_F \beta \;\Rightarrow\; F_\alpha < F_\beta. \tag{10.2}$$

By induction on $n<\omega$ we obtain $\psi_F(n)=n$. This is obvious for $n=0$ and in the successor case we have

$$\psi_F(n+1)=\max\{\psi_F(m)+1 \,|\, m\le n \wedge N(m)\le F(n+1)\}=n+1.$$

Hence

$$\psi_F\restriction\omega = Id_\omega. \tag{10.3}$$

The essential property of the functions $\psi_F$ is stated in the following lemma. We call an operator $F$ *strongly increasing* if it satisfies

$$F(x)+1<F(x+1).$$

For strongly increasing operators we obtain

$$n\neq 0 \;\Rightarrow\; n+F(x)<F(n+x) \tag{10.4}$$

by an easy induction on $n$.

**10.3.3 Lemma** *Let $F$ be a strongly increasing operator. Then*

$$\psi_F(\alpha+\psi_F(\beta))\le\psi_F(\alpha\,\#\,\beta). \tag{10.5}$$

*Proof* We prove the lemma by induction on $\beta$. For $\alpha=0$ or $\beta=0$ this follows from $\psi_F\restriction\omega=Id_\omega$. So assume $\alpha\neq 0$ and $\beta\neq 0$. Then $\psi_F(\beta)=\psi_F(\gamma)+1$ for some $\gamma\in O_F(\beta)\cap\beta$. Hence

$$\psi_F(\alpha+\psi_F(\beta))=\psi_F(\alpha+\psi_F(\gamma)+1)\le\psi_F(\alpha\,\#\,\gamma+1).$$

If $\alpha\,\#\,\beta=\alpha\,\#\,\gamma+1$ then we are done. So assume $\alpha\,\#\,\gamma+1<\alpha\,\#\,\beta$. But $N(\gamma)\le F(N(\beta))$ entails by (10.4) $N(\alpha\,\#\,\gamma+1)=N(\alpha)+N(\gamma)+1\le N(\alpha)+1+F(N(\beta))\le F(N(\alpha)+N(\beta))$. Hence $\alpha\,\#\,\gamma+1\in O_F(\alpha\,\#\,\beta)\cap(\alpha\,\#\,\beta)$ and therefore $\psi_F(\alpha\,\#\,\gamma+1)\le\psi_F(\alpha\,\#\,\beta)$. □

As a first consequence of Lemma 10.3.3 we obtain

$$F^n(x)\le\psi_F(\omega\cdot n+x), \tag{10.6}$$

where $F^n$ denotes the $n$-th iteration of the strongly increasing operator $F$.

We prove (10.6) by induction on $n$. First, we observe that $F(x)\ll_F\omega+x$, which implies $F(x)=\psi_F(F(x))<\psi_F(\omega+x)$. Hence $\omega\cdot n+F(x)\in O_F(\omega\cdot n+F(x))\cap\omega\cdot n+$

$\psi_F(\omega+x)$ by which we obtain $F^{n+1}(x) = F^n(F(x)) \le \psi_F(\omega{\cdot}n+F(x)) \le \psi_F(\omega{\cdot}n+\psi_F(\omega+x)) \le \psi_F(\omega{\cdot}(n+1)+x)$.

Conversely we obtain

$$\psi_F(\omega{\cdot}n+x) \le F^n(2{\cdot}n+x)+n \tag{10.7}$$

which is shown by main induction on $n$ and side induction on $x$. The case $n=0$ follows trivially from $\psi_F{\restriction}\omega = Id_\omega$. So assume that the claim is correct for $n$. Let

$$\eta_1 := \omega{\cdot}n+F(2n+x) \tag{i}$$

and

$$\eta_0 := \omega{\cdot}(n+1)+x-1 \tag{ii}$$

if $x \neq 0$. We claim

$$\psi_F(\omega{\cdot}(n+1)+x) = \max\{\psi_F(\eta_0), \psi_F(\eta_1)\}+1. \tag{iii}$$

Let $\alpha := \omega{\cdot}(n+1)+x$. First we show that $\eta_i \in O_F(\alpha)\cap\alpha$ holds true for $i=0,1$. We clearly have $\eta_i < \alpha$. We get $N(\eta_0) = 2n+x+1 \le F(2(n+1)+x) = F(N(\alpha))$ and $N(\eta_1) = 2n+F(2n+x) \le F(2(n+1)+x)$ since $F$ is strongly increasing. For $\omega{\cdot}(n+1) \le \xi \in O_F(\alpha)\cap\eta_0$ we get $N(\xi) \le 2{\cdot}(n+1)+x-1 \le F(2{\cdot}(n+1)+x-1) = F(N(\eta_0))$. Hence $\xi \in O_F(\eta_0)\cap\eta_0$ and thus $\psi_F(\xi) < \psi_F(\eta_0)$. For $\xi \in O_F(\alpha)\cap\omega{\cdot}(n+1)$ we get $N(\xi) \le F(N(\alpha)) = F(2(n+1)+x) \le F(2n+F(2n+x)) = F(N(\eta_1))$ and $\xi = \omega{\cdot}k+m$. For $k<n$ we obviously have $\xi < \eta_1$ and for $k=n$ we get $m \le F(2n+x)$, otherwise $N(\xi) \le F(N(\alpha))$ would fail. Hence $\xi = \eta_1$ or $\xi \ll_F \eta_1$ and thus $\psi_F(\xi) \le \psi_F(\eta_1)$.

If $\max\{\psi_F(\eta_0),\psi_F(\eta_1)\} = \psi_F(\eta_0)$ we get $\psi_F(\omega{\cdot}(n+1)+x) = \psi_F(\eta_0)+1 = \psi_F(\omega{\cdot}(n+1)+x-1)+1 \le F^{n+1}(2{\cdot}(n+1)+x-1)+n+2 \le F^{n+1}(2(n+1)+x)+n+1$ by the induction hypothesis for $x$.

If $\max\{\psi_F(\eta_0),\psi_F(\eta_1)\} = \psi_F(\eta_1)$ we get $\psi_F(\omega{\cdot}(n+1)+x) = \psi_F(\eta_1)+1 = \psi_F(\omega{\cdot}n+F(2n+x))+1 \le F^n(2n+F(2n+x))+n+1 \le F^n(F(2(n+1)+x))+n+1 = F^{n+1}(2(n+1)+x)+n+1$ by the induction hypothesis for $n$. This finishes the proof of (10.7).

Equations (10.6) and (10.7) show that the functions $\psi_F$ are closely connected to iterations of the "operator" $F$.

**10.3.4 Definition** In view of (10.6) and (10.7) it makes sense to define

$$F_\alpha := \lambda x.\, \psi_F(\alpha+x) \tag{10.8}$$

and to call $F_\alpha$ the $\alpha$-th "iteration" of $F$

However, Definition 10.3.4 is not completely stringent. So we get, for instance, $F_n(x) = n+x$ for finite $n$, which has nothing to do with iterations of the operator $F$. Defining $F_\alpha(x) := \psi_F(\omega{\cdot}\alpha+x)$ would have been more in the spirit of an iteration. This would have guaranteed $F_0 = Id$, $F_1 \approx F$ and $F_n \approx F^n$. However, in computing upper bounds for the provably recursive functions of NT we will very quickly reach ordinals beyond $\omega^\omega$. But $\omega \cdot \alpha = \alpha$ holds for all additively indecomposable ordinals $\alpha \geq \omega^\omega$. Therefore we have opted for the technically much simpler definition given in (10.8).

The next lemma is a corollary to (10.5).

**10.3.5 Lemma** *Let $F$ be a strongly increasing operator. Then $F_\alpha \circ F_\beta \leq F_{\alpha \# \beta}$.*

*Proof* We have $F_\alpha(F_\beta(x)) = \psi_F(\alpha + \psi_F(\beta + x)) \leq \psi_F(\alpha \# \beta + x) = F_{\alpha \# \beta}(x)$. □

We will be forced to iterate iterations of operators, i.e., to regard iterations $(F_\alpha)_\beta$. To find upper bounds for $(F_\alpha)_\beta$ we have to compute $\psi_{F_\alpha}(\beta)$. This computation needs the *symmetric product* of ordinals.

**10.3.6 Definition** We define the symmetric product $\alpha \times \beta$ in two steps. We start with additively indecomposable ordinals for which we put

$$\omega^\alpha \times \omega^\beta := \omega^{\alpha \# \beta}$$

where $\#$ denotes the symmetric sum (cf. Exercise 3.3.14). For ordinals $\alpha =_{NF} \alpha_1 + \cdots + \alpha_m$ and $\beta =_{NF} \beta_1 + \cdots + \beta_n$ in Cantor normal-form, we put

$$\alpha \times \beta := \sum_{i=1}^{m}{}^{\#} \left( \sum_{j=1}^{n}{}^{\#} \alpha_i \times \beta_j \right)$$

and define

$$\alpha \times 0 = 0 \times \alpha := 0.$$

First we have to show that the symmetric product behaves like a commutative product.

**10.3.7 Lemma** *For all ordinals $\alpha$, $\beta$ and $\gamma$ we have*

- $\alpha \times \beta = \beta \times \alpha$,
- $\alpha \times 1 = \alpha$,
- $\alpha \times (\beta \# \gamma) = (\alpha \times \beta) \# (\alpha \times \gamma)$,
- $\alpha \times \xi < \alpha \times \eta$ *for all ordinals* $\xi < \eta$

*and*

- $\alpha \times \lambda = \sup_{\eta < \lambda} (\alpha \times \eta)$ *for limit ordinals* $\lambda$.

*Proof* The first claim follows directly from the definition of the symmetric sum and Exercise 3.3.14. For the second claim, we first observe $\omega^\xi \times 1 = \omega^\xi \times \omega^0 = \omega^\xi$ and obtain $\alpha \times 1 = \#\sum_{i=1}^{n} \alpha_i \times 1 = \alpha$ for $\alpha =_{NF} \sum_{i=1}^{n} \alpha_i$. For the third claim, we assume $\alpha =_{NF} \sum_{i=1}^{l} \alpha_i$, $\beta =_{NF} \sum_{j=1}^{m} \beta_j$ and $\gamma =_{NF} \sum_{k=1}^{n} \gamma_k$. Then we obtain

$$\begin{aligned}\alpha \times (\beta \# \gamma) &= \sum_{i=1}^{l}{}^{\#} \left( \sum_{j=1}^{m}{}^{\#} (\alpha_i \times \beta_j) \# \sum_{k=1}^{n}{}^{\#} (\alpha_i \times \gamma_k) \right) \\ &= \sum_{i=1}^{l}{}^{\#} \sum_{j=1}^{m}{}^{\#} (\alpha_i \times \beta_j) \# \sum_{i=1}^{l}{}^{\#} \sum_{k=1}^{n}{}^{\#} (\alpha_i \times \gamma_k) = (\alpha \times \beta) \# (\alpha \times \gamma).\end{aligned}$$

For the fourth claim let $\alpha = \sum_{i=1}^{l}{}^{\#} \omega^{\alpha_i}$, $\xi =_{NF} \sum_{j=1}^{m} \omega^{\xi_j}$ and $\eta =_{NF} \sum_{k=1}^{n} \omega^{\eta_j}$. Since $\xi < \eta$ we either have $\xi_j = \eta_j$ for $j = 1,\ldots,m$ and $m < n$ or there is an $r < n$ such that $\xi_j = \eta_j$ for $1 \le j \le r$ and $\xi_j < \eta_{r+1}$ for $r+1 \le j \le m$. Then we obtain $\alpha \times \xi = \sum_{i=1}^{l}{}^{\#} \sum_{j=1}^{m}{}^{\#} \omega^{\alpha_i \# \xi_j} < \sum_{i=1}^{l}{}^{\#} \sum_{k=1}^{n}{}^{\#} \omega^{\alpha_i \# \eta_k} = \alpha \times \eta$ because either $\alpha_i \# \xi_j = \alpha_i \# \eta_j$ for $j = 1,\ldots,m$ and $m < n$ or $\alpha_i \# \xi_j = \alpha_i \# \eta_j$ for $1 \le j \le r$ and $\alpha_i \# \xi_j < \alpha_i \# \eta_{r+1}$ for $r+1 \le j \le m$.

For the last claim we have

$$\sup_{\eta<\lambda} (\alpha \times \eta) \le \alpha \times \lambda \tag{i}$$

by the previous claim. We first prove

$$\omega^\alpha \times \omega^{\xi+1} = \sup_{\eta<\omega^{\xi+1}} (\omega^\alpha \times \eta). \tag{ii}$$

If $\mu < \omega^\alpha \times \omega^{\xi+1} = \omega^{\alpha\#\xi+1}$ there is an $n < \omega$ such that $\mu < \sum_{i=1}^{n}{}^{\#}(\omega^{\alpha\#\xi}) = \sum_{i=1}^{n}{}^{\#}(\omega^\alpha \times \omega^\xi) = \omega^\alpha \times \sum_{i=1}^{n}{}^{\#}\omega^\xi < \sup_{\eta<\omega^{\xi+1}}(\omega^\alpha \times \eta)$ because $\sum_{i=1}^{n}{}^{\#}(\omega^\xi) < \sum_{i=1}^{n+1}{}^{\#}\omega^\xi < \omega^{\xi+1}$. Hence $\omega^\alpha \times \omega^{\xi+1} \le \sup_{\eta<\omega^{\xi+1}}(\omega^\alpha \times \eta)$. The converse inequality holds by (i).

Next we prove,

$$\omega^\alpha \times \omega^\lambda = \sup_{\eta<\omega^\lambda} (\omega^\alpha \times \eta) \tag{iii}$$

for limit ordinals $\lambda$. If $\mu < \omega^\alpha \times \omega^\lambda = \omega^{\alpha\#\lambda}$ there is a $\zeta < \lambda$ such that $\mu < \omega^{\alpha\#\zeta} = \omega^\alpha \times \omega^\zeta$. Since $\omega^\zeta < \omega^\lambda$ this implies $\omega^\alpha \times \omega^\lambda \le \sup_{\eta<\omega^\lambda}(\omega^\alpha \times \eta)$. The opposite inequality is again (i).

But (ii) and (iii) easily generalize to

$$\alpha \times \omega^\xi = \sup_{\eta<\omega^\xi} (\alpha \times \eta) \tag{iv}$$

for arbitrary ordinals $\alpha$ and $\xi \neq 0$. If $\lambda \in Lim$ we have $\lambda =_{NF} \sum_{j=1}^{n-1} \lambda_j + \lambda_n$ with $\lambda_n = \omega^\mu$ and $\mu \neq 0$. Then we obtain by (iv)

$$\alpha \times \lambda = \alpha \times \sum_{i=1}^{n}\!\!{}^{\#}\, \lambda_i = \sum_{i=1}^{n-1}\!\!{}^{\#}\, (\alpha \times \lambda_i) \# (\alpha \times \lambda_n) = \sum_{i=1}^{n-1}\!\!{}^{\#}\, (\alpha \times \lambda_i) \# \sup_{\xi<\lambda_n} (\alpha \times \xi)$$
$$= \sup_{\xi<\lambda_n}\left(\sum_{i=1}^{n-1}\!\!{}^{\#}\, (\alpha \times \lambda_i) \# (\alpha \times \xi)\right) = \sup_{\xi<\lambda_n}\left(\alpha \times \left(\sum_{i=1}^{n-1}\!\!{}^{\#}\, \lambda_i \# \xi\right)\right) \leq \sup_{\eta<\lambda}(\alpha \times \eta).$$

□

We need an upper bound for the norm of the symmetric product and want to show

$$N(\alpha)+N(\beta) < N(\alpha \times \beta) \leq N(\alpha)\cdot N(\beta). \tag{10.9}$$

For additively indecomposable ordinals $\alpha$ and $\beta$, we immediately get

$$N(\alpha \times \beta) = N(\alpha)+N(\beta)+1.$$

If $N(\alpha) \leq 2$ or $N(\beta) \leq 2$, we check $N(\alpha \times \beta) \leq N(\alpha)\cdot N(\beta)$ by looking at all possible cases. For $N(\alpha)$ and $N(\beta) > 2$, we obtain $N(\alpha)+N(\beta)+1 \leq N(\alpha)\cdot N(\beta)$ by induction on $N(\beta)$. Hence $N(\alpha \times \beta) \leq N(\alpha)\cdot N(\beta)$ for additively indecomposable ordinals. Therefore, we obtain for $\alpha =_{NF} \alpha_1 + \cdots \alpha_m$ and $\beta =_{NF} \beta_1 + \cdots \beta_n$

$$N(\alpha \times \beta) = \sum_{i=1}^{m}(\sum_{j=1}^{n} N(\alpha_i \times \beta_j)) \leq \sum_{i=1}^{m}\sum_{j=1}^{n}(N(\alpha_i)\cdot N(\beta_j))$$
$$= N(\alpha)\cdot N(\beta).$$

**10.3.8 Lemma** *Let $F$ be a strongly increasing operator satisfying also $x^2 \leq F(x)$. Then*

$$\psi_{F_\alpha}(\beta) \leq \psi_F(\alpha \times (\beta+1)+N(\beta)). \tag{10.10}$$

*Proof* We prove the lemma by induction on $\beta$. It is

$$\psi_{F_\alpha}(\beta) = \psi_{F_\alpha}(\eta)+1 \tag{i}$$

for some $\eta < \beta$ such that $N(\eta) \leq F_\alpha(N(\beta)) = \psi_F(\alpha+N(\beta))$. From (i) we obtain by the induction hypothesis

$$\psi_{F_\alpha}(\beta) \leq \psi_F(\alpha \times (\eta+1)+N(\eta))+1. \tag{ii}$$

If $\beta = \eta+1$ then (ii) directly implies $\psi_{F_\alpha}(\beta) \leq \psi_F(\alpha \times (\beta+1)+N(\beta))$ since clearly $\alpha \times \beta + N(\beta) \ll_F \alpha \times (\beta+1)+N(\beta)$. Otherwise we have

$$\alpha \times (\eta+1)+N(\eta) \ll_F \alpha \times \beta + \psi_F(\alpha+N(\beta)) \tag{iii}$$

since $\alpha \times (\eta+1)+N(\eta) < \alpha \times \beta + \psi_F(\alpha+N(\beta))$ and $N(\alpha \times (\eta+1)) + N(\eta) \leq N(\alpha)\cdot N(\eta)+N(\alpha)+N(\eta) \leq (N(\alpha)+N(\eta))^2 \leq F(N(\alpha)+N(\eta)) \leq F(N(\alpha \times \beta)+\psi_F(\alpha+N(\beta)))$.

But (iii) and (ii) imply

$$\psi_{F_\alpha}(\beta) \le \psi_F(\alpha\times(\eta+1)+N(\eta))+1 \le \psi_F(\alpha\times\beta+\psi_F(\alpha+N(\beta)))$$
$$\le \psi_F(\alpha\times\beta \# \alpha+N(\beta)) = \psi_F(\alpha\times(\beta+1)+N(\beta)). \tag{iv}$$

□

As a consequence of (10.10) we obtain an upper bound for $(F_\alpha)_\beta$.

**10.3.9 Lemma** *Let $F$ be a strongly increasing operator satisfying $x^2 \le F(x)$ and $\alpha$ be an infinite ordinal. Then*

$$(F_\alpha)_\beta(x) \le F_{\alpha\times(\beta\#\omega)}(x). \tag{10.11}$$

*Proof* By (10.10) we have

$$(F_\alpha)_\beta(x) = \psi_{F_\alpha}(\beta+x) \le \psi_F(\alpha\times(\beta+x+1)+N(\beta)+x). \tag{i}$$

We are done if we can show

$$\alpha\times(\beta+x+1)+N(\beta)+x \ll_F \alpha\times(\beta\#\omega)+x \tag{ii}$$

since then $(F_\alpha)_\beta(x) \le \psi_F(\alpha\times(\beta\#\omega)+x) = F_{\alpha\times(\beta\#\omega)}(x)$. We clearly have

$$\alpha\times(\beta+x+1)+N(\beta)+x < \alpha\times(\beta\#\omega)+x.$$

Moreover, we get

$$N(\alpha\times(\beta+x+1)+N(\beta)+x) \le N(\alpha)\cdot(N(\beta)+x+1)+N(\beta)+x$$
$$\le (N(\alpha)+N(\beta)+x)^2 \le F(N(\alpha)+N(\beta)+x) \le F(N(\alpha\times(\beta\#\omega)+x)).$$ □

The following theorem is a first application of Lemma 10.3.9.

**10.3.10 Theorem** *(Witnessing Theorem) Let $G$ be a $\exists^0_1$-sentence of $\mathcal{L}(\mathsf{NT})$, i.e., a sentence of the form $(\exists x)F_u(x)$ such that $F$ is quantifier free, and $F$ a strongly increasing operator satisfying $x^2 \le F(x)$. If $F \Big|\frac{\alpha}{0}\, (\exists x)F_u(x)$ then there is an $n < F_{\omega^{\alpha\cdot 2+1}}(\mathrm{par}(G))$ such that $\mathbb{N} \models F_u(\underline{n})$.*

*Proof* Induction on $\alpha$. The only possibility for $F \Big|\frac{\alpha}{0}\, (\exists x)F_u(x)$ is an inference according to $(\bigvee)$ whose premise is

$$F \Big|\frac{\alpha_0}{0}\, (\exists x)F_u(x), F_u(t) \tag{i}$$

for some term $t$ such that $\mathrm{par}(G, F_u(t)) \subseteq F(\mathrm{par}(G))$, hence $t^{\mathbb{N}} \le F(\mathrm{par}(G))$. In case of $\mathbb{N} \models F_u(t)$ we choose $n := t^{\mathbb{N}}$. Then $t^{\mathbb{N}} \le F(\mathrm{par}(G)) \le F_{\omega^{\alpha\cdot 2+1}}(\mathrm{par}(G))$ is true by the parameter condition for inferences with finitely many premises. Otherwise we get $F[\mathrm{par}(t)] \Big|\frac{\alpha_0}{0}\, (\exists x)F_u(x)$ by the Detachment Lemma. Since $\mathrm{par}(t) \le F(\mathrm{par}(G))$ this entails $F \circ F \Big|\frac{\alpha_0}{0}\, (\exists x)F_u(x)$. Hence $F_{\omega\cdot 2} \Big|\frac{\alpha_0}{0}\, (\exists x)F_u(x)$. By the induction hypothesis there is an $n \le (F_{\omega\cdot 2})_{\omega^{\alpha_0\cdot 2+1}}(\mathrm{par}(G))$ such that $\mathbb{N} \models F_u(\underline{n})$. By (10.11) we therefore obtain $n \le F_{\omega\cdot 2\times(\omega^{\alpha_0\cdot 2+1}\#\omega)}(\mathrm{par}(G))$. Putting $\beta := \omega\cdot 2\times(\omega^{\alpha_0\cdot 2+1}\#\omega) = \omega^{\alpha_0\cdot 2+2}\cdot 2 \# \omega^2\cdot 2$ it remains to check

$$F_\beta(\mathrm{par}(G)) \le F_{\omega^{\alpha\cdot 2+1}}(\mathrm{par}(G)). \tag{ii}$$

We certainly have $\beta < \omega^{\alpha\cdot 2+1}$. For $\alpha = \alpha_0 + 1$ we get $N(\beta) = 4\cdot N(\alpha_0) + 12 \le (2N(\alpha_0)+4)^2 \le F(2N(\alpha_0)+4) = F(N(\omega^{\alpha\cdot 2+1}))$, hence $\beta \ll_F \omega^{\alpha\cdot 2+1}$, which immediately entails (ii). Therefore, assume $\alpha_0 + 1 < \alpha$ and let $m := \max\mathrm{par}(G)$. Since $\alpha_0 \in \mathsf{O}_F(\mathrm{par}(G))$ we have $N(\alpha_0) \le \psi_F(\omega + m)$ and thus obtain $\beta \ll_F \omega^{\alpha\cdot 2} + 4\cdot\psi_F(\omega+m) + 12$. Hence

$$\begin{aligned} F_\beta(m) = \psi_F(\beta+m) &< \psi_F(\omega^{\alpha\cdot 2} + 4\cdot\psi_F(\omega+m) + 12 + m) \\ &\le \psi_F(\omega^{\alpha\cdot 2} \# \omega\cdot 4 + 12 + 5m). \end{aligned} \tag{iii}$$

But $\alpha \ge 1$ and $m \ge 1$ entail $\omega^{\alpha\cdot 2} \# \omega\cdot 4 + 12 + 5m \ll_F \omega^{\alpha\cdot 2+1} + m$ which together with (iii) implies (ii). □

It follows from the Witnessing Theorem and the Inversion Lemma that operator-controlled cut free derivations provide upper bounds for the Skolem-functions of $\forall^0_2$-sentences that are provable in the operator-controlled semi-formal system. Whenever we have $F \,|\!\frac{\alpha}{0}\, (\forall\vec{x})(\exists y)F(\vec{x},y,\underline{n}_1,\ldots,\underline{n}_k)$ for a quantifier free formula $F(\vec{x},y,\underline{n}_1,\ldots,\underline{n}_k)$ we obtain by inversion

$$F \,|\!\frac{\alpha}{0}\, (\exists y)F(\underline{m}_1,\ldots,\underline{m}_l,y,\underline{n}_1,\ldots,\underline{n}_k),$$

which by the Witnessing Theorem shows that there is a natural number $n < F_{\omega^{\alpha\cdot 2+1}}(\{m_1,\ldots,m_l,n_1,\ldots,n_k\})$ such that $\mathbb{N} \models F(\underline{m}_1,\ldots,\underline{m}_l,\underline{n},\underline{n}_1,\ldots,\underline{n}_k)$.

A Skolem function for a true $\forall^0_2$-sentence $(\forall x)(\exists y)F(x,y)$ is the recursive function $\mu y\,.\,F(x,y)$. If we succeed in characterizing the class of controlling operators for the $\forall^0_2$-sentences that are provable in NT we thus obtain a characterization of the recursive functions whose totality is provable in NT. Therefore, we have to prove cut-elimination for the semi-formal system with controlling operators.

**10.3.11 Exercise** The relation $\ll_F$ is not transitive. To make it transitive define

$$\begin{aligned} &\alpha \ll^0_F \beta \;:\Leftrightarrow\; \alpha = \beta \\ &\alpha \ll^{k+1}_F \beta \;:\Leftrightarrow\; (\exists\gamma)[\alpha \ll^k_F \gamma \wedge \gamma \ll_F \beta] \end{aligned}$$

and

$$\alpha \ll^*_F \beta \;:\Leftrightarrow\; (\exists k)[\alpha \ll^k_F \beta].$$

Prove $\psi_F(\alpha) = \max\{k \mid 0 \ll^k_F \alpha\}$.

## 10.4 Cut Elimination for Operator Controlled Derivations

We have now all the material to prove cut-elimination for operator controlled derivations in the semi-formal system. This will be the main tool in finding upper bounds for the provable recursive functions of NT. In this section we tacitly assume that all operators $F$ satisfy $x^3 \le F(x)$. This means no loss of generality because we will later see that the operators which result from the embedding procedure increase even stronger.

**10.4.1 Lemma** *(Elimination Lemma)* *Assume* $F \vdash^{\alpha}_{\rho+1} \Delta$. *Then* $F_{\omega^{\alpha+1}} \vdash^{\omega^{\alpha}}_{\rho} \Delta$.

*Proof* The proof is similar to the proofs of the analogous statements (Lemma 7.3.13, Theorem 9.3.15) proved before. We will, however, see that we have to be much more careful in calculating the controlling operators. In case that the last inference is not a cut of rank $\rho$, we have the premises

$$F \vdash^{\alpha_\iota}_{\rho+1} \Delta_\iota \text{ for } \iota \in I$$

and obtain

$$F_{\omega^{\alpha_\iota+1}} \vdash^{\omega^{\alpha_\iota}}_{\rho} \Delta_\iota \tag{i}$$

by the induction hypothesis. Now we show

$$F_{\omega^{\alpha_\iota+1}}[\mathrm{par}(\Delta_\iota)] \leq F_{\omega^{\alpha+1}}[\mathrm{par}(\Delta_\iota)]. \tag{ii}$$

Let $x \in \omega$, $d_\iota := \max \mathrm{par}(\Delta_\iota)$ and $e_\iota := \max\{d_\iota, x\}$. Then $F_{\omega^{\alpha_\iota+1}}[\mathrm{par}(\Delta_\iota)](x) = F_{\omega^{\alpha_\iota+1}}(e_\iota)$. From $\alpha_\iota \in \mathrm{O}_F(\mathrm{par}(\Delta_\iota))$ we obtain $N(\omega^{\alpha_\iota+1}) = N(\alpha_\iota)+2 \leq F(d_\iota)+2 \leq \psi_F(\omega+d_\iota)+2$. Therefore $\omega^{\alpha_\iota+1} \ll_F \omega^{\alpha}+\psi_F(\omega+d_\iota)+2$ and we obtain

$$\begin{aligned} F_{\omega^{\alpha_\iota+1}}(e_\iota) &< F_{\omega^{\alpha}+\psi_F(\omega+d_\iota)+2}(e_\iota) = \psi_F(\omega^{\alpha}+\psi_F(\omega+d_\iota)+2+e_\iota) \\ &\leq \psi_F(\omega^{\alpha} \# \omega+2e_\iota+2) < \psi_F(\omega^{\alpha+1}+e_\iota) = F_{\omega^{\alpha+1}}(e_\iota), \end{aligned} \tag{iii}$$

where we used $\omega^{\alpha} \# \omega+2e_\iota+2 \ll_F \omega^{\alpha+1}+e_\iota$, which is true because of $\omega^{\alpha} \# \omega+2e_\iota+2 < \omega^{\alpha+1}+e_\iota$ and $N(\omega^{\alpha} \# \omega+2e_\iota+2) = N(\alpha)+5+2e_\iota \leq (N(\alpha)+2+e_\iota)^2 < F(N(\omega^{\alpha+1}+e_\iota))$.

From (i) and (ii) we obtain

$$F_{\omega^{\alpha+1}} \vdash^{\omega^{\alpha_\iota}}_{\rho} \Delta_\iota \text{ for } \iota \in I \tag{iv}$$

by the Structural Lemma and from (iv)

$$F_{\omega^{\alpha+1}} \vdash^{\omega^{\alpha}}_{\rho} \Delta$$

with the same inference. In case of a finite index set $I$ it is obvious that $\mathrm{par}(\Delta_\iota) \subseteq F(\mathrm{par}(\Delta))$ also entails $\mathrm{par}(\Delta_\iota) \subseteq F_{\omega^{\alpha+1}}(\mathrm{par}(\Delta))$.

So assume that the last inference is a cut of rank $\rho$. Then, we have the premises

$$F \vdash^{\alpha_0}_{\rho+1} \Delta, F \text{ and } F \vdash^{\alpha_0}_{\rho+1} \Delta, \neg F \tag{v}$$

with $\mathrm{rnk}(F) = \rho$ and obtain by the induction hypothesis

$$F_{\omega^{\alpha_0+1}} \vdash^{\omega^{\alpha_0}}_{\rho} \Delta, F \text{ and } F_{\omega^{\alpha_0+1}} \vdash^{\omega^{\alpha_0}}_{\rho} \Delta, \neg F. \tag{vi}$$

Applying the Reduction Lemma (Lemma 10.2.6) in combination with Lemma 10.3.5 we obtain from (vi)

$$F_{(\omega^{\alpha_0+1})\cdot 3} \overset{\omega^{\alpha_0}\cdot 2}{\underset{\rho}{\vdash}} \Delta. \tag{vii}$$

Let $d := \max\text{par}(\Delta)$ and $x \geq d$. Since $N(\alpha_0) \leq F(\text{par}(\Delta, F))$ and $\text{par}(F) \leq F(\text{par}(\Delta)) = F(d)$ we obtain $N(\alpha_0) \leq F^2(d)$. Hence $N(\omega^{\alpha_0+1}) = N(\alpha_0) + 2 \leq F^2(d) + 2 \leq \psi_F(\omega\cdot 2 + d) + 2$, which in turn shows $\omega^{\alpha_0+1}\cdot 3 \ll_F \omega^{\alpha}\cdot 3 + \psi_F(\omega\cdot 2 + d)\cdot 3 + 6$. Therefore we obtain

$$\begin{aligned} F_{\omega^{\alpha_0+1}\cdot 3}(x) &< F_{\omega^{\alpha}\cdot 3+\psi_F(\omega\cdot 2+d)\cdot 3+6}(x) = \psi_F(\omega^{\alpha}\cdot 3 + \psi_F(\omega\cdot 2 + d)\cdot 3 + 6 + x) \\ &\leq \psi_F(\omega^{\alpha}\cdot 3 + \omega\cdot 6 + 4x + 6) \leq \psi_F(\omega^{\alpha+1} + x) = F_{\omega^{\alpha+1}}(x), \end{aligned} \tag{viii}$$

where in the last line we have used $\omega^{\alpha}\cdot 3 + \omega\cdot 6 + 4x + 6 \ll_F \omega^{\alpha+1} + x$, which follows because $N(\alpha) \geq 1$ entails $N(\omega^{\alpha}\cdot 3 + \omega\cdot 6 + 4x + 6) = N(\alpha)\cdot 3 + 4x + 21 < (N(\alpha) + 2 + x)^3 \leq F(N(\omega^{\alpha+1} + x))$. By (vii) and (viii) we obtain by the Structural Lemma

$$F_{\omega^{\alpha+1}} \overset{\omega^{\alpha}}{\underset{\rho}{\vdash}} \Delta. \qquad \square$$

Iterating the Elimination Lemma we obtain an Elimination Theorem which suffices to determine upper bounds for the provably recursive functions of NT.

**10.4.2 Theorem** *(Elimination Theorem) If* $F \overset{\alpha}{\underset{n}{\vdash}} \Delta$ *and* $n > 0$ *then* $F_{\varphi_0^n(\alpha+1)} \overset{\varphi_0^n(\alpha)}{\underset{0}{\vdash}} \Delta$.

*Proof* We induct on $n$. For $n = 1$ the claim follows from the Elimination Lemma (Lemma 10.4.1). For $n + 1 > 1$ we obtain $F_{\omega^{\alpha+1}} \overset{\omega^{\alpha}}{\underset{n}{\vdash}} \Delta$ from the Elimination Lemma. By the induction hypothesis it follows $(F_{\omega^{\alpha+1}})_{\varphi_0^n(\omega^{\alpha}+1)} \overset{\varphi_0^n(\omega^{\alpha})}{\underset{0}{\vdash}} \Delta$ which implies

$$F_{\omega^{\alpha+1}\times(\varphi_0^n(\omega^{\alpha}+1)\#\omega)} \overset{\varphi_0^{n+1}(\alpha)}{\underset{0}{\vdash}} \Delta \tag{i}$$

by Lemma 10.3.9. We prove

$$\beta := \omega^{\alpha+1} \times (\varphi_0^n(\omega^{\alpha} + 1) \# \omega) \ll_F \varphi_0^{n+1}(\alpha + 1) \tag{ii}$$

and get then the claim from (i) and (ii) by the Structural Lemma (Lemma 10.2.2).

We obtain $\beta < \varphi_0^{n+1}(\alpha + 1)$ since $\varphi_0^{n+1}(\alpha + 1)$ is multiplicatively indecomposable and $\omega^{\alpha+1} < \varphi_0^{n+1}(\alpha + 1)$ as well as $\varphi_0^n(\omega^{\alpha} + 1) \# \omega < \varphi_0^{n+1}(\alpha + 1)$. Since $n \geq 1$ we finally obtain $N(\beta) \leq (N(\alpha) + 2)\cdot(N(\alpha) + n + 4) \leq (N(\alpha) + n + 2)^3 \leq F(N(\varphi_0^{n+1}(\alpha + 1)))$ which finishes the proof of (ii). $\square$

**10.4.3 Exercise** Formulate and prove (ii) of Theorem 9.3.15 for $F \overset{\alpha}{\underset{\beta+\omega^{\rho}}{\vdash}} \Delta$.

## 10.5 The Embedding of NT

To obtain an embedding of formal derivations in NT into operator controlled derivations we have to majorize all primitive recursive functions by iterations of a suited strictly monotone operator. An essential role in the majorizing procedure is played by the diagonalizing operator $\lambda x.\, F^{x+1}(x)$.

**10.5.1 Lemma** *Let $F$ be a strictly increasing operator satisfying $x^2 \leq F(x)$. Then $\lambda x.\, F^{x+1}(x)$ is majorized by $F_{\omega^2}$.*

*Proof* We have $F^{x+1}(x) \leq \psi_F(\omega\cdot(x+1)+x)$. Since $\omega\cdot(x+1)+x \ll_F \omega^2 + x$ we obtain $F^{x+1}(x) \leq \psi_F(\omega^2+x) = F_{\omega^2}(x)$. □

**10.5.2 Lemma** *Let $F$ be a strongly increasing operator satisfying $x^2+2 \leq F(x)$. Then we obtain for every primitive recursive function $f$ a natural number $e_f$ such that*

$$f(x_1,\dots,x_n) \leq F_{\omega^{e_f}}(\{x_1,\dots,x_n\})$$

*Proof* Let $f$ be a primitive recursive function and $\underline{f}$ a primitive recursive function term representing $f$. We prove the lemma by induction on the definition of the primitive recursive function term $\underline{f}$.

If $\underline{f}$ is the successor symbol $S$ then $f(x) = x+1 \leq F_\omega(x)$.

If $\underline{f}$ is the symbol $C_k^n$ for the constant function with value $k$ we have $C_k^n(x_1,\dots,x_n) = k \leq F_{\omega^k}(\{x_1,\dots,x_n\})$.

If $\underline{f}$ is the symbol $P_k^n$ then $P_k^n(x_1,\dots,x_n) = x_k \leq F_\omega(\{x_1,\dots,x_n\})$.

Now assume that $\underline{f} = Sub(\underline{g},\underline{h}_1,\dots,\underline{h}_m)$. Then by the induction hypothesis, we have numbers $e_g$ and $e_{h_i}$ such that

$$g(x_1,\dots,x_m) \leq F_{\omega^{e_g}}(\{x_1,\dots,x_m\}) \tag{i}$$

and

$$h_i(x_1,\dots,x_n) \leq F_{\omega^{e_{h_i}}}(\{x_1,\dots,x_n\}) \quad \text{for} \quad i = 1,\dots,m. \tag{ii}$$

Then

$$\begin{aligned} g(h_1(x_1,\dots,x_n),\dots,h_m(x_1,\dots,x_n)) &\leq F_{\omega^{e_g}}(\{h_1(x_1,\dots,x_n),\dots,h_m(x_1,\dots,x_n)\}) \\ &\leq F_{\omega^{e_g}}(F_{\omega^{e_h}}(\{x_1,\dots,x_n\})) \leq F_{\omega^{e_g}\#\omega^{e_h}}(\{x_1,\dots,x_n\}) \leq F_{\omega^{e_f}}(\{x_1,\dots,x_n\}) \end{aligned}$$

for $e_h := \max\{e_{h_1},\dots,e_{h_m}\}$ and $e_f := e_g + e_h + 1$.

If $\underline{f}$ is the term $Rec(g,h)$ we have numbers $e_g$ and $e_h$ such that

$$g(x_1,\dots,x_n) \leq F_{\omega^{e_g}}(\{x_1,\dots,x_n\})$$

and

$$h(k,l,x_1,\dots,x_n) \leq F_{\omega^{e_h}}(\{k,l,x_1,\dots,x_n\}).$$

Let $e_0 := \max\{e_g,e_h\}$ and put $u_0 := g(x_1,\dots,x_n)$ and $u_{k+1} := h(k,u_k,x_1,\dots,x_n)$. Then $f(k,x_1,\dots,x_n) = u_k$ and we prove

$$u_k \leq F_{\omega^{e_0}}^{k+1}(k,x_1,\dots,x_n) \tag{iii}$$

by induction on $k$. We have

$$u_0 := g(x_1,\dots,x_n) \le F_{\omega^{e_0}}(\{x_1,\dots,x_n\})$$

and

$$\begin{aligned} u_{k+1} &:= h(k,u_k,x_1,\dots,x_n) \le F_{\omega^{e_0}}(\{k,u_k,x_1,\dots,x_n\}) \\ &\le F_{\omega^{e_0}}(\{k,F_{\omega^{e_0}}^{k+1}(\{k,x_1,\dots,x_n\}),x_1,\dots,x_n\}) \\ &\le F_{\omega^{e_0}}^{k+2}(\{k,x_1,\dots,x_n\}). \end{aligned}$$

Let $m := \max\{k,x_1,\dots,x_n\}$. Since $2{\cdot}x + 1 \le x^2 + 2 \le F(x)$, we obtain by Lemma 10.5.1 and Lemma 10.3.9

$$\begin{aligned} f(k,x_1,\dots,x_n) = u_k &\le F_{\omega^{e_0}}^{k+1}(m) \le F_{\omega^{e_0}}^{m+1}(m) \le (F_{\omega^{e_0}})_{\omega^2}(m) \\ &\le F_{\omega^{e_0+3}}(m) = F_{\omega^{e_0+3}}(\{k,x_1,\dots,x_n\}) \end{aligned}$$

and put $e_f := e_0 + 3$. □

**10.5.3 Theorem** *Let $F$ be a strongly increasing operator such that $x^2+2 \le F(x)$. Then every primitive recursive function is eventually majorized by $F_{\omega^\omega}$.*

*Proof* Let $f$ be an $n$-ary primitive recursive function. By Lemma 10.5.2, there is a natural number $e_f$ such that $f(x_1,\dots,x_n) \le F_{\omega^{e_f}}(\{x_1,\dots,x_n\})$. For all tuples $x_1,\dots,x_n$ such that $x := \max\{x_1,\dots,x_n\} \ge e_f$ we then obtain $\omega^{e_f} + x \ll_F \omega^\omega + x$, which implies $f(x_1,\dots,x_n) \le F_{\omega^{e_f}}(x) \le F_{\omega^\omega}(\{x_1,\dots,x_n\})$. □

Lemma 10.5.2 will be crucial in the proof of the following lemma which is based on the iterations of a strongly increasing operator $F$ fulfilling $x^2+2 \le F(x)$.

**10.5.4 Lemma** *Let $\Delta(x_1,\dots,x_n) \subseteq \Gamma(x_1,\dots,x_n)$ be a finite set of $\mathcal{L}(\mathsf{NT})$-formulas which contain at most the free variables $x_1,\dots,x_n$. If $\underset{\mathsf{T}}{\overset{m}{\vdash}} \Delta(x_1,\dots,x_n)$ then there is a $k < \omega$ such that $F_{\omega^k+m} \underset{0}{\overset{m}{\vdash}} \Gamma(\underline{z}_1,\dots,\underline{z}_n)$ for all $n$-tuples $\underline{z}_1,\dots,\underline{z}_n$ of numerals and all strongly increasing operators $F$ fulfilling $x^2+2 \le F(x)$.*

*Proof* Induction on $m$. We proceed by distinction on cases according to Definition 4.3.2.

If $\underset{\mathsf{T}}{\overset{m}{\vdash}} \Delta(x_1,\dots,x_n)$ holds by $(Ax)_L$ then there is an atomic formula $A(x_1,\dots,x_n)$ such that $\{A(x_1,\dots,x_n), \neg A(x_1,\dots,x_n)\} \subseteq \Delta(x_1,\dots,x_n)$. For an $n$-tuple $z_1,\dots,z_n$ of natural numbers either $A(\underline{z}_1,\dots,\underline{z}_n) \in Diag(\mathbb{N})$ or $\neg A(\underline{z}_1,\dots,\underline{z}_n) \in Diag(\mathbb{N})$. Hence $\Gamma(\underline{z}_1,\dots,\underline{z}_n) \cap Diag(\mathbb{N}) \ne \emptyset$, and we obtain $F_\omega \underset{0}{\overset{0}{\vdash}} \Gamma(\underline{z}_1,\dots,\underline{z}_n)$ by an inference $(\bigwedge)$ with empty premises.

In case of an inference according to $(\bigvee)$ we have the premise

$$\underset{\mathsf{T}}{\overset{m_0}{\vdash}} \Delta_0(x_1,\dots,x_n), A_i(x_1,\dots,x_n) \tag{i}$$

for some $i \in \{0,1\}$ and obtain

$$F_{\omega^k+m_0} \underset{0}{\overset{m_0}{\vdash}} \Gamma(\underline{z}_1,\dots,\underline{z}_n), A_i(\underline{z}_1,\dots,\underline{z}_n), (A_0 \vee A_1)(\underline{z}_1,\dots,\underline{z}_n) \tag{ii}$$

for any $n$-tuple $z_1,\dots,z_n$. From (ii), however, we obtain the claim by an inference $(\bigvee)$.

In case of an inference according to $(\wedge)$ we have the premises

$$\underset{\mathsf{T}}{\overset{m_i}{\models}} \Delta_0(x_1,\dots,x_n), A_i(x_1,\dots,x_n) \tag{iii}$$

for $i = 0, 1$ and obtain

$$F_{\omega^{k_0}+m_0} \underset{0}{\overset{m_0}{\models}} \Gamma(\underline{z}_1,\dots,\underline{z}_n), A_0(\underline{z}_1,\dots,\underline{z}_n), (A_0 \wedge A_1)(\underline{z}_1,\dots,\underline{z}_n) \tag{iv}$$

and

$$F_{\omega^{k_1}+m_1} \underset{0}{\overset{m_1}{\models}} \Gamma(\underline{z}_1,\dots,\underline{z}_n), A_1(\underline{z}_1,\dots,\underline{z}_n), (A_0 \wedge A_1)(\underline{z}_1,\dots,\underline{z}_n) \tag{v}$$

by the induction hypothesis. From (iv) and (v) we then obtain the claim by an inference $(\bigwedge)$ with $k := \max\{k_0, k_1\}$.

In case of an inference according to $(\exists)$ we have the premise

$$\underset{\mathsf{T}}{\overset{m_0}{\models}} \Delta_0(x_1,\dots,x_n), A_u(f(x_1,\dots,x_m), x_1,\dots,x_n). \tag{vi}$$

Without loss of generality, we may assume that $\{x_1,\dots,x_m\} \subseteq \{x_1,\dots,x_n\}$ because we can replace all variables that do not occur among $x_1,\dots,x_n$ by 0 without destroying the derivation. By the induction hypothesis we then obtain

$$F_{\omega^{k_0}+m_0} \underset{0}{\overset{m_0}{\models}} \begin{array}{l}\Gamma(\underline{z}_1,\dots,\underline{z}_n), A_u(f(\underline{z}_1,\dots,\underline{z}_n), \underline{z}_1,\dots,\underline{z}_n),\\ \qquad (\exists x)A_u(x, \underline{z}_1,\dots,\underline{z}_n).\end{array} \tag{vii}$$

By Lemma 10.5.2, there is a natural number $k_1$ such that $f(z_1,\dots,z_n) \leq F_{\omega^{k_1}}(\{z_1,\dots,z_n\}) < F_{\omega^{k_1}+m}(\mathrm{par}(\Gamma(\underline{z}_1,\dots,\underline{z}_n)))$. Thus

$$\mathrm{par}(\Gamma(\underline{z}_1,\dots,\underline{z}_n), A_u(f(\underline{z}_1,\dots,\underline{z}_n), \underline{z}_1,\dots,\underline{z}_n)) \subseteq F_{\omega^{k}+m}(\mathrm{par}(\Gamma(\underline{z}_1,\dots,\underline{z}_n)))$$

for $k = \max\{k_0, k_1\}$ and we obtain the claim from (vii) by an inference $(\bigvee)$.

The last case is an inference according to $(\forall)$. Then we have the premise

$$\underset{\mathsf{T}}{\overset{m_0}{\models}} \Delta_0(x_1,\dots,x_n), A(v, x_1,\dots,x_n) \tag{viii}$$

such that $v$ does not occur in $\Delta_0(x_1,\dots,x_n)$. Fix an $n$-tuple $z_1,\dots,z_n$ of natural numbers. Then we obtain from (viii)

$$F_{\omega^{k}+m_0} \underset{0}{\overset{m_0}{\models}} \Gamma(\underline{z}_1,\dots,\underline{z}_n), A_v(\underline{z}, \underline{z}_1,\dots,\underline{z}_n), (\forall x)A_u(x, \underline{z}_1,\dots,\underline{z}_n) \tag{ix}$$

for all natural numbers $z$ by the induction hypothesis. From (ix) we get the claim by an inference $(\bigwedge)$. □

It follows from Theorem 7.2.1 and Lemma 10.5.4 that for every $\mathscr{L}(\mathsf{NT})$-sentence that is provable in NT, there are axioms $A_1,\dots,A_n \in \mathsf{NT}$ and natural numbers $k$ and $m$ such that

$$F_{\omega \cdot k+m} \underset{0}{\overset{m}{\models}} \neg A_1,\dots,\neg A_n, F. \tag{10.12}$$

To obtain a characterization of the provably recursive functions of NT, we have to check if all the mathematical axioms of NT are also operator controlled derivable.

If $(\forall x_1)\dots(\forall x_n)F(x_1,\dots,x_n)$ is an identity axiom or a mathematical axiom of NT different from an instance of Mathematical Induction and the last identity axiom then we have

$$F_{x_1,\dots,x_n}(\underline{z}_1,\dots,\underline{z}_n) \in Diag(\mathbb{N})$$

for every tuple of numerals $\underline{z}_1,\dots,\underline{z}_n$. For any non trivial strictly increasing operator $\mathsf{F}$ and $n \geq 1$ we obtain $N(\omega) = 2 \leq \mathsf{F}(\mathrm{par}((\forall x_1)\dots(\forall x_n)F(x_1,\dots,x_n)))$. Therefore

$$\mathsf{F} \,\big|\!\frac{\omega}{0}\; (\forall x_1)\dots(\forall x_n)F(x_1,\dots,x_n) \tag{10.13}$$

is true for all strictly increasing operators $\mathsf{F}$ and all axioms different from instances of Mathematical Induction and the last identity axiom.

To check the operator controlled derivability of instances of Mathematical Induction, we need some preparations. The first is a operator controlled version of the Tautology Lemma.

**10.5.5 Lemma** *Let* $\mathsf{F}$ *be a strictly increasing operator which also satisfies* $2\cdot x \leq \mathsf{F}(x)$ *for all natural numbers x. Then* $\mathsf{F} \,\big|\!\frac{2\mathrm{rnk}(F)}{0}\; \Delta, F, \neg F'$ *holds for all numerical equivalent sentences F and* $F'$.

*Proof* We show the lemma by induction on $\mathrm{rnk}(F)$. Without loss of generality we may assume that $F \in \bigwedge$–type . If $F$ is atomic we obtain $\mathsf{F} \,\big|\!\frac{0}{0}\; \Delta, F, \neg F'$ by an inference $(\bigwedge)$ with empty premise. If $F$ is not atomic then $F$ is either a sentence $(F_0 \wedge F_1)$ or a sentence $(\forall x)G_u(x)$. In the first case, we obtain $\mathsf{F} \,\big|\!\frac{2\mathrm{rnk}(F_i)}{0}\; \Delta, F, \neg F', F_i, \neg F_i'$ for $i = 0, 1$ and obtain

$$\mathsf{F} \,\big|\!\frac{2\mathrm{rnk}(F_i)+1}{0}\; \Delta, F, \neg F', F_i \tag{i}$$

for $i = 0, 1$ by an inference $(\bigvee)$ and thence

$$\mathsf{F} \,\big|\!\frac{2\mathrm{rnk}(F)}{0}\; \Delta, F, \neg F' \tag{ii}$$

by an inference $(\bigwedge)$. Both inferences can be applied because $2\mathrm{rnk}(F_i) < 2\mathrm{rnk}(F_i) + 1 < 2\mathrm{rnk}(F) \leq \mathsf{F}(\mathrm{par}(\Delta, F, \neg F))$ and $\mathrm{par}(\Delta, F, F_i) \subseteq \mathrm{par}(\Delta, F)$.

If $F$ is a sentence $(\forall x)G_u(x)$ we obtain

$$\mathsf{F} \,\big|\!\frac{2\mathrm{rnk}(G)}{0}\; \Delta, F, \neg F', G_u(\underline{n}), \neg G_u'(\underline{n}) \tag{iii}$$

for all $n \in \omega$. By an inference $(\bigvee)$ we obtain from (iii)

$$\mathsf{F} \,\big|\!\frac{2\mathrm{rnk}(G)+1}{0}\; \Delta, F, \neg F', G_u(\underline{n}) \tag{iv}$$

for all $n \in \omega$. This inference can be applied because $2\mathrm{rnk}(G) < 2\mathrm{rnk}(G) + 1 < 2\mathrm{rnk}(F) \leq \mathsf{F}(\mathrm{par}(\Delta, F, G_u(\underline{n})))$ and $n \in \mathrm{par}(\Delta, F, G_u(\underline{n}))$. From (iv) we then obtain

$$F \overset{2\mathrm{rnk}(F)}{\underset{0}{\vdash}} \Delta,\, F,\, \neg F'$$

by an inference $(\bigwedge)$. □

As an immediate consequence of Lemma 10.5.5 we get $F \overset{\omega}{\underset{0}{\vdash}} A$ for all instances $A$ of the last identity axiom. Next we handle the scheme of Mathematical Induction.

**10.5.6 Lemma** *(Mathematical Induction Lemma) Let* $F$ *be a strictly increasing operator such that* $4{\cdot}n \le F(n)$*. Then*

$$F \overset{2\cdot(\mathrm{rnk}(F)+n)}{\underset{0}{\vdash}} \neg F_u(\underline{0})\,,\neg((\forall x)[F_u(x) \to F_u(x+1)]),\, F_u(\underline{n}).$$

*Proof* We prove the lemma by induction on $n$. The proof is literally that of Lemma 7.3.4. We have only to keep an eye on the occurring number parameters. First we have $2{\cdot}(\mathrm{rnk}(F)+n) \in O_F(\mathrm{par}(F_u(\underline{n})))$ because $2{\cdot}(\mathrm{rnk}(F)+n) \le 4{\cdot}\max\mathrm{par}(F_u(\underline{n})) \le F(\mathrm{par}(F_u(\underline{n})))$. Next we obtain

$$F \overset{2\cdot\mathrm{rnk}(F)}{\underset{0}{\vdash}} \neg F_u(\underline{0})\,,\neg((\forall x)[F_u(x) \to F_u(x+1)]),\, F_u(\underline{0}) \tag{i}$$

by Lemma 10.5.5. Assume now

$$F \overset{2\cdot(\mathrm{rnk}(F)+n)}{\underset{0}{\vdash}} \neg F_u(\underline{0})\,,\neg((\forall x)[F_u(x) \to F_u(x+1)]),\, F_u(\underline{n})$$

for induction hypothesis, i. e.

$$F \overset{2\cdot(\mathrm{rnk}(F)+n)}{\underset{0}{\vdash}} \neg F_u(\underline{0})\,,(\exists x)[F_u(x) \wedge \neg F_u(x+1)],\, F_u(\underline{n}). \tag{ii}$$

By Lemma 10.5.5 we have

$$F \overset{2\cdot\mathrm{rnk}(F)}{\underset{0}{\vdash}} \neg F_u(\underline{0})\,,(\exists x)[F_u(x) \wedge \neg F_u(x+1)],\, \neg F_u(\underline{n}+\underline{1}),\, F_u(\underline{n+1}). \tag{iii}$$

From (ii) and (iii) we obtain by an inference $(\bigwedge)$

$$F \overset{2\cdot(\mathrm{rnk}(F)+n)+1}{\underset{0}{\vdash}} \begin{array}{l}\neg F_u(\underline{0})\,,(\exists x)[F_u(x) \wedge \neg F_u(x+1)],\\ \quad F_u(\underline{n+1}),\, F_u(\underline{n}) \wedge \neg F_u(\underline{n}+\underline{1})\end{array} \tag{iv}$$

and from (iii) finally

$$F \overset{2\cdot(\mathrm{rnk}(F)+n)+2}{\underset{0}{\vdash}} F_u(\underline{0})\,,\neg(\forall x)[F_u(x) \to F_u(x+1)],\, F_u(\underline{n+1})$$

by an inference according to $(\bigvee)$. In both inferences the occurring number parameters are obviously controlled. □

From Lemma 10.5.6, we finally obtain all instances of Mathematical Induction.

**10.5.7 Lemma** *(Mathematical Induction) Let* $F$ *be a strictly increasing operator such that* $4{\cdot}n \le F(n)$*. Then*

$$F \overset{\omega+4}{\underset{0}{\vdash}} \neg F_u(\underline{0}) \vee \neg((\forall x)[F_u(x) \to F_u(x+1)]) \vee (\forall x)F_u(x))$$

*holds for any formula* $F$.

*Proof* This follows from Lemma 10.5.6 by an inference ($\bigwedge$) and inferences ($\bigvee$). □

Now we are prepared to prove the main theorem of this section.

**10.5.8 Theorem** *Let* $F$ *be a strongly increasing operator such that* $x^2+2 \le F(x)$ *for all natural numbers* $x$. *If* $\mathsf{NT} \vdash (\forall\vec{x})(\exists y)F(\vec{x},y)$ *for a quantifier free formula* $F(\vec{u},v)$ *then there is an* $\alpha < \varepsilon_0$ *such that*

$$\mathbb{N} \models (\forall x_1)\ldots(\forall x_n)(\exists y \in F_\alpha(\{x_1,\ldots,x_n\}))F(x_1,\ldots,x_n,y).$$

*Proof* Assume $\mathsf{NT} \vdash (\forall\vec{x})(\exists y)F(\vec{x},y)$. Then there are finitely many NT- axioms $A_1,\ldots,A_l$ such that

$$\frac{|m}{0}\ \neg A_1,\ldots,\neg A_l,\ (\forall\vec{x})(\exists y)F(\vec{x},y). \tag{i}$$

By Lemma 10.5.4 there is a $k \in \omega$ such that

$$F_{\omega^k+m} \frac{|m}{0}\ \neg A_1,\ldots,\neg A_l,\ (\forall\vec{x})(\exists y)F(\vec{x},y). \tag{ii}$$

Together with (10.13) (on p. 226) and Lemma 10.5.7 we obtain from (ii) finite ordinals $p,r < \omega$ such that

$$F_{\omega^p} \frac{|\omega\cdot 2}{r}\ (\forall\vec{x})(\exists y)F(\vec{x},y). \tag{iii}$$

We can choose $p$ so big that $x^3 \le F_{\omega^p}(x)$. Then the Elimination Theorem (Theorem 10.4.2) and (iii) imply

$$(F_{\omega^p})_{\varphi_0^r(\omega\cdot 2+1)} \frac{|\varphi_0^r(\omega\cdot 2)}{0}\ (\forall\vec{x})(\exists y)F(\vec{x},y). \tag{iv}$$

By Lemma 10.3.9 we obtain $(F_{\omega^p})_{\varphi_0^r(\omega\cdot 2+1)} \le F_{\omega^p\times(\varphi_0^r(\omega\cdot 2)\#\omega)} \le F_{\varphi_0^n(0)}$ for $n$ large enough. Then (iv) implies

$$F_{\varphi_0^n(0)} \frac{|\varphi_0^r(\omega\cdot 2)}{0}\ (\forall\vec{x})(\exists y)F(\vec{x},y). \tag{v}$$

by the Structural Rule (Lemma 10.2.2). By the Inversion Lemma (Lemma **??**) we get

$$F_{\varphi_0^n(0)}[\{x_1,\ldots,x_n\}] \frac{|\varphi_0^r(\omega\cdot 2)}{0}\ (\exists y)F(x_1,\ldots,x_n,y). \tag{vi}$$

Now let $n'$ so big that $\varphi_0^n(0) \times (\omega^{\varphi_0^r(\omega\cdot 2)\cdot 2+1} \# \omega) \ll_F \varphi_0^{n'}(0) =: \alpha$. Then $\alpha < \varepsilon_0$ and $\mathbb{N} \models (\exists y \in F_\alpha(\{x_1,\ldots,x_n\}))F(x_1,\ldots,x_n,y)$ by the Witnessing Theorem (Theorem 10.3.10) and Lemma 10.3.9. □

**10.5.9 Corollary** *The provably recursive functions of* NT *are eventually majorized by* $F_{\varepsilon_0}$ *for any strongly increasing operator satisfying* $x^2+2 \le F(x)$.

*Proof* Let $f$ be a provably recursive function of NT. Then there is an index $e$ such that $f(x_1,\ldots,x_n) \simeq U(\mu y\,.\,T(e,x_1,\ldots,x_n,y)$ and $\mathsf{NT} \vdash (\forall\vec{x})(\exists y)T(e,\vec{x},y)$. By Theorem 10.5.8 we obtain an $\alpha < \varepsilon_0$ such that $y < F_\alpha(\{x_1,\ldots,x_n\})$ holds true for the least $y$ satisfying $T(e,x_1,\dots,x_n,y)$. Since $f(\vec{x}) = U(y) \leq y$ we obtain $f(\vec{x}) \leq F_{\varepsilon_0}(\{x_1,\dots,x_n\})$ for all $\vec{x} = (x_1,\dots,x_n)$ such that $\max\{x_1,\dots,x_n\} \geq N(\alpha)$. □

On the other hand, we have shown in Sect 3.3.1 that we can talk about ordinals below $\varepsilon_0$ in $\mathcal{L}(\mathsf{NT})$. For the moment, it is useful to emphasize the difference between an ordinal $\alpha < \varepsilon_0$ and its code $\ulcorner\alpha\urcorner$. It is obvious that the function $N(\ulcorner\alpha\urcorner) := N(\alpha)$ is primitive recursive. For a finite ordinal $n$ we get according to Sect. 3.3.1

$$\ulcorner n\urcorner := \mu s\,.\,\{Seq(s) \wedge lh(s) = n \wedge (\forall i < n)[(s)_i = 0]\}$$

and the search operator can obviously be bounded. Hence $n \mapsto \ulcorner n\urcorner$ is primitive recursive. The inverse mapping $|\ulcorner n\urcorner|$ is then simply given by $|\ulcorner n\urcorner| = n = lh(\ulcorner n\urcorner)$, which shows that it is primitive recursive, too. For an ordinal $\alpha$ the function $\lambda n\,.\,\ulcorner\alpha + n\urcorner$ is primitive recursive because $\ulcorner\alpha + n\urcorner = \ulcorner\alpha\urcorner{}^\frown\ulcorner n\urcorner$.

For a strictly increasing function $F$ the function $\psi_F$ induces a mapping

$$\underline{\psi_F}\colon \mathrm{On} \longrightarrow \{\xi \in \mathrm{On} \,|\, \xi < \omega\} \text{ by putting } \underline{\psi_F}(\ulcorner\alpha\urcorner) := \ulcorner\psi_F(\alpha)\urcorner.$$

Hence

$$\psi_F(\alpha) = |\underline{\psi_F}(\ulcorner\alpha\urcorner)| = lh(\underline{\psi_F}(\ulcorner\alpha\urcorner)).$$

We will first prove in NT that the function $\underline{\psi_F}(\ulcorner\alpha\urcorner)$ is defined for all $\alpha < \varepsilon_0$ provided that $F$ is a provably recursive function of NT (which is the case for all primitive recursive functions). We use the Recursion Theorem (which is a theorem of NT) to obtain an index $e$ satisfying

$$\{e\}(\ulcorner\xi\urcorner) \simeq \max\nolimits_{\prec}\{\{e\}(\ulcorner\eta\urcorner)^\frown\langle 0\rangle \,|\, \ulcorner\eta\urcorner \prec \ulcorner\xi\urcorner \wedge N(\ulcorner\eta\urcorner) \leq F(N(\ulcorner\xi\urcorner))\}. \tag{10.14}$$

Then $e$ is an index for the function $\underline{\psi_F}$ and it remains to show that $\underline{\psi_F}$ is total. To this end we show by induction on $\ulcorner\beta\urcorner \prec \ulcorner\alpha\urcorner$, which is available in NT by Theorem 7.4.1,

$$(\forall\ulcorner\beta\urcorner \prec \ulcorner\alpha\urcorner)(\exists y)[\{e\}(\ulcorner\beta\urcorner) \simeq y \wedge y = \underline{\psi_F}(\ulcorner\beta\urcorner)]. \tag{10.15}$$

We have the induction hypothesis

$$(\forall\ulcorner\xi\urcorner \prec \ulcorner\beta\urcorner)(\exists y)[\{e\}(\ulcorner\xi\urcorner) \simeq y \wedge y = \underline{\psi_F}(\ulcorner\xi\urcorner)]$$

and obtain by (10.14)

$$\begin{aligned}\{e\}(\ulcorner\beta\urcorner) &\simeq \max\nolimits_{\prec}\{\{e\}(\ulcorner\xi\urcorner)^\frown\langle 0\rangle \,|\, \ulcorner\xi\urcorner \prec \ulcorner\beta\urcorner \wedge N(\ulcorner\xi\urcorner) \leq F(N(\ulcorner\beta\urcorner))\} \\ &\simeq \max\nolimits_{\prec}\{\ulcorner\psi_F(\xi)+1\urcorner \,|\, \ulcorner\xi\urcorner \prec \ulcorner\beta\urcorner \wedge N(\ulcorner\xi\urcorner) \leq F(N(\ulcorner\beta\urcorner))\} \\ &\simeq \underline{\psi_F}(\ulcorner\beta\urcorner).\end{aligned}$$

Since all $\{e\}(\ulcorner\xi\urcorner)$ are defined by induction hypothesis and NT proves that there are only finitely many $\ulcorner\xi\urcorner \prec \ulcorner\beta\urcorner$ such that $N(\ulcorner\xi\urcorner) \leq F(N(\ulcorner\beta\urcorner))$ their $\prec$-maximum exists and we obtain that $\{e\}(\ulcorner\beta\urcorner)$ is defined, too.

From (10.15) we obtain

$$(\exists y)[\underline{\psi_F}(\ulcorner\alpha\urcorner) = y] \tag{10.16}$$

for every ordinal $\alpha < \varepsilon_0$ and defining

$$\underline{F}_x(y) = lh(\underline{\psi_F}(x^\frown \ulcorner y \urcorner))$$

we obtain

$$F_\alpha(x) = \underline{F}_{\ulcorner\alpha\urcorner}(x) \tag{10.17}$$

since $F_\alpha(x) = \psi_F(\alpha+x) = |\underline{\psi_F}(\ulcorner\alpha+x\urcorner)| = lh(\underline{\psi_F}(\ulcorner\alpha+x\urcorner)) = lh(\underline{\psi_F}(\ulcorner\alpha\urcorner{}^\frown \ulcorner x\urcorner)) = \underline{F}_{\ulcorner\alpha\urcorner}(x)$. For $\alpha < \varepsilon_0$ we obtain $(\forall x)[\ulcorner\alpha + x\urcorner \prec \ulcorner\varepsilon_0\urcorner]$ by Mathematical Induction on $x$. So we have

$$(\forall x)(\exists y)[F_\alpha(x) \simeq y]$$

for any $\alpha < \varepsilon_0$ by (10.16) and (10.17). This shows that for $\alpha < \varepsilon_0$, every function $F_\alpha$ is a provably recursive function of NT. The provably recursive functions of NT are, however, clearly closed under "primitive recursive in." Therefore every function which is primitive recursive in some $F_\alpha$ with $\alpha < \varepsilon_0$ is a provably recursive function of NT. On the other hand, if $f$ is a provably recursive function of NT then there is an index $e$ such that $\mathsf{NT} \vdash (\forall \vec{x})(\exists y)\, T(e,\vec{x},y)$. Hence $\mathbb{N} \models (\forall x)(\exists y < F_\alpha(\{\vec{x}\}))\, T(e,\vec{x},y)$ for some $\alpha < \varepsilon_0$ by Theorem 10.5.8. This means that $f(\vec{x}) = \mu y < F_\alpha(\{\vec{x}\}) \,.\, T(e,\vec{x},y)$, i.e., that $f$ is primitive recursive in some $F_\alpha$. So we have the following theorem.

**10.5.10 Theorem** *The provably recursive functions of* NT *are exactly the functions which are primitive recursive in $F_\alpha$ for some $\alpha < \varepsilon_0$ and some strictly increasing operator $F$ which is provably recursive, e.g., primitive recursive, and satisfies $x^2 + 2 \leq F(x)$.*

**10.5.11 Corollary** *There is a true $\Pi^0_2$-sentence $(\forall x)(\exists y)F(x,y)$ such that* NT *proves $(\exists y)F(\underline{n},y)$ for all natural numbers n but* $\mathsf{NT} \nvdash (\forall x)(\exists y)F(x,y)$.

*Proof* Define $F(u,v) \;:\Leftrightarrow\; (u)_0 \in OT \to F_{(u)_0}((u)_1) = v$. □

## 10.6 Discussion

It follows from Theorems 10.5.8 and 10.5.10 that the Skolem functions for $\Pi^0_2$-sentences that are provable in NT form exactly the class of functions, which are primitive recursive in $\{F_\alpha \mid \alpha < \varepsilon_0\}$ for suitable "simple" operators $F$. We call the collection $\{F_\alpha \mid \alpha < \varepsilon_0\}$ *the subrecursive hierarchy* with basis $F$. For a collection $\mathscr{F}$ of functions let

$$PRH(\mathscr{F}) := \{f \mid f\colon \mathbb{N} \longrightarrow \mathbb{N} \text{ and } f \text{ is primitive recursive in some } g \in \mathscr{F}\}$$

denote the *primitive recursive hull* of the collection $\mathscr{F}$. By $SF_{\Pi_2^0}(T)$ we denote the class of Skolem functions of the provable $\Pi_2^0$-sentences of a theory $T$.

We say that the Skolem functions of the provable $\Pi_2^0$-sentences of a theory $T$ are *generated* by the subrecursive hierarchy $\{F_\alpha \mid \alpha < \kappa\}$ or, equivalently, that $\{F_\alpha \mid \alpha < \kappa\}$ form a *generating class* for the Skolem functions of the provable $\Pi_2^0$-sentences of the theory $T$ if $SF_{\Pi_2^0}(T) = PRH(\{F_\alpha \mid \alpha < \kappa\})$. The ordinal $\kappa$ is the *length of the generating class.*

It has become common to call the length of the shortest subrecursive hierarchy which is needed to generate the Skolem functions for the provable $\Pi_2^0$-sentences of a theory its $\Pi_2^0$*-ordinal.* The length of the generating subrecursive hierarchy for $SF_{\Pi_2^0}(\mathsf{NT})$ is pretty independent from its basic function as long as this function is a provably recursive function of NT with sufficient growth rate. Nevertheless the definition of the $\Pi_2^0$-ordinal of a theory is "less canonical" than the definition of its $\Pi_1^1$-ordinal. In defining the subrecursive hierarchy, we needed heavily the *norm of an ordinal*, i.e., some kind of unique term notation for ordinals. Therefore, the $\Pi_2^0$-ordinal of a theory is rather an ordinal equipped with some additional term structure than a pure ordinal in the set theoretic sense. Although we vaguely talk about the $\Pi_2^0$–ordinal of a theory, we actually mean an ordinal notation system equipped with a norm function. This causes no real problems for ordinals below $\Gamma_0$ because we have a relatively clear picture of a notation system for these ordinals. But even there we can observe surprising effects (cf. [107]). For bigger ordinals the situation becomes much worse. We have shown in Sect. 9.6 that there is a primitive recursive notation system for the Howard–Bachmann-ordinal $\psi(\varepsilon_{\Omega+1})$. Here, however, it is not at all clear why this notation system should be regarded as "canonical". A feasible characterization of "what makes a notation system canonical" is still one of the big challenges.

The question "which function is a suitable basis for a subrecursive hierarchy generating the Skolem functions for the provable $\Pi_2^0$-sentences for a theory?" is not so important for NT. In Theorem 10.5.10, we have just shown the fact that the subrecursive hierarchies which generate the provably recursive functions of NT are pretty independent of the choice of their generating functions. In determining the $\Pi_2^0$-ordinals of subtheories of NT or rather PA, however, the choice of the basis functions becomes important because these ordinals are not $\varepsilon$-number. A difference of one or even more $\omega$-powers in the length of the generating class makes a big difference there. But the analysis of such small theories is not among the aims of this book. It is, however, interesting in itself to find a basis for a generating class for the provably recursive functions of NT which is as simple as possible. To make "our"[1] theory of subrecursive hierarchies work we need a strongly increasing operator satisfying $x^2 + 2 \leq F(x)$. The function $\lambda x \,.\, x^2 + 2$ is strongly increasing except for $x = 0$ (which would be sufficient to develop the theory) but already $\lambda x \,.\, x^2 + 3$ is everywhere strongly increasing and can therefore serve as a basis. The requirements "strongly increasing" and $x^2 + 2 \leq F(x)$ are, however, somehow weird and at least

[1] This theory is based on the work of Weiermann, mainly published in [14] and [108].

the second condition is apparently only technically motivated. They exclude the simplest non trivial strictly increasing function, the successor function, as a basis. However, Arai in [2] and also Weiermann in [108] could show that the subrecursive hierarchy with basis $S$ (the operator induced by the successor function $S$) generates the Skolem functions of the provable $\Pi_2^0$-sentences of PA at level $\varepsilon_0$. This result, however, needs a detailed study of the properties of the subrecursive hierarchies which would lead us outside the scope of this book.

We will therefore briefly sketch an alternative approach to subrecursive hierarchies, which is historically the original one. In this discussion, we restrict ourselves to ordinals below $\varepsilon_0$ and assign *fundamental sequences* to the ordinals below $\varepsilon_0$. We do that in two steps and define first fundamental sequences for additively indecomposable ordinals.

**10.6.1 Definition** Let $\alpha < \varepsilon_0$ be an additively indecomposable ordinal and $x < \omega$. Put

$$\alpha[x] := \begin{cases} 0 & \text{if } \alpha = \omega^0 \\ \omega^\beta \cdot (x+1) & \text{if } \alpha = \omega^{\beta+1} \\ \omega^{\lambda[x]} & \text{if } \alpha = \omega^\lambda \text{ and } \lambda \in \mathit{Lim}. \end{cases}$$

For additively decomposable ordinals $\alpha =_{NF} \alpha_1 + \cdots + \alpha_n < \varepsilon_0$ let

$$\alpha[x] := \alpha_1 + \cdots + \alpha_n[x] \text{ and } 0[x] := 0.$$

We call $\langle \alpha[x] \,|\, x \in \omega \rangle$ the fundamental sequence for $\alpha$ and $\alpha[x]$ the $x$-th member of the fundamental sequence for $\alpha$.

Observe that for $\alpha \in \mathbb{H} \setminus \{1\}$ we obtain $\alpha = \sup_{x \in \omega} \alpha[x]$. Hence $\alpha = \sup_{x \in \omega} \alpha[x]$ for all $\alpha \in \mathit{Lim}$. If $\alpha \notin \mathit{Lim}$ we obtain $\alpha[x]$ as the predecessor of $\alpha$, independent from $x$.

We extend the definition of $\mathsf{F}^n$ to ordinals $< \varepsilon_0$ by putting

$$\mathsf{F}^\alpha(x) := \begin{cases} x & \text{if } \alpha = 0 \\ \mathsf{F}^{\alpha[x]}(\mathsf{F}(x)) & \text{if } 0 < \alpha. \end{cases}$$

**10.6.2 Lemma** *Let $\alpha =_{NF} \alpha_1 + \cdots + \alpha_n$. Then $\mathsf{F}^\alpha = \mathsf{F}^{\alpha_1} \circ (\cdots \circ \mathsf{F}^{\alpha_n} \cdots)$.*

*Proof* The proof is by induction on $\alpha$. For $\alpha_n = 1$ we obtain $\mathsf{F}^\alpha = \mathsf{F}^{\alpha_1 + \cdots + \alpha_{n-1}} \circ \mathsf{F}$ by definition which entails $\mathsf{F}^\alpha = \mathsf{F}^{\alpha_1} \circ (\cdots (\mathsf{F}^{\alpha_{n-1}} \circ \mathsf{F}) \cdots)$ by induction hypothesis. If $\alpha_n \in \mathit{Lim}$ we obtain $\mathsf{F}^\alpha(x) = \mathsf{F}^{\alpha[x]}(x) = \mathsf{F}^{\alpha_1 + \cdots + \alpha_{n-1} + \alpha_n[x]}(\mathsf{F}(x)) = \mathsf{F}^{\alpha_1}(\cdots \mathsf{F}^{\alpha_{n-1}}(\mathsf{F}^{\alpha_n[x]}(\mathsf{F}(x))) \cdots) = \mathsf{F}^{\alpha_1}(\cdots \mathsf{F}^{\alpha_{n-1}}(\mathsf{F}^{\alpha_n}(x)) \cdots)$. □

For the operator $\mathsf{S}$ we compute easily $\mathsf{S}^n(x) = n + x$, $\mathsf{S}^\omega(x) = 2 \cdot x + 2$, $\mathsf{S}^{\omega \cdot n}(x) = 2^n \cdot (x+2) - 2$ and $\mathsf{S}^{\omega^2}(x) = 2^{x+1} \cdot (x+3) - 2$.

Therefore we have $x^2+2 \le S^{\omega^2}(x)$ and may use $F := S^{\omega^2}$ as basis for a subrecursive hierarchy. We claim that this hierarchy is eventually majorized by the hierarchy $\{S^\alpha \mid \alpha < \varepsilon_0\}$. To prove the claim, we need a few observations. First, we remark that

$$\alpha \in Lim \;\Rightarrow\; x \le N(\alpha[x]), \tag{10.18}$$

which follows from the definition of $\alpha[x]$ by induction on $\alpha$. Next we observe

$$\alpha < \beta \;\Rightarrow\; \alpha \le \beta[N(\alpha)]. \tag{10.19}$$

To secure (10.19) let $\beta =_{NF} \beta_1 + \cdots + \beta_n$. If $\alpha \le \beta_1 + \cdots + \beta_{n-1} + \beta_n[0] \le \beta[N(\alpha)]$ we are done. Therefore assume $\beta_1 + \cdots + \beta_{n-1} + \beta_n[0] < \alpha < \beta_1 + \cdots + \beta_n$. Then there is a $k$ such that $\beta_1 + \cdots + \beta_{n-1} + \beta_n[k] < \alpha \le \beta_1 + \cdots + \beta_{n-1} + \beta_n[k+1] = \beta[k+1]$. But then $\alpha = \beta_1 + \cdots + \beta_{n-1} + \beta_n[k] + \gamma$ with $\gamma \ne 0$, which implies $N(\alpha) > k$ by (10.18). Hence $\alpha \le \beta[N(\alpha)]$. □

We show that the fundamental sequences have the *nesting property*

$$\alpha[x] < \beta < \alpha \;\Rightarrow\; \alpha[x] \le \beta[0]. \tag{10.20}$$

Let $\alpha =_{NF} \alpha_1 + \cdots + \alpha_n$. Then $\alpha[x] = \alpha_1 + \cdots + \alpha_{n-1} + \alpha_n[x] < \beta < \alpha_1 + \cdots + \alpha_n$. Hence $\beta =_{NF} \alpha_1 + \cdots + \alpha_{n-1} + \gamma_1 + \cdots + \gamma_m$ such that $\alpha_n[x] < \gamma := \gamma_1 + \cdots + \gamma_m$. Therefore there is a $\gamma_0 \ne 0$ such that $\gamma = \alpha_n[x] + \gamma_0$ and $\gamma[k] = \alpha_n[x] + \gamma_0[k]$ for all $k$. Hence $\alpha_n[x] \le \gamma[0]$, which implies $\alpha[x] \le \alpha_1 + \cdots + \alpha_{n-1} + \gamma[0] = \beta[0]$. □

We use the nesting property to show that the hierarchy $\{S^\alpha \mid \alpha < \varepsilon_0\}$ is strongly increasing. We prove

$$S^\alpha(x) + 1 < S^\alpha(x+1) \text{ for } \alpha \ne 0$$
and
$$\alpha[y] < \beta < \alpha \;\Rightarrow\; S^{\alpha[y]}(x) < S^\beta(x) \tag{10.21}$$

simultaneously by induction on $\alpha$ and side induction on $\beta$. If $\alpha = \alpha_0 + 1$, we obtain $S^\alpha(x) + 1 = S^{\alpha_0}(x+1) + 1 < S^{\alpha_0}(x+2) = S^\alpha(x+1)$ from the induction hypothesis for the first claim. If $\alpha \in Lim$, we use the nesting property to obtain $\alpha[x] < \alpha[x+1] \le \beta[0] < \beta < \alpha$ and the induction hypothesis for both claims to obtain $S^\alpha(x)+1 = S^{\alpha[x]}(x+1)+1 < S^{\alpha[x]}(x+2) < S^{\alpha[x+1]}(x+2) = S^\alpha(x+1)$. To prove the second claim we again use the nesting property to obtain $\alpha[y] \le \beta[x]$, which, together with the induction hypothesis for both claims, implies $S^{\alpha[y]}(x) \le S^{\beta[x]}(x) < S^{\beta[x]}(x+1) = S^\beta(x)$. □

The key property for the hierarchy is

$$\alpha < \beta \;\Rightarrow\; (\forall x)[N(\alpha) \le x \Rightarrow S^\alpha(x+1) \le S^\beta(x)]. \tag{10.22}$$

We prove (10.22) by induction on $\beta$. From (10.19), we obtain $\alpha \le \beta[N(\alpha)] \le \beta[x]$. For $\alpha = \beta[x]$ we obtain $S^\alpha(x+1) = S^{\beta[x]}(x+1) = S^\beta(x)$. If $\alpha < \beta[x]$ we obtain from the induction hypothesis $S^\alpha(x+1) \le S^{\beta[x]}(x) < S^{\beta[x]}(x+1) = S^\beta(x)$. □

**10.6.3 Lemma** *Let* $F = S^{\omega^2}$. *Then* $\psi_F(\alpha) \leq S^{\omega^2 \times (\alpha+1)}(N(\alpha)+2)$.

*Proof* We induct on $\alpha$. For $\alpha = 0$ the claim is obvious. So assume $\alpha > 0$. Then $\psi_F(\alpha) = \psi_F(\eta) + 1$ for some $\eta < \alpha$ such that $N(\eta) \leq F(N(\alpha))$. By the induction hypothesis we obtain $\psi_F(\eta) \leq S^{\omega^2 \times (\eta+1)}(N(\eta)+2)$. But $N(\omega^2 \times (\eta+1)) \leq 3{\cdot}N(\eta)+3$. Since $\omega^2 \times (\eta+1) \leq \omega^2 \times \alpha$ we obtain $\psi_F(\eta) < S^{\omega^2 \times \alpha}(3{\cdot}N(\eta)+3)$ by (10.22) and (10.21). Since $S^{\omega^2}(x) = 2^{x+1}(x+3) - 2$ we get $3{\cdot}S^{\omega^2}(x)+3 < S^{\omega^2}(x+2)$ and therefore $3{\cdot}N(\eta)+3 \leq 3{\cdot}S^{\omega^2}(N(\alpha))+3 < S^{\omega^2}(N(\alpha)+2)$ which implies $\psi_F(\alpha) = \psi_F(\eta)+1 \leq S^{\omega^2 \times \alpha}(3{\cdot}N(\eta)+3) \leq S^{\omega^2 \times \alpha}(S^{\omega^2}(N(\alpha)+2)) = S^{\omega^2 \times \alpha \# \omega^2}(N(\alpha)+2) = S^{\omega^2 \times (\alpha+1)}(N(\alpha)+2)$. □

**10.6.4 Theorem** *Let* $F = S^{\omega^2}$. *Then the subrecursive hierarchy* $\{F_\alpha \mid \alpha < \varepsilon_0\}$ *is eventually majorized by the hierarchy* $\{S^\alpha \mid \alpha < \varepsilon_0\}$.

*Proof* Because of $F_\alpha(x) = \psi_F(\alpha + x) \leq S^{\omega^2 \times (\alpha+x+1)}(N(\alpha)+x+2) \leq S^{\omega^2 \times (\alpha \# \omega)}(3{\cdot}N(\alpha)+3{\cdot}x+3) = S^{\omega^2 \times (\alpha \# \omega)+3{\cdot}N(\alpha)+3}(3x) \leq S^{\omega^2 \times (\alpha \# \omega)+3{\cdot}N(\alpha)+4}(x)$ we obtain that $\{F_\alpha \mid \alpha < \varepsilon_0\}$ is eventually majorized by $\{S^\alpha \mid \alpha < \varepsilon_0\}$. □

By Theorem 10.6.4, we have $SF_{\Pi^0_2}(\mathsf{NT}) = PRH(\{F_\alpha \mid \alpha < \varepsilon_0\}) \subseteq PRH(\{S^\alpha \mid \alpha < \varepsilon_0\})$. Therefore $\{S^\alpha \mid \alpha < \varepsilon_0\}$ generates at least all Skolem functions for the provable $\Pi^0_2$-sentences of NT. By induction on $\alpha < \varepsilon_0$ we obtain, however, that all the functions $S^\alpha$ are provably recursive functions of NT. Therefore $PRH(\{S^\alpha \mid \alpha < \varepsilon_0\}) \subseteq SF_{\Pi^0_2}(\mathsf{NT}) = PRH(\{F_\alpha \mid \alpha < \varepsilon_0\})$ and we have seen that the primitive recursive hulls of both hierarchies coincide. Therefore we obtain the following theorem.

**10.6.5 Theorem** *The hierarchy* $\{S^\alpha \mid \alpha < \varepsilon_0\}$ *generates the class of Skolem functions of the provable* $\Pi^0_2$*-sentences of* NT, *i.e.,*

$$SF_{\Pi^0_2}(\mathsf{NT}) = PRH(\{S^\alpha \mid \alpha < \varepsilon_0\}).$$

The hierarchy $\{S^\alpha \mid \alpha < \varepsilon_0\}$ is a slight variant of the Hardy-hierarchy. The original Hardy-hierarchy is defined by

$$H_\alpha(x) := \begin{cases} x & \text{if } \alpha = 0 \\ H_\beta(x+1) & \text{if } \alpha = \beta + 1 \\ H_{\alpha[x]}(x) & \text{if } \alpha \in Lim. \end{cases}$$

The difference is marginal and it is easy to show that the primitive recursive hulls of $S^\alpha$ and $H_\alpha$ coincide. It has become common to call hierarchies whose growth rate corresponds to the the growth rate of the Hardy-hierarchy "fast-growing" in contrast to the "slow-growing" hierarchy $G_\alpha$ which is defined by pointwise iteration of the successor function. Pointwise iteration of a function $P$ is defined by

$$P^p_\alpha(x) := \begin{cases} x & \text{if } \alpha = 0 \\ P(P^p_\beta(x)) & \text{if } \alpha = \beta + 1 \\ P^p_{\alpha[x]}(x) & \text{if } \alpha \in Lim. \end{cases}$$

So we obtain $G_{\alpha+1}(x) = G_\alpha(x) + 1$ for $G_\alpha(x) := S^p_\alpha(x)$. Girard [34] was the first who showed that the slow-growing hierarchy has to be iterated along the Howard–Bachmann-ordinal $\psi(\varepsilon_{\Omega+1})$ to generate $SF_{\Pi^0_2}(\mathsf{NT})$. This result was later reobtained by others. Careful studies by Weiermann, however, have shown that the slow-growing hierarchy is extremely sensible to variations in the definition of the fundamental sequences. There are assignments of fundamental sequences to limit ordinals which are sufficiently natural and leave the fast growing hierarchies unchanged but alter the slow growing hierarchies dramatically (cf. [107]). But these studies are widely outside the scope of this book.

# Chapter 11
# Ordinal Analysis for Kripke–Platek Set Theory with Infinity

The notion of "set" is the central notion of mathematics. All mathematical objects can be represented as sets. The set theoretical universe can therefore be regarded as *the mathematical universe.* The study of axiom systems for set theory should therefore be *the* central subject of proof theory. In the following section, we will present the first step into the study of axiom systems for set theory. The pioneering work in this direction is mainly due to Gerhard Jäger [48].

## 11.1 Naive Set Theory

We want to start with a rough heuristic. Set theory has been introduced by Georg Cantor on the turn of the nineteenth to the twentieth century. He gave a definition of a set as a "collection of well-distinguished objects into a new object." This is of course not a mathematical definition as we understand it today. The difficulties in defining "what is a set" are similar to the difficulties in defining "what is a vector." Instead of defining vectors, we define vector spaces by their closure properties and say that a vector is an element of a vector space. In the same way, we will describe the set theoretical universe and declare a set as a "member of the universe." Following this line we try to describe *the universe* $\mathsf{U}$.

The basic notion of set theory is the membership relation $a \in b$ stating that the object $a$ is a member of the object $b$. The formula $a \in \mathsf{U}$ then expresses that $a$ is a member of the universe, which is synonymous to saying "$a$ is a set." The universe is then a "collection" of sets. A collection of sets, i.e., an object $A$ having the property $(\forall x \in A)[x \in \mathsf{U}]$, is commonly called a *class*. So we have to deal with two types of objects, *classes* and *sets*. One of the basic properties of the universe is that it is closed under the membership relation. All the members of a set are supposed to be sets themselves. Therfore, we postulate the *axiom of transitivity*

(Tran) $\quad a \in \mathsf{U} \;\Rightarrow\; (\forall x \in a)[x \in \mathsf{U}]$

which says that every set is also a class. The next thing to do is to define equality of classes. We say that two classes are equal if and only if they have the same elements, i.e., we postulate the *axiom of extensionality*

(Ext) $A = B \;\Leftrightarrow\; (\forall x \in A)[x \in B] \wedge (\forall y \in B)[x \in A]$.

If $F(x)$ is some property assigned to the set $x$, we can comprehend all these sets having this property into a class. As an abbreviation we write $\{x \mid F(x)\}$ for this class and postulate the *axiom of class comprehension*

(Comp) $a \in \{x \mid F(x)\} \;:\Leftrightarrow\; a \in \mathsf{U} \wedge F(a)$.

From (Comp) we see immediately that not all classes are sets. The famous counterexample is Russell's class $R := \{x \mid x \notin x\}$. The assumption $R \in \mathsf{U}$ leads to the contradiction $R \in R \;\Leftrightarrow\; R \notin R$. So we have $R \notin \mathsf{U}$. This shows that the class $R$ is somehow too big as to be a set. Sets are classes, which – in what sense so ever – are "smaller than the universe." So a subclass of a set – which is already a "small class" – should again be a set. Separating sets out of a set should again lead to a "small class." Formulating this as an axiom gives the *separation axiom*

(Sep) $a \in \mathsf{U} \;\Rightarrow\; \{x \in a \mid F(x)\} \in \mathsf{U}$.

Another axiom in this guise is the *collection axiom*

(Coll) $(\forall a)\big[(\forall x \in a)(\exists y)F(x,y) \Rightarrow (\exists z)(\forall x \in a)(\exists y \in z)F(x,y)\big]$

saying that given as set $a$ and to every element $x \in a$, a witness $y$ having the property $F(x,y)$ then there should already exist a "small class," i.e., a set, $z$ which contains such a witness for every $x \in a$.

There are simple set operations such as pairing, i.e., forming the class $\{a,b\} := \{x \mid x = a \vee x = b\}$ from sets $a,b$, and union $\bigcup a := \{x \mid (\exists y \in a)[x \in y]\}$, which should not lead outside the universe. So we introduce also the axioms

(Pair) $a \in \mathsf{U} \wedge b \in \mathsf{U} \;\Rightarrow\; \{a,b\} \in \mathsf{U}$

and

(Union) $a \in \mathsf{U} \;\Rightarrow\; \bigcup a \in \mathsf{U}$.

Until now we do not have an axiom that postulates that the universe is inhabited. One possibility is to postulate that the empty class $\emptyset := \{x \mid x \neq x\}$ is a set. But what we really are interested in are universes that contain infinite sets. Therefore, we introduce the *axiom of infinity*

(Inf) $(\exists x \in \mathsf{U})\big[x \neq \emptyset \wedge (\forall y \in x)(\exists z \in x)[y \in z]\big]$

requiring the existence of a set $x$ which is not empty and internally unbounded, i.e., for every element $y$ of $x$ there is still an element $z \in x$, which "majorizes $y$" in the sense of the $\in$-relation.

There is another requirement which is perhaps a bit harder to motivate. We want to exclude infinitely descendent $\in$ sequences, i.e., sequences of the form

$a_0 \ni a_1 \ni \cdots \ni a_i \ni a_{n+1} \ni \cdots$. This requirement is equivalent to the requirement that the $\in$ relation is well-founded on $\mathsf{U}$. So we have another axiom, the axiom of foundation

(FOUND*) $A \subseteq \mathsf{U} \wedge (\exists x)[x \in A] \Rightarrow (\exists y)[y \in A \wedge (\forall z \in y)[z \notin A]]$.

The foundation axiom especially excludes the existence of set that are "self-referential", i.e., the existence of a set $a$ such that $a \in a$ holds true.

The axiomatization of set theory is connected with the names Ernst Zermelo and Abraham Fraenkel. We have essentially (still informally) introduced a part of the axioms, which today are known as Zermelo–Fraenkel set theory. However, two axioms are still lacking. The first is the *power–set axiom.* The class $\mathrm{Pow}(a) := \{x \mid (\forall y \in x)[y \in a]\}$ of all subsets of a set $a$ is the *power-class of a.* The *power-set axiom* postulates that $\mathrm{Pow}(a)$ is again a set, i.e.,

(Pow) $a \in \mathsf{U} \;\Rightarrow\; \mathrm{Pow}(a) \in \mathsf{U}$.

From a foundational point of view the power set axiom is more problematic. It is by far not clear why $\mathrm{Pow}(a)$ should again be a "small" class.[1] But already in Cantor's work $\mathrm{Pow}(a)$ was considered as a set.

The second axiom – which for other reasons is somewhat problematic from a foundational point of view – is the *axiom of choice* postulating the existence of a choice function for every family of nonempty sets. Its formulation requires the definition of a function from sets into sets.

(AC) $A \subseteq \mathsf{U} \wedge (\forall x \in A)(\exists y)[y \in x] \Rightarrow (\exists f)\big[Fun(f) \wedge (\forall x \in A)[f(x) \in x]\big]$.

The formalization of the axioms (Ext), (Pair), (Union), (Sep), (Inf), (Pow), and (FOUND) in first-order logic (as presented in the forthcoming section) are commonly called Zermelo set theory. If (AC) is included one talks about Zermelo set theory with choice ZC. In Zermelo–Fraenkel set theory the axiom of separation (Sep) is replaced by (the formalization of) the *axiom of replacement*

(Repl) $Fun(f) \wedge a \in \mathsf{U} \;\Rightarrow\; \{f(x) \mid x \in a\} \in \mathsf{U}$

which again has the clear meaning that the image of a set $a$ under a function $f$ cannot be too big. By ZF we commonly denote Zermelo–Fraenkel set theory without the axiom of choice while ZFC stands for Zermelo–Fraenkel set theory with choice. If we want to state something for both ZF and ZFC we denote that by ZF(C). We want to mention that the axioms of ZFC are equivalent to the system of axioms (Ext), (Pair), (Union), (Sep), (Coll), (Pow), (Inf), (AC), and (Found), i.e., that the axiom of replacement is – on the basis of the other axioms – equivalent to separation plus collection.

---

[1] Even ZFC is not sufficient to decide the size of the class $\mathrm{Pow}(a)$. It is still open which axioms are the right ones to decide the size of $\mathrm{Pow}(\mathbb{N})$ [113, 114]. The fact that $\mathrm{Pow}(a)$ it is again a set is, however, not a debatable item among set theorists.

## 11.2 The Language of Set Theory

Formalization of naive set theory as presented in the previous section would require a two-sorted language with a sort for classes and another sort for sets. However, classes play a role different from that of sets. Talking about all classes would mean to introduce a "superuniverse" containing all classes and thus "superclasses" as objects which collect classes but are not classes, i.e., elements of the superuniverse, themselves, etc. This shows that quantification over classes may cause problems. There are formalizations of set theory, which involve quantifications over classes but the most common axiomatizations avoid classes as basic objects. In fact there is no need to talk about classes. We already introduced the notion $\{x \mid F(x)\}$ as an "abbreviation" for the formula $F(x)$. So it becomes possible to talk at least about definable classes and to formalize nearly all of the axioms of the previous section in terms of a one-sorted first-order logic. By Zermelo–Fraenkel set theory, one usually understands the axiomatization of set theory in an one-sorted first-order logic with identity. The only nonlogical symbol that is needed to formalize ZF(C) in a first-order language is a binary relation symbol $\in$ for membership.

**11.2.1 Definition** The language $\mathcal{L}(\in)$ of set theory is the first-order language with identity not containing relation variables whose only nonlogical symbol is the binary relation constant $\in$.

In the language $\mathcal{L}(\in)$ the axioms of the previous section take the following form:

(Ext) $(\forall x)(\forall y)\big[x = y \leftrightarrow (\forall z \in x)[z \in y] \wedge (\forall z \in y)[z \in x]\big]$

(Pair) $(\forall x)(\forall y)(\exists z)(\forall u)[u \in z \leftrightarrow u = x \vee u = y]$

(Union) $(\forall x)(\exists y)(\forall u)\big[u \in y \leftrightarrow (\exists z \in x)[u \in z]\big]$

(Sep) $(\forall x)(\exists y)(\forall u)[u \in y \leftrightarrow u \in x \wedge F(u)]$

(Coll) $(\forall a)[(\forall x \in a)(\exists y)F(x,y) \to (\exists z)(\forall x \in a)(\exists y \in z)F(x,y)]$

where $F$ is a formula in the language of set theory. If $F$ contains further free variables we assume that (Sep) and (Coll) are the universal closure of these schemes.

(Inf) $(\exists x)\big[x \neq \emptyset \wedge (\forall y \in x)(\exists z \in x)[y \in z]\big]$

where $x \neq \emptyset$ can be expressed by $(\exists y)[y \in x]$ and

(FOUND) $(\exists x)F(x) \quad \to \quad (\exists y)\big[F(y) \wedge (\forall z \in y)[\neg F(z)]\big].$

We leave the formulation of the power-set axiom and the axiom of choice in terms of $\mathcal{L}(\in)$ to the reader since we will not deal with these axioms in this book.

Observe that the sacrifice of classes forced us to replace the axioms (Sep) and (Coll) by schemes. This is unavoidable. It can be shown that ZF(C) is not finitely axiomatizable. In full ZF(C) it is possible to replace the foundation scheme (FOUND) by a single axiom

(Found) $(\forall a)\big[(\exists y)(y \in a) \to (\exists z \in a)[(\forall y \in z)(y \notin a)]\big]$.

This is in general not true for subsystems of ZF(C).

The remarks on the limits of first-order logic in Sect. 4.1 of course also apply to models of ZF(C). If $M$ is a model of ZF(C) then $M$ will have all the properties required in the informal Sect. 11.1 possibly except (Tran) and (FOUND*). Even if $M \models$ (FOUND) we will not know that $M$ is in fact well-founded (this cannot be secured in first-order logic). To remedy that one usually only regards well-founded and transitive models of ZF(C).

It is no surprise that, similar to the situation in arithmetic, there is again no first-order axiomatization which fixes the set theoretical universe up to isomorphisms. In the case of set theory, the situation is even worse because we do not have a picture of the "standard" set theoretical universe which is comparably clear to that of the standard structure of natural numbers.[2] But there is a similarity. Every model of the axioms of NT contains $\mathbb{N}$ as an initial segment. Likewise every model of ZFC contains a least inner class model, i.e., a model $L$ which contains all ordinals. This inner model – the constructible hierarchy invented by Gödel – will overtake the role of the standard natural numbers in the proof-theoretical analysis of axiom systems for set theory.

## 11.3 Constructible Sets

Whenever we have a set theoretical universe U, we can define a hierarchy $V$ by the following clauses

$$V_0 := \emptyset$$
$$V_{\alpha+1} := \mathrm{Pow}(V_\alpha)$$
$$V_\lambda := \bigcup_{\xi<\lambda} V_\xi \text{ for } \lambda \in Lim$$

and

$$V := \bigcup_{\xi \in On} V_\xi.$$

This construction is due to von Neumann. The von Neumann-hierarchy is *cumulative* in the sense that $\alpha < \beta$ implies $V_\alpha \subseteq V_\beta$ and $V_\lambda$ is the collection of all previous stages at limit ordinals $\lambda$. It is not too difficult to check that $V$, when constructed within a set theoretical universe U, is again a transitive model of

[2] This is of course only true as long as we restrict ourselves to the "first-order part" of the structure of natural number. That our intuition of its second-order part, i.e., the reals, is likewise unclear is often overlooked.

ZF(C). It is moreover also easy to see that $\mathsf{U} = V$ if and only if $\mathsf{U} \models$ (FOUND). So every well-founded universe can be visualized as a von Neumann-hierarchy. But due to the unclear meaning of the power-set, we do not know too much about the von Neumann-hierarchy. Therefore, Gödel invented a hierarchy whose stages are more carefully built. Instead of taking the full power-set at the successor stage he only allows those members of the power-set which are constructible from the already obtained part of the universe. The key here is the notion of *definability*. If $M$ is a set and $F$ an $\mathcal{L}(\in)$-sentence we write shortly $M \models F$ for $(M, \in) \models F$.

**11.3.1 Definition** Let $M$ be a transitive set. A set $a$ is *definable from $M$* if there is an $\mathcal{L}(\in)$ formula $F(x, y_1, \ldots, y_n)$ and elements $a_1, \ldots, a_n \in M$ such that $a = \{x \in M \mid M \models F(x, a_1, \ldots, a_n)\}$.

Let $Def(M)$ be the collection of all sets which are definable from $M$.

**11.3.2 Definition** Using definability we introduce the constructible hierarchy $L$ by the following clauses.

$$L_0 := \emptyset$$
$$L_{\alpha+1} := Def(L_\alpha)$$
$$L_\lambda := \bigcup_{\xi<\lambda} L_\xi \text{ for } \lambda \in Lim$$

and

$$L := \bigcup_{\xi \in On} L_\xi.$$

Although we will not deal too much with the set theoretical properties of the constructible hierarchy, we are going to state some basic facts. For a more profound study, we recommend any text book on set theory (e.g. [59, 21]).

**11.3.3 Lemma** *For the stages of the constructible hierarchy we have*

*(A)* $\quad (\forall \alpha \in On)[Tran(L_\alpha)]$

*and*

*(B)* $\quad \alpha \leq \beta \Rightarrow L_\alpha \subseteq L_\beta.$

*Proof* We prove (A) and

(C) $\quad Tran(L_\alpha) \;\Rightarrow\; L_\alpha \subseteq L_{\alpha+1}.$

Property (B) then follows by induction on $\beta$. It is obvious for $\beta = 0$ and $\beta \in Lim$. For successor ordinals it follows from (A) and (C).

If $Tran(L_\alpha)$ and $x \in L_\alpha$ we immediately get $x = \{z \in L_\alpha \mid z \in x\} \in L_{\alpha+1}$. Hence $L_\alpha \subseteq L_{\alpha+1}$. We show (A) by induction on $\alpha$. It is trivial for $\alpha = 0$ and follows directly from the induction hypothesis for $\alpha \in Lim$. If $x \in y \in L_{\alpha+1}$ we have the induction hypothesis $Tran(L_\alpha)$ and obtain $y = \{z \in L_\alpha \mid L_\alpha \models F(z,\vec{a})\}$ for some $\mathcal{L}(\in)$-formula $F$ and a tuple $\vec{a}$ of sets in $L_\alpha$. Hence $x \in L_\alpha \subseteq L_{\alpha+1}$ by (C). □

Once we have defined a set at stage $\alpha$ in the constructible hierarchy, we want to be sure that it will not change its meaning during the process of expanding $L$. This requires the notion of absoluteness. We call a formula *absolute* if it keeps its meaning in all higher stages. More generally we define absoluteness as follows.

**11.3.4 Definition** We call an $\mathcal{L}(\in)$-formula $F(x_1,\dots,x_n)$ *upwards persistent* if

$$M \models F(a_1,\dots,a_n) \;\Rightarrow\; N \models F(a_1,\dots,a_n),$$

for all transitive sets $M, N$ such that $M \subseteq N$. We call $F(x_1,\dots,x_n)$ *downwards persistent* if

$$N \models F(a_1,\dots,a_n) \;\Rightarrow\; M \models F(a_1,\dots,a_n)$$

and *absolute* if

$$M \models F(a_1,\dots,a_n) \;\Leftrightarrow\; N \models F(a_1,\dots,a_n)$$

is true for all transitive sets $M, N$ with $M \subseteq N$. In all cases $a_1,\dots,a_n$ is an arbitrary tuple of elements in $M$.

Persistency and absoluteness are model-theoretically defined properties of formulas. Therefore, it would be convenient to have syntactical criteria for a formula being persistent or absolute. In the next definition, we introduce sufficient criteria for persistency and absoluteness in terms of the Levy-hierarchy of $\mathcal{L}(\in)$-formulas. For these purposes, we consider *bounded* and *unbounded quantifiers* as different basic symbols. A quantifier Q is bounded if it only occurs in the form $(Qx \in u)$. The meaning of $(\forall x \in u)[\cdots]$ is of course $(\forall x)[x \in u \rightarrow \cdots]$ while $(\exists x \in u)[\cdots]$ means $(\exists x)[x \in u \wedge \cdots]$.

**11.3.5 Definition** The class of $\Delta_0$-formulas is the smallest class of $\mathcal{L}(\in)$-formulas, which contains all atomic formulas $(u = v)$ and $(u \in v)$ and is closed under the propositional connectives $\neg$, $\wedge$, $\vee$, and bounded quantification.

The class of $\Sigma$-formulas is the smallest class which comprises all $\Delta_0$-formulas and is closed under the positive propositional connectives $\wedge$ and $\vee$, bounded quantification and unbounded $\exists$-quantification.

Dually the $\Pi$-formulas form the smallest class, which comprises all $\Delta_0$-formulas and is closed under the positive propositional connectives $\wedge$ and $\vee$, bounded quantification and unbounded $\forall$-quantification.

An $\mathcal{L}(\in)$-formula is $\Sigma_1$ if it has the form $(\exists x)F_u(x)$ for a $\Delta_0$-formula $F$.

Dually a formula is $\Pi_1$ if it has the shape $(\forall x)F_u(x)$ with $F$ a $\Delta_0$-formula.

It is obvious from the definition that every $\Sigma_1$-formula is also a $\Sigma$-formula and every $\Pi_1$-formula is also a $\Pi$-formula. The converse direction is in general false. A list of $\Delta_0$-notions is displayed in Fig. 11.1. Recall that the ordered pair is definable by $(x,y) := \{\{x\},\{x,y\}\}$.

| Notion | Abbreviation | $\Delta_0$–definition |
|---|---|---|
| subset | $a \subseteq b$ | $(\forall x \in a)[x \in b]$ |
| $a$ is the pair $\{u,v\}$ | $a = \{u,v\}$ | $(\forall x \in a)[x = u \vee x = v] \wedge u \in a \wedge v \in a$ |
| $a$ is the ordered pair $(u,v)$ | $a = (u,v)$ | $(\exists x \in a)(\exists y \in a)[x = \{u\} \wedge y = \{u,v\} \wedge a = \{x,y\}]$ |
| $a = (x,y)$ for some $y$ | $x = P_0(a)$ | $(\exists u \in a)(\exists y \in u)[a = (x,y)]$ |
| $a = (x,y)$ for some $x$ | $y = P_1(a)$ | $(\exists u \in a)(\exists x \in u)[a = (x,y)]$ |
| $a$ is an ordered pair | $Pair(a)$ | $(\exists u \in a)(\exists x \in u)(\exists y \in u)[a = (x,y)]$ |
| $a$ is a relation | $Rel(a)$ | $(\forall x \in a)[Pair(x)]$ |
| $a$ is a function | $Fun(a)$ | $Rel(a) \wedge (\forall x \in a)(\forall y \in a)[P_0(x) = P_0(y) \rightarrow P_1(x) = P_1(y)]$ |
| domain of a relation | $dom(f) = a$ | $Rel(f) \wedge (\forall x \in f)[P_0(x) \in a] \wedge (\forall x \in a)(\exists y \in f)[P_0(y) = x]$ |
| range of a relation | $rng(f) = a$ | $Rel(f) \wedge (\forall x \in f)[P_1(x) \in a] \wedge (\forall x \in a)(\exists y \in f)[P_1(y) = x]$ |
| field of a relation | $field(f) = a$ | $Rel(f) \wedge a = dom(f) \cup rng(f)$ |
| image of an element | $f(x) = a$ | $Fun(f) \wedge (x,a) \in f$ |
| union of a set | $a = \bigcup b$ | $(\forall x \in a)(\exists y \in b)[x \in y] \wedge (\forall y \in b)(\forall x \in y)[x \in a]$ |
| $a$ is not empty | $a \neq \emptyset$ | $(\exists x \in a)[x \in a]$ |
| $a$ is transitive | $Tran(a)$ | $(\forall x \in a)(\forall y \in x)[y \in a]$ |
| $a$ is an ordinal | $a \in On$ | $Tran(a) \wedge (\forall x \in a)Tran(x)$ |
| $a$ is a limit ordinal | $a \in Lim$ | $a \neq \emptyset \wedge a \in On \wedge (\forall x \in a)(\exists y \in a)[x \in y]$ |
| $a$ belongs to a finite set | $a \in \{a_1, \dots, a_n\}$ | $a = a_1 \vee \cdots \vee a = a_n$ |

**Fig. 11.1** Some $\Delta_0$–notions

**11.3.6 Lemma** *Every $\Sigma$-formula is upwards persistent. Dually every $\Pi$-formula is downwards persistent. The $\Delta_0$-formulas are therefore absolute.*

*Proof* It suffices to prove the lemma for $\Sigma$-formulas. The rest follows by dualization. So let $F(x_1,\dots,x_n)$ be a $\Sigma$-formula, $M$, $N$ transitive sets and $a_1,\dots,a_n \in M$. We show

$$M \models F(a_1,\dots,a_n) \;\Rightarrow\; N \models F(a_1,\dots,a_n) \tag{i}$$

by induction on the complexity of the formula $F(x_1,\dots,x_n)$. This is obvious for atomic formulas and follows directly from the induction hypothesis if $F(x_1,\dots,x_n)$

is a boolean combination. So let $F(x_1,\dots,x_n)$ be a formula $(\forall y \in x_1)F_0(y,x_1,\dots,x_n)$ and assume

$$M \models F_0(b,a_1,\dots,a_n) \text{ for all } b \in a_1 \in M. \tag{ii}$$

Since $M$ is transitive we obtain from $c \in a_1 \cap N$ already $c \in M$. Hence

$$N \models F_0(c,a_1,\dots,a_n) \text{ for all } c \in a_1 \cap N \tag{iii}$$

by the induction hypothesis which means $N \models (\forall x \in a_1)F_0(x,a_1,\dots,a_n)$.

Finally assume that $F(x_1,\dots,x_n)$ is a formula $(\exists y)F_0(y,x_1,\dots,x_n)$ and

$$M \models (\exists y)F_0(y,a_1,\dots,a_n). \tag{iv}$$

Then there is a $b \in M \subseteq N$ such that $M \models F_0(b,a_1,\dots,a_n)$ and we obtain

$$N \models F_0(b,a_1,\dots,a_n), \text{ i.e., } N \models (\exists y)F_0(y,a_1,\dots,a_n) \tag{v}$$

directly by the induction hypothesis. This includes of course also the case of a bounded existential quantifier. □

**11.3.7 Definition** Let $M$ be a set. We call $F(x_1,\dots,x_n)$ a $\Delta$-formula in $M$ if there is a $\Sigma$-formula $F_\Sigma(x_1,\dots,x_n)$ and a $\Pi$-formula $F_\Pi(x_1,\dots,x_n)$ such that

$$M \models (\forall x_1)\dots(\forall x_n)[F(x_1,\dots,x_n) \leftrightarrow F_\Sigma(x_1,\dots,x_n)]$$

and

$$M \models (\forall x_1)\dots(\forall x_n)[F(x_1,\dots,x_n) \leftrightarrow F_\Pi(x_1,\dots,x_n)].$$

Observe that the concept of a $\Delta$-formula is not longer purely syntactical. It needs – in contrast to the purely syntactical concept of a $\Delta_0$-formula – a proof to see that $F$ is a $\Delta$-formula in some set $M$.

**11.3.8 Corollary** *Let $M$ and $N$ be transitive sets, $M \subseteq N$ and $F$ a $\Delta$-formula in $M$ as well as in $N$. Then $F$ is absolute for $M$ and $N$, i.e., we have $M \models F(a_1,\dots,a_n)$ if $N \models F(a_1,\dots,a_n)$ for all $a_1,\dots,a_n \in M$.*

*Proof* This follows immediately from Lemma 11.3.6. □

For the stage $L_\alpha$ in the constructible hierarchy we obtain

$$L_\alpha = \{x \in L_\alpha \mid L_\alpha \models x = x\}.$$

Hence $L_\alpha \in L_{\alpha+1}$ and, since $x = x$ is a $\Delta_0$ formula, this definition of $L_\alpha$ is absolute for all $L_\beta$ with $\beta > \alpha$.

For a set $a \in L$ we define

$$\operatorname{rnk}_L(a) := \min\{\eta \mid a \in L_\eta\} \tag{11.1}$$

and call $\operatorname{rnk}_L(a)$ the $L$-rank of $a$.

In Sect. 3, we have shown that in the presence of the foundation scheme (FOUND) we can define

$$\alpha \in On \;:\Leftrightarrow\; Tran(\alpha) \wedge (\forall x \in \alpha) Tran(x).$$

Being an ordinal is thus expressible by a $\Delta_0$-formula. The notion "$\alpha$ is an ordinal" is therefore absolute. We prove

$$L_\alpha \cap On = \alpha \text{ and } \mathrm{rnk}_L(\alpha) = \alpha + 1 \tag{11.2}$$

by induction on $\alpha$. Both claims are clear for $\alpha = \emptyset$. Next assume $\alpha \in Lim$. Then

$$L_\alpha \cap On = \bigcup_{\xi<\alpha} (L_\xi \cap On) = \bigcup_{\xi<\alpha} \xi = \alpha$$

by the induction hypothesis, and, since $x \in On$ is absolute,

$$\alpha = \{x \in L_\alpha \mid x \in On\} = \{x \in L_\alpha \mid L_\alpha \models x \in On\} \in L_{\alpha+1}.$$

Hence

$$\mathrm{rnk}_L(\alpha) \leq \alpha + 1$$

and $\mathrm{rnk}_L(\alpha) \leq \alpha$ would imply $\alpha \in L_\eta$ for some $\eta \leq \alpha$ contradicting the just proved first claim.

In the successor case $\alpha = \alpha_0 + 1$ we have the induction hypothesis $L_{\alpha_0} \cap On = \alpha_0$ and $\mathrm{rnk}_L(\alpha_0) = \alpha_0 + 1$, i.e., $\alpha_0 \in L_\alpha$. Hence $\alpha = \alpha_0 \cup \{\alpha_0\} \subseteq L_\alpha \cap On$. Conversely, $\xi \in L_\alpha \cap On$ implies $\xi \subseteq L_{\alpha_0} \cap On = \alpha_0$ which in turn implies $\xi \leq \alpha_0 < \alpha$. Hence $L_\alpha \cap On \subseteq \alpha$ and we have $L_\alpha \cap On = \alpha$ which is the first claim and entails $\alpha < \mathrm{rnk}_L(\alpha)$. But by the first claim and the absoluteness of $On$ we again have

$$\alpha = \{x \in L_\alpha \mid x \in On\} = \{x \in L_\alpha \mid L_\alpha \models x \in On\} \in L_{\alpha+1}.$$

Hence $\mathrm{rnk}_L(\alpha) \leq \alpha + 1$, which shows also the second claim. □

The constructible hierarchy is obiously well-founded. Therefore, it follows immediately that all stages $L_\alpha$ satisfy the scheme (FOUND). But we have even more.

**11.3.9 Theorem** *Let $\alpha > \omega$ be a limit ordinal. Then*

$$L_\alpha \models (\mathrm{Ext}) \wedge (\mathrm{Pair}) \wedge (\mathrm{Union}) \wedge (\mathrm{Inf}) \wedge (\mathrm{FOUND}).$$

*Proof* Assume $a, b \in L_\alpha$ and $a \neq b$. Then there is an $x \in a$ such that $x \notin b$ or there is an $x \in b$ such that $x \notin a$. Since $L_\alpha$ is transitive we get $(\exists x \in L_\alpha \cap a)[x \notin b]$ or $(\exists x \in L_\alpha \cap b)[x \notin a]$. Hence $L_\alpha \models a \neq b \to (\exists x \in a)[x \notin b] \vee (\exists x \in b)[x \notin a]$, i.e., $L_\alpha \models (\mathrm{Ext})$.

If $a, b \in L_\alpha$ then there is an $\eta < \alpha$ such that $a, b \in L_\eta$. But then

$$\{a, b\} = \{x \in L_\eta \mid L_\eta \models (x = a \vee x = b)\} \in L_{\eta+1} \subseteq L_\alpha.$$

Hence $L_\alpha \models (\mathrm{Pair})$.

If $a \in L_\alpha$ then there is an $\eta < \alpha$ such that $a \in L_\eta$ and $\bigcup a = \{x \mid (\exists y \in a)[x \in y]\}$. Since $L_\eta$ is transitive and $(\exists y \in a)[x \in y]$ is a $\Delta_0$-formula we obtain

$$\bigcup a = \{x \in L_\eta \mid L_\eta \models (\exists y \in a)[x \in y]\} \in L_{\eta+1} \subseteq L_\alpha.$$

Hence $L_\alpha \models$ (Union).

By (11.2) we have $\omega \in L_{\omega+1} \subseteq L_\alpha$ for every limit ordinal $\alpha > \omega$. Hence $L_\alpha \models$ (Inf). We already mentioned that $L_\alpha \models$ (FOUND) holds for all $\alpha$. □

Defining a set $c := \{x \in a \mid F(x, b_1, \ldots, b_n)\}$ by separation makes it dependent on the property expressed by the formula $F(x, b_1, \ldots, b_n)$. To make sure that this set remains unaltered during the expansion process of the universe, we must only allow absolute properties in defining sets by separation. Therefore, we restrict the separation scheme to properties which can be expressed by $\Delta_0$-formulas, a notion which is still syntactically checkable. Let us denote the scheme (Sep) in which the formulas $F(u)$ are restricted to $\Delta_0$-formulas by ($\Delta_0$–Sep). Analogously we denote by ($\Delta_0$–Coll) the collection scheme restricted to $\Delta_0$-formulas $F(x, y)$.

**11.3.10 Theorem** *For any limit ordinal $\alpha$ we have $L_\alpha \models$ ($\Delta_0$–Sep).*

*Proof* Let $a \in L_\alpha$, $F(x, y_1, \ldots, y_n)$ be a $\Delta_0$-formula and $b_1, \ldots, b_n \in L_\alpha$. Again there is an $\eta < \alpha$ such that $a, b_1, \ldots, b_n \in L_\eta$. By absoluteness and transitivity of $L_\eta$ we obtain

$$\{x \in a \mid F(x, b_1, \ldots, b_n)\} = \{x \in L_\eta \mid L_\eta \models x \in a \land F(x, b_1, \ldots, b_n)\} \in L_{\eta+1}.$$

But $\alpha \in Lim$ implies $L_{\eta+1} \subseteq L_\alpha$. □

## 11.4 Kripke–Platek Set Theory

We have seen that the constructible hierarchy at limit levels already satisfies a certain amount of closure conditions. Axiomatizing these closure conditions we obtain a set BST of basic axioms, which are satisfied by any rudimentarily closed universe. The acronym BST stands for Basic Set Theory. Our aim is to analyze these axiom proof-theoretically. Therefore, we reformulate these axioms more parsimoniously. Since $\Delta_0$-separation is among the axioms of basic set theory, it suffices to require the existence of supersets of pair, union, etc., to obtain the axioms of pairing and union. The precise sets can then be separated by $\Delta_0$-separation. The axioms of BST comprise, therefore, the following sentences and schemes.

(Ext') $(\forall u)(\forall v)\big[(\forall x \in u)[x \in v] \land (\forall y \in v)[y \in u] \to u = v\big]$

(Pair') $(\forall u)(\forall v)(\exists z)[u \in z \land v \in z]$

(Union') $(\forall u)(\exists z)[(\forall x \in u)[x \subseteq z]]$

$$(\Delta_0\text{–Sep})\quad (\forall v_1)\dots(\forall v_n)(\forall u)(\exists z)[(\forall x\in z)[x\in u\wedge F(x,v_1,\dots,v_n)] \\ \wedge (\forall x\in u)[F(x,v_1,\dots,v_n)\to x\in z]],$$

where $F(x,v_1,\dots,v_n)$ is a $\Delta_0$-formula, and the scheme of foundation

$$(\text{FOUND})\quad (\forall\vec{v})\big[(\exists x)F(x,\vec{v})\to(\exists x)[F(x,\vec{v})\wedge(\forall y\in x)[\neg F(y,\vec{v})]]\big].$$

We have not included the axiom of infinity into BST. Therefore, we need an axiom that secures the existence of at least one set. We add the axiom

$$(\text{Nullset})\quad (\exists x)(\forall y)[y\notin x]$$

securing the existence of the empty set.

Augmenting BST by absolute collection, i.e.,

$$(\Delta_0\text{–Coll})\quad (\forall\vec{v})(\forall u)[(\forall x\in u)(\exists y)F(x,y,\vec{v})\to(\exists z)(\forall x\in u)(\exists y\in z)F(x,y,\vec{v})]$$

for $\Delta_0$-formulas $F(x,\vec{v})$, we obtain the axiom system KP of Kripke–Platek set theory. Adding also the axiom

$$(\text{Inf'})\quad (\exists z)[z\in Lim]$$

which replaces the axiom of infinity (and makes axiom (Nullset) superfluous) we obtain the system KP$\omega$ of Kripke–Platek set theory with infinity. These systems are profoundly studied in Barwise's book [4]. A transitive set $\mathbb{A}$ is called *admissible* if $\mathbb{A}\models$ KP, if $\mathbb{A}\models$ KP$\omega$ we call $\mathbb{A}$ *admissible* above $\omega$. An ordinal $\alpha$ is called *admissible* if $L_\alpha\models$ KP and *admissible above* $\omega$ if $L_\alpha\models$ KP$\omega$. Admissible sets and ordinals are important because they possess enough closure properties to develop an abstract recursion theory.

The theory of admissible sets and ordinals is not the topic of this book. Those who are interested in learning more about admissible sets and recursion theory on admissible ordinals are recommended to consult Barwise's and Hinman's books [4] and [45]. We will restrict ourselves to sketch the main features which are needed for our coming studies.

First we check the closure properties of admissible sets.

**11.4.1 Lemma** *Any admissible set is closed under pairs, ordered pairs, unions, intersections, and Cartesian products.*

*Proof* Let $\mathbb{A}$ be an admissible set and $u,v\in\mathbb{A}$. By (Pair') there is an $a\in\mathbb{A}$ such that $u,v\in a$. By ($\Delta_0$–Sep) we obtain $\{x\in a\,|\,x=u\vee x=v\}\in\mathbb{A}$. Hence $\{u,v\}\in\mathbb{A}$.

If $u\in\mathbb{A}$ we obtain by (Union') a $v\in\mathbb{A}$ such that $(\forall x\in u)[x\subseteq v]$. By ($\Delta_0$-Sep) we get $\bigcup u=\{x\in v\,|\,(\exists y\in u)[x\in y]\}\in\mathbb{A}$.

Since $(u,v)=\{\{u\},\{u,v\}\}$ we obtain for $u,v\in\mathbb{A}$ also $(u,v)\in\mathbb{A}$. So we have closure under ordered pairs.

For sets $a\in\mathbb{A}$ such that $a\neq\emptyset$ we get

$$\bigcap a=\{x\,|\,(\forall y\in a)[x\in y]\}=\{x\in\bigcup a\,|\,(\forall y\in a)[x\in y]\}$$

which exists by closure under unions and ($\Delta_0$–Sep). So we have closure under intersections.

Observe that for $a, b \in \mathbb{A}$ also $a \cup b = \bigcup\{a,b\} \in \mathbb{A}$ and $a \cap b = \bigcap\{a,b\} \in \mathbb{A}$, which shows that we also have closure under finite unions and intersections.

Finally we want to show that admissible sets are closed under Cartesian products. This is not immediately clear from the defining axioms, since there is no power set axiom in KP. For $a, b \in \mathbb{A}$ we have

$$a \times b = \{(u,v) \mid u \in a \wedge v \in b\}. \tag{i}$$

If we succeed in finding a set $c \in \mathbb{A}$ such that

$$a \times b = \{x \in c \mid x = (u,v) \wedge u \in a \wedge v \in b\} \tag{ii}$$

we obtain $a \times b \in \mathbb{A}$ by ($\Delta_0$-Sep). Since we have closure under ordered pairs we obtain for every $x \in a$

$$(\forall y \in b)(\exists z)[z = (x,y)] \tag{iii}$$

which by ($\Delta_0$–Coll) entails

$$(\forall x \in a)(\exists w_x)(\forall y \in b)(\exists z \in w_x)[z = (x,y)]. \tag{iv}$$

Applying ($\Delta_0$–Coll) again we obtain

$$(\exists c_0)(\forall x \in a)(\exists w \in c_0)(\forall y \in b)(\exists z \in w)[z = (x,y)]. \tag{v}$$

Since $\mathbb{A}$ is closed under unions we have $c := \bigcup c_0 \in \mathbb{A}$ and obtain from (v)

$$(\forall x \in a)(\forall y \in b)[(x,y) \in c].$$

Hence

$$a \times b = \{x \in c \mid x = (u,v) \wedge u \in a \wedge v \in b\} \in \mathbb{A}. \qquad \square$$

For a $\mathscr{L}(\in)$-formula $F$ and a set $a$ we denote by $F^a$ the formula, which is obtained from $F$ by restricting all quantifiers to $a$. It is obvious that $F^a$ is always a $\Delta_0$-formula.

The fact that $\Sigma$-formulas are upwards - and $\Pi$-formulas downwards persistent can be formally expressed by

$$F^a \wedge a \subseteq b \;\rightarrow\; F^b \tag{11.3}$$

and

$$F^a \;\rightarrow\; F \tag{11.4}$$

for $\Sigma$-formulas $F$ and

$$F^b \wedge a \subseteq b \;\rightarrow\; F^a \tag{11.5}$$

and

$$F \;\rightarrow\; F^a \tag{11.6}$$

for $\Pi$-formulas. Observe that (11.3) through (11.6) are theorems of pure logic. The proof is a straightforward induction on the complexity of the formula $F$.

**11.4.2 Theorem** *($\Sigma$-reflection) Let $F(\vec{v})$ be a $\Sigma$-formula with only the shown variables free. Then*

$$\mathsf{KP} \vdash (\forall \vec{v})(\exists a)[F(\vec{v}) \leftrightarrow F(\vec{v})^a].$$

*Proof* As just stated the direction from right to left holds for logical reasons. We prove the opposite direction by induction on the complexity of the $\Sigma$-formula $F$ and argue in an arbitrary model $\mathbb{A}$ of $\mathsf{KP}$.

If $F(\vec{v})$ is $\Delta_0$ then $F(\vec{v})^a$ is equal to $F(\vec{v})$ and we have nothing to show. If $F(\vec{v})$ is a Boolean combination, say $F(\vec{v}) \equiv F_0(\vec{v}) \wedge F_1(\vec{v})$, we obtain by the induction hypothesis sets $a_0$ and $a_1$ in $\mathbb{A}$ such that $F_0(\vec{v}) \leftrightarrow F_0(\vec{v})^{a_0}$ and $F_1(\vec{v}) \leftrightarrow F_1(\vec{v})^{a_1}$. But then $a_i \subseteq a := a_0 \cup a_1 \in \mathbb{A}$ and the claim follows from (11.3).

So assume that $F(\vec{v})$ is a formula $(\forall x \in v_1)F_0(x,\vec{v})$. Let $\vec{v} \in \mathbb{A}$ such that $F(\vec{v})$ is true in $\mathbb{A}$. Then we have by induction hypothesis

$$(\forall x \in v_1)(\exists a_0)F_0(x,\vec{v})^{a_0}. \tag{i}$$

By ($\Delta_0$–Coll) we obtain from (i)

$$(\exists z)(\forall x \in v_1)(\exists a_0 \in z)F_0(x,\vec{v})^{a_0}. \tag{ii}$$

But then $a_0 \subseteq a := \bigcup z \in \mathbb{A}$ and together with (11.3) we get

$$(\exists a)(\forall x \in v_1)F(x,\vec{v})^a. \tag{iii}$$

Let finally $F(\vec{v}) \equiv (\exists y)F_0(y,\vec{v})$ and $\vec{v} \in \mathbb{A}$ such that $\mathbb{A} \models F(\vec{v})$. By induction hypothesis we have

$$(\exists y)F_0(y,\vec{v}) \to (\exists y)(\exists a_0)F_0(y,\vec{v})^{a_0}. \tag{iv}$$

Let $y \in \mathbb{A}$ be a witness for $(\exists y)F_0(y,\vec{v})$ and $a := a_0 \cup \{y\}$. Then we have $a_0 \subseteq a \in \mathbb{A}$ and $y \in a$ which imply

$$(\exists a)(\exists y \in a)F_0(y,\vec{v})^a.$$ □

**11.4.3 Theorem** *($\Sigma$-Collection) Let $F(x,y,\vec{v})$ be a $\Sigma$-formula. Then* $\mathsf{KP}$ *proves*

$$(\forall \vec{v})\big[(\forall x \in a)(\exists y)[F(x,y,\vec{v})] \to (\exists b)[(\forall x \in a)(\exists y \in b)[F(x,y,\vec{v})] \\ \wedge (\forall y \in b)(\exists x \in a)F(x,y,\vec{v})]\big].$$

*Proof* We argue in an arbitrary model $\mathbb{A}$ of $\mathsf{KP}$. Since $(\forall x \in a)(\exists y)[F(x,y,\vec{v})]$ is a $\Sigma$-formula we find a set $c$ such that $(\forall x \in a)(\exists y \in c)[F(x,y,\vec{v})^c]$ by $\Sigma$-reflection. Put

$$b := \{y \in c \mid (\exists x \in a)[F(x,y,\vec{v})^c]\}.$$

Then $b \in \mathbb{A}$ by ($\Delta_0$–Sep). Since $F(x,y,\vec{v})^c \to F(x,y,\vec{v})$ by (11.14) we finally get $(\forall x \in a)(\exists y \in b)F(x,y,\vec{v})$ and $(\forall y \in b)(\exists x \in a)F(x,y,\vec{v})$. □

**11.4.4 Theorem** *($\Delta$-Separation) Let $F(x,\vec{v})$ be a $\Sigma$- and $G(x,\vec{v})$ a $\Pi$-formula such that* $\mathsf{KP} \vdash (\forall\vec{v})(\forall x)[F(x,\vec{v}) \leftrightarrow G(x,\vec{v})]$*. Then* KP *proves that $\{x \in u \mid F(x,\vec{v})\}$ is a set. I.e., more precisely,*

$$\mathsf{KP} \vdash (\forall\vec{v})(\forall u)(\exists z)[z = \{x \in u \mid F(x,\vec{v})\}].$$

*Proof* We argue in an arbitrary model $\mathbb{A}$ of KP. Let $d := \{x \in u \mid F(x,\vec{v})\}$. By $\Sigma$-reflection we find a set $c \in \mathbb{A}$ such that

$$(\forall x \in u)[F(x,\vec{v})^c \vee \neg G(x,\vec{v})^c]. \tag{i}$$

Then

$$z := \{x \in u \mid F(x,\vec{v})^c\} \in \mathbb{A} \tag{ii}$$

by ($\Delta_0$-Sep). If $x \in z$ we get $F(x,\vec{v})^c$ which by upwards persistency entails $F(x,\vec{v})$. Hence $x \in d$. If $x \in d$ we have $x \in u$ and $F(x,\vec{v})$ which by hypothesis entails $G(x,\vec{v})$. By downward persistency it follows $G(x,\vec{v})^c$ which by (ii) entails $F(x,\vec{v})^c$. Hence $x \in z$. So $d = z \in \mathbb{A}$. □

**11.4.5 Theorem** *($\Sigma$-replacement) Let $F(x,y,\vec{v})$ be a $\Sigma$-formula, then* KP *proves*

$$(\forall\vec{v})\big[(\forall x \in a)(\exists! y)[F(x,y,\vec{v})] \;\rightarrow\; (\exists f)[Fun(f) \wedge dom(f) = a \wedge (\forall x \in a)[F(x,f(x),\vec{v})]]\big].$$

*Proof* We argue in an arbitrary model $\mathbb{A}$ of KP. Assume $(\forall x \in a)(\exists! y)[F(x,y,\vec{v})]$. By $\Sigma$-collection there is a $b \in \mathbb{A}$ such that $(\forall x \in y)(\exists y \in b)[F(x,y,\vec{v})]$. Because of the uniqueness condition we obtain

$$\begin{aligned} f &:= \{(x,y) \in a \times b \mid F(x,y,\vec{v})\} \\ &= \{(x,y) \in a \times b \mid (\forall z)[F(x,z,\vec{v}) \rightarrow z = y].\} \end{aligned}$$

Hence $f \in \mathbb{A}$ by $\Delta$-Separation and $f$ is a function with domain $a$ which satisfies the claim. □

**11.4.6 Theorem** *(Strong $\Sigma$-replacement) Let $F(x,y,\vec{z})$ be a $\Sigma$-formula. Then* KP *proves*

$$(\forall\vec{v})\big[(\forall x \in a)(\exists y)[F(x,y,\vec{v})] \rightarrow (\exists f)[Fun(f) \wedge dom(f) = a \wedge (\forall x \in a)[f(x) \neq \emptyset] \wedge (\forall x \in a)(\forall y \in f(x))[F(x,y,\vec{v})]]\big]$$

*Proof* We argue in an arbitrary KP model $\mathbb{A}$. Assume $(\forall x \in a)(\exists y)[F(x,y,\vec{v})]$. By $\Sigma$-collection there is a set $b \in \mathbb{A}$ such that

$$(\forall x \in a)(\exists y \in b)[F(x,y,\vec{v})] \tag{i}$$

and

$$(\forall y \in b)(\exists x \in a)[F(x,y,\vec{v})]. \tag{ii}$$

By $\Sigma$-reflection we find a $c \in \mathbb{A}$ such that

$$(\forall x \in a)(\exists y \in b)[F(x,y,\vec{v})]^c \text{ and } (\forall y \in b)(\exists x \in a)[F(x,y,\vec{v})]^c. \tag{iii}$$

For $x \in a$ we obtain $d_x := \{y \in b \mid F(x,y,\vec{b})^c\} \in \mathbb{A}$. Since $d_x$ is uniquely determined by $x \in a$ we obtain by $\Sigma$-replacement a function $f \in \mathbb{A}$ such that $dom(f) = a$ and $f(x) = d_x$. The function $f$ then obviously satisfies the claim. □

It is a useful observation that we may extend the language of $\mathscr{L}(\in)$ by new relation and function symbols without altering the class of $\Delta$- and $\Sigma$-formulas, provided that the new relation symbols have a $\Delta$-definition and the new function symbols a $\Sigma$-definition. We state the theorem omitting its (more or less standard) proof, which can be found in [4].

**11.4.7 Theorem** *Let $F(\vec{v})$ be a $\Sigma$-formula and $G(\vec{v})$ be a $\Pi$-formula such that* KP *proves $(\forall\vec{v})[F(\vec{v}) \leftrightarrow G(\vec{v})]$. Moreover let $F'(\vec{v},y)$ be a $\Sigma$-formula such that* KP *proves $(\forall\vec{v})(\exists! y)F'(\vec{v},y)$. We extend the language $\mathscr{L}(\in)$ to the language $\mathscr{L}(\in)^*$ by adding new relation symbols $\mathsf{R}_F$ and new function symbols $\mathsf{F}_{F'}$. Let $\mathsf{KP}^*$ be the theory in the language $\mathscr{L}(\in)^*$ which comprises all axioms of* KP *together with the defining axioms*

$$(\forall\vec{v})[\mathsf{R}_F(\vec{v}) \leftrightarrow F(\vec{v})]$$

*and*

$$(\forall\vec{v})F'(\vec{v},\mathsf{F}_{F'}(\vec{v}))$$

*for the new symbols. We call $\mathsf{R}_F$ a $\Delta$-relation symbol and $\mathsf{F}_{F'}$ a $\Sigma$-function symbol. The theory $\mathsf{KP}^*$ is an extention by definitions of* KP*, i.e., if $\mathsf{KP}^* \vdash H$ for an $\mathscr{L}(\in)$-formula $H$ then also $\mathsf{KP} \vdash H$ and for every $\mathscr{L}(\in)^*$-formula $H$ there is an $\mathscr{L}(\in)$-formula $H'$ such that $\mathsf{KP}^* \vdash H \leftrightarrow H'$. Moreover, if $H$ is a $\Sigma$-formula then $H'$ is again a $\Sigma$-formula and if $H$ is a $\Delta$-formula then $H'$ is also a $\Delta$-formula, i.e., the classes of $\Sigma$-, $\Pi$- and $\Delta$-formulas are preserved under such extensions.*

Although class terms are not part of the language of $\mathsf{KP}\omega$ we will sometimes use class terms of the form $\{x \mid F(x)\}$. The formula $s \,\varepsilon\, \{x \mid F(x)\}$ will then be an "abbreviation" for the formula $F(s)$. We will sometimes even write $s \in \{x \mid F(x)\}$.

The next theorem we are aiming at is the $\Sigma$-recursion theorem. It is the $\Sigma$-recursion theorem that accounts for the fact that admissible sets are good candidates for generalizations of recursion theory. The first step is to show that $\mathsf{KP}\omega$ proves the existence of the transitive closure of a set.[3] By the transitive closure $TC(a)$ of a set $a$ we mean the least transitive set $b$ such that $a \subseteq b$.

**11.4.8 Theorem** *The theory $\mathsf{KP}\omega$ proves that every set possesses a transitive closure.*

[3] This is even true for KP. The proof there needs, however, a bit more care [4].

*Proof* We work informally in $\mathsf{KP}\omega$. Let

$$\alpha = \omega \;:\Leftrightarrow\; \alpha \in Lim \wedge (\forall \xi \in \alpha)[\xi \notin Lim].$$

Using (Inf') we obtain $(\exists! \alpha)[\alpha = \omega]$, i.e., $\mathsf{KP}\omega$ proves the existence of $\omega$. Let

$$P(f,n,a) \;:\Leftrightarrow\; Fun(f) \wedge n \in \omega \wedge dom(f) = n+1 \\ \wedge f(0) = a \wedge (\forall m \in n)[f(m+1) = \bigcup f(m)].$$

Then we obtain

$$P(f,n,a) \wedge P(g,n,a) \;\Rightarrow\; f = g \tag{i}$$

by $\in$-induction and again by $\in$-induction

$$(\forall n \in \omega)(\exists f) P(f,n,a) \tag{ii}$$

expanding $f$ with domain $n+1$ to $f \cup \{(n+1, \bigcup f(n))\}$. By (i) and (ii) we have $(\forall n \in \omega)(\exists! f) P(f,n,a)$ and by $\Sigma$-replacement (Theorem 11.4.5) we obtain a function $F$ such that $dom(F) = \omega$, $(\forall n \in \omega) P(F(n),n,a)$ and $F(m) = F(n) \restriction (m+1)$ for $m < n$. Now define $TC(a) := \bigcup_{n \in \omega} F(n)(n)$. Then we obtain $a = F(0)(0) \subseteq F(1)(1) = \bigcup F(0)(0)$. Hence $a \subseteq TC(a)$. If $x \in y \in TC(a)$ there is a $n \in \omega$ such that $y \in F(n)(n)$ and we obtain $x \in \bigcup F(n)(n) = F(n+1)(n+1)$, which shows $x \in \bigcup_{n \in \omega} F(n)(n)$. Hence $Tran(TC(a))$. If $a \subseteq b$ and $Tran(b)$ we show

$$F(n)(n) \subseteq b \tag{iii}$$

by induction on $n$. $F(0)(0) = a \subseteq b$ holds by hypothesis. By induction hypothesis we have $F(n)(n) \subseteq b$ which entails $F(n+1)(n+1) = \bigcup F(n)(n) \subseteq \bigcup b \subseteq b$ by transitivity of $b$. This proves that $TC(a)$ is the least transitive set which comprises $a$. □

**11.4.9 Theorem** *(Induction along the transitive closure) The theory* $\mathsf{KP}$ *proves*

$$(\forall \vec{v})\big[(\forall x)[(\forall y \in TC(x)) F(y,\vec{v}) \to F(x,\vec{v})] \;\to\; (\forall x) F(x,\vec{v})\big].$$

*Proof* By $\in$-induction we prove

$$(\forall x)(\forall y \in TC(x)) F(y,\vec{v}). \tag{i}$$

The claim follows from (i) because $x \in TC(\{x\})$. To prove (i) we have the induction hypothesis

$$(\forall z \in x)(\forall y \in TC(z)) F(y,\vec{v}). \tag{ii}$$

From (ii) we get $(\forall z \in x) F(z,\vec{v})$. Therefore, we have $F(y,\vec{v})$ for all $y$ in the set $x \cup \bigcup \{TC(z) \mid z \in x\} = TC(x)$. □

The next theorem is the most important theorem of Kripke–Platek set theory.

**11.4.10 Theorem** *($\Sigma$-recursion theorem). Let G be an $n+2$-ary $\Sigma$-function symbol. Then* $\mathsf{KP}^*$ *proves the existence of an $n+1$-ary $\Sigma$-function symbol F satisfying*

$$F(\vec{v},x) = G(\vec{v},x,F{\restriction}TC(x)),$$

*where* $F{\restriction}TC(x) := \{(z,F(\vec{v},z)) \mid z \in TC(x)\}$.

*Proof* We work informally in $\mathsf{KP}^*$. The proof is essentially the familiar proof of transfinite recursion with attention paid to the restricted means of $\mathsf{KP}^*$ and to the complexity of the involved definitions. Let

$$P(f,y,\vec{v},z) \;:\Leftrightarrow\; Fun(f) \wedge dom(f) = TC(y) \wedge (\forall u \in dom(f))[f(u) = G(\vec{v},u,f{\restriction}TC(u))] \wedge z = G(\vec{v},y,f).$$

First we show by induction on $TC(y)$

$$P(f,y,\vec{v},z) \wedge P(f',y,\vec{v},z') \;\Rightarrow\; f = f' \wedge z = z'. \tag{i}$$

We have the induction hypothesis

$$(\forall u \in TC(y))(\forall f)(\forall f')(\forall z)(\forall z')[P(f,u,\vec{v},z) \wedge P(f',u,\vec{v},z') \;\Rightarrow\; f = f' \wedge z = z']. \tag{ii}$$

From $u \in TC(y) \wedge P(f,y,\vec{v},z)$ we obtain $P(f{\restriction}TC(u),u,\vec{v},f(u))$ by definition of $P$. Therefore, we obtain by the induction hypothesis $f{\restriction}TC(u) = f'{\restriction}TC(u)$ and $f(u) = f'(u)$ for all $u \in TC(y)$. Hence $f = f'$ which implies $z = G(\vec{v},y,f) = G(\vec{v},y,f') = z'$.

Next we show

$$(\forall y)(\exists f)(\exists z)P(f,y,\vec{v},z) \tag{iii}$$

by induction on $TC(y)$. We have the induction hypothesis

$$(\forall u \in TC(y))(\exists f_u)(\exists z_u)P(f_u,u,\vec{v},z_u) \tag{iv}$$

which by (i) entails

$$(\forall u \in TC(y))(\exists! f_u)(\exists! z_u)P(f_u,u,\vec{v},z_u). \tag{v}$$

Using $\Sigma$-replacement we obtain the function $f := \{(u,z_u) \mid u \in TC(y)\}$. Let $u \in TC(y)$. For $v \in TC(u) \subseteq TC(y)$ we have $P(f_v,v,\vec{v},z_v)$ and obtain $f_v = f_u{\restriction}TC(v)$ by induction on $TC(v)$ because $x \in TC(v)$ implies $f_v(x) = G(\vec{v},x,f_v{\restriction}TC(x)) = G(\vec{v},x,f_u{\restriction}TC(x)) = f_u(x)$. By (v) we now obtain $P(f_u{\restriction}TC(v),v,\vec{v},z_v)$, i.e., $f_u(v) = z_v = f(v)$. Hence $f{\restriction}TC(u) = f_u$. Again by (v) we have $P(f{\restriction}TC(u),u,\vec{v},z_u)$ which implies $f(u) = z_u = G(\vec{v},u,f{\restriction}TC(u))$. Therefore, $f(u) = z_u = G(\vec{v},u,f{\restriction}TC(u))$ is true for all $u \in TC(y)$ which shows $P(f,y,\vec{v},G(\vec{v},y,f))$ and terminates the proof of (iii).

By (iii) and (i) we have

$$(\forall y)(\exists! z)(\exists f)P(f,y,\vec{v},z) \tag{vi}$$

and we may introduce a $\Sigma$-function symbol $F$ together with the defining axiom

$$F(\vec{v},y) = z \;\Leftrightarrow\; (\exists f)P(f,y,\vec{v},z). \tag{vii}$$

Then we have

$$F(\vec{v},y) = G(\vec{v},y,f) \text{ if } P(f,y,\vec{v},G(\vec{v},y,f)). \tag{viii}$$

But $P(f,y,\vec{v},G(\vec{v},y,f))$ implies $P(f\restriction TC(u),u,\vec{v},f(u))$ for $u \in TC(y)$. Hence $f(u) = F(\vec{v},u)$ for all $u \in TC(y)$, i.e., $f = \{(u,F(\vec{v},u)) \mid u \in TC(y)\}$. So we have

$$F(\vec{v},y) = G(\vec{v},y,\{(u,F(\vec{v},u)) \mid u \in TC(y)\})$$

which finishes the proof. □

There are variations and consequences of the $\Sigma$-recursion theorem.

**11.4.11 Theorem** *Let $G$ be an $n+2$-ary $\Sigma$-function symbol. Then* KP$^*$ *proves the existence of an $n+1$-ary $\Sigma$-function symbol $F$ such that*

$$F(\vec{v},y) = G(\vec{v},y,\{(z,F(\vec{v},z)) \mid z \in y\}).$$

*Proof* Define $G'(\vec{v},y,f) := G(\vec{v},y,f\restriction y)$. Then we obtain by Theorem 11.4.10 a $\Sigma$-function $F$ such that

$$F(\vec{v},y) = G'(\vec{v},y,\{(z,F(\vec{v},z)) \mid z \in TC(y)\}) = G(\vec{v},y,\{(z,F(\vec{v},z)) \mid z \in y\}). \quad \square$$

Another special form is the definition of $\Sigma$-functions on ordinals. Here we start with a $\Sigma$-function $G$ and define

$$F(\vec{v},\alpha) = G(\vec{v},\alpha,F\restriction\alpha). \tag{11.7}$$

Assertion (11.7) is mostly applied in the following form.

**11.4.12 Theorem** *Let $G_0$, $G_S$, and $G_L$ be $\Sigma$-function symbols. Then* KP$^*$ *proves the existence of a $\Sigma$-function symbol $F$ such that $dom(F) = On$ and*

*(0)* $F(\vec{v},0) = G_0(\vec{v})$,

*(S)* $F(\vec{v},\alpha+1) = G_S(\vec{v},\alpha+1,F(\alpha))$,

*(L)* $F(\vec{v},\lambda) = G_L(\vec{v},\lambda,F\restriction\lambda)$ *for* $\lambda \in Lim$.

*Proof* Working informally in KP$^*$ we define a function symbol $G'$ by

$$G'(\vec{v},\alpha,f) := \begin{cases} G_0(\vec{v}) & \text{if } \alpha = 0 \\ G_S(\vec{v},\alpha,f(\beta)) & \text{if } \alpha = \beta+1 \\ G_L(\vec{v},\alpha,f) & \text{if } \alpha \in Lim. \end{cases}$$

It is obvious that $G'$ is again a $\Sigma$-function symbol. Applying (11.7) to $G'$ yields the claim. □

It is also obvious that we can apply the $\Sigma$-recursion Theorem in the form that for a $\Sigma$-function symbol $G$, we obtain a new $\Sigma$-function symbol $F$ satisfying

$$F(y,a_1,\dots,a_n) = G(F[y],y,a_1,\dots,a_n) \tag{11.8}$$

where $F[y] := \{F(x,a_1,\dots,a_n) \mid x \in y\}$. Just define

$$G'(a_1,\dots,a_n,y,f) := G(a_1,\dots,a_n,y,rng(f \restriction y))$$

and apply the $\Sigma$-recursion theorem.

An important application of the $\Sigma$-recursion theorem is the implicit definition of $\Delta$-predicate symbols.

**11.4.13 Theorem** *Let $P$ be a $\Delta$-predicate symbol. Then $\mathsf{KP}^*$ proves the existence of a $\Delta$-predicate symbol $R$ satisfying*

$$R(\vec{v},y) \;\Leftrightarrow\; P(\vec{v},y,\{z \in TC(y) \mid R(\vec{v},z)\}).$$

*Proof* Let

$$G(\vec{v},y,f) := \begin{cases} 0 & \text{if } P(\vec{v},y,\{x \in TC(y) \mid f(x) = 0\}) \\ 1 & \text{if } \neg P(\vec{v},y,\{x \in TC(y) \mid f(x) = 0\}). \end{cases}$$

Then

$$\begin{aligned} G(\vec{v},y,f) = z \;\Leftrightarrow\; (\exists u)\big[&(\forall v \in u)(f(v) = 0 \wedge v \in TC(y)) \\ &\wedge (\forall v \in TC(y))(f(v) = 0 \rightarrow v \in u) \\ &\wedge \big((z = 0 \wedge P(\vec{v},y,u)) \vee (z = 1 \wedge \neg P(\vec{v},y,u))\big)\big]. \end{aligned}$$

Hence $G$ is a $\Sigma$-function symbol. By the $\Sigma$-recursion Theorem we obtain a $\Sigma$-function symbol $F$ such that $F(\vec{v},y) \in 2$ and

$$F(\vec{v},y) = 0 \;\Leftrightarrow\; G(\vec{v},y,F \restriction TC(y)) = 0 \;\Leftrightarrow\; P(\vec{v},y,\{z \in TC(y) \mid F(\vec{v},z) = 0\}). \quad \text{(i)}$$

Defining

$$R(\vec{v},y) \;:\Leftrightarrow\; F(\vec{v},y) = 0$$

we see by (i) that $R$ satisfies the recursion condition of the theorem. Since

$$\neg R(\vec{v},y) \;\Leftrightarrow\; F(\vec{v},y) = 1$$

we see that $R$ is a $\Delta$-relation symbol. □

An important consequence of the $\Sigma$-recursion theorem is that the constructible hierarchy can be developed within the framework on $\mathsf{KP}$. Ruminating Definition 11.3.2 we see that the stages $L_\alpha$ of the constructible hierarchy are definable by $\Sigma$-recursion as soon as it is secured that $a \mapsto Def(a)$ is a $\Sigma$-function. There are different methods to secure that. One is using Gödel-functions, another to goedelize the language of set theory, i.e., to code every $\mathcal{L}(\in)$-formula $F$ by a set $\ulcorner F \urcorner$, and then to show that the satisfaction predicate

$$Sat(a,\ulcorner F \urcorner,\vec{b}) \;:\Leftrightarrow\; (a,\in) \models F[\vec{b}]$$

is a $\Delta$-predicate. We will not give the details here. They can be found in the standard literature, e.g. [4] or [21]. The function $\alpha \mapsto L_\alpha$ is thus a $\Sigma$-function of $\mathsf{KP}$.

## 11.5 $\mathsf{ID}_1$ as a Subtheory of $\mathsf{KP}\omega$

In this section, we want to sketch that $\mathsf{ID}_1$ may be regarded as a subtheory of $\mathsf{KP}\omega$. The first step is to ensure that all primitive recursive functions are sets in $\mathsf{KP}\omega$. We take finite ordinals, i.e., ordinals $< \omega$, to represent natural numbers. Remember that $\omega$ is a set in $\mathsf{KP}\omega$. Likewise all finite ordinals $k < \omega$ are sets in $\mathsf{KP}\omega$, we put $0 := \emptyset$ and $k+1 := \{0,\dots,k\}$. In the remainder of this section $m$, $n$, $k$, $n_0$,... denote finite ordinals.

We define $n$-tuples in the usual way by

$$(x_1,\dots,x_{n+1}) := ((x_1,\dots,x_n),x_{n+1})$$

using the pairing function of set theory. We leave it as an exercise to check that the predicates

$$a \text{ is an } n\text{-tuple} \quad :\Leftrightarrow \quad a = (x_1,\dots,x_n) \text{ for some } x_1,\dots,x_n$$

and

$$P^n_i(a) = x_i \quad :\Leftrightarrow \quad a = (x_1,\dots,x_n) \text{ for some } x_1,\dots,x_{i-1},x_{i+1},\dots,x_n$$

are again $\Delta_0$-notions. Likewise we define the $k$-fold Cartesian product of sets by

$$a_1 \times \dots \times a_k := \{(x_1,\dots,x_k) \mid \bigwedge_{i=1}^{k} x_i \in a_i\} = (a_1 \times \dots \times a_{k-1}) \times a_k.$$

It follows by a meta-induction on $k$ that $\mathsf{KP}$ also proves closure under $k$-fold Cartesian products.

We are going to show that all primitive recursive functions are sets in $\mathsf{KP}\omega$. We obtain

$$S = \{z \in \omega \times \omega \mid (\exists x \in \omega)[z = (x, x \cup \{x\})]\}.$$

So the successor function $S$ on natural numbers is a set by $\Delta_0$-separation. Put

$$C^n_k := \{z \in \omega^{n+1} \mid (\exists x_1 \in \omega)\dots(\exists x_n \in \omega)[z = (x_1,\dots,x_n,k)]\}$$

and

$$P^n_k := \{z \in \omega^{n+1} \mid (\exists x_1 \in \omega)\dots(\exists x_n \in \omega)[z = (x_1,\dots,x_n,x_k)]\}$$

showing that $C^n_k$ and $P^n_k$ are sets by $\Delta_0$-separation. Now assume that we have an $m$-ary function $g$ and $n$-ary functions $h_1,\dots,h_m$ on natural numbers as sets. Then let

$$Sub(g,h_1,\dots,h_m) := \{z \in \omega^{n+1} \mid (\exists x_1 \in \omega)\dots(\exists x_n \in \omega)(\exists z_1 \in \omega)\dots(\exists z_m \in \omega)(\exists y \in \omega)$$
$$[\bigwedge_{i=1}^{m} h_i(x_1,\dots,x_n) = z_i \wedge g(z_1,\dots,z_m) = y \wedge z = (x_1,\dots,x_n,y)]\}.$$

So $Sub(g,h_1,\dots,h_m)$ is a set by $\Delta_0$-separation. Finally assume that we have an $n$-ary function $g$ and an $n+2$-ary function $h$ on the natural numbers. Then put

$$\begin{aligned}Rec(g,h) := \{z \in \omega^{n+2} \mid (\exists u \in \omega)(\exists x_1 \in \omega)\dots(\exists x_n \in \omega)(\forall f)[&Fun(f) \wedge \\ &dom(f) = \omega \wedge f(0) = g(x_1,\dots,x_n) \wedge \\ &(\forall n \in \omega)[f(n+1) = h(n,f(n),x_1,\dots,x_n)] \wedge \\ &z = (u,x_1,\dots,x_n,f(u))\}\end{aligned}$$

Since then also

$$\begin{aligned}Rec(g,h) = \{z \in \omega^{n+2} \mid (\exists u \in \omega)(\exists x_1 \in \omega)\dots(\exists x_n \in \omega)(\forall f)\big[&Fun(f) \wedge \\ &dom(f) = \omega \wedge f(0) = g(x_1,\dots,x_n) \wedge \\ &(\forall n \in \omega)[f(n+1) = h(n,f(n),x_1,\dots,x_n)] \\ &\quad \to z = (u,x_1,\dots,x_n,f(u))\big]\}\end{aligned}$$

we obtain $Rec(g,h)$ as a set by $\Delta$-separation. Therefore, all primitive recursive functions are available in $\mathsf{KP}\omega$. This makes it easy to embed the first-order language $\mathscr{L}(\mathsf{NT})$ of number theory into the language of set theory. Atomic formulas $\underline{f}(x_1,\dots,x_n) = y$ and $\underline{f}(x_1,\dots,x_n) \neq y$ are translated by replacing the function symbol $\underline{f}$ by the primitive recursive function $f$, which is available as set in $\mathsf{KP}\omega$. One asymmetry is caused by the fact that we have free predicate variables in $\mathscr{L}(\mathsf{NT})$, which we did not introduce in the language of set theory. We will therefore, for the moment, augment the language of set theory by free predicate variables, which we denote by capital letters $X, Y, \dots$. Again we often write $(a_1,\dots,a_n) \;\varepsilon\; X$ instead of $(Xa_1,\dots,a_n)$. In case that $X$ is a set we will also often just write $a \in X$. The new atomic formulas $(Xa_1,\dots,a_n)$ are counted among the $\Delta_0$-formulas of $\mathsf{KP}\omega(\vec{X})$.[4] Let us denote by $\mathsf{KP}\omega(\vec{X})$ the theory in the extended language. Since there are no defining axioms for predicate variables, it is obvious that $\mathsf{KP}\omega(\vec{X})$ is a conservative extension of $\mathsf{KP}\omega$. So far we obtain $\mathsf{NT}$ as a subtheory of $\mathsf{KP}\omega$.

**11.5.1 Theorem** *Let $F(\vec{X})$ be an $\mathscr{L}(\mathsf{NT})$-formula such that* $\mathsf{NT} \vdash F(\vec{X})$. *Then* $\mathsf{KP}\omega \vdash F(\vec{X})^\omega$.

*Proof* It suffices to check that all nonlogical axioms of $\mathsf{NT}$ are provable in $\mathsf{KP}\omega$. Obviously, we always have $S(x) \neq 0$ and from $S(x) = x \cup \{x\} = y \cup \{y\} = S(y)$ we obtain $(\forall z \in x)[z \in y]$ and $(\forall z \in y)[z \in x]$, i.e., $x = y$. So the successor axioms are provable in $\mathsf{KP}\omega$. For $x \in \omega$ we obtain $S(x) = x \cup \{x\} = x + 1 \in \omega$ by definition of $S$. Likewise we can prove that $x_1,\dots,x_n \in \omega$ implies $C_k^n(x_1,\dots,x_n) = k \in \omega$ as well as $P_k^n(x_1,\dots,x_n) = x_k \in \omega$. If $g$ and $h_1,\dots,h_m$ are given we obtain $Sub(g,h_1,\dots,h_m)(x_1,\dots,x_n) = g(h_1(x_1,\dots,x_n),\dots,h_m(x_1,\dots,x_n))$ by definition of $Sub(g,h_1,\dots,h_m)$. For given functions $g$ and $h$ we obtain by definition of $Rec(g,h)$ also $Rec(g,h)(u,x_1,\dots,x_n) = f(u)$ where $f$ is the uniquely determined function used in the definition of $Rec(g,h)$. Hence $Rec(g,h)(0,x_1,\dots,x_n) = f(0) = g(x_1,\dots,x_n)$ and

$$\begin{aligned}Rec(g,h)(k+1,x_1,\dots,x_n) &= f(k+1) = h(k,f(k),x_1,\dots,x_n) \\ &= h(k,Rec(g,h)(k,x_1,\dots,x_n),x_1,\dots,x_n).\end{aligned}$$

[4] The predicate variables should rather be viewed as predicate constants. The substitution rule $\vdash F(X) \Rightarrow \vdash F(S)$ for any "class term" $S = \{x \mid G(x)\}$ does not hold for $\mathsf{KP}\omega(\vec{X})$. It is only true for class terms that are $\Delta$-definable.

So all defining axioms for primitive recursive functions are provable in $\mathsf{KP}\omega$. It remains to check the scheme of mathematical induction. So assume $F(0,\vec{X})^\omega$ and $(\forall x\in\omega)[F(x,\vec{X})^\omega \to F(x+1,\vec{X})^\omega]$. If $(\exists x\in\omega)[\neg F(x,\vec{X})^\omega]$ then we obtain by (FOUND) a least $k<\omega$ such that $\neg F(k,\vec{X})^\omega$. The hypothesis $F(0,k)^\omega$ excludes $k=0$. Since $\omega$ is the least limit ordinal it follows $k=k_0+1$ which implies $F(k_0,\vec{X})^\omega$. The hypothesis $(\forall x\in\omega)[F(x,\vec{X})^\omega \to F(x+1,\vec{X})^\omega]$, however, leads to the contradiction $F(k,\vec{X})^\omega$. Hence $(\forall x\in\omega)F(k,\vec{X})^\omega$. □

To show that also $\mathsf{ID}_1$ is a subtheory of $\mathsf{KP}\omega$ we have to represent fixed-points of arithmetically definable inductive definitions in $\mathsf{KP}\omega$. This is prepared by the following lemma.

**11.5.2 Lemma** *Let $B(X,x,a_1,\dots,a_n)$ be a $\Delta$-formula of* $\mathsf{KP}\omega$. *Then there is a $\Sigma$-function symbol $I_B$ of* $\mathsf{KP}\omega$ *such that*

$$I_B(\alpha,a_1,\dots,a_n)=\{x\in a_1 \mid B(I_B[\alpha],x,a_1,\dots,a_n)\}$$

*where*

$$I_B[\alpha]:=\{z\in a_1 \mid (\exists\beta\in\alpha)[z\in I_B(\beta,a_1,\dots,a_n)]\}$$

*Proof* Let

$$\mathsf{G}_B(X,a_1,\dots,a_n):=\{x\in a_1 \mid B(X,x,a_1,\dots,a_n)\}. \tag{i}$$

Then

$$\mathsf{G}_B(X,a_1,\dots,a_n)=y \;\Leftrightarrow\; (\forall z\in y)[z\in a_1 \wedge B(X,z,a_1,\dots,a_n)] \wedge (\forall z\in a_1)[B(X,z,a_1,\dots,a_n)\to z\in y]$$

which shows that $\mathsf{G}_B$ is a $\Sigma$-function. By the $\Sigma$-recursion theorem (11.8) we obtain therefore a $\Sigma$-function $I_B$ such that

$$I_B(\alpha,a_1,\dots,a_n)=\mathsf{G}_B(I_B[\alpha],a_1,\dots,a_n)=\{z\in a_1 \mid B(I_B[\alpha],z,a_1,\dots,a_n)\}.\ \square$$

If $X$ occurs positively in an $\mathcal{L}(\in)$-formula $B(X,x,\vec{a})$ then the associated operator

$$\mathsf{G}_{B,\vec{a}}(X):=\{z\in a_1 \mid B(X,z,\vec{a})\}$$

is monotone. We have seen in the proof of the previous lemma that $\mathsf{G}_{B,\vec{a}}$ is a $\Sigma$-operator of $\mathsf{KP}\omega$ if $B(X,x,\vec{a})$ is $\Delta$-formula. The fact that a class $S\subseteq a_1$ is closed under the operator $\mathsf{G}_{B,\vec{a}}$ is abbreviated by

$$Cl_{B,\vec{a}}(S) \;:\Leftrightarrow\; (\forall z\in a_1)[B(S,z,\vec{a})\to z\,\varepsilon\,S].$$

Next we show that the least fixed point of $\mathsf{G}_{B,\vec{a}}$ is $\Sigma$-definable in $\mathsf{KP}\omega$. Put

$$\boldsymbol{I}_{B,\vec{a}}:=\{z\in a_1 \mid (\exists\xi)[z\in I_B(\xi,\vec{a})]\}.$$

Notice that $\boldsymbol{I}_{B,\vec{a}}$ is in general not a set in $\mathsf{KP}\omega$. The notion $z\,\varepsilon\,\boldsymbol{I}_{B,\vec{a}}$ is supposed to be an "abbreviation" for $(\exists\xi)[z\in I_B(\xi,\vec{a})]\wedge z\in a_1$.

**11.5.3 Theorem** *Let $B(x,X,\vec{a})$ be an $X$-positive $\Delta$-formula of* $\mathsf{KP}\omega$. *Then* $\mathsf{KP}\omega$ *proves*

(A) $\quad Cl_{B,\vec{a}}(\boldsymbol{I}_{B,\vec{a}})$

*and*

(B) $\quad Cl_{B,\vec{a}}(S) \rightarrow \boldsymbol{I}_{B,\vec{a}} \subseteq S.$

*Proof* Working in $\mathsf{KP}\omega$ assume $B(\boldsymbol{I}_{B,\vec{a}},x,\vec{a})$ and $x \in a_1$. Since $B(X,x,\vec{a})$ is $X$-positive, $B(\boldsymbol{I}_{B,\vec{a}},x,\vec{a})$ is a $\Sigma$-formula. By $\Sigma$-reflection there is a $c$ such that

$$B(\{z \in a_1 \mid (\exists \xi \in c)[z \in I_B(\xi,\vec{a})]\},x,\vec{a}). \tag{i}$$

Now define

$$\beta := \bigcup\{\eta \in c \mid \eta \in On\} \text{ and } \alpha = \beta \cup \{\beta\}.$$

Then $\alpha$ is a set by $\Delta_0$–separation, (Union) and (Pair) such that $c \cap On \subseteq \alpha$. By upwards persistency and monotonicity we obtain

$$B(\{z \in a_1 \mid (\exists \xi \in \alpha)[z \in I_B(\xi,\vec{a})]\},x,\vec{a}),$$

which by Lemma 11.5.2 entails $x \in I_B(\alpha,\vec{a})$ and thus $x \,\varepsilon\, \boldsymbol{I}_{B,\vec{a}}$. This proves part (A).

For the proof of part (B) assume $Cl_{B,\vec{a}}(S)$. We prove

$$I_B(\alpha,\vec{a}) \subseteq S \tag{ii}$$

by induction on $\alpha$. From the induction hypothesis we have

$$I_B[\alpha] \subseteq S$$

which by $X$-positivity implies

$$B(I_B[\alpha],x,\vec{a}) \rightarrow B(S,x,\vec{a}),$$

i.e.,

$$x \in I_B(\alpha,\vec{a}) \rightarrow B(S,x,\vec{a}). \tag{iii}$$

From (iii) and $Cl_{B,\vec{a}}(S)$ we finally obtain

$$(\forall x)[x \in I_B(\alpha,\vec{a}) \rightarrow x \,\varepsilon\, S].$$

Hence

$$\boldsymbol{I}_{B,\vec{a}} \subseteq S.$$

□

If $F(X,x,\vec{x})$ is an $X$-positive arithmetical formula then $F(X,x,\vec{x})^{\omega}$ becomes an $X$-positive $\Delta$-formula in $\mathcal{L}(\in)$, which defines a monotone operator

$$\mathsf{G}_{F,\vec{x}}\colon \mathrm{Pow}(\omega^n) \longrightarrow \mathrm{Pow}(\omega^n).$$

It follows from Theorem 11.5.3 that $\boldsymbol{I}_{F,\vec{x}}$ is the least fixed point of $\mathsf{G}_{F,\vec{x}}$. This shows that we can embed the theory $\mathsf{ID}_1$ into $\mathsf{KP}\omega$. So we have the following theorem.

**11.5.4 Theorem** *The theory* $\mathsf{ID}_1$ *is a subtheory of* KP$\omega$.

*Proof* We have already seen that all axioms of NT are provable in KP$\omega$. It follows from Theorem 11.5.3 that for any $X$-positive arithmetical formula $F(X,x,\vec{x})$, we have a $\Sigma$-definable class $I_{F^\omega,\vec{x}}$ satisfying axioms $ID_1^1$ and $ID_1^2$. □

## 11.6 Variations of KP$\omega$ and Axiom $\beta$

Among others this section prepares an ordinal analysis of the theory $(\mathsf{ATR})_0$ which has been introduced in Exercise 8.5.32.[5] The ordinal analysis will be accomplished in Chap. 12. Since $(\mathsf{ATR})_0$ is a theory that is not in the focus of this book, we will be rather sketchy here and leave proofs as exercises (with extensive hints).

**11.6.1 Exercise** Extend the language $\mathcal{L}(\in)$ of set theory to the language $\mathcal{L}^*(\in)$ by adding constants for all primitive recursive functions. Interpreting natural numbers as ordinals less than $\omega$ we obtain $\mathcal{L}(\mathsf{NT})$ as a sublanguage of $\mathcal{L}^*(\in)$. Denote by KPN the theory in the language $\mathcal{L}^*(\in)$ which comprises KP$\omega$, whose schemes are all extended to the new language, and all defining axioms for primitive recursive functions restricted to $\omega$. The theory $\mathsf{KPN}_0$ is obtained from KPN replacing the scheme of foundation by axiom (Ind).

(a) Show that NT $\vdash F$ implies KPN $\vdash F^\omega$

(b) Show that KPN is an extension by definitions of KP$\omega$.

To calibrate theories by the amount of induction they provide we define $<_{\mathbb{N}} := \{(\alpha,\beta) \in \omega \times \omega \mid \alpha \in \beta\}$ and introduce the axiom

$$(\mathrm{Ind})_\omega \quad (\forall u)\,[(\forall x)[(\forall y)(y <_{\mathbb{N}} x \to y \in u) \to x \in u)] \;\to\; (\forall x \in \omega)[x \in u]]$$

of Mathematical Induction and the corresponding scheme

$$(\mathrm{IND})_\omega \quad (\forall x \in \omega)[(\forall y)(y <_{\mathbb{N}} x \to F(y)) \to F(x)] \;\to\; (\forall x \in \omega)F(x).$$

Let $\mathsf{BST}\omega^-$ be the theory $\mathsf{BST} + (\mathrm{Inf}')$ without the foundation scheme (FOUND) and (Union') replaced by the axiom

$$(\mathrm{TranC}) \quad (\forall x)(\exists y)[\mathit{Tran}(y) \land x \subseteq y]$$

which entails that every set possesses a transitive hull. This axiom is necessary since we need foundation to prove the existence of the transitive hull. Because of $\bigcup a \subseteq$

---

[5] This section will only be needed in Chap. 12 in which we revisit predicative proof theory. Readers who are only interested in the first step into impredicativity may therefore omit this section in a first reading.

$TC(a)$, we obtain unions by $\Delta_0$-separation. The axiom (TranC) makes therefore axiom (Union) superfluous.

**11.6.2 Exercise** Show $\mathsf{BST}\omega^- \vdash (\forall x)(\exists y)[y = TC(x)]$.

Hint: By (TranC) there is a $y$ such that $Tran(y)$ and $u \subseteq y$. By (Inf') you get the existence of $\omega$. Define

$$TC(u) := \{x \in y \,|\, (\exists n \in \omega)(\exists f)[Fun(f) \wedge dom(f) = n+1] \\ \wedge f(0) \in u \wedge (\forall k \in n)[f(k+1) \in f(k) \wedge f(n) = x]\}$$

and show that the existential quantifier can be bounded.

By $\mathsf{KP}\omega^0$ we denote the theory $\mathsf{KP}\omega$ in which the scheme of foundation is replaced by the axiom $(\mathrm{Ind})_\omega$ of Mathematical Induction. By $\mathsf{KP}\omega^r$ we denote the theory $\mathsf{KP}\omega$ in which the scheme (FOUND) is replaced by the axiom (Found).

A key role in the ordinal analysis of $(\mathsf{ATR})_0$ plays Axiom $\beta$ which we are going to formulate. Express by $Wf(a,r)$ that $r$ is a well-founded binary relation on $a$, i.e.,

$$Wf(a,r) \;:\Leftrightarrow\; Rel(r) \wedge field(r) = a \wedge \\ (\forall x)\big[x \subseteq a \wedge x \neq \emptyset \rightarrow (\exists y \in x)[(\forall z \in x)[(z,y) \notin r]]\big].$$

Axiom $\beta$ postulates that every transitive well-founded relation possesses an order type, i.e., a bit more general,

$$(\mathrm{Ax}\beta) \quad (\forall a)(\forall r)\big[Wf(a,r) \rightarrow (\exists f)(\exists b)[Fun(f) \wedge dom(f) = a \wedge rng(f) = b \wedge \\ (\forall x \in a)(\forall y \in a)[(x,y) \in r \rightarrow f(x) \in f(y)]]\big].$$

The function $f$ in Axiom $\beta$ is uniquely defined and commonly called the *collapsing function* of the well-founded relation $r$. Its range is called the Mostowski-collapse of the well-founded relation $r$ (cf. Definition 3.2.9). The Mostowski-collapse of a transitive well-founded relation is an ordinal which is the order type of the relation.

To formalize admissible sets, we introduce an unary relation symbol $Ad$ with the defining axioms

$(Ad\ 1)$ $\quad (\forall x)[Ad(x) \rightarrow Tran(x)]$

$(Ad\ 2)$ $\quad (\forall x)(\forall y)[Ad(x) \wedge Ad(y) \rightarrow x \in y \vee x = y \vee y \in x]$

$(Ad\ 3)$ $\quad (\forall x)[Ad(x) \rightarrow (\mathrm{Pair'})^x \wedge (\mathrm{Union'})^x \wedge (\Delta_0\text{–Sep})^x \wedge (\Delta_0\text{–Coll})^x]$.

Of course we have to replace $(\mathrm{Union'})^x$ by $(\mathrm{TranC})^x$ for theories with restricted foundation. The formula $Ad(a)$ axiomatizes the fact that $a$ is an admissible set. Let $\mathsf{Ad}$ be the set $\{Ad\ 1,\dots,Ad\ 3\}$.

By $\mathsf{KPl}$ we denote the theory which comprises $\mathsf{BST}$, the axioms in $\mathsf{Ad}$ , and the limit axiom

(Lim) $\quad (\forall x)(\exists y)[Ad(y) \wedge x \in y]$

which postulates that the universe is a union of admissible sets. Even stronger is the theory $\mathsf{KPi}$ , which is $\mathsf{KP} + \mathsf{Ad} + \mathrm{(Lim)}$ and axiomatizes an admissible union of admissible universes, i.e., a recursively inaccessible universe. An ordinal analysis for these theories is outside a first step into predicativity. However, restricting the amount of induction – or equivalently foundation – we obtain much weaker theories.

**11.6.3 Exercise** Let

$$u = \mathbb{A}_0 \;:\Leftrightarrow\; Ad(u) \wedge (\forall x \in u)[\neg Ad(x)].$$

(a) Show that KPl proves $(\exists! u)[u = \mathbb{A}_0]$.
(b) The formula $x \in \mathbb{A}_0$ stands for $(\exists u)[u = \mathbb{A}_0 \wedge x \in u]$. Show that KPl proves $(\forall x)\big[x \in \mathbb{A}_0 \leftrightarrow (\forall u)[u = \mathbb{A}_0 \rightarrow x \in u]\big]$.
(c) Show that KPl proves $(\forall x)\big[x \in \mathbb{A}_0 \leftrightarrow (\forall u)[Ad(u) \rightarrow x \in u]\big]$.

Check if you need foundation for these results.

Hint: Observe that even in absence of foundation you have $Ad(a) \rightarrow a \notin a$ because otherwise you would get $r := \{x \in a \,|\, x \notin x\} \in a$ by $\Delta_0$–separation leading to the contradiction $r \in r \;\Leftrightarrow\; r \notin r$.

In the standard interpretation $\mathbb{A}_0$ is interpreted as the set $L_\omega$ of hereditarily finite sets.

The axiom

$$(\forall u)\big[(\forall x \in X)[(\forall y \in x)[y \in X \rightarrow y \in u] \rightarrow x \in u] \rightarrow (\forall x \in X)[x \in u]\big]$$

of $\in$-induction restricted to a class $X$ is equivalent to the axiom

$$(\text{Found}(X)) \quad (\forall u)\big[(\exists y \in u)[y \in X] \rightarrow (\exists y \in u)[y \in X \wedge (\forall z \in y)[z \in X \rightarrow z \notin u]]\big]$$

saying that $\in$ restricted to $X$ is well-founded on sets. Let $\mathsf{KPl}_0$ and $\mathsf{KPi}_0$be the theories in which the scheme of foundation is replaced by the axiom $(\text{Found}(\mathbb{A}_0))$. This is a slight generalization of the axiom of Mathematical Induction (and is in fact equivalent to it). Observe that you did not need foundation to solve Exercise 11.6.3. Therefore $\mathbb{A}_0$ is available in $\mathsf{KPl}_0$ and thus also in $\mathsf{KPi}_0$. In contrast to KPl and KPi, which are much stronger than KP$\omega$, the theories $\mathsf{KPl}_0$ and $\mathsf{KPi}_0$ are reducible to predicative theories. This will be handled in Sect. 12.7.

**11.6.4 Exercise** (a) Show that $\mathsf{KPl}_0$ proves Axiom $\beta$.[6] Show moreover that $\mathsf{KPl}_0$ proves that the "collapsing function" $f$ for a well-founded relation $r$ on a set $a$ belongs to an admissible $u$ whenever $a$ and $r$ are in $u$.
(b) Show that the MOSTOWSKI collapse of a well-ordering is an ordinal.

Hint: Assume $Wf(a,r)$. By (Lim) you find an admissible set $u$ such that $a, r \in u$. Let

$$\begin{aligned} TC_{a,r}(v) := \{x \in a \,|\, (\exists n \in \omega)(\exists f)[&Fun(f) \wedge dom(f) = n+1 \wedge f(0) = x \\ &\wedge (\forall k \in n)[(f(k+1), f(k)) \in r] \wedge f(n) \in v]\} \end{aligned}$$

denote the transitive closure of the $r$ predecessors of $v$. Define the Mostowski collapse of $r$ by the formula

[6] Observe that in $\mathsf{KPl}_0$ the existence of a collapsing function does not necessarily imply that $r$ is well-founded. The Mostowski collapse of $r$ is just a hereditarily transitive set. Only the presence of foundation makes it an ordinal. The "collapse" of a well-founded relation is, however, always an ordinal.

$$A(x,g) \;:\Leftrightarrow\; Fun(g) \wedge dom(g) = TC_{a,r}(\{x\}) \wedge (\forall y \in dom(g))[g(y) = \{g(z) \mid (z,y) \in r\}].$$

Show by induction on $r$ that there is an uniquely determined $g$ with $A(x,g)$ and find a set $c \in u$ such that

$$f := \{(x,y) \in a \times c \mid (\exists g \in c)[A(x,g) \wedge y = g(x)]\}.$$

The ordinal $\omega = \mathbb{A}_0 \cap On$ is a set in $\mathsf{KPI}_0$. Since all primitive recursive functions are definable as sets in $\mathsf{KPI}_0$ a canonical translation of the formulas of second-order arithmetic into the language of set theory is obtained by restricting all first-order quantifiers to $\omega$ and all second-order quantifiers to $\mathrm{Pow}(\omega)$. The second-order language of $\mathcal{L}(\mathsf{NT})$ becomes so a sublanguage of $\mathcal{L}(\in)$.

**11.6.5 Exercise** Show that $\mathsf{KPI}_0$ comprises $(\mathsf{ATR})_0$ (cf. Exercise 8.5.32).

Hint: The only axiom that needs checking is *(ATR)*. Let $A(Z,x)$ be an arithmetical formula, assume $WO(\omega,X)$, i.e., that $X$ is a well-ordering on $\omega$, and put $H(a,Y) \;:\Leftrightarrow\; Hier_A(X\restriction a, Y)$ where $X\restriction a := \{\langle x,y\rangle \mid \langle x,y\rangle \,\varepsilon\, X \wedge \langle y,a\rangle \,\varepsilon\, X\}$. Use induction on $X$ to show that there is a uniquely determined $Y$ satisfying $H(a,Y)$.

In $\mathsf{KPI}_0$ we can define all finite levels of the admissible hierarchy. Define

$$u = \mathbb{A}_{k+1}(u) \;:\Leftrightarrow\; Ad(u) \wedge (\forall x \in u)[\neg Ad(x) \vee \bigvee_{i=1}^{k} x = \mathbb{A}_k].$$

**11.6.6 Exercise** (a) Show that $\mathsf{KPI}_0$ proves $(\exists! y)[y = \mathbb{A}_k]$ for all natural numbers $k$.
(b) Define $x \in \mathbb{A}_k \;:\Leftrightarrow\; (\exists u)[u = \mathbb{A}_k \wedge x \in u]$ and show that $\mathsf{KPI}_0$ proves

$$(\forall x)\big[x \in \mathbb{A}_{n+1} \leftrightarrow (\forall u)[Ad(u) \wedge \bigwedge_{i=0}^{n} \mathbb{A}_i \in u \rightarrow x \in u]\big].$$

**11.6.7 Remark** The term Axiom $\beta$ has its origin in the theory of models of second-order arithmetic. A structure for second-order arithmetic has the form $\mathfrak{M} = (N, S, \cdots)$, where $N$ is the domain of the first-order quantifiers and $S \subseteq \mathrm{Pow}(N)$ the domain of the second-order quantifiers (Sect. 4.2). A structure for second-order arithmetic is a $\beta$-structure if it is absolute for well-orderings[7], i.e., $\mathfrak{M} \models (\forall X)TI(\prec, X)$ iff $\prec$ is a well-ordering in standard structure. Any model $\mathfrak{M}$ of second-order arithmetic that satisfies the axiom or the scheme of Mathematical Induction thinks that $N$ is well-ordered by the *"less than"*-relation on $N$. If $\mathfrak{M}$ is a $\beta$-structure then *"less than"* is a well-ordering in the real world, which implies that $N$ is isomorphic to $\mathbb{N}$, the standard natural numbers, i.e., that $\mathfrak{M}$ is an $\omega$-model. However, Mostowski has shown that in general there are $\omega$-models, which are not $\beta$-structures (cf. [64]).

If $\mathfrak{M}$ is a transitive model of $\mathsf{KP}\omega+$ Axiom $\beta$ then $\mathfrak{M}$ is absolute for well-orderings. This is obvious because the well-foundedness of a relation $\prec$ can then be expressed not only by the familiar $\Pi$-formula but also by saying that there is a

[7] bon-ordre is French for well-ordering, hence $\beta$.

collapsing function for $\prec$, which is a $\Sigma$-formula. So the notion of well-foundedness of relations (on sets) is expressible by a $\Delta$-formula, hence absolute.

It is a theorem of abstract recursion theory that a model of second-order arithmetic is a $\beta$-model if and only if it is closed under hyperjumps, i.e., if it is closed under $\Pi^1_1$-comprehension. On the side of set theory one hyperjump corresponds to the passage to the next admissible set. Therefore, KPl axiomatizes an universe that is "closed under hyperjumps." This is the recursion theoretic background for the fact that KPl proves Axiom $\beta$.

## 11.7 The $\Sigma$–Ordinal of KP$\omega$

In this section, we study the connection between the proof-theoretic ordinal of a theory $T$ in the language of set theory and the minimal constructible model for the $\Sigma_1$-sentences provable in $T$.

Let $a$ be a set and $\prec$ a linear ordering with $\mathit{field}(\prec) \subseteq a$. Let

$$Prog^a(\prec,S) \;:\Leftrightarrow\; (\forall x\in a)[(\forall y\in a)[y\prec x \to y\,\varepsilon\, S] \to x\,\varepsilon\, S],$$

expressing that the class term $S$ is "progressive" with respect to $\prec$ inside of $a$, and

$$TI^a(\prec,S) \;:\Leftrightarrow\; \mathit{field}(\prec)\subseteq a \wedge (Prog^a(\prec,S) \to (\forall x\in \mathit{field}(\prec))[x\,\varepsilon\, S]),$$

expressing the scheme of induction along $\prec$.

For theories in the language of set theory, we therefore modify the definition of the proof-theoretic ordinal as defined in Sect. 6.7 to

$$\|T\| := \sup\{\mathrm{otyp}(\prec) \mid \prec\subseteq \omega\times\omega \text{ is } \Delta_0\text{-definable and } T\vdash TI^{\omega}(\prec,S)\}. \tag{11.9}$$

Now put

$$\mathrm{Acc}_a(X,x,\prec) \;:\Leftrightarrow\; x\in a \wedge (\forall y\in a)[y\prec x \to y\in X]$$

Then $\mathrm{Acc}_a(X,x,\prec)$ defines an $X$-positive operator. We obtain $Cl_{\mathrm{Acc}_{a,\prec}}(S) \Leftrightarrow Prog^a(\prec,S)$. If we assume that KP$\omega$ proves the scheme of induction along $\prec$, i.e., KP$\omega \vdash TI^a(\prec,S)$ for any class term $S$, then KP$\omega$ proves especially $TI^a(\prec,\boldsymbol{I}_{\mathrm{Acc}_{a,\prec}})$. From Theorem 11.5.3 we then obtain

$$\mathsf{KP}\omega \vdash (\forall x\in a)[x\in \mathit{field}(\prec) \to (\exists\beta)[x\in I_{\mathrm{Acc}_{a,\prec}}(\beta)]]. \tag{11.10}$$

But $I_{\mathrm{Acc}_{a,\prec}}(\beta)$ is exactly the $\beta$th stage in the accessible part of the well-ordering $\prec$. We have shown in Sect. 6.5 that the order type of $x\in \mathit{field}(\prec)$ is equal to its inductive norm in the accessible part (c.f. Lemma 6.5.2). If we succeed in finding an upper bound for the ordinals $\beta$ in (11.10), we will also have an upper bound for the order type of $\prec$ and thus an upper bound for the proof-theoretic ordinal of KP$\omega$ as defined in Sect. 6.7. This is the background of the following definition.

**11.7.1 Definition** Let

$$\|\mathsf{KP}\omega\|_\Sigma := \min\{\alpha \mid L_\alpha \models F \text{ for all } \Sigma\text{-sentences } F \text{ such that } \mathsf{KP}\omega \vdash F\}.$$

We call $\|\mathsf{KP}\omega\|_\Sigma$ the $\Sigma$-ordinal of $\mathsf{KP}\omega$.

We have just explained that the $\Sigma$-ordinal of $\mathsf{KP}\omega$ is an upper bound for the order types of well-orderings for which the scheme of transfinite induction is provable in $\mathsf{KP}\omega$. Therefore, we have the following theorem.

**11.7.2 Theorem** *The $\Sigma$-ordinal of $\mathsf{KP}\omega$ is an upper bound for its proof-theoretic ordinal.*

**11.7.3 Remark** Theorem 11.7.2 is the first step in showing $\|\mathsf{KP}\omega\| = \|\mathsf{KP}\omega\|_\Sigma$. As a word of warning we want to emphasize that this situation is singular for $\mathsf{KP}\omega$. The special status of $\mathsf{KP}\omega$ is due to the fact that $\omega_1^{CK}$ is the least admissible ordinal above $\omega$, i.e., that $L_{\omega_1^{CK}}$ is the least model of $\mathsf{KP}\omega$ in the constructible hierarchy. If $F$ is a $\Sigma$-sentence such that $\mathsf{KP}\omega \vdash F$ then $L_{\omega_1^{CK}} \models F$ and by $\Sigma$-reflection there is an ordinal $\alpha_F < \omega_1^{CK}$ such that $L_{\alpha_F} \models F$. Since the function $\alpha \mapsto L_\alpha$ is $\omega_1^{CK}$-recursive, i.e., it is definable by a $\Sigma$-formula in $L_{\omega_1^{CK}}$ (see below), and $L_\alpha \models F$ is expressible as $\Delta$-relation, we obtain $F \mapsto \alpha_F$ as a $\omega_1^{CK}$-recursive function. The set $\{F \mid \mathsf{KP}\omega \vdash F\}$ is recursively enumerable. Therefore, $\sup\{\alpha_F \mid F \text{ is a } \Sigma\text{-formula and } \mathsf{KP}\omega \vdash F\} < \omega_1^{CK}$ and we obtain

$$\|\mathsf{KP}\omega\|_\Sigma < \omega_1^{CK}. \tag{11.11}$$

For stronger theories $T$, however, the $\Sigma$-ordinal is bigger than $\omega_1^{CK}$. There we have to introduce the notion of a $\Sigma^{\omega_1^{CK}}$-ordinal $\|T\|_{\Sigma^{\omega_1^{CK}}}$, which is the least stage in the constructible hierarchy, which models the provable $\Sigma^{\omega_1^{CK}}$-sentences, i.e., sentences of the form $(\exists x \in L_{\omega_1^{CK}})F(x)$ where $F(x)$ is $\Delta_0$. This requires that $L_{\omega_1^{CK}}$ is definable in these theories. For such theories $T$, we obtain $\|T\| = \|T\|_{\Sigma^{\omega_1^{CK}}}$.[8] The abstract background is the hyperarithmetical quantifier theorem which states that every $\Pi^1_1$-relation is uniformly equivalent to a relation which is $\Sigma$-definable on $L_{\omega_1^{CK}}$. This theorem becomes provable in stronger theories such as $\mathsf{KPl}$ (cf. [52, 74]). It can even be extracted from the $\omega$-completeness theorem (Theorem 5.4.9) as sketched in [8]. In the case of $\mathsf{KP}\omega$ the $\Sigma$- and $\Sigma^{\omega_1^{CK}}$-ordinals coincide.

**11.7.4 Theorem** $\|\mathsf{KP}\omega\|_\Sigma \geq \psi(\varepsilon_{\Omega+1})$.

*Proof* We have shown in Theorem 9.6.19 that $\mathsf{ID}_1$ proves that for every $\alpha < \psi(\varepsilon_{\Omega+1})$ there is an $n \in \omega$ such that $n \in \mathrm{Acc}_\prec$ and $|n|_{\mathrm{Acc}_\prec} \geq \alpha$ for a primitive recursively definable ordering $\prec$ on $\omega$. By Theorem 11.5.4 we therefore obtain

8 Provided that enough foundation is available in $T$. Cf. Theorem 12.6.14 for an example in which these ordinals differ.

KP$\omega \vdash (\exists\beta)[n \in I_{\mathrm{Acc}_{\omega,\prec}}(\beta)]$. But then $L_\alpha \not\models (\exists\beta)[n \in I_{\mathrm{Acc}_{\omega,\prec}}(\beta)]$ because otherwise $|n|_{\mathrm{Acc}_\prec} < \alpha$. Hence $\alpha < \|\mathsf{KP}\omega\|_\Sigma$. □

For $\alpha := \|\mathsf{KP}\omega\|_\Sigma$ the set $L_\alpha$ is by definition the least stage in the constructible hierarchy at which we have a partial model for the $\Sigma$-sentences which are provable in KP$\omega$. We are going to show that $L_\alpha$ is closed under the $\omega_1^{CK}$-recursive functions whose totality is provable in KP$\omega$. A partial function $F : L^n_{\omega_1^{CK}} \longrightarrow_p L_{\omega_1^{CK}}$ is $\omega_1^{CK}$-recursive if its graph

$$\mathsf{G}_F(a_1,\dots,a_n,b) \;:\Leftrightarrow\; F(a_1,\dots,a_n) \simeq b$$

is $\Sigma$-definable in $L_{\omega_1^{CK}}$. A partial function $F$ is provably total in KP$\omega$ if

$$\mathsf{KP}\omega \vdash (\forall x_1)\dots(\forall x_n)(\exists y)\mathsf{G}_F(x_1,\dots,x_n,y).$$

A $\Pi_2$-formula is a formula of the shape $(\forall x)(\exists y)G(x,y,\vec{v})$ where $G(x,y,\vec{v})$ is a $\Delta_0$-formula. The next theorem states that the least model for the provable $\Sigma$-sentences of KP$\omega$ is already the least model for the provable $\Pi_2$-sentences of KP$\omega$.

**11.7.5 Theorem** *Let $\alpha = \|\mathsf{KP}\omega\|_\Sigma$ and $F$ be a $\Pi_2$-sentence such that* KP$\omega \vdash F$. *Then $L_\alpha \models (\forall x)(\exists y)F(x,y)$.*

*Proof* For the proof we borrow from [4] that KP$\omega$ proves that the constructible hierarchy is an inner class model of KP$\omega$. Therefore we obtain $\Sigma$-reflection in the form

$$\mathsf{KP}\omega \vdash G \rightarrow (\exists u)(\exists\xi)[u = L_\xi \wedge G^u] \tag{i}$$

for $\Sigma$-formulas $G$. Let $(\forall x)(\exists y)F(x,y)$ be a provable $\Pi_2$-sentence of KP$\omega$, i.e., assume KP$\omega \vdash (\forall x)(\exists y)F(x,y)$ and let $a \in L_\alpha$. We have to show $L_\alpha \models (\exists y)F(a,y)$. Since $\alpha \in Lim$ there is a $\beta < \alpha$ such that $a \in L_\beta$. Since $\beta < \|\mathsf{KP}\omega\|_\Sigma$ there is a $\Sigma$-sentence $G$ such that KP$\omega \vdash G$ but $L_\beta \not\models G$. By (i) we obtain

$$(\exists\xi)(\exists u)[u = L_\xi \wedge G^u \wedge (\forall x \in u)(\exists y)F(x,y)]. \tag{ii}$$

Applying $\Sigma$-reflection again we obtain

$$(\exists z)(\exists\xi)(\exists u)[u = L_\xi \wedge G^u \wedge (\forall x \in u)(\exists y \in z)F(x,y)] \tag{iii}$$

and thus

$$L_\alpha \models (\exists z)(\exists\xi)(\exists u)[u = L_\xi \wedge G^u \wedge (\forall x \in u)(\exists y \in z)F(x,y)]. \tag{iv}$$

So there is a $\xi < \alpha$ such that $L_\xi \models G$ and $(\forall x \in L_\xi)(\exists y \in L_\alpha)F(x,y)$. Since $L_\beta \not\models G$ we have $\beta < \xi$ by upwards persistency. Hence $a \in L_\xi$ and we are done. □

**11.7.6 Corollary** *Let $\alpha := \|\mathsf{KP}\omega\|_\Sigma$. Then $L_\alpha$ is closed under the provably $\omega_1^{CK}$-recursive functions of* KP$\omega$.

*Proof* Let $f$ be a provably $\omega_1^{CK}$-recursive function of $\mathsf{KP}\omega$. Let $F(x,y)$ be the $\Sigma_1$-formula which defines $f$. Then $\mathsf{KP}\omega \vdash (\forall x)(\exists y)F(x,y)$. Since $(\forall x)(\exists y)F(x,y)$ is a $\Pi_2$-sentence we obtain $L_\alpha \models (\forall x)(\exists y)F(x,y)$ by Theorem 11.7.5. Hence $f(x) \in L_\alpha$ for all $x \in L_\alpha$. □

## 11.8 The Theory ($\Pi_2$–REF)

We have seen that KP proves the reflection principle for $\Sigma$-formulas. We will now briefly introduce a theory that seems to be stronger but has – as we will see in the following sections – the same proof-theoretical strength. To formulate the theory, we first introduce the scheme of $\Pi_2$-reflection.

($\Pi_2$–*REF*) $(\forall \vec{v})\big[F(\vec{v}) \to (\exists a)[a \neq \emptyset \wedge Tran(a) \wedge \vec{v} \in a \wedge F(\vec{v})^a]\big]$,

where $F(\vec{v})$ is a $\Pi_2$ formula and $\vec{v} \in a$ stands for $v_i \in a$ for $i = 1,\ldots,n$ if $\vec{v} = (v_1,\ldots,v_n)$.

To simplify notations we use the shorthand

$$a \models F \;:\Leftrightarrow\; a \neq \emptyset \wedge Tran(a) \wedge F^a.$$

There are admissible sets that do not satisfy ($\Pi_2$–*REF*). The following theorem, however, shows that for constructible sets this is always the case.

**11.8.1 Theorem** *Let $\alpha$ be an admissible ordinal. Then $L_\alpha \models$ ($\Pi_2$–REF).*

*Proof* Let $\mathsf{G} :\equiv (\forall x)(\exists y)F(x,y,\vec{v})$ be a $\Pi_2$-formula and $\vec{b}$ a tuple in $L_\alpha$ such that $L_\alpha \models \mathsf{G}[\vec{b}]$. Since $\alpha \in Lim$ there is a $\beta_0 < \alpha$ such that $\vec{b} \in L_{\beta_0}$. Then

$$L_\alpha \models (\forall x \in L_{\beta_0})(\exists y)F(x,y,\vec{b}). \tag{i}$$

Since $\alpha$ is admissible we obtain by $\Sigma$-reflection a set $b_0 \in L_\alpha$ such that

$$L_\alpha \models (\forall x \in L_{\beta_0})(\exists y \in b_0)F(x,y,\vec{b})^{b_0}.$$

But then there is a $\beta_1 < \alpha$ such that $b_0 \in L_{\beta_1}$. Because of $Tran(L_{\beta_1})$ we also have $b_0 \subseteq L_{\beta_1}$ which implies

$$L_\alpha \models (\forall x \in L_{\beta_0})(\exists y \in L_{\beta_1})F(x,y,\vec{b})^{L_{\beta_1}}. \tag{ii}$$

We may now choose $\beta_1 > \beta_0$ minimal satisfying (ii). Then we get

$$L_\alpha \models (\forall x \in L_{\beta_1})(\exists y)F(x,y,\vec{b}) \tag{iii}$$

and construct a $\beta_2 > \beta_1$ such that

$$L_\alpha \models (\forall x \in L_{\beta_1})(\exists y \in L_{\beta_2})F(x,y,\vec{b})^{L_{\beta_2}}$$

exactly as before. Iterating the procedure we obtain a sequence $\beta_0, \beta_1, \ldots$ of ordinals $< \omega$. We show that this function is $\Sigma$-definable. Having defined $\beta_n$ we put

$$\beta_{n+1} = z \;\Leftrightarrow\; z \in On \wedge \beta_n < z \wedge (\exists u)\big[u = L_z \wedge (\forall x \in L_{\beta_n})(\exists y \in u)F(x,y,\vec{b})^u \wedge (\forall \rho < z)(\forall v \in u)[v \neq L_\rho \vee (\exists x \in L_{\beta_n})(\forall y \in v)\neg F(x,y,\vec{b})^v]\big].$$

So we have $(\forall x \in \omega)(\exists z)[z = \beta_x]$ and obtain by $\Sigma$-collection and union that $\beta := \sup_{n\in\omega} \beta_n < \alpha$. But for $x \in L_\beta$ we find an $n < \omega$ such that $x \in L_{\beta_n}$. So there is a $y \in L_{\beta_{n+1}}$ such that $F(x,y,\vec{b})^{L_{\beta_{n+1}}}$. Since $L_{\beta_{n+1}} \subseteq L_\beta$ we obtain finally

$$(\forall x \in L_\beta)(\exists y \in L_\beta)F(x,y,\vec{b})^{L_\beta}.$$

Since $L_\beta \in L_\alpha$ we have a reflection point. □

It is obvious that $\Delta_0$-collection follows from *($\Pi_2$–REF)*. If $(\forall x \in a)(\exists y)F(x,y,\vec{b})$ for a $\Delta_0$-formula $F(x,y,\vec{b})$ we reformulate it as $(\forall x)(\exists y)[x \in a \rightarrow F(x,y,\vec{b})]$ and obtain by *($\Pi_2$–REF)* a transitive set $c$ such that $(\forall x \in c)(\exists y \in c)[x \in a \rightarrow F(x,y,\vec{b})^c]$ and $a \in c$. From $Tran(c)$ we obtain $a \subseteq c$ which entails $(\forall x \in a)(\exists y \in c)[F(x,y,\vec{b})]$ by absoluteness of $F(x,y,\vec{b})$.

**11.8.2 Definition** The theory ($\Pi_2$–REF) comprises the axioms of BST together with the scheme *($\Pi_2$–REF)*.

We have already seen that ($\Pi_2$–REF) proves $\Delta_0$-collection. Therefore, ($\Pi_2$–REF) is at least as strong as KP. It is, however, easy to see that in ($\Pi_2$–REF) we can also prove the axiom of infinity. Let $x$ be an ordinal. Then obviously $x \cup \{x\}$ is again a hereditarily transitive set, i.e., an ordinal, such that $x \in x \cup \{x\}$. So ($\Pi_2$–REF) proves $(\forall x)(\exists y)[x \in On \rightarrow y \in On \wedge x \in y]$. By *($\Pi_2$–REF)* there is a transitive set $a$ such that $(\forall x \in a)(\exists y \in a)[x \in On \rightarrow y \in On \wedge x \in y]$. By $\Delta_0$-separation the set $\alpha := \{x \in a \,|\, x \in On\}$ exists and we have $(\forall x \in \alpha)(\exists y \in \alpha)[x \in y]$, i.e., $\alpha \in Lim$. Therefore, we have the following theorem.

**11.8.3 Theorem** *The theory* KP$\omega$ *is a subtheory of* ($\Pi_2$–REF).

## 11.9 An Infinitary Verification Calculus for the Constructible Hierarchy

In analogy to the infinitary verification calculus for $\Pi^1_1$-sentences in the language of arithmetic, we develop an infinitary verification calculus for the $\Sigma$-sentences in the language of the constructible hierarchy. The first step is to introduce a language for the constructible hierarchy.

As a peculiarity of this and the following sections, we do not count the equality symbol among the basic symbols of $\mathcal{L}(\in)$ but view equations $s = t$ as abbreviations. We put

$$s = t \;:\Leftrightarrow\; (\forall x \in t)[x \in s] \wedge (\forall x \in s)[x \in t]. \tag{11.12}$$

This has some technical advantages. In analogy to Sect. 5 we introduce the Tait language of $\mathscr{L}(\in)$. Since we regard equality as defined, the only relation symbols of the Tait language of $\mathscr{L}(\in)$ are the binary relations $\in$ and $\notin$. Formulas are built from the atomic formulas $(t \in s)$ and $(t \notin s)$ by the boolean connectives $\wedge$ and $\vee$, bounded and unbounded quantification. Negation is not among the logical symbols. Instead we define $\sim F$ by

- $\sim(t \in s) :\equiv (t \notin s)$ and $\sim(t \notin s) :\equiv (t \in s)$
- $\sim(F_0 \vee F_1) :\equiv (\sim F_0 \wedge \sim F_1)$ and $\sim(F_0 \wedge F_1) :\equiv (\sim F_0 \vee \sim F_1)$
- $\sim((\exists x \in t)F(x)) :\equiv (\forall x \in t)[\sim F(x)]$ and $\sim((\forall x \in t)F(x)) :\equiv (\exists x \in t)[\sim F(x)]$
- $\sim((\exists x)F(x)) :\equiv (\forall x)[\sim F(x)]$ and $\sim((\forall x)F(x)) :\equiv (\exists t)[\sim F(x)]$

Again we have $\mathscr{S} \models \sim F \;\Leftrightarrow\; \mathscr{S} \models \neg F$ for any $\mathscr{L}(\in)$–structure $\mathscr{S}$. Therefore, we will mostly use $\sim F$ and $\neg F$ synonymously, although negation is not among the basic symbols.

For the following definition of $\mathscr{L}_{RS}$-terms and sentences we assume that $\mathscr{L}(\in)$ is given as Tait language.

**11.9.1 Definition** (Terms of the language $\mathscr{L}_{RS}$ of ramified set theory) We define the terms of $\mathscr{L}_{RS}$ together with their stages inductively by the following clauses.

- For every ordinal $\alpha$ the symbol $\mathsf{L}_\alpha$ is an (atomic) $\mathscr{L}_{RS}$-term of stage $\alpha$.
- If $a_1,\ldots,a_n$ are $\mathscr{L}_{RS}$-terms of stages $< \alpha$ and $F(v_1,\ldots,v_n)$ is an $\mathscr{L}(\in)$-formula then $\{x \in \mathsf{L}_\alpha \mid F(x,a_1,\ldots,a_n)^{\mathsf{L}_\alpha}\}$ is a (composed) $\mathscr{L}_{RS}$-term of stage $\alpha$.

By $stg(t)$ we denote the stage of an $\mathscr{L}_{RS}$-term $t$.

By $\mathscr{T}_\alpha$ we denote the set of $\mathscr{L}_{RS}$-terms of stages less than $\alpha$. We assume an ordering $<_{ST}$ on $\mathscr{T}_\alpha$ such that all terms in $\mathscr{T}_\alpha$ come before the terms in $\mathscr{T}_\beta \setminus \mathscr{T}_\alpha$ if $\alpha < \beta$.

**11.9.2 Definition** If $F(v_1,\ldots,v_n)$ is an $\mathscr{L}(\in)$-formula which contains at most the shown free variables $v_1,\ldots,v_n$ and $a_1,\ldots,a_n$ are $\mathscr{L}_{RS}$-terms then $F_{v_1,\ldots,v_n}(a_1,\ldots,a_n)$ is an $\mathscr{L}_{RS}$-sentence.

The stage $stg(F)$ of an $\mathscr{L}_{RS}$-formula is the maximum of the stages of $\mathscr{L}_{RS}$-terms occurring in $F$.

The notions of $\Delta_0$-, $\Sigma$-, $\Pi$-, $\Sigma_1$-, $\cdots$ formulas carry over to $\mathscr{L}_{RS}$.

The semantics for $\mathscr{L}_{RS}$ is defined in the obvious way. We are only interested in the standard meaning of $\mathscr{L}_{RS}$-terms and sentences and will therefore only define their meaning in the constructible hierarchy $L$. We put

- $\mathsf{L}_\alpha^L = L_\alpha$,
- $\{x \in \mathsf{L}_\alpha \,|\, F(x, a_1, \dots, a_n)^{\mathsf{L}_\alpha}\}^L := \{x \in L_\alpha \,|\, L_\alpha \models F[x, a_1^L, \dots, a_n^L]\}$

and

- $L \models F(a_1, \dots, a_n) \;:\Leftrightarrow\; L \models F[a_1^L, \dots, a_n^L]$.

**11.9.3 Lemma** *For every constructible set $a \in L$ there is an $\mathcal{L}_{RS}$-term $t$ such that $a = t^L$.*

*Proof* We prove the lemma by induction on $\mathrm{rnk}_L(a)$. If $\mathrm{rnk}_L(a) = \alpha + 1$ we have $a = \{x \in L_\alpha \,|\, L_\alpha \models F[x, a_1, \dots, a_n]\}$ for constructible sets $a_1, \dots, a_n$. Since $\mathrm{rnk}_L(a_i) < \mathrm{rnk}_L(a)$ for $i = 1, \dots, n$ we have by induction hypothesis $\mathcal{L}_{RS}$-terms $t_1, \dots, t_n$ such that $a_i = t_i^L$. Then $t := \{x \in \mathsf{L}_\alpha \,|\, F(x, t_1, \dots, t_n)^{\mathsf{L}_\alpha}\}$ is an $\mathcal{L}_{RS}$-term and we obtain $t^L = \{x \in L_\alpha \,|\, L_\alpha \models F[x, t_1^L, \dots, t_n^L]\} = \{x \in L_\alpha \,|\, L_\alpha \models F[x, a_1, \dots, a_n]\} = a$. □

The following corollary follows immediately from the proof of Lemma 11.9.3 and the definition of $t^L$.

**11.9.4 Corollary** *For every $a \in L_\alpha$ there is an $\mathcal{L}_{RS}$-term $t \in \mathcal{T}_\alpha$ such that $s = t^L$. Vice versa $t^L \in L_\alpha$ is true for all $t \in \mathcal{T}_\alpha$.*

We divide the $\mathcal{L}_{RS}$-sentences into two types.

**11.9.5 Definition** The $\bigvee$–type comprises

- Sentences of the form $(s \in t)$ for $\mathcal{L}_{RS}$-terms $s$ and $t$
- Sentences of the form $(F_0 \vee F_1)$
- Sentences of the form $(\exists x \in t)F(x)$
- Sentences of the form $(\exists x)F(x)$.

Dually the $\bigwedge$–type comprises

- Sentences of the form $(s \notin t)$ for $\mathcal{L}_{RS}$-terms $s$ and $t$
- Sentences of the form $(F_0 \wedge F_1)$
- Sentences of the form $(\forall x \in t)F(x)$
- Sentences of the form $(\forall x)F(x)$.

It is obvious that an $\mathcal{L}_{RS}$-sentence $G$ belongs to $\bigwedge$–type if and only if $\sim G$ belongs to $\bigvee$–type and vice versa.

**11.9.6 Definition** We define the *characteristic sequence* for sentences in $\bigvee$–type by the following clauses

$$\mathrm{CS}(s \in t) := \begin{cases} \langle s = s' \mid s' \in \mathcal{T}_\alpha \rangle & \text{if } t = \mathsf{L}_\alpha \\ \langle s = s' \wedge F(s') \mid s' \in \mathcal{T}_\alpha \rangle & \text{if } t = \{x \in \mathsf{L}_\alpha \mid F(x)\}, \end{cases}$$

$$\mathrm{CS}(F_0 \vee F_1) = \langle F_0, F_1 \rangle,$$

$$\mathrm{CS}((\exists x \in t)F(x)) = \begin{cases} \langle F(s) \mid s \in \mathcal{T}_\alpha \rangle & \text{if } t = \mathsf{L}_\alpha \\ \langle F(s) \wedge G(s) \mid s \in \mathcal{T}_\alpha \rangle & \text{if } t = \{x \in \mathsf{L}_\alpha \mid G(x)\}, \end{cases}$$

$$\mathrm{CS}((\exists x)F(x)) = \langle F(t) \mid t \in \bigcup_{\alpha \in On} \mathcal{T}_\alpha \rangle.$$

Dually we define for sentences in $\bigwedge$–type

$$\mathrm{CS}(s \notin t) := \begin{cases} \langle \sim(s = s') \mid s' \in \mathcal{T}_\alpha \rangle & \text{if } t = \mathsf{L}_\alpha \\ \langle \sim(s = s') \vee \sim F(s') \mid s' \in \mathcal{T}_\alpha \rangle & \text{if } t = \{x \in \mathsf{L}_\alpha \mid F(x)\}, \end{cases}$$

$$\mathrm{CS}(F_0 \wedge F_1) = \langle F_0, F_1 \rangle,$$

$$\mathrm{CS}((\forall x \in t)F(x)) = \begin{cases} \langle F(s) \mid s \in \mathcal{T}_\alpha \rangle & \text{if } t = \mathsf{L}_\alpha \\ \langle \sim G(s) \vee F(s) \mid s \in \mathcal{T}_\alpha \rangle & \text{if } t = \{x \in \mathsf{L}_\alpha \mid G(x)\} \end{cases}$$

$$\mathrm{CS}((\forall x)F(x)) = \langle F(t) \mid t \in \bigcup_{\alpha \in On} \mathcal{T}_\alpha \rangle.$$

For every member $G \in \mathrm{CS}(F)$ let $i(G) \in \mathcal{T}_\alpha$ denote its index. In case of $F \equiv (F_0 \circ F_1)$ for $\circ \in \{\wedge, \vee\}$ put $i(F_0) = 0$ and $i(F_1) = 1$.

Observe that

$$F \in \bigvee\text{–type} \quad \Leftrightarrow \quad \sim F \in \bigwedge\text{–type}$$

and

$$\mathrm{CS}(\sim F) = \langle \sim G \mid G \in \mathrm{CS}(F) \rangle$$

holds true for all $\mathcal{L}_{RS}$-sentences $F$.

**11.9.7 Remark** According to (11.12) we regard equations and inequalities as defined by extensionality. An alternative approach is to put $s = t$ into $\bigwedge$–type , $s \neq t$ into $\bigvee$–type and to define $\mathrm{CS}(s = t) := \langle (\forall x \in s)[x \in t], (\forall x \in t)[x \in s] \rangle$ and dually $\mathrm{CS}(s \neq t) := \langle (\exists x \in s)[x \notin t], (\exists x \in t)[x \notin s] \rangle$. To reduce the number of cases we opted for the approach via extensionality.

**11.9.8 Lemma** *Let $F$ be an $\mathcal{L}_{RS}$-sentence. Then*

$$F \in \bigvee\text{–type} \quad \Rightarrow \quad (L \models F \quad \Leftrightarrow \quad (\exists G \in \mathrm{CS}(F)) L \models G)$$

$$F \in \bigwedge\text{–type} \;\Rightarrow\; (L \models F \;\Leftrightarrow\; (\forall G \in \mathrm{CS}(F))L \models G).$$

*Proof* We prove the lemma only for sentences in $\bigvee$–type. This suffices because for $F \in \bigwedge$–type we we have $\sim F \in \bigvee$–type and

$$L \models F \;\Leftrightarrow\; L \not\models \sim F \;\Leftrightarrow\; (\forall G \in \mathrm{CS}(\sim F))[L \not\models G]$$
$$\Leftrightarrow\; (\forall G \in \mathrm{CS}(\sim F))[L \models \sim G] \;\Leftrightarrow\; (\forall G \in \mathrm{CS}(F))[L \models G].$$

If $F$ is a sentence $s \in \mathsf{L}_\alpha$ we have $L \models (s \in \mathsf{L}_\alpha)$ if and only if $s^L \in L_\alpha$ which holds if and only if $s^L = a$ for some set $a \in L_\alpha$. But then there is an $\mathcal{L}_{RS}$-term $t \in \mathcal{T}_\alpha$ such that $s^L = t^L$. By absoluteness this holds if and only if $L \models s = t$ and obviously $(s = t) \in \mathrm{CS}(s \in \mathsf{L}_\alpha)$.

If $F$ is a sentence $s \in \{x \in \mathsf{L}_\alpha \mid F(x, a_1, \dots, a_n)^{\mathsf{L}_\alpha}\}$ we get

$$L \models F \text{ if and only if } s^L \in \{x \in L_\alpha \mid L_\alpha \models F[x, a_1^L, \dots, a_n^L]\}.$$

This is the case if and only if

$$s^L \in L_\alpha \text{ and } L_\alpha \models F[s^L, a_1^L, \dots, a_n^L].$$

By Corollary 11.9.4 there is a term $t \in \mathcal{T}_\alpha$ such that $s^L = t^L$. Then also $L \models s^L = t^L$ and $L_\alpha \models F[t^L, a_1^L, \dots, a_n^L]$ by absoluteness. This entails

$$L \models s = t \wedge F(t, a_1, \dots, a_n)^{\mathsf{L}_\alpha}$$

and $(s = t \wedge F(t, a_1, \dots, a_n)^{\mathsf{L}_\alpha}) \in \mathrm{CS}(F)$.

If vice versa $L \models s = t \wedge F(t, a_1, \dots, a_n)^{\mathsf{L}_\alpha}$ for some $t \in \mathcal{T}_\alpha$ and terms $a_i \in \mathcal{T}_\alpha$ then $s^L = t^L \in L_\alpha$ and, since $a_i^L \in L_\alpha$ for $i = 1, \dots, n$, we get by absoluteness also $L \models s^L \in \{x \in L_\alpha \mid L_\alpha \models F[x, a_1^L, \dots, a_n^L]\}$.

If $F$ is a sentence $F_0 \vee F_1$ we immediately obtain

$$L \models F \;\Leftrightarrow\; L \models F_i \text{ for some } i \in \{0, 1\}.$$

Now let $F$ be a sentence $(\exists x \in t)F_0(x, a_1, \dots, a_n)$. First assume that $t = \mathsf{L}_\alpha$. Then $L \models F$ if and only if there is a set $a \in L_\alpha$ such that $L \models F_0[a, a_1^L, \dots, a_n^L]$. By Corollary 11.9.4 there is a term $t \in \mathcal{T}_\alpha$ such that $a = t^L$ and we obtain

$$L \models F_0[t^L, a_1^L, \dots, a_n^L], \text{ i.e., } L \models F_0(t, a_1, \dots, a_n)$$

and $F_0(t, a_1, \dots, a_n) \in \mathrm{CS}(F)$. If conversely $L \models F_0(t, a_1, \dots, a_n)$ for some $t \in \mathcal{T}_\alpha$ then $t^L \in L_\alpha$ which implies $L \models (\exists x \in \mathsf{L}_\alpha)F_0(x, a_1, \dots, a_n)$.

If $t = \{x \in \mathsf{L}_\alpha \mid G(x, b_1, \dots, b_m)^{\mathsf{L}_\alpha}\}$ we have $L \models F$ if and only if there is a set $a \in L_\alpha$ such that $L \models G[a, b_1^L, \dots, b_m^L] \wedge F_0[a, a_1^L, \dots, a_n^L]$. But then there is a term $t \in \mathcal{T}_\alpha$ such that $a = t^L$ and we obtain

$$L \models G(t, b_1, \dots, b_m) \wedge F_0(t, a_1, \dots, a_n).$$

If conversely

$$L \models G(t, b_1, \dots, b_m) \wedge F_0(t, a_1, \dots, a_n)$$

for some $t \in \mathcal{T}_\alpha$ then we obtain $t^L \in L_\alpha$ which in turn implies

$$L \models (\exists x \in L_\alpha)\big[G[x, b_1^L, \ldots, b_m^L] \wedge F_0[x, a_1^L, \ldots, a_n^L]\big].$$

Hence $L \models F$.

The case that $F$ is a sentence $(\exists x)F(x_0, a_1, \ldots, a_n)$ is even simpler and treated analogously. □

Having defined $\bigvee$–type and $\bigwedge$–type and the characteristic sequences, we obtain a verification relation $\models^{\alpha} \Delta$ for finite sets $\Delta$ of $\Sigma$-sentences in the language $\mathcal{L}_{RS}$ according to Definition 5.4.3. Since there are no free predicate variables in $\mathcal{L}_{RS}$ we can, however, dispense with clause *(Ax)*.

Therefore there are only the rules

($\bigwedge$) If $F \in \Delta \cap \bigwedge$–type and $\models^{\alpha_G} \Delta, G$ as well as $\alpha_G < \alpha$ for all $G \in \mathrm{CS}(F)$ then $\models^{\alpha} \Delta$.

($\bigvee$) If $F \in \Delta \cap \bigvee$–type and $\models^{\alpha_0} \Delta, G$ for some $G \in \mathrm{CS}(F)$ then $\models^{\alpha} \Delta$ for all $\alpha > \alpha_0$ such that $stg(i(G)) < \alpha$.

**11.9.9 Theorem** *Let $\Delta$ be a finite set of $\Sigma$-sentences in the language $\mathcal{L}_{RS}$. If $\models^{\alpha} \Delta$ then $L \models \bigvee \Delta^{L_\alpha}$.*

*Proof* We induct on $\alpha$. If the last inference is an inference according to ($\bigwedge$) then there is a sentence $F \in \Delta \cap \bigwedge$–type such that $\models^{\alpha_G} \Delta, G$ for all $G \in \mathrm{CS}(F)$. If $L \models \bigvee \Delta^{L_\alpha}$ we are done. Otherwise we obtain by persistency also $L \not\models \bigvee \Delta^{L_{\alpha_G}}$ which by the induction hypothesis entails $L \models G^{L_{\alpha_G}}$ for all $G \in \mathrm{CS}(F)$. All sentences in $\mathrm{CS}(F)$ are again $\Sigma$-sentences. By persistency we get therefore $L \models G^{L_\alpha}$ for all $G \in \mathrm{CS}(F)$. But $\mathrm{CS}(F^{L_\alpha}) = \langle G^{L_\alpha} \mid G \in \mathrm{CS}(F)\rangle$. By Lemma 11.9.8, however, we then obtain $L \models F^{L_\alpha}$ contradicting the assumption $L \not\models \bigvee \Delta^{L_\alpha}$.

Now assume that the last inference is an $\bigvee$-inference. Then there is a sentence $F \in \bigvee$–type $\cap \Delta$, an $\mathcal{L}_{RS}$-term $t$ such that $stg(t) < \alpha$, a sentence $G \in \mathrm{CS}(F)$ such that $i(G) = t$ and an ordinal $\alpha_0 < \alpha$ such that $\models^{\alpha_0} \Delta, G$. If $L \not\models \bigvee \Delta^{L_\alpha}$ we obtain by persistency and the induction hypothesis $L \models G^{L_{\alpha_0}}$. Since $stg(t) < \alpha$ we get in any case $G^{L_\alpha} \in \mathrm{CS}(F^{L_\alpha})$ and obtain by Lemma 11.9.8 $L \models F^{L_\alpha}$ contradicting $L \not\models \bigvee \Delta^{L_\alpha}$. □

**11.9.10 Definition** Let $\Delta$ be a set of $\Sigma$-sentences of $\mathcal{L}_{RS}$. Put

$$\mathrm{par}(\Delta) = \{\alpha \mid \mathsf{L}_\alpha \text{ occurs in some formula of } \Delta\}.$$

**11.9.11 Corollary** *Assume $\models^{\alpha} F$ for a $\Sigma$-sentence $F$ such that* $\mathrm{par}(F) \subseteq \alpha$. *Then $L_\alpha \models F$.*

*Proof* By Theorem 11.9.9 we obtain $L \models F^{L_\alpha}$. Since $\mathrm{par}(F) \subseteq \alpha$ we obtain $t^L \in L_\alpha$ for all terms occurring in $F$. Hence $L_\alpha \models F$ by absoluteness. □

**11.9.12 Definition** (Truth complexity for $\Sigma$-sentences) For a $\Sigma$-sentence $F$ we define

$$\mathrm{tc}(F) := \min(\{\alpha \mid \models^{\alpha} F\} \cup \{\infty\})$$

where $\infty$ is a symbol such that $\alpha < \infty$ and $\alpha + \infty = \infty$ as well as $\infty + \alpha = \infty$ hold true for all ordinals $\alpha$.

**11.9.13 Theorem** *If $F$ is a $\Sigma$-formula such that $\alpha := \mathrm{tc}(F) < \infty$ and $\mathrm{par}(F) \subseteq \mathrm{tc}(F)$ then $L_\alpha \models F$.*

*Proof* This is immediate from Corollary 11.9.11 and the definition of $\mathrm{tc}(F)$. □

The next aim is to show that the sentences in the characteristic sequence of a sentence $F$ have lower complexity than $F$. By an $\mathscr{L}_{RS}$-expression we mean either an $\mathscr{L}_{RS}$-term or an $\mathscr{L}_{RS}$-sentence.

**11.9.14 Definition** (The rank of an $\mathscr{L}_{RS}$-expression) We define the rank of an $\mathscr{L}_{RS}$-expression by the following clauses

$$\mathrm{rnk}(\mathsf{L}_\alpha) := \omega \cdot \alpha$$

$$\mathrm{rnk}(\{x \in \mathsf{L}_\alpha \mid F\}) := \max\{\mathrm{rnk}(\mathsf{L}_\alpha), \mathrm{rnk}(F_x(\mathsf{L}_0)) + 2\}$$

$$\mathrm{rnk}(s \in t) := \mathrm{rnk}(s \notin t) := \max\{\mathrm{rnk}(s) + 6, \mathrm{rnk}(t) + 1\}$$

$$\mathrm{rnk}(A \vee B) := \mathrm{rnk}(A \wedge B) := \max\{\mathrm{rnk}(A), \mathrm{rnk}(B)\} + 1$$

$$\mathrm{rnk}((\exists x \in s)F(x)) := \mathrm{rnk}((\forall x \in s)F(x)) := \max\{\mathrm{rnk}(s), \mathrm{rnk}(F(\mathsf{L}_0)) + 2\}$$

$$\mathrm{rnk}((\exists x)F(x)) = \mathrm{rnk}((\forall x)F(x)) := \infty.$$

**11.9.15 Lemma** *Let $a$ be an $\mathscr{L}_{RS}$-expression. If $\mathrm{rnk}(a) < \infty$ there is an $n < \omega$ such that $\mathrm{rnk}(a) = \omega \cdot stg(a) + n$.*

*Proof* We prove the lemma by induction on $\mathrm{rnk}(a) < \infty$. If $a$ is a term $\mathsf{L}_\alpha$ then $n :=0$. If $a$ is a composed term $\{x \in \mathsf{L}_\alpha \mid F\}$ we get $\mathrm{rnk}(a) = \max\{\mathrm{rnk}(\mathsf{L}_\alpha), \mathrm{rnk}(F_x(\mathsf{L}_0)) + 2\}$. Then $stg(F_x(\mathsf{L}_0)) \leq \alpha$ and by induction hypothesis there is an $n_0 < \omega$ such that $\mathrm{rnk}(F_x(\mathsf{L}_0)) = \omega \cdot stg(F_x(\mathsf{L}_0)) + n_0$. If $stg(F_x(\mathsf{L}_0)) < \alpha$ put $n := 0$ otherwise put $n := n_0 + 2$.

If $a$ is a formula $(s \in t)$ or $(s \notin t)$ then $\mathrm{rnk}(a) = \max\{\mathrm{rnk}(s) + 6, \mathrm{rnk}(t) + 1\}$. By induction hypothesis there are $n_0, n_1 < \omega$ such that $\mathrm{rnk}(s) = \omega \cdot stg(s) + n_0$ and $\mathrm{rnk}(t) = \omega \cdot stg(t) + n_1$. If $stg(s) < stg(t)$ put $n := n_1 + 1$, if $stg(t) < stg(s)$ put $n := n_0 + 6$, if $stg(s) = stg(t)$ put $n := \max\{n_0 + 6, n_1 + 1\}$.

If $a$ is a formula $(A \vee B)$ or $(A \wedge B)$ then $\mathrm{rnk}(a) = \max\{\mathrm{rnk}(A), \mathrm{rnk}(B)\} + 1$. By induction hypothesis we have $n_0, n_1 < \omega$ such that $\mathrm{rnk}(A) = \omega \cdot stg(A) + n_0$ and $\mathrm{rnk}(B) = \omega \cdot stg(B) + n_1$. If $stg(A) < stg(B)$ put $n := n_1 + 1$, if $stg(B) < stg(A)$ put $n := n_0 + 1$ and if $stg(A) = stg(B)$ put $n := \max\{n_0 + 1, n_1 + 1\}$.

If $a$ is a sentence of the form $(\exists x \in s)F(x)$ or $(\forall x \in s)F(x)$ then we obtain $\mathrm{rnk}(a) = \max\{\mathrm{rnk}(s), \mathrm{rnk}(F(\mathsf{L}_0)) + 2\}$. We have $\mathrm{rnk}(s) = \omega \cdot stg(s) + n_0$ and $\mathrm{rnk}(F(\mathsf{L}_0)) = \omega \cdot stg(F(\mathsf{L}_0)) + n_1$ for $n_0, n_1 < \omega$ by induction hypothesis. If $stg(F(\mathsf{L}_0)) < stg(s)$ put $n := n_0$, if $stg(s) < stg(F(\mathsf{L}_0))$ put $n := n_1 + 2$ if $stg(s) = stg(F(\mathsf{L}_0))$ put $n := \max\{n_0, n_1 + 2\}$.

The cases that $a$ is a formula $(\exists x)F(x)$ or $(\forall x)F(x)$ are excluded by the hypothesis $\mathrm{rnk}(a) < \infty$. □

**11.9.16 Lemma** *Let $a$ and $b_x(\mathsf{L}_0)$ be $\mathscr{L}_{RS}$-expressions. Then*

*(A)* $stg(a) < stg(b_x(\mathsf{L}_0))$ *implies* $\mathrm{rnk}(b_x(a)) = \mathrm{rnk}(b_x(\mathsf{L}_0))$

*and*

*(B)* $stg(a) < \alpha$ *implies* $\mathrm{rnk}(b_x(a)) < \max\{\omega \cdot \alpha, \mathrm{rnk}(b_x(\mathsf{L}_0)) + 1\}$.

*Proof* We show (A) by induction on $\mathrm{rnk}(b_x(\mathsf{L}_0))$. The case that $b_x(\mathsf{L}_0)$ is $\mathsf{L}_0$ is excluded. So assume $b_x(\mathsf{L}_0)$ is the term $\{y \in \mathsf{L}_\alpha \mid F_x(\mathsf{L}_0)\}$. First assume $stg(b_x(\mathsf{L}_0)) = stg(F_x(\mathsf{L}_0))$. Then $stg(a) < stg(F_x(\mathsf{L}_0))$ and we obtain $\mathrm{rnk}(F_x(a)) = \mathrm{rnk}(F_x(\mathsf{L}_0))$ by the induction hypothesis. But this immediately implies $\mathrm{rnk}(b_x(a)) = \mathrm{rnk}(b_x(\mathsf{L}_0))$. Next assume $stg(F_x(\mathsf{L}_0)) < stg(b_x(\mathsf{L}_0))$. Because of $stg(a) < stg(b_x(\mathsf{L}_0)) = \alpha$ we get $stg(F_x(a)) < \alpha$ and thus $\mathrm{rnk}(b_x(a)) = \omega \cdot \alpha = \mathrm{rnk}(b_x(\mathsf{L}_0))$.

Now assume that $b_x(\mathsf{L}_0)$ is a formula $(s_x(\mathsf{L}_0) \in t_x(\mathsf{L}_0))$ or $(s_x(\mathsf{L}_0) \notin t_x(\mathsf{L}_0))$. If $stg(s_x(\mathsf{L}_0)) < stg(t_x(\mathsf{L}_0))$ then $\mathrm{rnk}(b_x(\mathsf{L}_0)) = \mathrm{rnk}(t_x(\mathsf{L}_0)) + 1$ and $stg(a) < stg(t_x(\mathsf{L}_0))$. By induction hypothesis it follows $\mathrm{rnk}(t_x(a)) = \mathrm{rnk}(t_x(\mathsf{L}_0))$ and, since still $stg(s_x(a)) < stg(t_x(a))$, also $\mathrm{rnk}(b_x(a)) = \mathrm{rnk}(b_x(\mathsf{L}_0))$. If $stg(t_x(\mathsf{L}_0)) < stg(s_x(\mathsf{L}_0))$ we obtain the claim by a similar argument. If $stg(s_x(\mathsf{L}_0)) = stg(t_x(\mathsf{L}_0))$ we get the claim directly from the induction hypotheses. The remaining cases are all similar and left as an exercise.

If $stg(a) < stg(b_x(\mathsf{L}_0))$ we obtain claim (B) directly from (A). If $stg(b_x(\mathsf{L}_0)) \leq stg(a) < \alpha$ we have $stg(b_x(a)) < \alpha$ which implies $\mathrm{rnk}(b_x(a)) < \omega \cdot \alpha$ by Lemma 11.9.15. □

**11.9.17 Lemma** *Let $s$ and $t$ be $\mathscr{L}_{RS}$-terms. Then*

$$\mathrm{rnk}(s = t) = \mathrm{rnk}(s \neq t) = \max\{9, \mathrm{rnk}(s) + 4, \mathrm{rnk}(t) + 4\}.$$

*Proof* The sentence $s = t$ is a shorthand for

$$(\forall x \in s)[x \in t] \wedge (\forall x \in t)[x \in s]. \tag{i}$$

According to Definition 11.9.14 we get

$$\mathrm{rnk}((\forall x \in s)[x \in t]) = \max\{\mathrm{rnk}(s), 8, \mathrm{rnk}(t) + 3\}.$$

Hence $\mathrm{rnk}(s=t) = \max\{8, \mathrm{rnk}(t)+3, \mathrm{rnk}(s)+3\}+1$.

The case of a sentence $s \neq t$ is completely analogous. □

**11.9.18 Lemma** *Let $F$ be a $\Delta_0$-sentence of $\mathscr{L}_{RS}$ and $G \in \mathrm{CS}(F)$. Then* $\mathrm{rnk}(G) < \mathrm{rnk}(F)$.

*Proof* We run through the cases. First let $F$ be a formula $(s \in \mathsf{L}_\alpha)$ or $(s \notin \mathsf{L}_\alpha)$. Then $\alpha \neq 0$ and $G \in \mathrm{CS}(F)$ is of the form $s = s'$ for some $s' \in \mathscr{T}_\alpha$. Hence $\mathrm{rnk}(G) = \max\{9, \mathrm{rnk}(s)+4, \mathrm{rnk}(s')+4\} < \max\{\mathrm{rnk}(s)+6, \omega\cdot\alpha+1\} = \mathrm{rnk}(F)$.

If $F$ is the sentence $s \in \{x \in \mathsf{L}_\alpha \,|\, A\}$ or $s \notin \{x \in \mathsf{L}_\alpha \,|\, A\}$ then $\alpha \neq 0$ and $G$ is a sentence $s = s' \wedge A_x(s')$ or $\sim(s=s') \vee \sim A_x(s')$ for some $s' \in \mathscr{T}_\alpha$. Hence

$$\begin{aligned}\mathrm{rnk}(G) &= \max\{10, \mathrm{rnk}(s)+5, \mathrm{rnk}(s')+5, \mathrm{rnk}(A_x(s'))+1\}\\ &< \max\{\mathrm{rnk}(s)+6, \omega\cdot\alpha+1, \mathrm{rnk}(A_x(\mathsf{L}_0)+3)\} = \mathrm{rnk}(F)\end{aligned}$$

by Lemma 11.9.16.

The claim follows directly from Definition 11.9.14 if $F$ is a sentence $A \vee B$ or $A \wedge B$.

So assume that $F$ is of the form $(\forall x \in \mathsf{L}_\alpha)A$ or $(\exists x \in \mathsf{L}_\alpha)A$. Then $\alpha \neq 0$ and $G \equiv A_x(s')$ for some $s' \in \mathscr{T}_\alpha$. By Lemma 11.9.16 *(B)* we obtain $\mathrm{rnk}(A_x(s')) < \max\{\omega\cdot\alpha, \mathrm{rnk}(A_x(\mathsf{L}_0))+2\} = \mathrm{rnk}(F)$.

Finally assume that $F$ is a formula of the shape $(\exists x \in \{y \in \mathsf{L}_\alpha \,|\, H\})A$ or a formula $(\forall x \in \{y \in \mathsf{L}_\alpha \,|\, H\})A$. Then $\alpha \neq 0$ and $G$ is a sentence $H_y(s') \wedge A_x(s')$ or $\sim H_y(s') \vee A_x(s')$. Then we get by Lemma 11.9.16 *(B)* $\mathrm{rnk}(G) = \max\{\mathrm{rnk}(H_y(s'))+1, \mathrm{rnk}(A_x(s')+1)\} < \max\{\omega\cdot\alpha, \mathrm{rnk}(H_y(\mathsf{L}_0))+2, \mathrm{rnk}(A_x(\mathsf{L}_0))+2\} = \mathrm{rnk}(F)$. □

**11.9.19 Corollary** *For a $\Delta_0$-sentence $A$ which is true in the constructible hierarchy we have* $\mathrm{tc}(A) \leq \mathrm{rnk}(A)$.

*Proof* Using Lemma 11.9.18 and Lemma 11.9.8 we obtain by induction on $\mathrm{rnk}(A)$

$$L \models A \;\Rightarrow\; \overset{\mathrm{rnk}(A)}{\models\!\!=} \Delta, A$$

for any finite set $\Delta$ of $\mathscr{L}_{RS}$-sentences. Hence $\overset{\mathrm{rnk}(A)}{\models\!\!=} A$ which entails $\mathrm{tc}(A) \leq \mathrm{rnk}(A)$.

□

## 11.10 A Semiformal System for Ramified Set Theory

Our aim is the computation of the $\Sigma$-ordinal of $\mathsf{KP}\omega$ or even $(\Pi_2\text{–}\mathsf{REF})$. We have already computed a lower bound for $\|\mathsf{KP}\omega\|_\Sigma$ in Theorem 11.7.4. According to Theorem 11.9.13 it suffices to compute an upper bound for the truth-complexities of the $\Sigma$-sentences which are provable in $(\Pi_2\text{–}\mathsf{REF})$ to obtain also an upper bound for $\|(\Pi_2\text{–}\mathsf{REF})\|_\Sigma$. We follow the pattern of Sect. 9.3 and design a semiformal system with operator controlled infinitary derivations. We repeat the definition of the semi-formal system.

**11.10.1 Definition** We define the proof relation $\frac{|\alpha}{\rho}\ \Delta$ inductively by the following clauses:

($\bigwedge$) If $F \in \bigwedge$–type $\cap \Delta$ and $\frac{|\alpha_G}{\rho}\ \Delta, G$ as well as $\alpha_G < \alpha$ for all $G \in \mathrm{CS}(F)$ then $\frac{|\alpha}{\rho}\ \Delta$.

($\bigvee$) If $F \in \bigvee$–type $\cap \Delta$ and $\frac{|\alpha_0}{\rho}\ \Delta, G$ for some $G \in \mathrm{CS}(F)$ then $\frac{|\alpha}{\rho}\ \Delta$ for all $\alpha > \alpha_0$ such that $stg\,(i(G)) < \alpha$.

(cut) If $\frac{|\alpha_0}{\rho}\ \Delta, F$ and $\frac{|\alpha_0}{\rho}\ \Delta, {\sim}F$ for some $\alpha_0 < \alpha$ and sentence $F$ such that $\mathrm{rnk}(F) < \rho$ then $\frac{|\alpha}{\rho}\ \Delta$.

First we obtain a theorem corresponding to Theorem 9.3.2.

**11.10.2 Theorem** *Let $\Gamma$ be a finite set of $\mathscr{L}_{RS}$-sentences. Then*

$$\frac{|\alpha}{\rho}\ \Gamma \;\Rightarrow\; \models^{\alpha} \Gamma.$$

*Proof* The proof is exactly that of Theorem 9.3.2 in which $\mathbb{N} \models F$ is replaced by $L \models F$. □

We copy the following definition from Sect. 9.3.

**11.10.3 Definition** Let $\mathscr{H}$ be a Skolem–hull operator. We define the relation $\mathscr{H}\,\frac{|\alpha}{\rho}\ \Delta$ by the clauses ($\bigwedge$), ($\bigvee$) and (cut) of Definition 11.10.1 with the additional conditions

- $\alpha \in \mathscr{H}(\mathrm{par}(\Delta))$

and for an inference

$$\mathscr{H}\,\frac{|\alpha_\iota}{\rho}\ \Delta_\iota \text{ for } \iota \in I \;\Rightarrow\; \mathscr{H}\,\frac{|\alpha}{\rho}\ \Delta$$

with finite $I$ also

- $\mathrm{par}(\Delta_\iota) \subseteq \mathscr{H}(\mathrm{par}(\Delta))$.

In spite of the slight modification in the ($\bigvee$)-rule the calculus $\mathscr{H}\,\frac{|\alpha}{\rho}\ \Delta$ coincides with the calculus defined in Sect. 9.3. The properties 9.3.9 through 9.3.15 still hold true for the new calculus. For Lemma 9.3.13, however, observe that the only sentences of $\bigwedge$–type with empty characteristic sequences are formulas of the form $s \notin t$ with $stg\,(t) = 0$, i.e., with terms $t$ representing the empty set. The first difference is in Lemma 9.3.16 which is replaced by the following boundedness lemma.

**11.10.4 Lemma** *(Boundedness) Assume* $\mathcal{H} \overset{\alpha}{\underset{\rho}{\vdash}} \Delta, F$ *for a* $\Sigma$*-sentence* $F$. *Then* $\mathcal{H} \overset{\alpha}{\underset{\rho}{\vdash}} \Delta, F^{L_\beta}$ *for all* $\beta \geq \alpha$.

*Proof* First we observe that $\text{par}(\Delta, F) \subseteq \text{par}(\Delta, F^{L_\beta})$. This, in turn, implies $\alpha \in \mathcal{H}(\text{par}(\Delta, F^{L_\beta}))$. We show the lemma by induction on $\alpha$. Assume that $F$ is not a sentence of the form $(\exists x)G(x)$ which is simultaneously also the critical formula of the last inference

(I) $\quad \mathcal{H} \overset{\alpha_\iota}{\underset{\rho}{\vdash}} \Delta_\iota, F, G_\iota \text{ for } \iota \in I \;\Rightarrow\; \mathcal{H} \overset{\alpha}{\underset{\rho}{\vdash}} \Delta, F.$

Then we obtain

$$\mathcal{H} \overset{\alpha_\iota}{\underset{\rho}{\vdash}} \Delta_\iota, F^{L_\beta}, G_\iota^{L_\beta} \tag{i}$$

for all $\beta \geq \alpha$ by the induction hypothesis. In case of a finite index set $I$ we also have $\text{par}(\Delta_\iota, F^{L_\beta}, G_\iota^{L_\beta}) = \text{par}(\Delta_\iota, F, G_\iota) \cup \{\beta\} \subseteq \mathcal{H}(\text{par}(\Delta, F)) \cup \{\beta\} \subseteq \mathcal{H}(\text{par}(\Delta, F^{L_\beta}))$. Therefore, we obtain $\mathcal{H} \overset{\alpha}{\underset{\rho}{\vdash}} \Delta, F^{L_\beta}$ from (i) by an inference (I).

If $F$ is a $\Sigma$-sentence $(\exists x)G$ which is the critical formula of the last inference then this is an inference according to ($\bigvee$) whose premise is

$$\mathcal{H} \overset{\alpha_0}{\underset{\rho}{\vdash}} \Delta, F, G_x(t) \tag{ii}$$

for a term $t$ such that $stg(t) < \alpha$. But then $G_x(t) \in \text{CS}(F^{L_\beta})$ for $\beta \geq \alpha$ and $\text{par}(\Delta, F, G_x(t)) \subseteq \mathcal{H}(\text{par}(\Delta, F)) \subseteq \mathcal{H}(\text{par}(\Delta, F^{L_\beta}))$ and we obtain $\mathcal{H} \overset{\alpha}{\underset{\rho}{\vdash}} \Delta, F^{L_\beta}$ by an inference ($\bigvee$). □

Derivations in the semiformal systems are bothersome because of the ordinal bounds and the controlling operators. But we have seen in Corollary 11.9.19 that an upper bound for the truth complexity of a sentence can be computed from its rank. In most cases also the extended identity operator will suffice to control the parameters. The extended identity operator is the operator defined by

$$\mathcal{I}(X) := \text{closure of } X \cup \{0, \Omega\} \text{ under the successor function and } \lambda\xi\,.\,\omega^\xi\,.$$

We introduce an auxiliary calculus $\overset{\circ}{\vdash} \Delta$ for multisets $\Delta$ of $\Delta_0$-sentences of $\mathcal{L}_{RS}$. A multiset is a finite unordered sequence. Two multisets $\langle F_1, \ldots, F_n \rangle$ and $\langle G_1, \ldots, G_m \rangle$ are equal if $n = m$ and there is a permutation $\pi$ of the numbers $1, \ldots, n$ such that $F_i = G_{\pi(i)}$ for $i = 1, \ldots, n$. If $\Delta := \langle F_1, \ldots, F_n \rangle$ and $\Gamma$ are multisets we write $\Delta \subseteq \Gamma$ if there is a permutation $\pi$ such that $F_{\pi(1)}, \ldots, F_{\pi(n)}$ is an initial sequence of $\Gamma$, i.e., all $F_i$ occur in $\Gamma$ with at least the same multiplicity in which they occur in $\Delta$.

**11.10.5 Definition** The calculus $\overset{\circ}{\vdash} \Delta$ is defined by the two clauses:

($\bigvee$)° If $F \in \bigvee\text{–type}$ and $\overset{\circ}{\vdash} \Delta, \Gamma$ for some nonvoid set $\Gamma \subseteq \text{CS}(F)$ then $\overset{\circ}{\vdash} \Lambda$ holds true for all $\Lambda \supseteq \Delta, F$ (viewed as multisets)

and

($\bigwedge$)° If $F \in \bigwedge\text{–type}$ and $\overset{\circ}{\vdash} \Delta, G$ for all $G \in \text{CS}(F)$ then $\overset{\circ}{\vdash} \Lambda$ for all $\Lambda \supseteq \Delta, F$ (viewed as multisets) with the proviso that in case of a finite number of premises $\overset{\circ}{\vdash} \Delta,\Gamma$ or $\overset{\circ}{\vdash} \Delta,G$ we have $\mathrm{par}(\Delta,\Gamma) \subseteq \mathscr{I}(\mathrm{par}(\Lambda))$ or $\mathrm{par}(\Delta,G) \subseteq \mathscr{I}(\mathrm{par}(\Lambda))$, respectively. We call the formula $F$ in clauses $(\bigwedge)^\circ$ and $(\bigvee)^\circ$ the *critical sentence* of the inference.

For a multiset $\Delta = \langle F_1,\dots,F_n\rangle$ of $\Delta_0$-sentences we define its derivation rank

$$\sum_{G\in\Gamma}{}^{\#}\, \omega^{\mathrm{rnk}(G)} + \omega < \omega^{\mathrm{rnk}(F)}.$$

**11.10.6 Lemma** *If* $\overset{\circ}{\vdash} \Lambda$ *for a multiset* $\Lambda$ *then* $\mathscr{H} \overset{\mathrm{drnk}(\Lambda)}{\underset{0}{\vdash}} \Lambda$ *holds for any Cantorian closed Skolem–hull operator* $\mathscr{H}$.

*Proof* We induct on the definition of $\overset{\circ}{\vdash} \Lambda$. First we observe that for a Cantorian closed operator $\mathscr{H}$ we always have $\mathrm{drnk}(\Lambda) \in \mathscr{H}(\mathrm{par}(\Lambda))$. Let $F$ be the critical sentence of the last inference

(I) $\quad \overset{\circ}{\vdash} \Delta,\Gamma \;\Rightarrow\; \overset{\circ}{\vdash} \Lambda.$

If (I) is an inference according to $(\bigvee)^\circ$ we have the premise $\overset{\circ}{\vdash} \Delta,\Gamma$ for some non-void set $\Gamma \subseteq \mathrm{CS}(F)$ and $\Delta,F \subseteq \Lambda$. By the induction hypothesis we obtain

$$\mathscr{H} \overset{\mathrm{drnk}(\Delta,\Gamma)}{\underset{0}{\vdash}} \Delta,\Gamma. \tag{i}$$

But then $\mathrm{rnk}(G) < \mathrm{rnk}(F)$ for all $G \in \Gamma$ by Lemma 11.9.18. This, however, implies

$$\sum_{G\in\Gamma}{}^{\#}\, \omega^{\mathrm{rnk}(G)} + \omega < \omega^{\mathrm{rnk}(F)}.$$

and thus $\mathrm{drnk}(\Delta,\Gamma) + \omega < \mathrm{drnk}(\Delta,F) \leq \mathrm{drnk}(\Lambda)$. Here we see why it is important to have multisets, because $F$ may occur in $\Delta$ and thus has to be counted with multiplicity 2 in $\Delta,F$ to ensure $\mathrm{drnk}(\Delta,\Gamma) + \omega < \mathrm{drnk}(\Delta,F)$. From (i), however, we obtain

$$\mathscr{H} \overset{\mathrm{drnk}(\Delta,F)}{\underset{0}{\vdash}} \Delta,F \text{ and therefore also } \mathscr{H} \overset{\mathrm{drnk}(\Lambda)}{\underset{0}{\vdash}} \Lambda$$

by the structural rule and finitely many applications of inferences $(\bigvee)$ which are applicable because $\mathrm{par}(\Delta,\Gamma) \subseteq \mathscr{I}(\mathrm{par}(\Lambda)) \subseteq \mathscr{H}(\mathrm{par}(\Lambda))$ holds for any Cantorian closed Skolem–hull operator.

The case of an inference according to $(\bigwedge)^\circ$ is analogous with the simplification that we only need one application of an inference $(\bigwedge)$ and do not have to bother about the parameters if there are infinitely many premises. □

Another awkwardness of ramified set theory is the fact that sentences of the form $a \in b$, $a \notin b$, $(\exists x \in b)F(x)$ and $(\forall x \in b)F(x)$ have different characteristic sequences according to the shape of $b$. To unify the notation we introduce a modified membership relation[9] $a \mathrel{\underline{\in}} b$ as well as its negation $a \mathrel{\underline{\notin}} b$ and assume that $a \mathrel{\underline{\in}} b$ and $a \mathrel{\underline{\notin}} b$ are empty sentences in case that $b$ is an atomic term $\mathsf{L}_\alpha$ and denote the formula $F(a)$

9 This simplification as well as the simplified calculus $\overset{\circ}{\vdash} \Delta$ are due to Buchholz [13].

or $\sim F(a)$, respectively, in case that $b$ is a composed term $\{x \in \mathsf{L}_\alpha \,|\, F(x)\}$. More precisely we define

$$a \mathrel{\underline{\in}} b \wedge G(s) \;:\Leftrightarrow\; \begin{cases} G(s) & \text{if } b = \mathsf{L}_\alpha \\ F(a) \wedge G(s) & \text{if } b = \{x \in \mathsf{L}_\alpha \,|\, F(x)\} \end{cases}$$

$$a \mathrel{\underline{\notin}} b \vee G(s) \;:\Leftrightarrow\; \begin{cases} G(s) & \text{if } b = \mathsf{L}_\alpha \\ \sim F(a) \vee G(s) & \text{if } b = \{x \in \mathsf{L}_\alpha \,|\, F(x)\} \end{cases}$$

and for multisets also

$$\cdots, a \mathrel{\underline{\notin}} b, G(s), \cdots \;:\Leftrightarrow\; \begin{cases} \cdots, G(s), \cdots & \text{if } b = \mathsf{L}_\alpha \\ \cdots, \sim F(a), G(s), \cdots & \text{if } b = \{x \in \mathsf{L}_\alpha \,|\, F(x)\}. \end{cases}$$

This has the advantage that we obtain

$$\mathrm{CS}(a \in b) = \langle t \mathrel{\underline{\in}} b \wedge t = a \,|\, t \in \mathcal{T}_{stg(b)} \rangle,$$

$$\mathrm{CS}(a \notin b) = \langle t \mathrel{\underline{\notin}} b \vee t \neq a \,|\, t \in \mathcal{T}_{stg(b)} \rangle,$$

$$\mathrm{CS}((\exists x \in b)F(x)) = \langle t \mathrel{\underline{\in}} b \wedge F(t) \,|\, t \in \mathcal{T}_{stg(b)} \rangle$$

as well as

$$\mathrm{CS}((\forall x \in b)F(x)) = \langle t \mathrel{\underline{\notin}} b \vee F(t) \,|\, t \in \mathcal{T}_{stg(b)} \rangle$$

independent of the shape of $b$.

There are some derived rules using this new notation which we will use quite frequently.

(Str) If $\overset{\circ}{\vdash} \Delta$ and $\Delta \subseteq \Gamma$ then $\overset{\circ}{\vdash} \Gamma$

(Taut) $\overset{\circ}{\vdash} \Delta, A, \sim A$

(Sent) If $\overset{\circ}{\vdash} \Delta, A$ then $\overset{\circ}{\vdash} \Delta, \sim B, A \wedge B$

($\in$) If $\overset{\circ}{\vdash} \Delta, t \mathrel{\underline{\in}} b \wedge t = a$ and $\mathrm{par}(t) \subseteq \mathrm{par}(\Delta, a \in b)$ for some $t \in \mathcal{T}_{stg(b)}$ then $\overset{\circ}{\vdash} \Delta, a \in b$

($\notin$) If $\overset{\circ}{\vdash} \Delta, t \mathrel{\underline{\notin}} b, t \neq a$ for all $t \in \mathcal{T}_{stg(b)}$ then $\overset{\circ}{\vdash} \Delta, a \notin b$

($\exists^b$) If $\overset{\circ}{\vdash} \Delta, t \mathrel{\underline{\in}} b \wedge F(t)$ and $\mathrm{par}(t) \subseteq \mathrm{par}(\Delta, (\exists x \in b)F(x))$ for some $t \in \mathcal{T}_{stg(b)}$ then $\overset{\circ}{\vdash} \Delta, (\exists x \in b)F(x)$

($\forall^b$) If $\overset{\circ}{\vdash} \Delta, t \mathrel{\underline{\notin}} b, F(t)$ for all $t \in \mathcal{T}_{stg(b)}$ then $\overset{\circ}{\vdash} \Delta, (\forall x \in b)F(x)$

The structural rule (Str) has been built into the calculus $\overset{\circ}{\vdash}$. The tautology rule (Taut) is proved by induction on the rank of the sentence $A$. The sentential rule (Sent) follows from the tautology rule $\overset{\circ}{\vdash} \Delta, \sim B, B$ and the hypothesis $\overset{\circ}{\vdash} \Delta, A$ by an inference $(\bigwedge)^\circ$ since $\mathrm{par}(\Delta, A) \cup \mathrm{par}(\Delta, \sim B, B) \subseteq \mathrm{par}(\Delta, \sim B, A \wedge B)$. The remaining rules are special cases of $(\bigvee)^\circ$ or $(\bigwedge)^\circ$.

We will now prove a series of sentences in the simplified calculus.

$$\overset{\circ}{\vdash} a \notin a \text{ for every } \mathscr{L}_{RS}\text{-term a.} \tag{11.13}$$

We show (11.13) by induction on $\mathrm{rnk}(a)$. Let $\alpha := stg(a)$. For $\alpha = 0$ we obtain the claim by an inference $(\bigwedge)^\circ$ because then $\mathrm{CS}(a \in a) = \emptyset$. If $\alpha > 0$ we have $\mathrm{rnk}(b) < \mathrm{rnk}(a)$ for all $b \in \mathscr{T}_\alpha$ and obtain

$$\overset{\circ}{\vdash} b \notin b \tag{i}$$

for all $b \in \mathscr{T}_\alpha$ by induction hypothesis and from (i)

$$\overset{\circ}{\vdash} b \not\mathrel{\underline{\in}} a, b \mathrel{\underline{\in}} a \wedge b \notin b \tag{ii}$$

by (Sent). By $(\exists^a)$ it then follows

$$\overset{\circ}{\vdash} b \not\mathrel{\underline{\in}} a, (\exists x \in a)[x \notin b], \text{ i.e., } \overset{\circ}{\vdash} b \not\mathrel{\underline{\in}} a, b \neq a \tag{iii}$$

for all $b \in \mathscr{T}_\alpha$. By $(\notin)$ this entails $\overset{\circ}{\vdash} a \notin a$. □

$$\overset{\circ}{\vdash} a \subseteq a, \text{ hence also } \overset{\circ}{\vdash} a = a, \text{ for all } \mathscr{L}_{RS}\text{-terms } a. \tag{11.14}$$

We prove (11.14) by induction on $\alpha := \mathrm{rnk}(a)$. For $\alpha = 0$ the characteristic sequence of $(\forall x \in a)[x \in a]$ is empty. Therefore, we obtain $\overset{\circ}{\vdash} a \subseteq a$ by an inference $(\bigwedge)^\circ$. For $\alpha > 0$ we obtain $\overset{\circ}{\vdash} b \subseteq b$ and, by symmetry, therefore also

$$\overset{\circ}{\vdash} b = b \tag{i}$$

for all $b \in \mathscr{T}_\alpha$ by induction hypothesis. From (i) we obtain

$$\overset{\circ}{\vdash} b \not\mathrel{\underline{\in}} a, b \mathrel{\underline{\in}} a \wedge b = b \tag{ii}$$

by (Sent) and from (ii) by $(\in)$

$$\overset{\circ}{\vdash} b \not\mathrel{\underline{\in}} a, b \in a \text{ for all } b \in \mathscr{T}_\alpha. \tag{11.15}$$

By $(\forall^a)$ we get $\overset{\circ}{\vdash} a \subseteq a$ from (11.15). □

As a corollary to (11.14) we get

$$\overset{\circ}{\vdash} a \in \mathsf{L}_\alpha \text{ for all } a \in \mathscr{T}_\alpha \tag{11.16}$$

and thus

$$\alpha \neq 0 \Rightarrow \overset{\circ}{\vdash} \mathsf{L}_\alpha \neq \emptyset. \tag{11.17}$$

By (Taut) we also have

$$\stackrel{\circ}{\vdash} a \neq b, a = b. \tag{11.18}$$

An important observation is

$$\stackrel{\circ}{\vdash} Tran(\mathsf{L}_\alpha). \tag{11.19}$$

Let $a \in \mathscr{T}_\alpha$. If $b \in \mathscr{T}_{stg(a)}$ then $stg(b) < \alpha$ and by (11.15) we get

$$\stackrel{\circ}{\vdash} b \notin a, b \in \mathsf{L}_\alpha \tag{i}$$

for all $b \in \mathscr{T}_{stg(a)}$. By $(\forall^a)$ (i) implies

$$\stackrel{\circ}{\vdash} (\forall y \in a)[y \in \mathsf{L}_\alpha] \text{ i.e. } \stackrel{\circ}{\vdash} a \subseteq \mathsf{L}_\alpha \tag{11.20}$$

for all $a \in \mathscr{T}_\alpha$. But from (11.20) we obtain again by $(\forall^{\mathsf{L}_\alpha})$

$$\stackrel{\circ}{\vdash} (\forall x \in \mathsf{L}_\alpha)(\forall y \in x)[y \in \mathsf{L}_\alpha]$$

which is the claim. □

This is all we need for the collapsing theorem. We will return to the simplified calculus $\stackrel{\circ}{\vdash}$ in Sect. 11.12 where we prove the axioms of $\mathsf{KP}\omega$ and $(\Pi_2\text{–REF})$ in ramified set theory.

## 11.11 The Collapsing Theorem for Ramified Set Theory

This section is in principle a remake of Sect. 9.4. We will, therefore, use the notions of this section and also often rely on the results obtained there. The main difference is that we will prove the Collapsing Lemma for *($\Pi_2$–REF)* while the direct conversion of Sect. 9.4 would only give us $\Sigma$-reflection. Recall the scheme

$$(\Pi_2\text{–}REF) \quad (\forall \vec{v})\big[F(\vec{v}) \ \rightarrow \ (\exists a)[a \models F(\vec{v})]\big]$$

for $\Pi_2$-formulas $F(\vec{v}) \equiv (\forall x)(\exists y)G(x,y,\vec{v})$. We start with a simple technical remark.

**11.11.1 Remark** If $\mathscr{H} \,\big|\!\frac{\alpha}{\rho}\, \Delta,\ F^{\mathsf{L}_\Omega}$ for a $\Sigma_1$-sentence $F$ then $\mathscr{H} \,\big|\!\frac{\alpha}{\rho}\, \Delta, F$.

The proof is an obvious induction on $\alpha$ using $\mathscr{H}(\mathrm{par}(\Delta, F^{\mathsf{L}_\Omega})) = \mathscr{H}(\mathrm{par}(\Delta, F))$. □

**11.11.2 Theorem** *(Collapsing Theorem) Let $\Gamma$ be a finite set of instances of the scheme ($\Pi_2$–REF), $\Delta$ a finite set of $\Sigma$-sentences with* $\mathrm{par}(\Delta) \subset \Omega$*. Let moreover $\mathscr{H}$ be a Cantorian closed transitive Skolem–hull operator such that* $\mathrm{par}(\Gamma, \Delta) \subseteq \mathscr{H}(\emptyset)$.
*Then $\mathscr{H} \,\big|\!\frac{\beta}{\Omega}\, \neg\Gamma^{\mathsf{L}_\Omega}, \Delta^{\mathsf{L}_\Omega}$ entails $\mathscr{H}_{\omega^{\omega^\beta}+1} \,\big|\!\frac{\psi_{\mathscr{H}}(\omega^{\omega^\beta})}{\Omega}\, \Delta^{\mathsf{L}_\Omega}$.*

*Proof* We adapt the key properties in the proof of Lemma 9.4.5. First we obtain

$$\beta \in \mathscr{H}(\emptyset) \text{ and } \omega^{\omega^\beta} < \gamma \;\Rightarrow\; \psi_{\mathscr{H}}(\omega^{\omega^\beta}) < \psi_{\mathscr{H}}(\gamma) \tag{i}$$

which is again obvious by (9.20) (on p. 173) since we have $\omega^{\omega^\beta} \in \mathscr{H}(\emptyset) \cap \gamma \subseteq \mathscr{H}_\gamma(\emptyset) \cap \gamma$. From $\mathrm{par}(\Gamma, \Delta) \subseteq \mathscr{H}(\emptyset)$ we get

$$\mathscr{H}(\mathrm{par}(\Delta)) = \mathscr{H}(\emptyset) \tag{ii}$$

and finally we have

$$\beta \in \mathscr{H}(\emptyset) \;\Rightarrow\; \omega^{\omega^\beta} \in \mathscr{H}_{\omega^{\omega^\beta}+1}(\emptyset) \text{ and } \psi_{\mathscr{H}}(\omega^{\omega^\beta}) \in \mathscr{H}_{\omega^{\omega^\beta}+1}(\emptyset) \tag{iii}$$

by (9.19) and the closure properties of $\mathscr{H}_{\omega^{\omega^\beta}+1}(\emptyset)$.

Let us first assume that the last inference

$$\text{(J)} \qquad \mathscr{H} \,\Big|\frac{\beta_\iota}{\Omega}\, \neg\Gamma^{\llcorner\Omega}, \Delta_\iota^{\llcorner\Omega} \text{ for } \iota \in I \;\Rightarrow\; \mathscr{H} \,\Big|\frac{\beta}{\Omega}\, \neg\Gamma^{\llcorner\Omega}, \Delta^{\llcorner\Omega}$$

does not have a critical sentence which belongs to $\neg\Gamma^{\llcorner\Omega}$. We show first:

$$\text{"All sentences in } \Delta_\iota \text{ are } \Sigma\text{-sentences"} \tag{iv}$$

and

$$\mathrm{par}(\Delta_\iota) \subseteq \mathscr{H}(\emptyset). \tag{v}$$

The only case in which there are sentences in $\Delta_\iota$ which are not subsentences of sentences in $\Delta$ is a cut with cut-formula $H$, say. If $H$ contains unbounded quantifiers then $\mathrm{rnk}(H^{\llcorner\Omega}) \geq \Omega$. So (iv) follows because of cut rank $\Omega$. In the case of a finite index set $I$ we obtain (v) from $\mathrm{par}(\Delta_\iota^{\llcorner\Omega}) \subseteq \mathscr{H}(\mathrm{par}(\Gamma^{\llcorner\Omega}, \Delta^{\llcorner\Omega})) \subseteq \mathscr{H}(\emptyset)$. The only possibility for an infinite index set $I$ is an inference according to $(\bigwedge)$. Then we have the critical sentence $(\forall x \in a)G(x)$ in $\Delta^{\llcorner\Omega}$ and for every $\iota \in I$ there is an $\mathscr{L}_{RS}$-term $t \in \mathscr{T}_{stg(a)}$ such that $\mathrm{par}(\Delta_\iota) \subseteq \mathrm{par}(\Delta) \cup \mathrm{par}(t)$. But then $\mathrm{par}(t) \subseteq stg(a) \in \mathscr{H}(\emptyset) \cap \Omega$. Since $\mathscr{H}$ is transitive it follows $\mathrm{par}(t) \subseteq \mathscr{H}(\emptyset)$. Because of (iv) and (v) the induction hypothesis applies to the premise(s) of (J). So we have

$$\mathscr{H}_{\omega^{\omega^{\beta_\iota}}+1} \,\Big|\frac{\psi_{\mathscr{H}}(\omega^{\omega^{\beta_\iota}})}{\Omega}\, \Delta_\iota^{\llcorner\Omega} \tag{vi}$$

and obtain with the aid of (i) and (iii)

$$\mathscr{H}_{\omega^{\omega^\beta}+1} \,\Big|\frac{\psi_{\mathscr{H}}(\omega^{\omega^\beta})}{\Omega}\, \Delta^{\llcorner\Omega} \tag{vii}$$

from (vi) by an inference (J).

Now assume that the critical formula of (J) belongs to $\neg\Gamma^{\llcorner\Omega}$. All formulas in $\neg\Gamma$ have the shape

$$\begin{aligned} H(t_1,\dots,t_k) \;:\Leftrightarrow\; & (\exists v_{k+1})\dots(\exists v_n)\big[(\forall x)(\exists y)G(x,y,t_1,\dots,t_k,v_{k+1},\dots,v_n) \\ & \wedge (\forall a)[a = \emptyset \vee \neg Tran(a) \vee t_1,\dots,t_n \notin a \vee v_{k+1},\dots,v_n \notin a \\ & \vee \neg(\forall x \in a)(\exists y \in a)G(x,y,\vec{t},\vec{v}\,)]\big] \end{aligned}$$

for a $\Delta_0$-formula $G(x,y,\vec{t},\vec{v}\,)$. Therefore, we have to distinguish the cases that (J) is an inference according to $(\bigvee)$ (the case that $k < n$ in the critical sentence) or an inference according to $(\bigwedge)$ (the case that $k = n$ in the critical sentence). We start with the first case. Then we have the premise

$$\mathscr{H} \,\Big|\frac{\beta_0}{\Omega}\, \neg\Gamma^{\mathsf{L}_\Omega}, \neg H(t_1,\ldots,t_{k+1})^{\mathsf{L}_\Omega}, \Delta^{\mathsf{L}_\Omega}. \tag{viii}$$

By the parameter condition for $(\bigvee)$ inferences we obtain $\mathrm{par}(H(t_1,\ldots,t_{k+1})) \subseteq \mathscr{H}(\mathrm{par}(\Gamma,\Delta)) \subseteq \mathscr{H}(\emptyset)$ and can therefore apply the induction hypothesis to (viii). So we have

$$\mathscr{H}_{\omega^{\omega^{\beta_0}}+1} \,\Big|\frac{\psi_{\mathscr{H}}(\omega^{\omega^{\beta_0}})}{\Omega}\, \Delta^{\mathsf{L}_\Omega} \tag{ix}$$

and obtain $\mathscr{H}_{\omega^{\omega^{\beta}}+1} \,\Big|\frac{\psi_{\mathscr{H}}(\omega^{\omega^{\beta}})}{\Omega}\, \Delta^{\mathsf{L}_\Omega}$ from (ix) by the structural rule.

The real interesting case is that of an inference according to $(\bigwedge)$. There we have the premises

$$\mathscr{H} \,\Big|\frac{\beta_0}{\Omega}\, \neg\Gamma^{\mathsf{L}_\Omega}, (\forall x \in \mathsf{L}_\Omega)(\exists y \in \mathsf{L}_\Omega)G(x,y,\vec{t}\,), \Delta^{\mathsf{L}_\Omega} \tag{x}$$

and

$$\mathscr{H} \,\Big|\frac{\beta_0}{\Omega}\, \begin{array}{l}\neg\Gamma^{\mathsf{L}_\Omega}, (\forall a \in \mathsf{L}_\Omega)[a = \emptyset \vee \neg Tran(a) \vee \vec{t} \notin a \vee \\ \qquad \neg(\forall x \in a)(\exists y \in a)G(x,y,\vec{t}\,)], \Delta^{\mathsf{L}_\Omega}\end{array} \tag{xi}$$

for some $\beta_0 < \beta$. By $(\bigwedge)$ inversion we obtain from (x)

$$\mathscr{H} \,\Big|\frac{\beta_0}{\Omega}\, \neg\Gamma^{\mathsf{L}_\Omega}, (\exists y \in \mathsf{L}_\Omega)G(t,y,\vec{t}\,), \Delta^{\mathsf{L}_\Omega} \tag{xii}$$

for every term $t \in \mathscr{T}_\Omega$. Let $\eta_n := \omega^{\omega^{\beta_0}} \cdot (n+1)$, $\eta_\omega := \omega^{\omega^{\beta_0}} \cdot \omega$ and $\eta := \psi_{\mathscr{H}}(\eta_\omega)$. Then (xii) is true for all $t \in \mathscr{T}_\eta$. We have $\mathscr{H}_{\eta_\omega}(\emptyset) = \bigcup_{n\in\omega} \mathscr{H}_{\eta_n}(\emptyset)$ because both sets are closed under $\mathscr{H}$ and $\psi_{\mathscr{H}} \restriction \eta_\omega$. Since $\eta = \mathscr{H}(\eta_\omega) \cap \Omega$ we obtain for every $t \in \mathscr{T}_\eta$ an $n \in \omega$ such that $\mathrm{par}(t) \subseteq \mathscr{H}_{\eta_n}(\emptyset) \cap \Omega = \psi_{\mathscr{H}}(\eta_n)$. From (xii) we get

$$\mathscr{H}_{\eta_n} \,\Big|\frac{\beta_0}{\Omega}\, \neg\Gamma^{\mathsf{L}_\Omega},\; (\exists y \in \mathsf{L}_\Omega)G(t,y,\vec{t}\,), \Delta^{\mathsf{L}_\Omega} \tag{xiii}$$

by the structural rule and can now apply the induction hypothesis to (xiii) to obtain

$$\mathscr{H}_{\eta_{n+1}+1} \,\Big|\frac{\psi_{\mathscr{H}}(\eta_{n+1})}{\Omega}\, (\exists y \in \mathsf{L}_\Omega)G(t,y,\vec{t}\,), \Delta^{\mathsf{L}_\Omega}. \tag{xiv}$$

By Remark 11.11.1 and the Boundedness Lemma (Lemma 11.10.4) we get from (xiv)

$$\mathscr{H}_{\eta_{n+1}+1} \,\Big|\frac{\psi_{\mathscr{H}}(\eta_{n+1})}{\Omega}\, (\exists y \in \mathsf{L}_\eta)G(t,y,\vec{t}\,), \Delta^{\mathsf{L}_\Omega}.$$

This shows that for every $t \in \mathscr{T}_\eta$ there is an $n \in \omega$ such that

$$\mathscr{H}_{\eta_\omega+1} \,\Big|\frac{\psi_{\mathscr{H}}(\eta_n)}{\Omega}\, (\exists y \in \mathsf{L}_\eta)G(t,y,\vec{t}\,), \Delta^{\mathsf{L}_\Omega}$$

and this entails

$$\mathcal{H}_{\eta_\omega+1} \left|\frac{\psi_{\mathcal{H}}(\eta_\omega)}{\Omega}\right. (\forall x \in \mathsf{L}_\eta)(\exists y \in \mathsf{L}_\eta)G(t,y,\vec{t}\,),\Delta^{\mathsf{L}_\Omega} \tag{xv}$$

by an inference $(\bigwedge)$. From (xi) we get by $(\bigwedge)$-inversion and $\bigvee$-exportation

$$\mathcal{H}_{\eta_\omega+1} \left|\frac{\beta_0}{\Omega}\right. \begin{array}{l}\neg\Gamma^{\mathsf{L}_\Omega}, \mathsf{L}_\eta = \emptyset,\ \neg Tran(\mathsf{L}_\eta),\ t_1 \notin \mathsf{L}_\eta, \ldots, t_n \notin \mathsf{L}_\eta, \\ \neg(\forall x \in \mathsf{L}_\eta)(\exists y \in \mathsf{L}_\eta)G(x,y,\vec{t}\,),\Delta^{\mathsf{L}_\Omega}.\end{array} \tag{xvi}$$

Since $\mathrm{par}(\mathsf{L}_\eta) = \{\eta\} \subseteq \mathcal{H}_{\eta_\omega+1}(\emptyset)$ we can apply the induction hypothesis to (xvi) to obtain

$$\mathcal{H}_{\eta_\omega+\omega^{\omega^{\beta_0}}+1} \left|\frac{\psi_{\mathcal{H}}(\eta_\omega+\omega^{\omega^{\beta_0}})}{\Omega}\right. \begin{array}{l}\mathsf{L}_\eta = \emptyset,\ \neg Tran(\mathsf{L}_\eta),\ t_1 \notin \mathsf{L}_\eta, \ldots, t_n \notin \mathsf{L}_\eta, \\ \neg(\forall x \in \mathsf{L}_\eta)(\exists y \in \mathsf{L}_\eta)G(x,y,\vec{t}\,),\Delta^{\mathsf{L}_\Omega}.\end{array} \tag{xvii}$$

Since $\mathrm{rnk}(Tran(\mathsf{L}_\eta)) = \eta + 6$ and $\mathrm{rnk}(\mathsf{L}_\eta = \emptyset) = \eta + 3$ we obtain by (11.19) and (11.16) in p. 282 and Lemma 11.10.6

$$\mathcal{H} \left|\frac{\eta \cdot \omega^3}{0}\right. \mathsf{L}_\eta \neq \emptyset \tag{xviii}$$

and

$$\mathcal{H} \left|\frac{\eta \cdot \omega^6}{0}\right. Tran(\mathsf{L}_\eta). \tag{xix}$$

Since $stg(t_i) \in \mathcal{H}(\emptyset) \cap \Omega \subseteq \mathcal{H}_{\eta_\omega}(\emptyset) \cap \Omega = \eta$ we obtain $t_i \in \mathcal{T}_\eta$. By (11.16) we therefore have $\overset{\circ}{\models} t_i \in \mathsf{L}_\eta$ and thus

$$\mathcal{H} \left|\frac{\eta \cdot \omega}{0}\right. t_i \in \mathsf{L}_\eta \tag{xx}$$

for $i = 1, \ldots, n$ by Lemma 11.10.6. Cutting (xvii), (xviii), (xix), (xx), and (xv) gives

$$\mathcal{H}_{\omega^{\omega^\beta}+1} \left|\frac{\psi_{\mathcal{H}}(\omega^{\omega^\beta})}{\Omega}\right. \Delta^{\mathsf{L}_\Omega}. \tag{xxi}$$

To ensure that these cuts are correct it suffices to check

$$\{\eta \cdot \omega^k, \eta_\omega + \omega^{\omega^{\beta_0}} + 1\} \subseteq \mathcal{H}_{\omega^{\omega^\beta}}(\emptyset) \cap \omega^{\omega^\beta} \tag{xxii}$$

because from (xxii) we obtain $\{\eta \cdot \omega^k, \psi_{\mathcal{H}}(\eta_\omega + \omega^{\omega^{\beta_0}})\} \subseteq \mathcal{H}_{\omega^{\omega^\beta}}(\emptyset) \cap \Omega = \psi_{\mathcal{H}}(\omega^{\omega^\beta})$ as well as $\mathcal{H}_{\eta_\omega+\omega^{\omega^{\beta_0}}+1} \subseteq \mathcal{H}_{\omega^{\omega^\beta}+1}$ which also imply that the parameters of all cut formulas are controlled by $\mathcal{H}_{\omega^{\omega^\beta}+1}(\emptyset) = \mathcal{H}_{\omega^{\omega^\beta}+1}(\Delta^{\mathsf{L}_\Omega})$.

But (xxii) is obvious because $\beta_0 \in \mathcal{H}(\emptyset) \subseteq \mathcal{H}_{\omega^{\omega^\beta}}(\emptyset)$ and thus also $\eta_\omega + \omega^{\omega^{\beta_0}} = \omega^{\omega^{\beta_0}+1} + \omega^{\omega^{\beta_0}} \in \mathcal{H}_{\omega^{\omega^\beta}}(\emptyset) \cap \omega^{\omega^\beta}$. This in turn also implies $\eta = \psi_{\mathcal{H}}(\eta_\omega) \in \mathcal{H}_{\omega^{\omega^\beta}}(\emptyset)$ and $\eta \cdot \omega^k = \omega^{\eta+k} < \omega^{\psi_{\mathcal{H}}(\omega^{\omega^\beta})} = \psi_{\mathcal{H}}(\omega^{\omega^\beta})$. Since $\beta \in \mathcal{H}(\emptyset)$ we finally also obtain $\psi_{\mathcal{H}}(\omega^{\omega^\beta}) \in \mathcal{H}_{\omega^{\omega^\beta}+1}(\emptyset)$. □

## 11.12 Ordinal Analysis for Kripke–Platek Set Theory

Recent proof-theoretical research has shown that, via the semiformal system for ramified set theory, the constructible hierarchy is the best suited instrument for ordinal analyses of stronger impredicative systems. In our first step into impredicativity, this does not yet become visible. The ordinal analysis of $\mathsf{ID}_1$ is still simpler than that of $\mathsf{KP}\omega$ or $(\Pi_2\text{–}\mathsf{REF})$. One of the reasons is the extensionality of sets which makes the proof of the identity axioms cumbersome in ramified set theory. Their proof will be therefore the first step in the ordinal analysis of $\mathsf{KP}\omega$.

Let $A((v_1,\dots,v_n))$ denote that every free variable $v_1,\dots,v_n$ occurs at most once in the formula $A$.

**11.12.1 Lemma** *(Identity Lemma) Let $A((v_1,\dots,v_n))$ be a $\Delta_0$-formula of $\mathcal{L}(\in)$. Then*

$$\overset{\circ}{\vdash} \neg(s_1 \subseteq t_1), \neg(t_1 \subseteq s_1), \dots, \neg(s_n \subseteq t_n), \neg(t_n \subseteq s_n), \neg A(s_1,\dots,s_n), A(t_1,\dots,t_n).$$

*Proof* We prove the lemma by induction on $\operatorname{rnk}(A(s_1,\dots,s_n)) \# \operatorname{rnk}(A(t_1,\dots,t_n))$. First let $A((v_1,\dots,v_n)) \equiv v_1 \in v_2$. If $stg(s_2) = 0$ we obtain $\overset{\circ}{\vdash} s_1 \notin s_2$ by an inference $(\notin)$ and thus the claim by (Str). So assume $stg(s_2) > 0$. If $stg(t_2) = 0$ we obtain $\overset{\circ}{\vdash} b \notin t_2$ and by (11.15) on p. 282 also $\overset{\circ}{\vdash} b \not\mathrel{\underline{\in}} s_2, b \neq s_1, b \in s_2$ for all $b \in \mathscr{T}_{stg(s_2)}$. Hence $\overset{\circ}{\vdash} b \not\mathrel{\underline{\in}} s_2, b \neq s_1, b \in s_2 \wedge b \notin t_2$ for all $b \in \mathscr{T}_{stg(s_2)}$. By an inference $(\bigvee)^\circ$ and $(\notin)$ we then get $\overset{\circ}{\vdash} \neg(s_2 \subseteq t_2), s_1 \notin s_2$ which again implies the claim by (Str). Therefore assume $stg(s_2) \neq 0$ as well as $stg(t_2) \neq 0$. For $s \in \mathscr{T}_{stg(s_2)}$ and $t \in \mathscr{T}_{stg(t_2)}$ we obtain $\operatorname{rnk}(s = t_1) = \max\{9, \operatorname{rnk}(s)+4, \operatorname{rnk}(s_1)+4\} < \max\{\operatorname{rnk}(s_1)+6, \operatorname{rnk}(s_2)\} = \operatorname{rnk}(s_1 \in s_2)$ and analogously $\operatorname{rnk}(t = t_1) < \operatorname{rnk}(t_1 \in t_2)$. So we obtain by the induction hypothesis

$$\overset{\circ}{\vdash} \neg(s \subseteq t), \neg(t \subseteq s), \neg(s_1 \subseteq t_1), \neg(t_1 \subseteq s_1), s \neq s_1, t = t_1. \tag{i}$$

From (i) we obtain by (Sent)

$$\overset{\circ}{\vdash} \neg(s \subseteq t), \neg(t \subseteq s), \neg(s_1 \subseteq t_1), \neg(t_1 \subseteq s_1), t \not\mathrel{\underline{\in}} t_2, s \neq s_1, t \mathrel{\underline{\in}} t_2 \wedge t = t_1. \tag{ii}$$

By $(\in)$ we obtain from (ii)

$$\overset{\circ}{\vdash} \neg(s \subseteq t), \neg(t \subseteq s), \neg(s_1 \subseteq t_1), \neg(t_1 \subseteq s_1), t \not\mathrel{\underline{\in}} t_2, s \neq s_1, t_1 \in t_2 \tag{iii}$$

and from (iii)

$$\overset{\circ}{\vdash} \neg(s_1 \subseteq t_1), \neg(t_1 \subseteq s_1), s \neq t, t \not\mathrel{\underline{\in}} t_2, s \neq s_1, t_1 \in t_2 \tag{iv}$$

for all $t \in \mathscr{T}_{stg(t_2)}$ by $(\bigvee)^\circ$. From (iv) we obtain

$$\overset{\circ}{\vdash} \neg(s_1 \subseteq t_1), \neg(t_1 \subseteq s_1), s \notin t_2, s \neq s_1, t_1 \in t_2 \tag{v}$$

by an inference $(\notin)$. By (v) we obtain by (Sent)

$$\overset{\circ}{\vdash} \neg(s_1 \subseteq t_1), \neg(t_1 \subseteq s_1), s \mathrel{\dot\in} s_2 \wedge s \notin t_2, s \mathrel{\dot\notin} s_2, s \neq s_1, t_1 \in t_2 \tag{vi}$$

which by $(\exists^{s_2})$ implies

$$\overset{\circ}{\vdash} \neg(s_1 \subseteq t_1), \neg(t_1 \subseteq s_1), (\exists x \in s_2)[x \notin t_2], s \mathrel{\dot\notin} s_2, s \neq s_1, t_1 \in t_2 \tag{vii}$$

for all $s \in \mathscr{T}_{stg(s_2)}$. From (vii), however, we get finally

$$\overset{\circ}{\vdash} \neg(s_1 \subseteq t_1), \neg(t_1 \subseteq s_1), \neg(s_2 \subseteq t_2), \neg(t_2 \subseteq s_2), s_1 \notin s_2, t_1 \in t_2$$

by an inference $(\notin)$ and (Str).

Next let $A((v_1,\dots,v_n))$ be a conjunction $A_0((v_1,\dots,v_n)) \wedge A_1((v_1,\dots,v_n))$. By the induction hypothesis we obtain

$$\overset{\circ}{\vdash} \neg(s_1 \subseteq t_1), \neg(t_1 \subseteq s_1), \dots, \neg(s_n \subseteq t_n), \neg(t_n \subseteq s_n), \neg A_i(s_1,\dots,s_n), A_i(t_1,\dots,t_n)$$

for $i = 0, 1$. By $(\bigvee)^\circ$ we then obtain

$$\overset{\circ}{\vdash} \neg(s_1 \subseteq t_1), \neg(t_1 \subseteq s_1), \dots, \neg(s_n \subseteq t_n), \neg(t_n \subseteq s_n), \neg A_0(\vec{s}\,) \vee \neg A_1(\vec{s}\,), A_i(\vec{t}\,)$$

for $i = 0, 1$ and finally by an inference $(\bigwedge)^\circ$

$$\overset{\circ}{\vdash} \neg(s_1 \subseteq t_1), \neg(t_1 \subseteq s_1), \dots, \neg(s_n \subseteq t_n), \neg(t_n \subseteq s_n), \neg A(\vec{s}\,), A(\vec{t}\,).$$

Now assume $A((v_1,\dots,v_n)) \equiv (\exists x \in v_1)B((v_2,\dots,v_n))$. Let $\vec{s} := s_2,\dots,s_n$ and $\vec{t} := t_2,\dots,t_n$. For $s \in \mathscr{T}_{stg(s_1)}$ and $t \in \mathscr{T}_{stg(t_1)}$ we obtain

$$\overset{\circ}{\vdash} \begin{array}{r} \neg(s_2 \subseteq t_2), \neg(t_2 \subseteq s_2), \dots, \neg(s_n \subseteq t_n), \neg(t_n \subseteq s_n), \neg(s \subseteq t), \neg(t \subseteq s), \\ \neg B(s,\vec{s}\,), B(t,\vec{t}\,) \end{array} \tag{viii}$$

by the induction hypothesis. Putting

$$\Delta := \neg(s_2 \subseteq t_2), \neg(t_2 \subseteq s_2), \dots, \neg(s_n \subseteq t_n), \neg(t_n \subseteq s_n)$$

we obtain from (viii)

$$\overset{\circ}{\vdash} \Delta, \neg(s \subseteq t), \neg(t \subseteq s), t \mathrel{\dot\notin} t_1, \neg B(s,\vec{s}\,), t \mathrel{\dot\in} t_1 \wedge B(t,\vec{t}\,) \tag{ix}$$

by (Sent) and from (ix)

$$\overset{\circ}{\vdash} \Delta, s \neq t, t \mathrel{\dot\notin} t_1, \neg B(s,\vec{s}\,), (\exists x \in t_1)B(x,\vec{t}\,) \tag{x}$$

for all $t \in \mathscr{T}_{stg(t_1)}$ by an inference $(\exists^{t_1})$ and an inference $(\bigvee)^\circ$. From (x) we obtain by $(\notin)$

$$\overset{\circ}{\vdash} \Delta, s \notin t_1, \neg B(s,\vec{s}\,), (\exists x \in t_1)B(x,\vec{t}\,). \tag{xi}$$

By an inference (Sent) we infer

$$\overset{\circ}{\vdash} \Delta, s \notin t_1 \wedge s \mathrel{\dot\in} s_1, s \mathrel{\dot\notin} s_1, \neg B(s,\vec{s}\,), (\exists x \in t_1)B(x,\vec{t}\,)$$

from (xi) and by $(\exists^{s_1})$

$$\overset{\circ}{\vdash} \Delta, \neg(s_1 \subseteq t_1), s \mathrel{\dot\notin} s_1, \neg B(s,\vec{s}\,), (\exists x \in t_1)B(s,\vec{t}\,) \tag{xii}$$

for all $s \in \mathcal{T}_{stg(s_1)}$. From (xii) we get finally

$$\overset{\circ}{\vdash} \Delta, \neg(s_1 \subseteq t_1), \neg(t_1 \subseteq s_1), \neg((\exists x \in s_1)B(x,\vec{s})), (\exists x \in t_1)B(x,\vec{t})$$

by $(\forall^{s_1})$ and (Str). This, however, is the claim.

The cases that $A((v_1,\dots,v_n))$ is a formula $A_0((v_1,\dots,v_n)) \vee A_1((v_1,\dots,v_n))$ or a formula $(\forall x \in v_1)B((v_2,\dots,v_n))$ are symmetrical to the already treated cases. Therefore the lemma is proved. □

The following identity theorem is now a corollary to Lemma 11.12.1.

**11.12.2 Theorem** *Let $A(v_1,\dots,v_n)$ be an $\mathcal{L}(\in)$-formula. Then*

$$\overset{\circ}{\vdash} s_1 \neq t_1,\dots,s_n \neq t_n, \neg A(s_1,\dots,s_n)^{\mathsf{L}_\Omega}, A(t_1,\dots,t_n)^{\mathsf{L}_\Omega}$$

*holds for all $\mathcal{L}_{RS}$-terms $s_1,\dots,s_n$ and $t_1,\dots,t_n$.*

*Proof* Let $B((u_1,\dots,u_k))$ be a formula such that

$$A(v_1,\dots,v_n) = B_{u_1,\dots,u_k}(v_1,-,v_1,v_2,-,v_2,\ \dots\ ,v_n,-,v_n).$$

Then apply the identity lemma to obtain

$$\overset{\circ}{\vdash} \begin{array}{l} \neg(\tilde{s}_1 \subseteq \tilde{t}_1), \neg(\tilde{t}_1 \subseteq \tilde{s}_1),\dots,\neg(\tilde{s}_k \subseteq \tilde{t}_k), \neg(\tilde{t}_k \subseteq \tilde{s}_k), \\ \qquad \neg B(\tilde{s}_1,\dots,\tilde{s}_k)^{\mathsf{L}_\Omega}, B(\tilde{t}_1,\dots,\tilde{t}_k)^{\mathsf{L}_\Omega} \end{array} \tag{i}$$

where $\tilde{s}_1,\dots,\tilde{s}_k = s_1,-,s_1,\dots,s_n,-,s_n$ and analogously $\tilde{t}_1,\dots,\tilde{t}_k = t_1,-,t_1,\dots,t_n,-,t_n$. From (i) we infer the claim by successively applying $(\bigvee)^\circ$-inferences. □

Next we show how logically valid formulas are translated into ramified set theory. Since we treat bounded quantifiers as basic symbols in the language of $\mathsf{KP}\omega$, we have to extend the Tait calculus introduced in Definition 4.3.2 by rules for these basic symbols. We define

$(\exists)^b$ If $\underset{\mathsf{T}}{\overset{m_0}{\vdash}} \Delta, v \in u \wedge A(v)$, then $\underset{\mathsf{T}}{\overset{m}{\vdash}} \Delta, (\exists x \in u)A_v(x)$ for all $m > m_0$.

$(\forall)^b$ If $\underset{\mathsf{T}}{\overset{m_0}{\vdash}} \Delta, v \notin u, A(v)$ and the variable $v$ is not free in any of the formulas in $\Delta, (\forall x \in u)A_v(x)$, then $\underset{\mathsf{T}}{\overset{m}{\vdash}} \Delta, (\forall x \in u)A_v(x)$ for all $m > m_0$.

**11.12.3 Theorem** *Let $\Delta(x_1,\dots,x_k)$ be a finite set of $\mathcal{L}(\in)$-formulas which contain at most the variables $x_1,\dots,x_k$ free. If $\underset{\mathsf{T}}{\overset{m}{\vdash}} \Delta(x_1,\dots,x_k)$ we obtain*

$$(\forall \alpha \le \Omega)(\exists n \in \omega)(\exists r \in \omega)(\forall a_1 \in \mathcal{T}_\alpha)\dots(\forall a_k \in \mathcal{T}_\alpha)\mathcal{H} \,\overset{\omega^{\omega\cdot\alpha+n}}{\underset{\omega\cdot\alpha+r}{\vdash}}\, \Delta(a_1,\dots,a_k)^{\mathsf{L}_\alpha}$$

*for every Cantorian closed Skolem–hull operator $\mathcal{H}$.*

*Proof* We prove the theorem by induction on $m$. In case of an axiom according to $(Ax)_L$ we get $\overset{\circ}{\vdash} \Delta(a_1,\dots,a_k)^{\mathsf{L}_\alpha}$ by (Taut). Since $\mathrm{drnk}(\Delta(a_1,\dots,a_k)^{\mathsf{L}_\alpha}) \le \omega^{\omega\cdot\alpha+n}$ for some $n < \omega$ we obtain the claim.

The cases of inferences according to $(\wedge)$ or $(\vee)$ follow immediately from the induction hypotheses.

In the case of an inference according to $(\exists)$ or $(\exists)^b$ we have the premise

$$\frac{m_0}{\mathsf{T}}\ \Delta(x_1,\dots,x_k), A(u,x_1,\dots,x_k) \tag{i}$$

or

$$\frac{m_0}{\mathsf{T}}\ \Delta(x_1,\dots,x_k), u \in x_i \wedge A(u,x_1,\dots,x_k) \tag{ii}$$

and distinguish the following cases.

1. The variable $u$ is different from all variables $x_1,\dots,x_k$. Then we obtain

$$\mathscr{H} \left|\frac{\omega^{\omega\cdot\alpha+n_0}}{\omega\cdot\alpha+r}\right. \Delta(a_1,\dots,a_k)^{\mathsf{L}_\alpha}, A_u(\mathsf{L}_0,a_1,\dots,a_k)^{\mathsf{L}_\alpha} \tag{iii}$$

by the induction hypothesis and obtain by an inference according to $(\bigvee)$

$$\mathscr{H} \left|\frac{\omega^{\omega\cdot\alpha+n_0+1}}{\omega\cdot\alpha+r}\right. \Delta(a_1,\dots,a_k)^{\mathsf{L}_\alpha}, (\exists x \in \mathsf{L}_\alpha)A_u(x,a_1,\dots,a_k)^{\mathsf{L}_\alpha}$$

in case of $(\exists)$ or

$$\mathscr{H} \left|\frac{\omega^{\omega\cdot\alpha+n_0+1}}{\omega\cdot\alpha+r}\right. \Delta(a_1,\dots,a_k)^{\mathsf{L}_\alpha}, (\exists x \in a_i)A_u(x,a_1,\dots,a_k)^{\mathsf{L}_\alpha}$$

in case of $(\exists)^b$. Since the parameters of $\mathsf{L}_0$ are trivial we don't have to bother about them.

2. The variable $u$ occurs in the list $x_1,\dots,x_k$. In case of $(\exists)$ we get

$$\mathscr{H} \left|\frac{\omega^{\omega\cdot\alpha+n_0}}{\omega\cdot\alpha+r}\right. \Delta(a_1,\dots,a_k)^{\mathsf{L}_\alpha}, A_u(a_i,a_1,\dots,a_k)^{\mathsf{L}_\alpha} \tag{iv}$$

by the induction hypothesis. Since $a_i \in \mathscr{T}_\alpha$ and $a_i$ occurs in the conclusion we obtain

$$\mathscr{H} \left|\frac{\omega^{\omega\cdot\alpha+n_0+1}}{\omega\cdot\alpha+r}\right. \Delta(a_1,\dots,a_k), (\exists x \in \mathsf{L}_\alpha)A_u(x,a_1,\dots,a_k)$$

by an inference $(\bigvee)$. In case of $(\exists)^b$ we obtain

$$\mathscr{H} \left|\frac{\omega^{\omega\cdot\alpha+n_0}}{\omega\cdot\alpha+r}\right. \Delta(a_1,\dots,a_k)^{\mathsf{L}_\alpha}, a_i \in a_j \wedge A_u(a_i,a_1,\dots,a_k)^{\mathsf{L}_\alpha} \tag{v}$$

by the induction hypothesis. We have

$$\models^{\circ} \Delta(a_1,\dots,a_k)^{\mathsf{L}_\alpha}, t \not\subseteq a_j, t \neq a_i, \neg A_u(a_i,a_1,\dots,a_k)^{\mathsf{L}_\alpha}, \\ t \subseteq a_j \wedge A_u(t,a_1,\dots,a_k)^{\mathsf{L}_\alpha} \tag{vi}$$

for all $t \in \mathscr{T}_{stg(a_j)}$ by the Identity Theorem (Theorem 11.12.2). This implies

$$\mathscr{H} \left|\frac{\omega^{\omega\cdot\alpha+n_1}}{0}\right. \Delta(a_1,\dots,a_k)^{\mathsf{L}_\alpha}, t \not\subseteq a_j, t \neq a_i, \neg A_u(a_i,a_1,\dots,a_k)^{\mathsf{L}_\alpha}, \\ t \subseteq a_j \wedge A_u(t,a_1,\dots,a_k)^{\mathsf{L}_\alpha} \tag{vii}$$

for some $n_1 < \omega$ by Lemma 11.9.15 and Lemma 11.10.6. From (v) we get

$$\mathscr{H} \left|\frac{\omega^{\omega\cdot\alpha+n_0}}{\omega\cdot\alpha+r}\right. \Delta(a_1,\dots,a_k)^{\mathsf{L}_\alpha}, a_i \in a_j \tag{viii}$$

and

$$\mathscr{H} \left|\frac{\omega^{\omega\cdot\alpha+n_0}}{\omega\cdot\alpha+r}\right. \Delta(a_1,\dots,a_k)^{\mathsf{L}_\alpha}, A_u(a_i,a_1,\dots,a_k)^{\mathsf{L}_\alpha} \tag{ix}$$

by inversion. Cutting (vii) and (ix) yields natural numbers $n_1$ and $r_1$ such that

$$\mathcal{H} \overset{\omega^{\omega\cdot\alpha+n_2}}{\underset{\omega\cdot\alpha+r_1}{\vdash}} \Delta(a_1,\dots,a_k)^{\mathsf{L}_\alpha}, t \not\subseteq a_j, t \neq a_i, t \subseteq a_j \wedge A_u(t,a_1,\dots,a_k)^{\mathsf{L}_\alpha} \tag{x}$$

for all $t \in \mathcal{T}_{stg(a_j)}$. By an inference $(\bigvee)$ we obtain

$$\mathcal{H} \overset{\omega^{\omega\cdot\alpha+n_2+1}}{\underset{\omega\cdot\alpha+r_1}{\vdash}} \Delta(a_1,\dots,a_k)^{\mathsf{L}_\alpha}, t \not\subseteq a_j, t \neq a_i, (\exists x \in a_j)A_u(x,a_1,\dots,a_k)^{\mathsf{L}_\alpha} \tag{xi}$$

for all $t \in \mathcal{T}_{stg(a_j)}$ and by inferences $(\bigvee)$ and $(\bigwedge)$

$$\mathcal{H} \overset{\omega^{\omega\cdot\alpha+n_3}}{\underset{\omega\cdot\alpha+r_1}{\vdash}} \Delta(a_1,\dots,a_k)^{\mathsf{L}_\alpha}, a_i \notin a_j, (\exists x \in a_j)A_u(x,a_1,\dots,a_k)^{\mathsf{L}_\alpha}. \tag{xii}$$

Cutting (viii) and (xii) yields the claim.

In the case according to an inference $(\forall)$ or $(\forall)^b$ we have the premises

$$\overset{m_0}{\underset{\mathsf{T}}{\vdash}} \Delta(x_1,\dots,x_k), A(u,x_1,\dots,x_k) \tag{xiii}$$

or

$$\overset{m_0}{\underset{\mathsf{T}}{\vdash}} \Delta(x_1,\dots,x_k), u \notin x_j, A(u,x_1,\dots,x_k), \tag{xiv}$$

respectively, such that $u$ is not among the variables $x_1,\dots,x_k$. From (xiii) and (xiv) we obtain by the induction hypothesis

$$\mathcal{H} \overset{\omega^{\omega\cdot\alpha+n_0}}{\underset{\omega\cdot\alpha+r}{\vdash}} \Delta(a_1,\dots,a_k)^{\mathsf{L}_\alpha}, A_u(t,a_1,\dots,a_k)^{\mathsf{L}_\alpha} \tag{xv}$$

for all $t \in \mathcal{T}_\alpha$ or

$$\mathcal{H} \overset{\omega^{\omega\cdot\alpha+n_0}}{\underset{\omega\cdot\alpha+r}{\vdash}} \Delta(a_1,\dots,a_k), t \notin a_j, A_u(t,a_1,\dots,a_k) \tag{xvi}$$

for all $t \in \mathcal{T}_{stg(a_j)}$, respectively. From (xv) we obtain by an inference $(\bigwedge)$ directly

$$\mathcal{H} \overset{\omega^{\omega\cdot\alpha+n}}{\underset{\omega\cdot\alpha+r}{\vdash}} \Delta(a_1,\dots,a_k)^{\mathsf{L}_\alpha}, (\forall x \in \mathsf{L}_\alpha)A_u(x,a_1,\dots,a_k)^{\mathsf{L}_\alpha}.$$

By (11.15) (in p. 282) we have

$$\overset{\circ}{\vdash} t \not\subseteq a_j, t \in a_j \tag{xvii}$$

for all $t \in \mathcal{T}_{stg(a_j)}$. Cutting (xvii) and (xvi) yields $n_1 < \omega$ and $r_0 < \omega$ such that

$$\mathcal{H} \overset{\omega^{\omega\cdot\alpha+n_1}}{\underset{\omega\cdot\alpha+r_0}{\vdash}} \Delta(a_1,\dots,a_k), t \not\subseteq a_j \vee A_u(t,a_1,\dots,a_k) \tag{xviii}$$

for all $t \in \mathcal{T}_{stg(a_j)}$. By an inference $(\bigwedge)$ we finally get from (xviii)

$$\mathcal{H} \overset{\omega^{\omega\cdot\alpha+n_2}}{\underset{\omega\cdot\alpha+r_0}{\vdash}} \Delta(a_1,\dots,a_k), (\forall x \in a_j)A_u(x,a_1,\dots,a_k). \qquad \square$$

Now we derive the axioms of $(\Pi_2\text{–}\mathsf{REF})$. We begin with the equality axioms $\mathsf{IDEN}$. From (11.14) (in p. 282) we obtain

$$\overset{\circ}{\vdash} (\forall x \in \mathsf{L}_\alpha)[x = x] \tag{11.21}$$

for all $\alpha$ and from (11.18) also

$$\overset{\circ}{\vdash} (\forall x \in \mathsf{L}_\alpha)(\forall y \in \mathsf{L}_\alpha)[x = y \to y = x] \tag{11.22}$$

and from the Identity Theorem (Theorem 11.12.2) also

$$\overset{\circ}{\vdash} (\forall x \in \mathsf{L}_\alpha)(\forall y \in \mathsf{L}_\alpha)(\forall z \in \mathsf{L}_\alpha)[x = y \land y = z \to x = z] \tag{11.23}$$

and

$$\overset{\circ}{\vdash} (\forall x \in \mathsf{L}_\alpha)(\forall y \in \mathsf{L}_\alpha)(\forall a \in \mathsf{L}_\alpha)(\forall b \in \mathsf{L}_\alpha)[x = y \land a = b \to x \in a \to y \in b]. \tag{11.24}$$

These are all the identity axioms.

Since $a = b$ is an abbreviation for the formula $(\forall x \in a)[x \in b] \land (\forall x \in b)[x \in a]$ we obtain

$$\overset{\circ}{\vdash} (\text{Ext'})^{\mathsf{L}_\alpha} \tag{11.25}$$

from (Taut).

**11.12.4 Lemma** *Let $\alpha$ be a limit ordinal. Then* $\overset{\circ}{\vdash} (\text{Pair'})^{\mathsf{L}_\alpha}$.

*Proof* Let $a, b \in \mathscr{T}_\alpha$ and put $\beta := \max\{stg(a), stg(b)\} + 1$. Then

$$\overset{\circ}{\vdash} a \in \mathsf{L}_\beta \land b \in \mathsf{L}_\beta \tag{i}$$

by (11.16). Since $\alpha \in Lim$ we have $\beta < \alpha$ and obtain

$$\overset{\circ}{\vdash} (\exists z \in \mathsf{L}_\alpha)[a \in z \land b \in z] \tag{ii}$$

for all $a, b \in \mathscr{T}_\alpha$ by an inference $(\bigvee)^\circ$. From (ii), however, we obtain

$$\overset{\circ}{\vdash} (\forall x \in \mathsf{L}_\alpha)(\forall y \in \mathsf{L}_\alpha)(\exists z \in \mathsf{L}_\alpha)[x \in z \land y \in z]. \qquad \square$$

**11.12.5 Lemma** *Let $\alpha$ be a limit ordinal. Then* $\overset{\circ}{\vdash} (\text{Union'})^{\mathsf{L}_\alpha}$.

*Proof* Let $a \in \mathscr{T}_\alpha$ and $\beta = stg(a)$. For $t \in \mathscr{T}_\beta$ we obtain by (11.16) and (Str)

$$\overset{\circ}{\vdash} s \notin t, s \in \mathsf{L}_\beta \tag{i}$$

for all $s \in \mathscr{T}_{stg(t)}$ and thus

$$\overset{\circ}{\vdash} (\forall y \in t)[y \in \mathsf{L}_\beta] \tag{ii}$$

by an inference $(\bigwedge)^\circ$. From (ii) we obtain by (Str)

$$\overset{\circ}{\vdash} t \notin a, (\forall y \in t)[y \in \mathsf{L}_\beta] \tag{iii}$$

and from (iii) by $(\forall^a)$

$$\overset{\circ}{\vdash} (\forall x \in a)(\forall y \in x)[y \in \mathsf{L}_\beta]. \tag{iv}$$

Hence

$$\overset{\circ}{\vdash} (\exists z \in \mathsf{L}_\alpha)(\forall x \in a)(\forall y \in x)[y \in z] \tag{v}$$

by $(\exists)^{\mathsf{L}_\alpha}$ with $\mathsf{L}_\beta$ as witness. Since this holds for all $a \in \mathscr{T}_\alpha$ we finally obtain

$$\overset{\circ}{\vdash} (\forall u \in \mathsf{L}_\alpha)(\exists z \in \mathsf{L}_\alpha)(\forall x \in u)(\forall y \in x)[y \in z]$$

by an inference $(\forall^{\mathsf{L}_\alpha})$. □

**11.12.6 Lemma** *Let $\alpha$ be a limit ordinal. Then* $\overset{\circ}{\vdash} (\Delta_0\text{–Sep})^{\mathsf{L}_\alpha}$.

*Proof* Let $F(x,x_1,\dots,x_n)$ be a $\Delta_0$-formula. Choose $a,a_1,\dots,a_n \in \mathscr{T}_\alpha$ and define $\beta := \max\{stg(a), stg(a_1),\dots,stg(a_n)\}+1$. Put

$$b := \{x \in \mathsf{L}_\beta \mid x \in a \wedge F(x,a_1,\dots,a_n)\}.$$

From (Taut) we get

$$\overset{\circ}{\vdash} t \mathrel{\not\in} b, t \in a \wedge F(t,a_1,\dots,a_n) \tag{i}$$

for all $t \in \mathscr{T}_\beta$. Hence

$$\overset{\circ}{\vdash} (\forall x \in b)[x \in a \wedge F(x,a_1,\dots,a_n)] \tag{ii}$$

by $(\forall^b)$. Conversely we have

$$\overset{\circ}{\vdash} t \mathrel{\not\in} a, \neg F(t,a_1,\dots,a_n), t \in a \wedge F(t,a_1,\dots,a_n) \wedge t = t$$

for all $t \in \mathscr{T}_{stg(a)}$ by (11.15), (11.14) and (Sent). Using the definition of $t \mathrel{\underline{\in}} b$ this means

$$\overset{\circ}{\vdash} t \mathrel{\not\in} a, \neg F(t,a_1,\dots,a_n), t \mathrel{\underline{\in}} b \wedge t = t. \tag{iii}$$

Since $stg(t) < stg(b)$ we obtain

$$\overset{\circ}{\vdash} t \mathrel{\not\in} a, \neg F(t,a_1,\dots,a_n) \vee t \in b \tag{iv}$$

by $(\in)$ and $(\bigvee)^\circ$. Hence

$$\overset{\circ}{\vdash} (\forall x \in a)[F(x,a_1,\dots,a_n) \rightarrow x \in b]. \tag{v}$$

By (ii) and (v) we therefore have

$$\overset{\circ}{\vdash} (\forall x \in b)[x \in a \wedge F(x,a_1,\dots,a_n)] \wedge (\forall x \in a)[F(x,a_1,\dots,a_n) \rightarrow x \in b]. \tag{vi}$$

Since $b \in \mathscr{T}_\alpha$ this yields

$$\overset{\circ}{\vdash} \begin{array}{l}(\exists z \in \mathsf{L}_\alpha)\big[(\forall x \in z)[x \in a \wedge F(x,a_1,\dots,a_n)] \wedge \\ \qquad (\forall x \in a)[F(x,a_1,\dots,a_n) \rightarrow x \in z]\big]\end{array} \tag{vii}$$

by an inference $(\bigvee)^\circ$. By iterated applications of $(\forall^{\mathsf{L}_\alpha})$ we finally obtain the provability of

$$(\forall x_1 \in \mathsf{L}_\alpha)\dots,(\forall x_n \in \mathsf{L}_\alpha)(\forall x \in \mathsf{L}_\alpha)(\exists z \in \mathsf{L}_\alpha)\big[z = \{u \in x \mid F(x,x_1,\dots,x_n)\}\big]. \square$$

To deal also with the foundation scheme we prove a foundation lemma.

**11.12.7 Lemma** *(Foundation Lemma). Let $\mathscr{H}$ be a Cantorian closed Skolem–hull operator and $F(x,x_1,\dots,x_n)$ an arbitrary $\mathscr{L}(\in)$-sentence. Then*

$$\mathscr{H} \,\Big|\frac{2\cdot\rho+5\cdot(stg\,(a)+1)}{0}\; \begin{array}{l}\neg F(a,a_1,\dots,a_n)^{\mathsf{L}_\alpha},\\ (\exists x\in \mathsf{L}_\alpha)\big[F(x,a_1,\dots,a_n)^{\mathsf{L}_\alpha} \wedge (\forall y\in x)[\neg F(y,a_1,\dots,a_n)^{\mathsf{L}_\alpha}]\big]\end{array}$$

*for all $\alpha$ and $\mathscr{L}_{RS}$-terms $a,a_1,\dots,a_n \in \mathscr{T}_\alpha$ with $\rho := \mathrm{rnk}(F(a,a_1,\dots,a_n)^{\mathsf{L}_\alpha})$.*

*Proof* Choose $a,a_1,\dots,a_n \in \mathscr{T}_\alpha$. To improve readability we suppress mentioning the parameters $a_1,\dots,a_n$. We prove the lemma by induction on $stg\,(a)$ and put $\rho_a := 2{\cdot}\mathrm{rnk}(F(a)^{\mathsf{L}_\alpha})$. For $b \in \mathscr{T}_{stg\,(a)}$ we get

$$\mathscr{H} \,\Big|\frac{\rho_b+5\cdot(stg\,(b)+1)}{0}\; \neg F(b)^{\mathsf{L}_\alpha}, (\exists x\in \mathsf{L}_\alpha)\big[F(x)^{\mathsf{L}_\alpha} \wedge (\forall y\in x)[\neg F(y)^{\mathsf{L}_\alpha}]\big] \tag{i}$$

by the induction hypothesis. From (i) we obtain structurally

$$\mathscr{H} \,\Big|\frac{\rho_b+5\cdot(stg\,(b)+1)}{0}\; b \notin a, \neg F(b)^{\mathsf{L}_\alpha}, (\exists x\in \mathsf{L}_\alpha)\big[F(x)^{\mathsf{L}_\alpha} \wedge (\forall y\in x)[\neg F(y)^{\mathsf{L}_\alpha}]\big] \tag{ii}$$

which by two inferences $(\bigvee)$ and an inference $(\bigwedge)$ implies

$$\mathscr{H} \,\Big|\frac{\rho_b+5\cdot stg\,(a)+3}{0}\; (\forall y\in a)\neg F(y)^{\mathsf{L}_\alpha}, (\exists x\in \mathsf{L}_\alpha)\big[F(x)^{\mathsf{L}_\alpha} \wedge (\forall y\in x)[\neg F(y)^{\mathsf{L}_\alpha}]\big]. \tag{iii}$$

By tautology we obtain

$$\mathscr{H} \,\Big|\frac{\rho_a}{0}\; \neg F(a)^{\mathsf{L}_\alpha}, F(a)^{\mathsf{L}_\alpha}, (\exists x\in \mathsf{L}_\alpha)\big[F(x)^{\mathsf{L}_\alpha} \wedge (\forall y\in x)[\neg F(y)^{\mathsf{L}_\alpha}]\big]. \tag{iv}$$

Since $\rho_b \le \rho_a$ we obtain from (iii) and (iv) by $(\bigwedge)$

$$\mathscr{H} \,\Big|\frac{\rho_a+5\cdot stg\,(a)+4}{0}\; \begin{array}{l}\neg F(a)^{\mathsf{L}_\alpha}, F(a)^{\mathsf{L}_\alpha} \wedge (\forall y\in a)\neg F(y)^{\mathsf{L}_\alpha},\\ (\exists x\in \mathsf{L}_\alpha)\big[F(x)^{\mathsf{L}_\alpha} \wedge (\forall y\in x)[\neg F(y)^{\mathsf{L}_\alpha}]\big]\end{array}. \tag{v}$$

Since $a \in \mathscr{T}_\alpha$ we obtain from (v)

$$\mathscr{H} \,\Big|\frac{\rho_a+5\cdot stg\,(a)+5}{0}\; \neg F(a)^{\mathsf{L}_\alpha}, (\exists x\in \mathsf{L}_\alpha)\big[F(x)^{\mathsf{L}_\alpha} \wedge (\forall y\in x)[\neg F(y)^{\mathsf{L}_\alpha}]\big] \tag{vi}$$

by an inference $(\bigvee)$. The parameter conditions are satisfied because $a$ occurs in the conclusion. Since $\rho_a + 5{\cdot}stg\,(b)+5 \le \rho_a + 5{\cdot}stg\,(a)$, we obtain the claim from (vi). □

As an immediate consequence of the Foundation Lemma we obtain the following foundation theorem.

**11.12.8 Theorem** *(Foundation Theorem) Let $\mathscr{H}$ be a Cantorian closed Skolem–hull operator, $\alpha$ a limit ordinal and $F(x,x_1,\dots,x_n)$ an $\mathscr{L}(\in)$–formula. Then*

$$\mathscr{H} \,\Big|\frac{2\cdot\omega\cdot\alpha+\alpha}{0}\; \begin{array}{l}(\forall \vec{x}\in \mathsf{L}_\alpha)\big[(\exists z\in \mathsf{L}_\alpha)F(z,\vec{x})^{\mathsf{L}_\alpha} \to\\ \qquad (\exists z\in \mathsf{L}_\alpha)[F(z,\vec{x})^{\mathsf{L}_\alpha} \wedge (\forall y\in z)\neg F(y,\vec{x})^{\mathsf{L}_\alpha}]\big].\end{array}$$

Now we have collected all the facts which are needed to compute an upper bound for the $\Sigma$-ordinal of ($\Pi_2$–REF).

**11.12.9 Theorem** $\|(\Pi_2\text{–REF})\|_\Sigma \leq \psi(\varepsilon_{\Omega+1})$.

*Proof* Assume $(\Pi_2\text{–REF}) \vdash F$ for a $\Sigma$-sentence $F$. Then there are finitely many axioms $A_1,\dots,A_n$ of $(\Pi_2\text{–REF})$ such that

$$\underset{\mathsf{T}}{\overset{m}{\vdash}} \neg A_1,\dots,\neg A_n, F \tag{i}$$

in the Tait calculus augmented by the rules for the bounded quantifiers. According to Theorem 11.12.3 we obtain natural numbers $n$ and $r$ such that

$$\mathscr{H} \underset{\Omega+r}{\overset{\Omega\cdot\omega^n}{\vdash}} \neg A_1^{\mathsf{L}_\Omega},\dots,\neg A_n^{\mathsf{L}_\Omega}, F^{\mathsf{L}_\Omega}. \tag{ii}$$

For axioms $A_i$ which are different from instances of *($\Pi_2$–REF)* we obtain by equations (11.21) through (11.25), Lemma 11.12.4 through 11.12.6 and Theorem 11.12.8 natural numbers $n_i$ such that

$$\mathscr{H} \underset{0}{\overset{\Omega\cdot\omega^{n_i}}{\vdash}} A_i^{\mathsf{L}_\Omega}. \tag{iii}$$

By cuts we therefore obtain natural numbers $m$ and $s$ such that

$$\mathscr{H} \underset{\Omega+s}{\overset{\Omega\cdot\omega^m}{\vdash}} \neg B_1^{\mathsf{L}_\Omega},\dots,\neg B_m^{\mathsf{L}_\Omega}, F^{\mathsf{L}_\Omega} \tag{iv}$$

where the $B_i$'s are all the instances of the scheme *($\Pi_2$–REF)*, which were needed in the formal proof of $F$. This holds for any Cantorian closed Skolem–hull operator and is especially true for the minimal operator $\mathscr{B}$ defined in Chap. 9. Applying cut elimination we obtain an ordinal $\alpha \in \mathscr{B}_{\varepsilon_{\Omega+1}}(\emptyset) \cap \varepsilon_{\Omega+1}$ such that

$$\mathscr{B} \underset{\Omega}{\overset{\alpha}{\vdash}} \neg B_1^{\mathsf{L}_\Omega},\dots,\neg B_m^{\mathsf{L}_\Omega}, F^{\mathsf{L}_\Omega}. \tag{v}$$

By the Collapsing Theorem (Theorem 11.11.2) we obtain

$$\mathscr{B}_{\omega^{\omega^\alpha}+1} \underset{\Omega}{\overset{\psi(\omega^{\omega^\alpha})}{\vdash}} F^{\mathsf{L}_\Omega}. \tag{vi}$$

By semantical cut elimination (Theorem 11.10.2) we then obtain

$$\overset{\psi(\omega^{\omega^\alpha})}{\Vdash\!=\!=} F^{\mathsf{L}_\Omega} \tag{vii}$$

which implies by Corollary 11.9.11 $L_{\psi(\omega^{\omega^\alpha})} \models F$. So we have $L_{\psi(\varepsilon_{\Omega+1})} \models F$ for all $\Sigma$-sentences provable in $(\Pi_2\text{–REF})$. Hence $\|(\Pi_2\text{–REF})\|_\Sigma \leq \psi(\varepsilon_{\Omega+1})$. □

In combination with Theorem 11.7.4 we finally obtain the ordinal analysis of $(\Pi_2\text{–REF})$.

**11.12.10 Theorem** *(Ordinal analysis of Kripke–Platek set theory)* $\|\mathsf{KP}\omega\| = \|\mathsf{KP}\omega\|_\Sigma = \psi(\varepsilon_{\Omega+1})$ *and* $\|(\Pi_2\text{–REF})\| = \|(\Pi_2\text{–REF})\|_\Sigma = \psi(\varepsilon_{\Omega+1})$.

*Proof* By Theorem 11.5.4 the theory $\mathsf{ID}_1$ is a subtheory of $\mathsf{KP}\omega$ which in turn is a subtheory of $(\Pi_2\text{–REF})$. Therefore we obtain $\psi(\varepsilon_{\Omega+1}) = \|\mathsf{ID}_1\| \leq \|\mathsf{KP}\omega\| \leq \|\mathsf{KP}\omega\|_\Sigma \leq \|(\Pi_2\text{–REF})\|_\Sigma \leq \psi(\varepsilon_{\Omega+1})$ and therefore also $\psi(\varepsilon_{\Omega+1}) = \|\mathsf{KP}\omega\| \leq \|(\Pi_2\text{–REF})\| \leq \|(\Pi_2\text{–REF})\|_\Sigma = \psi(\varepsilon_{\Omega+1})$. □

# Chapter 12
# Predicativity Revisited

The ordinal analyses of $\mathsf{ID}_1$ and $(\Pi_2\text{–}\mathsf{REF})$ show that their crucial "impredicative" axioms are $ID_1^1$ and the reflection scheme. Their presence forced us to develop the collapsing machinery in the ordinal analysis. This, however, is not the complete truth. The impredicative character of these axioms comes only in connection with foundation.[1] This observation is due to Jäger (cf. [48] and [52]). He has shown that theories which are considerably stronger[2] than $\mathsf{KP}\omega$ become reducible to predicative theories as soon as the foundation scheme is removed or restricted. The methods of predicative proof theory are not in the center of this book. However, we have introduced most of the main notions and techniques needed in Jäger's work. Therefore, we sketch Jäger's approach leaving most of the proofs as exercises. One aim of this chapter is to compute $\Gamma_0$ as an upper bound for proof-theoretic ordinal of the theories $\mathsf{KPi}_0$ and $\mathsf{KPl}_0$ introduced in Sect. 11.6. Together with Exercise 11.6.5 this also yields $\Gamma_0$ as upper bound for the proof-theoretic ordinal of the theory $(\mathsf{ATR})_0$ which in turn with Exercise 8.5.31 implies that $\Gamma_0$ is the exact proof-theoretic ordinal for all these theories.

## 12.1 Admissible Extension

The recursion theoretic background of Jäger's research is the theory of the next admissible set. Let $T$ be a theory in a language $\mathscr{L}(T)$ and $\mathfrak{M}$ a model of $T$. Then we can build an admissible universe $\mathbb{A}_{\mathfrak{M}}$ above $\mathfrak{M}$ in which the elements of $\mathfrak{M}$ act as "urelements", i.e., objects without elements such that the domain $M$ of $\mathfrak{M}$ is an element of $\mathbb{A}_{\mathfrak{M}}$. If $T$ is a theory in the language of set theory and $\mathfrak{M}$ a transitive model of $T$ we may extend $\mathfrak{M}$ to an admissible set that contains $M$ as an element. Here, we will mainly deal with theories in the language of set theory and introduce the

[1] This chapter is intended to give some background information and is not needed in the following sections. It may therefore be omitted in a first reading.

[2] Their ordinal analysis needs a second or even further steps into impredicativity.

*admissible extension* $T^+$ of a theory $T$.[3] Since we are going to analyze set theories without full foundation we cannot longer rely on the $\Sigma$-recursion theorem which we used to make clear that the $\Sigma$-ordinal of a set theory is an upper bound for its proof-theoretic ordinal. Therefore, we may be forced to use pseudo $\Pi^1_1$-sentences to define the proof-theoretic ordinal of theories which do not contain enough foundation. To provide pseudo $\Pi^1_1$-sentences let us assume that there are free second-order variables in the language of set theory. Since there are no defining axioms or rules for second-order variables they are harmless. However, if we count formulas $t \in X$ or $t \notin X$ among $\Delta_0$-formulas, we must be aware that $X$ stands only for classes which are $\Delta_0$-definable. Any of our theories formulated with second-order variables is a conservative extension of its first-order part. For the definition of the admissible extension, recall the theory $\mathsf{BST}^-$ (cf. p. 261) which contains the axioms of basic set theory $\mathsf{BST}$ (cf. p. 247) but the foundation scheme and (Union') replaced by (TranC). Let $\mathsf{KP}^-$ be $\mathsf{BST}^- + (\Delta_0\text{–Coll})$ and $\mathsf{KP}^r$ be $\mathsf{BST}^- + (\text{Found}) + (\Delta_0\text{–Coll})$.

**12.1.1 Definition** *(Admissible extension)* Let $T$ be a theory in the language of set theory. To obtain the language $\mathcal{L}(T^+)$, we augment the language $\mathcal{L}(T)$ by a constant $\mathsf{U}$ whose intended meaning is a universe for the theory $T$.

The axioms of $T^+$ comprise:

- For all axioms $A \in T$ the sentence $A^{\mathsf{U}}$.
- The sentence $Tran(\mathsf{U})$ together with all sentences $t \in \mathsf{U}$ for closed $\mathcal{L}(T)$-terms $t$.
- All axioms in $\mathsf{KP}^-$.

Observe that $T^+ \vdash (\exists y)[y = \mathsf{U}]$ and $T^+ \vdash (\forall x \in \mathsf{U})(\exists y)[x \in y]$, i.e., $\mathsf{U}$ is a set and a transitive subset of the universe. If a structure $\mathbb{A} := (A, U, E)$ is a model of $T^+$ then $U = \mathsf{U}^{\mathbb{A}}$ is an element of $A$ and $\mathfrak{A} := (U, E)$ is an $\mathcal{L}(T)$-structure such that $\mathfrak{A} \models T$. If $\mathbb{A}$ is well-founded and transitive and $E$ the standard membership relation then $\mathbb{A}$ is an admissible structure above $\mathfrak{A}$ in the sense of [4]. Obviously $\mathbb{A}$ is an end-extension of $\mathfrak{A}$. However, observe that we have not included the scheme of foundation in the axioms of $T^+$. As we will see, this makes the theory proof-theoretically much weaker.

## 12.2 $\mathfrak{M}$-Logic

Let $T$ be a theory and $\mathfrak{M}$ be the intended standard structure for $\mathcal{L}(T)$. Our plan is to study $T^+$ by capturing an admissible segment of the constructible hierarchy above the "urelement structure" $\mathfrak{M}$ by a semi-formal system $RS_{\mathfrak{M}}$. In Definition 7.3.5 we designed a semi-formal system which is $\omega$-complete, i.e., complete with respect to the standard model $\mathbb{N}$. To obtain a semi-formal system which is complete with

[3] This differs from Jäger's notation who called it $T^e$.

respect to a countable $\mathscr{L}(T)$-structure $\mathfrak{M}$ we have to introduce a variant of this semi-formal system.

**12.2.1 Definition** Assume that $\mathfrak{M}$ is a countable $\mathscr{L}(T)$-structure such that for every element $m$ in the domain of $\mathfrak{M}$ there is an $\mathscr{L}(T)$-term $\underline{m}$ representing $m$, i.e., we require $\underline{m}^{\mathfrak{M}} = m$. We put all sentences in the diagram of $\mathfrak{M}$ into $\bigwedge$–type and alter Definition 5.3.3 by defining

$$\mathrm{CS}^{\mathfrak{M}}((\mathsf{Q}x)F(x)) := \langle F_x(\underline{m}) \,|\, m \in dom(\mathfrak{M})\rangle.$$

To get a verification calculus $\mathfrak{M} \models^{\alpha} \Delta$ similar to that in Definition 5.4.3 we have to modify *(Ax)* to *(Ax)*$_{\mathfrak{M}}$ below.

$(Ax)_{\mathfrak{M}}$ If $t^{\mathfrak{M}} = s^{\mathfrak{M}}$ then $\mathfrak{M} \models^{\alpha} \Delta, s \not\varepsilon X, t \varepsilon X$ for all ordinals $\alpha$.

Finally, we define $\mathfrak{M} \models^{\alpha} \Delta$ using the rules $(Ax)_{\mathfrak{M}}$, $(\bigwedge)$ and $(\bigvee)$. Then $\mathfrak{M} \models^{\alpha}$ satisfies a counterpart of the $\omega$-completeness theorem (Theorem 5.4.9). This is known as $\mathfrak{M}$-completeness theorem.

Extending the verification calculus as in Definition 7.3.5 by a cut-rule, we obtain a semi-formal system $\mathfrak{M} \vdash^{\alpha}_{\rho} \Delta$ for $\mathscr{L}(T)$. This semi-formal system is uniquely determined by the structure $\mathfrak{M}$. Therefore, we commonly do not distinguish the structure $\mathfrak{M}$ and its associated semi-formal system.

**12.2.2 Exercise** (*$\mathfrak{M}$-completeness theorem* ) Let $\mathfrak{M}$ be a countable $\mathscr{L}(T)$-structure. Show that $\mathfrak{M} \models (\forall X)F(X)$ holds true iff there is a countable ordinal $\alpha$ such that $\mathfrak{M} \models^{\alpha} F(X)$.

Hint: Modify the proof of the $\omega$-completeness theorem.

**12.2.3 Definition** Let $T$ be a theory, $\mathfrak{M}$ a countable $\mathscr{L}(T)$-structure. For a finite set $\Delta$ of $\mathscr{L}(T)$-formulas, we denote by $T_{\mathfrak{M}} \vdash^{\alpha}_{\rho} \Delta$ that there is a finite subset $\Gamma \subseteq T$ such that $\mathfrak{M} \vdash^{\alpha}_{\rho} \neg\Gamma, \Delta$.

**12.2.4 Exercise** Let $T$ be a theory and $\mathfrak{M}$ a countable $\mathscr{L}(T)$-structure. Show that $T \vdash F$ implies $T_{\mathfrak{M}} \vdash^{<\omega}_{0} F$ for all $\mathscr{L}(T)$-sentences $F$.

Hint: This is the standard embedding argument. Show by induction on $n$ that $\vdash^{n}_{\mathsf{T}} \Delta(\vec{x})$ implies $\mathfrak{M} \vdash^{n}_{0} \Delta(\vec{\underline{m}})$ for all tuples $\vec{m}$ of elements in the domain of $\mathfrak{M}$.

**12.2.5 Exercise** Show that $\mathfrak{M} \vdash^{\alpha}_{\rho} \Delta, F$ for an $\mathscr{L}(\mathfrak{M})$-sentence $F$ which is false in $\mathfrak{M}$ implies $\mathfrak{M} \vdash^{\alpha}_{\rho} \Delta$.

Hint: This is an easy induction on $\alpha$. In case that $F$ is not the critical formula of the last inference you get the claim immediately from the induction hypothesis. In case that $F$ is the critical formula use that $\mathfrak{M} \not\models F$ implies $\mathfrak{M} \not\models G$ for all (some) $G \in \mathrm{CS}(F)$ if $F \in \bigvee$–type $(F \in \bigwedge$–type ).

## 12.3 Extending Semi-formal Systems

The main tools in studying the admissible extension of a theory are semi-formal systems for ramified set theory. To this end we will extend a given semi-formal system $\mathbb{S}$ into two different directions. One direction is to extend $\mathbb{S}$ to a semi-formal system $RS_{\mathbb{S}}$ which formalizes a set universe above the domain of $\mathbb{S}$. The other extension is a system $\mathbb{S}^{\mathsf{U}}$ which engrafts first-order logic into $\mathbb{S}$. We start by fixing an abstract notion of semi-formal systems.

**12.3.1 Definition** A semi-formal system[4] $\mathbb{S}$ over a countable structure $\mathfrak{M}$ is given by a language $\mathscr{L}(\mathbb{S})$, comprising the language of $\mathfrak{M}$, whose formulas are arranged in $\bigwedge$–type , $\bigvee$–type , and possibly also atomic formulas without types such that every formula $F \in (\bigvee\text{–type} \cup \bigwedge\text{–type})$ is equipped with a characteristic sequence $\mathrm{CS}_{\mathbb{S}}(F)$. Moreover, we assume that there is a well-defined rank function $\mathrm{rnk}_{\mathbb{S}}(F)$ for the formulas in the language of $\mathbb{S}$ satisfying $\mathrm{rnk}_{\mathbb{S}}(G) < \mathrm{rnk}_{\mathbb{S}}(F)$ for $G \in \mathrm{CS}_{\mathbb{S}}(F)$. The derivability relation $\mathbb{S} \mathop{|\!\!\frac{\alpha}{\rho}} \Delta$ for finite $\mathscr{L}(\mathbb{S})$-formulas is given by the axioms

$(Ax)_L$ If $A$ is an atomic formula not in $\bigwedge$–type $\cup \bigvee$–type and $\{A, \neg A\} \subseteq \Delta$ then $\mathop{|\!\!\frac{\alpha}{\rho}} \Delta$ holds true for all ordinals $\alpha$ and $\rho$

and

$(Ax)_{\mathfrak{M}}$ If $s$ and $t$ are $\mathfrak{M}$-terms such that $s^{\mathfrak{M}} = t^{\mathfrak{M}}$ then $\mathop{|\!\!\frac{\alpha}{\rho}} \Delta, s \in X, t \notin X$ holds true for all ordinals $\alpha$ and $\rho$

together with the familiar rules $(\bigwedge)$, $(\bigvee)$ and (cut).

If $T$ is a theory in the language of $\mathbb{S}$ then we denote by $T_{\mathbb{S}} \mathop{|\!\!\frac{\alpha}{\rho}} \Delta$ that there is a finite set $\Lambda \subseteq T \cup \mathsf{IDEN}$ such that $\mathbb{S} \mathop{|\!\!\frac{\alpha}{\rho}} \neg\Lambda, \Delta$ where $\mathsf{IDEN}$ is the set of identity axioms (cf. p. 61). We say that $\Delta$ is a consequence of $T$ in the framework of $\mathbb{S}$-logic. We view $T_{\mathbb{S}}$ as a semi-formal system with additional axioms. When talking about semi-formal systems, we include semi-formal systems with additional axioms.

For convenience we require that there are no function symbols in the language of $\mathbb{S}$. This means no restriction for our studies.

**12.3.2 Remark** We did not, in general, require that the language of $\mathscr{L}(\mathbb{S})$ includes the membership symbol. To avoid tedious case distinctions, we assume that $\mathscr{L}(\mathbb{S})$ always contains $\in$ (and thus also $\notin$) and put $\mathrm{CS}_{\mathbb{S}}(s \in t) := \mathrm{CS}_{\mathbb{S}}(s \notin t) := \emptyset$ for $\mathscr{L}(\mathbb{S})$-terms $s$ and $t$ if $\in$ is not among the regular symbols of $\mathscr{L}(\mathbb{S})$. Observe that in languages without membership symbols $s \notin t$ is a sentence in $\bigwedge$–type with empty characteristic sequence. We therefore obtain $\mathbb{S} \mathop{|\!\!\frac{0}{0}} s \notin t, (\exists x \in t)(\forall y \in x)[y \notin t]$, for all $\mathscr{L}(\mathbb{S})$-terms $s$ and $t$ and thus $\mathbb{S} \mathop{|\!\!\frac{<\omega}{0}}$ (Found).

[4] Here we understand semi-formal systems in the liberated sense of Note 7.3.17.

Our plan is now to enhance $\mathbb{S}$-logic by additional set theoretic axioms. These additional axioms are formulated in first-order logic. Therefore, we have to engraft first-order logic into $\mathbb{S}$-logic. This is done in the following definition, where we allow $\mathbb{S}$ to be a semi-formal systems with additional axioms.

**12.3.3 Definition** *(The extension $\mathbb{S}^{\mathsf{U}}$)* Let $\mathbb{S}$ be a semi-formal system over a countable structure $\mathfrak{M}$. We augment the language $\mathcal{L}(\mathbb{S})$ by a new constant $\mathsf{U}$ (for urelements) and free variables (if not already present in the language $\mathcal{L}(\mathbb{S})$) to the language $\mathcal{L}(\mathbb{S}^{\mathsf{U}})$. Terms and formulas of $\mathcal{L}(\mathbb{S}^{\mathsf{U}})$ are defined by the following clauses:

- Every $\mathcal{L}(\mathbb{S})$-term is an $\mathcal{L}(\mathbb{S}^{\mathsf{U}})$-term.
- Every new free variable and the constant $\mathsf{U}$ are $\mathcal{L}(\mathbb{S}^{\mathsf{U}})$-terms.
- If $F$ is an $\mathcal{L}(\mathbb{S})$-formula then $F^{\mathsf{U}}$ is an $\mathcal{L}(\mathbb{S}^{\mathsf{U}})$-formula. We put $F^{\mathsf{U}}$ in $\bigwedge$–type ($\bigvee$–type) iff $F$ is in $\bigvee$–type ($\bigwedge$–type) and define $\mathrm{CS}^{\mathbb{S}^{\mathsf{U}}}(F^{\mathsf{U}}) := \langle G^{\mathsf{U}} \,|\, G \in \mathrm{CS}^{\mathbb{S}}(F)\rangle$.
- We put $\mathsf{U} \in \mathsf{U}$ as well as $\mathsf{U} \neq \mathsf{U}$ in $\bigvee$–type and dually $\mathsf{U} \notin \mathsf{U}$ and $\mathsf{U} = \mathsf{U}$ in $\bigwedge$–type. All these sentences have empty characteristic sequences.
- If $t$ is an $\mathcal{L}(\mathbb{S})$-term then $t \in \mathsf{U}$ is in $\bigvee$–type and we put $\mathrm{CS}(t \in \mathsf{U}) = \big\langle t = s \,|\, s \text{ is an } \mathcal{L}(\mathbb{S})\text{-term}\big\rangle$. Dually $t \notin \mathsf{U}$ is in $\bigwedge$–type with $\mathrm{CS}(t \notin \mathsf{U}) = \big\langle t \neq s \,|\, s \text{ is an } \mathcal{L}(\mathbb{S})\text{-term}\big\rangle$.
- If $s$ and $t$ are $\mathcal{L}(\mathbb{S}^{\mathsf{U}})$-terms different from $\mathsf{U}$ then $s = t$, $s \neq t$, $s \in t$ and $s \notin t$ are atomic $\mathcal{L}(\mathbb{S}^{\mathsf{U}})$-formulas. Every atomic $\mathcal{L}(\mathbb{S}^{\mathsf{U}})$-formula is an $\mathcal{L}(\mathbb{S}^{\mathsf{U}})$-formula.

  Atomic formulas which do not belong to $\mathcal{L}(\mathbb{S})$ and do not contain $\mathsf{U}$ are neither in $\bigvee$–type nor in $\bigwedge$–type.
- The $\mathcal{L}(\mathbb{S}^{\mathsf{U}})$-formulas are closed under the positive boolean operations $\wedge$ and $\vee$. Conjunctions belong to $\bigwedge$–type while disjunctions are in $\bigvee$–type. The characteristic sequences of conjunctions and disjunctions are defined in the obvious way.
- If $F(u)$ is an $\mathcal{L}(\mathbb{S}^{\mathsf{U}})$-formula where $u$ is a free variable then $(\forall x)F_u(x)$ and $(\forall x \in \mathsf{U})F_u(x)$ are $\mathcal{L}(\mathbb{S}^{\mathsf{U}})$-formulas in $\bigwedge$–type and $(\exists x)F_u(x)$ and $(\exists x \in \mathsf{U})F_u(x)$ are $\mathcal{L}(\mathbb{S}^{\mathsf{U}})$-formula in $\bigvee$–type. Let

  $$\mathrm{CS}^{\mathbb{S}^{\mathsf{U}}}((\mathsf{Q}x \in \mathsf{U})F_u(x)) := \langle F_u(t) \,|\, t \text{ is an } \mathcal{L}(\mathbb{S})\text{-term}\rangle$$

  and

  $$\mathrm{CS}^{\mathbb{S}^{\mathsf{U}}}((\mathsf{Q}x)F_u(x)) := \langle F_u(t) \,|\, t \text{ is an } \mathcal{L}(\mathbb{S}^{\mathsf{U}})\text{-term}\rangle.$$

We define

- $\text{rnk}_{\mathsf{U}}(F) := \begin{cases} 0 & \text{if } F \notin \bigvee\text{–type} \cup \bigwedge\text{–type} \\ \sup\{\text{rnk}_{\mathsf{U}}(G)+1 \mid G \in \text{CS}^{\mathbb{S}^{\mathsf{U}}}(F)\} & \text{if } F \in \bigvee\text{–type} \cup \bigwedge\text{–type}. \end{cases}$

Generalizing Definition 11.10.1, we obtain the proof relation $\mathbb{S}^{\mathsf{U}} \vdash^{\alpha}_{\rho} \Delta$ for a finite set $\Delta$ of $\mathscr{L}(\mathbb{S}^{\mathsf{U}})$-formulas using $(Ax)_L$, $(Ax)_{\mathfrak{M}}$, $(\bigwedge)$, $(\bigvee)$ and (cut) where the cut rank $\rho$ is computed according to $\text{rnk}_{\mathsf{U}}$.

If $\mathbb{S}$ is a semi-formal system with additional axioms $T$, then the sentences $T^{\mathsf{U}}$ become additional axioms of $\mathbb{S}^{\mathsf{U}}$. The obvious abbreviation $\mathbb{S}^{\mathsf{U}} \vdash^{<\alpha}_{<\rho} \Delta$ denotes that there are ordinals $\alpha_0 < \alpha$ and $\rho_0 < \rho$ such that $\mathbb{S}^{\mathsf{U}} \vdash^{\alpha_0}_{\rho_0} \Delta$. Similarly, we use $\mathbb{S}^{\mathsf{U}} \vdash^{<\alpha}_{\rho_0} \Delta$.

Observe that $\mathbb{S}^{\mathsf{U}}$ is again a semi-formal system over $\mathfrak{M}$.

**12.3.4 Exercise** (a) Check that $\text{rnk}_{\mathsf{U}}(F)$ is well defined.
(b) Show that the Predicative Elimination Lemma (Lemma 7.3.14) and the Elimination Theorem (Theorem 7.3.15) remain true for the semi-formal system $\mathbb{S}^{\mathsf{U}}$.

Hint: (a) Pick an additively indecomposable ordinal $\eta$ such that $\text{rnk}_{\mathbb{S}}(F) < \eta$ holds true for all $\mathscr{L}(\mathbb{S})$-formulas $F$. Then define $\widehat{\text{rnk}}(F^{\mathsf{U}}) = \text{rnk}_{\mathbb{S}}(F)$ for $\mathscr{L}(\mathbb{S})$-formulas $F$, $\widehat{\text{rnk}}(F) := \eta$ for all atomic formulas which are not $\mathscr{L}(\mathbb{S})$-formulas, $\widehat{\text{rnk}}(A \circ B) := \max\{\widehat{\text{rnk}}(A), \widehat{\text{rnk}}(B)\} + 1$ and $\widehat{\text{rnk}}((\mathsf{Q}x)F(x)) = \widehat{\text{rnk}}(F(u)) + 1$ and show that $G \in \text{CS}^{\mathbb{S}^{\mathsf{U}}}(F)$ implies $\widehat{\text{rnk}}(G) < \widehat{\text{rnk}}(F)$.
(b) Just modify the proofs of 7.3.14 and 7.3.15.

**12.3.5 Exercise** Let $\Delta$ be a finite set of $\mathscr{L}(\mathbb{S})$-formulas.
(a) Show that $\mathbb{S} \vdash^{\alpha}_{\rho} \Delta$ implies $\mathbb{S}^{\mathsf{U}} \vdash^{\alpha}_{\rho} \Delta^{\mathsf{U}}$ and conclude that $T_{\mathbb{S}} \vdash^{\alpha}_{\rho} \Delta$ implies $T^{\mathsf{U}}_{\mathbb{S}^{\mathsf{U}}} \vdash^{\alpha}_{\rho} \Delta^{\mathsf{U}}$.
(b) Show that $\mathbb{S}^{\mathsf{U}} \vdash^{\alpha}_{0} \Delta^{\mathsf{U}}$ implies $\mathbb{S} \vdash^{\alpha}_{0} \Delta$. Hence $T^{\mathsf{U}}_{\mathbb{S}^{\mathsf{U}}} \vdash^{\alpha}_{0} \Delta^{\mathsf{U}}$ entails $T_{\mathbb{S}} \vdash^{\alpha}_{0} \Delta$.

Hint: (a) holds essentially by definition.
(b) This is an easy induction on $\alpha$.

The next step is to resolve the additional set theoretic axioms enhancing a semi-formal system $\mathbb{S}$ into a ramified set theory above $\mathsf{U}$. The ramified set theory above $\mathbb{S}$ is formalized by the semi-formal system $RS_{\mathbb{S}}$ introduced below.

**12.3.6 Definition** *(The language $\mathscr{L}(RS_{\mathbb{S}})$)* Assume again that $\mathbb{S}$ is a semi-formal system over a countable structure $\mathfrak{M}$, possibly with additional axioms. We define the $\mathscr{L}(RS_{\mathbb{S}})$-terms inductively by the following clauses.

- Every $\mathscr{L}(\mathbb{S})$-term is an $\mathscr{L}(RS_{\mathbb{S}})$-term of stage $-1$.

- Every constant $\mathsf{L}^{\mathbb{S}}_{\alpha}$ is an $\mathscr{L}(RS_{\mathbb{S}})$-term of stage $\alpha$.

- Let $G(u,\vec{u})$ be an $\mathscr{L}(\mathbb{S}^{\mathsf{U}})$-formula such that $u,\vec{u}$ comprises all its free variables and $\vec{a}$ a tuple of closed $\mathscr{L}(RS_{\mathbb{S}})$-terms of stages less than $\alpha$. By $G^{\alpha}$ we denote the formula which is obtained from $G$ replacing $\mathsf{U}$ by $\mathsf{L}_0^{\mathbb{S}}$ and restricting all unbounded quantifiers in $G$ to $\mathsf{L}_{\alpha}^{\mathbb{S}}$. Then $\{x \in \mathsf{L}_{\alpha}^{\mathbb{S}} \mid G_{u,\vec{u}}^{\alpha}(x,\vec{a})\}$ is an $\mathscr{L}(RS_{\mathbb{S}})$-term of stage $\alpha$.

We use $stg_{\mathbb{S}}^{RS}(t)$ as a token for the stage of an $\mathscr{L}(RS_{\mathbb{S}})$-term and denote by $\mathscr{T}_{\alpha}^{\mathbb{S}}$ the set of all $\mathscr{L}(RS_{\mathbb{S}})$-terms of stages less than $\alpha$.

Formulas are defined by the following clause.

- If $F(u_1,\dots,u_n)$ is an $\mathscr{L}(\mathbb{S}^{\mathsf{U}})$-formula whose free variables occur all in the list $u_1,\dots,u_n$ then $F_{u_1,\dots,u_n}^{\alpha}(a_1,\dots,a_n)$ is an $\mathscr{L}(RS_{\mathbb{S}})$-formula for all $n$-tuples $a_1,\dots,a_n$ of $\mathscr{L}(RS_{\mathbb{S}})$-terms.

Observe that for an $\mathscr{L}(\mathbb{S})$-formula $F$ we obtain $F^{\mathsf{U}}$ as an $\mathscr{L}(\mathbb{S}^{\mathsf{U}})$-formula and thus $(F^{\mathsf{U}})^{\alpha}$, i.e., $F^{\mathsf{L}_0^{\mathbb{S}}}$, as an $\mathscr{L}(RS_{\mathbb{S}})$-formula.

Since we (may) have equations in the "basis language" $\mathscr{L}(\mathbb{S})$ we need the equality symbol as basic symbol anyway. Therefore, we regard equations $s = t$ and $s \neq t$ of $\mathscr{L}(RS_{\mathbb{S}})$ no longer as defined but as separate atomic formulas. This is in contrast to the language of ramified set theory as introduced in Sect. 11.9 and forces us to define also the type and the characteristic sequences for equations and inequalities. However, it becomes immediately clear that the difference is only of technical nature.

**12.3.7 Definition** *(The $\bigvee$–type and $\bigwedge$–type of $\mathscr{L}(RS_{\mathbb{S}})$ and the characteristic sequences of $\mathscr{L}(RS_{\mathbb{S}})$-sentences.)* We define $\bigwedge$–type, $\bigvee$–type and $\mathrm{CS}_{\mathfrak{M}}^{\mathrm{RS}}(F)$ for $\mathscr{L}(RS_{\mathbb{S}})$-sentences by the following clauses:

- If $F$ is an formula $F_0^{\mathsf{L}_0^{\mathbb{S}}}$ where $F_0$ is an $\mathscr{L}(\mathbb{S})$-formula we put $F$ in $\bigwedge$–type ($\bigvee$–type) iff $F_0$ is in $\bigwedge$–type ($\bigvee$–type) of the semi-formal system $\mathbb{S}$ and define $\mathrm{CS}_{\mathbb{S}}^{\mathrm{RS}}(F) = \langle G^{\mathsf{L}_0^{\mathbb{S}}} \mid G \in \mathrm{CS}_{\mathbb{S}}(F_0)\rangle$.

For the following clauses, assume that $F$ is not of the form $F_0^{\mathsf{L}_0}$ for an $\mathscr{L}(\mathbb{S})$-formula $F_0$.

- Formulas $s = t$ belong to $\bigwedge$–type with

$$\mathrm{CS}_{\mathbb{S}}^{\mathrm{RS}}(s = t) := \langle (\forall x \in s)(x \in t), (\forall x \in t)(x \in s)\rangle.$$

- Formulas $s \neq t$ belong to $\bigvee$–type with

$$\mathrm{CS}_{\mathbb{S}}^{\mathrm{RS}}(s \neq t) := \langle (\exists x \in s)(x \notin t), (\exists x \in t)(x \notin s)\rangle.$$

For simplicity we put $\mathrm{CS}_{\mathbb{S}}^{\mathrm{RS}}(\mathsf{L}_0^{\mathbb{S}} = \mathsf{L}_0^{\mathbb{S}}) = \mathrm{CS}_{\mathbb{S}}^{\mathrm{RS}}(\mathsf{L}_0 \neq \mathsf{L}_0) := \emptyset$.

Atomic formulas $s \in t$ belong to $\bigvee$–type and we define their characteristic sequences as

$$\mathrm{CS}^{\mathrm{RS}}_{\mathbb{S}}(s \in t) := \begin{cases} \emptyset & \text{if } s \text{ is } \mathsf{L}^{\mathbb{S}}_0 \text{ and } t \in \mathscr{T}^{\mathbb{S}}_0 \\ \emptyset & \text{if } t \text{ and } s \text{ are both the term } \mathsf{L}^{\mathbb{S}}_0 \\ \langle s = a \,|\, a \in \mathscr{T}^{\mathbb{S}}_\alpha \rangle & \text{if } t \text{ is a term } \mathsf{L}^{\mathbb{S}}_\alpha \\ \langle s = a \wedge a \in t \,|\, a \in \mathscr{T}^{\mathbb{S}}_0 \rangle & \text{if } t \in \mathscr{T}^{\mathbb{S}}_0 \text{ and } s \text{ is not } \mathsf{L}^{\mathbb{S}}_0 {}^{5} \\ \langle s = a \wedge F(a) \,|\, a \in \mathscr{T}^{\mathbb{S}}_\alpha \rangle & \text{if } t = \{x \in \mathsf{L}^{\mathbb{S}}_\alpha \,|\, F(x)\}. \end{cases}$$

Dually, we count formulas $s \notin t$ among $\bigwedge$–type and define
$\mathrm{CS}^{\mathrm{RS}}_{\mathbb{S}}(s \notin t) := \langle \neg F \,|\, F \in \mathrm{CS}^{\mathrm{RS}}_{\mathbb{S}}(s \in t) \rangle$.

Quantifiers which are bounded by terms of stage $-1$ are regarded as defined. That is, for $t \in \mathscr{T}^{\mathbb{S}}_0$ we put

$$\mathrm{CS}^{\mathrm{RS}}_{\mathbb{S}}((\mathsf{Q}x \in t)F(x)) = \begin{cases} \langle s \in t \wedge F(s) \,|\, s \in \mathscr{T}^{\mathbb{S}}_0 \rangle & \text{if } \mathsf{Q} = \exists \\ \langle s \notin t \vee F(s) \,|\, s \in \mathscr{T}^{\mathbb{S}}_0 \rangle & \text{if } \mathsf{Q} = \forall \end{cases}.$$

In the definition of the characteristic sequences $\mathrm{CS}^{\mathrm{RS}}_{\mathbb{S}}(F)$ of the remaining sentences $F$ we follow the rules in Definition 11.9.6 where we replace $\mathscr{T}_\alpha$ by $\mathscr{T}^{\mathbb{S}}_\alpha$.

We define

$$\mathrm{rnk}^{RS}_{\mathbb{S}}(F) := \begin{cases} 0 & \text{if } F \notin (\bigvee\text{–type} \cup \bigwedge\text{–type}) \\ \sup\{\mathrm{rnk}^{RS}_{\mathbb{S}}(G) + 1 \,|\, G \in \mathrm{CS}^{\mathrm{RS}}_{\mathbb{S}}(F)\} & \text{if } F \in (\bigvee\text{–type} \cup \bigwedge\text{–type}). \end{cases}$$

We will, however, mostly omit the subscript and just write $\mathrm{rnk}(F)$ instead of $\mathrm{rnk}^{RS}_{\mathbb{S}}(F)$ if there is no danger of confusion. It is mostly clear from the context (or inessential) to which rank we refer.

Finally, we obtain the derivability relation for $RS_{\mathbb{S}} \,|\!\frac{\alpha}{\rho}\, \Delta$ using axioms $(Ax)_{\mathfrak{M}}$, $(Ax)_L$ and the rules $(\bigwedge)$, $(\bigvee)$ and (cut).

**12.3.8 Exercise** Show that $\mathrm{rnk}^{RS}_{\mathbb{S}}(F)$ is well-defined.

Hint: Let $\eta$ be an additively indecomposable ordinal such that $\mathrm{rnk}_{\mathbb{S}}(F) \le \eta$ for all formulas in $\mathscr{L}(\mathbb{S})$. Define $\widehat{\mathrm{rnk}}(F^{\mathsf{L}^{\mathbb{S}}_0}) = \mathrm{rnk}_{\mathbb{S}}(F)$ for all formulas $F$ in $\mathscr{L}(\mathbb{S})$. For $\mathscr{L}(\mathbb{S})$-terms $t$ put $\widehat{\mathrm{rnk}}(t) = \mathrm{rnk}_{\mathbb{S}}(t)$ or $\widehat{\mathrm{rnk}}(t) = 0$ in case that $\mathrm{rnk}_{\mathbb{S}}(t)$ is undefined. Then define $\widehat{\mathrm{rnk}}(\mathsf{L}^{\mathbb{S}}_\alpha) := \eta + \omega{\cdot}\alpha$, put, according to Lemma 11.9.17,

$$\widehat{\mathrm{rnk}}(s = t) = \widehat{\mathrm{rnk}}(s \neq t) := \max\{9, \widehat{\mathrm{rnk}}(s) + 4, \widehat{\mathrm{rnk}}(t) + 4\}$$

and then follow the clauses in Definition 11.9.14 to define $\widehat{\mathrm{rnk}}(E)$ for all $\mathscr{L}(RS_{\mathbb{S}})$-expressions $E$. Finally, show that $\widehat{\mathrm{rnk}}(G) < \widehat{\mathrm{rnk}}(F)$ holds true for all $G \in \mathrm{CS}^{\mathrm{RS}}_{\mathbb{S}}(F)$.

**12.3.9 Exercise** (a) Show that for finite sets $\Delta$ of $\mathscr{L}(\mathbb{S})$-formulas $RS_{\mathbb{S}} \,|\!\frac{\alpha}{0}\, \Delta^{\mathsf{L}^{\mathbb{S}}_0}$ implies $\mathbb{S} \,|\!\frac{\alpha}{0}\, \Delta$.

[5] Strictly speaking, we should distinguish between the membership relation in the language $\mathscr{L}(\mathbb{S})$ and $\mathscr{L}(RS_{\mathbb{S}})$. It will, however, always be clear from the context which membership relation is meant.

(b) Show that $\mathbb{S} \vdash_{\rho}^{\alpha} \Delta$ implies $RS_{\mathbb{S}} \vdash_{\rho}^{\alpha} \Delta^{\mathsf{L}_0^{\mathbb{S}}}$.

Hint: Use induction on $\alpha$.

The semi-formal system $RS_{\mathbb{S}}$ is a generalization of ramified set theory $RS$ as introduced in Sect. 11.9. The system $RS$ represents the "pure part" of $RS_{\mathbb{S}}$. According to [4] Chapter II 1, the pure part of an admissible set $\mathbb{A}$ consists of the elements $a \in \mathbb{A}$ with empty support, i.e., of those elements for which $TC(a)$ contains no urelements. Since $\mathcal{T}_0 = \emptyset$ we get $RS \vdash_0^0 s \notin \mathsf{L}_0$ for all $\mathcal{L}_{RS}$-terms $s$, i.e., $\mathsf{L}_0$ represents the empty set in $RS$. In contrast to that, we get $RS_{\mathbb{S}} \vdash_{\rho}^{\alpha} s \in \mathsf{L}_0^{\mathbb{S}}$ for those $s \in \mathcal{T}_0^{\mathbb{S}}$ for which $\mathbb{S} \vdash_{\rho}^{\beta} s = s$ holds for some $\beta < \alpha$. Therefore $\mathsf{L}_0^{\mathbb{S}}$ in $RS_{\mathbb{S}}$ represents the set of all "urelements" (i.e., $\mathcal{L}(\mathbb{S})$-terms) for which $\mathbb{S}$ proves $s = s$. If we assume $\mathbb{S} \vdash_0^0 s = s$ for all $\mathcal{L}(\mathbb{S})$-terms $s$ then $\mathsf{L}_0^{\mathbb{S}}$ contains all "urelements". Defining $\emptyset := \{x \in \mathsf{L}_0^{\mathbb{S}} \mid x \neq x\}$ we get $RS_{\mathbb{S}} \vdash_0^1 t \notin \emptyset$ for all $\mathcal{L}(RS_{\mathbb{S}})$-terms. The role of $\mathsf{L}_0$ in $RS$ is therefore played by $\emptyset$ in $RS_{\mathbb{S}}$. The elements in the "pure part" of $RS_{\mathbb{S}}$ are those whose transitive closures intersected with $\mathsf{L}_0^{\mathbb{S}}$ is empty (cf. Fig. 12.1).

We want to transfer wide parts of the results of Sects. 11.10 and 11.12 to the system $RS_{\mathbb{S}}$. This means to ruminate all the proofs given there. The situation is, however, not so bad since we do not need controlling operators, i.e., we do not have to care about the parameters occurring in formulas.

In checking the properties of the empty set as defined in $RS_{\mathbb{S}}$ we have already seen that the underlying semi-formal system $\mathbb{S}$ should fulfill some prerequisites, e.g., $\mathbb{S} \vdash_0^0 s = s$ for every $\mathcal{L}(\mathbb{S})$-term $s$. This is the background of the following definition.

**12.3.10 Definition** Call a semi-formal system $\mathbb{S}$ (possibly with additional axioms) *chaste* iff

- $\mathbb{S} \vdash_0^{<\omega} s \neq t, s = t$, $\mathbb{S} \vdash_0^{<\omega} s \in t, s \notin t$ and $\mathbb{S} \vdash_0^{<\omega} s \notin s$ hold true for all $\mathcal{L}(\mathbb{S})$-terms $s$ and $t$

and

- $\mathbb{S} \vdash_0^{<\omega} F$ for all instances $F$ of identity axioms in $\mathsf{IDEN}$.

**12.3.11 Exercise** Let $\mathfrak{M}$ be a countable $\mathcal{L}(\in)$-structure which is well-founded and transitive.

(a) Show that $\mathfrak{M}$-logic is chaste.

(b) Show that for any chaste semi-formal system $\mathbb{S}$ and any theory $T$ in the language of $\mathbb{S}^{\mathsf{U}}$ the system $T_{\mathbb{S}^{\mathsf{U}}}$ is chaste.

Hint: (a) follows from the $\Pi_1^1$-completeness of $\mathfrak{M}$-logic. For (b) recall the convention that all theories are supposed to comprehend the identity axioms.

So assume that $\mathbb{S}$ is a chaste semi-formal system. First, we observe that the calculus $\overset{\circ}{\vdash} \Delta$ defined in Definition 11.10.5 carries over to the system $RS_{\mathbb{S}}$ where we may add the rule

($\mathbb{S}$) $\mathbb{S} \vdash_0^{<\omega} \Delta$ implies $\overset{\circ}{\vdash} \Delta$

and drop the parameter conditions. Also the derived rules (Str) through ($\forall^b$) stay valid where in the case of (Taut) we need the chasteness of $\mathbb{S}$ in case that $A$ is a formula $s \in t$ for $\mathscr{L}(\mathbb{S})$-term $s$ and $t$. We easily check that Lemma 11.10.6 modifies to the semi-formal system $RS_{\mathbb{S}}$, i.e., we have

$$RS_{\mathbb{S}} \overset{\circ}{\vdash} \Delta \;\Rightarrow\; RS_{\mathbb{S}} \vdash_0^{\omega + \mathrm{drnk}(\Delta)} \Delta. \tag{12.1}$$

Next, we check equation (11.13) on p. 282, i.e., $(a \notin a)$ for all $RS_{\mathbb{S}}$-terms $a$. Here we have the additional case that $a$ is in $\mathscr{T}_0^{\mathbb{S}}$. But then $RS_{\mathbb{S}} \overset{\circ}{\vdash} a \notin a$ follows from the chasteness of $\mathbb{S}$. Similarly we get (11.14), i.e., $(a = a)$, in the case of $a \in \mathscr{T}_0^{\mathbb{S}}$ by the chasteness of $\mathbb{S}$.

The lacking case $a \in \mathscr{T}_0^{\mathbb{S}}$ in (11.15) (on p. 282) is also covered by the chasteness of $\mathbb{S}$. Equations (11.16) and (11.17) follow in the same way. The restriction $\alpha \neq 0$ in (11.17) can be dropped. In (11.19), we have to supplement $Tran(\mathsf{L}_0^{\mathbb{S}})$. To get it observe that $RS_{\mathbb{S}} \vdash_0^{<\omega} s = s$ implies $RS_{\mathbb{S}} \vdash_0^{<\omega} s \in \mathsf{L}_0^{\mathbb{S}}$ which in turn entails $RS_{\mathbb{S}} \vdash_0^{<\omega} s \notin t \vee s \in \mathsf{L}_0^{\mathbb{S}}$, for all $\mathscr{L}(\mathbb{S})$-terms $t$ and $s$. Hence $RS_{\mathbb{S}} \vdash_0^{<\omega} (\forall y \in t)[y \in \mathsf{L}_0^{\mathbb{S}}]$ and thus $RS_{\mathbb{S}} \vdash_0^{<\omega} (\forall x \in \mathsf{L}_0^{\mathbb{S}})(\forall y \in x)[x \in \mathsf{L}_0^{\mathbb{S}}]$, i.e., $RS_{\mathbb{S}} \vdash_0^{<\omega} Tran(\mathsf{L}_0^{\mathbb{S}})$. Here, we have also shown that (11.20) transfers to $RS_{\mathbb{S}}$.

A bit more tedious to show is the counterpart of Theorem 11.12.2

$$RS_{\mathbb{S}} \overset{\circ}{\vdash} t_1 \neq s_1, \ldots, t_n \neq s_n, \neg A(t_1, \ldots, t_n), A(s_1, \ldots, s_n).$$

Here we show first the counterpart of Lemma 11.12.1, where we need the chasteness of $\mathbb{S}$ to treat the case of terms of negative stages.

Now it is easy to check that (11.21) – (11.25), Lemmas 11.12.4 – 11.12.6 (pp. 291 and 293) become provable in $RS_{\mathbb{S}}$. In the coming text, we refer to these results as properties of $RS_{\mathbb{S}}$.

Observe, however, that in general, we cannot transfer the Foundation Lemma 11.12.7. The reason is that $\mathscr{T}_0^{\mathbb{S}}$ is no longer empty. We will study in Sect. 12.5 what is needed to handle the axiom of foundation. Exercise 12.3.14 below displays a special situation in which $RS_{\mathbb{S}}$ proves foundation.

**12.3.12 Remark** Another difference is that the system $RS_{\mathbb{S}}$ may include free second-order parameters which are supposed to represent subsets of the domain of the urelement structure. According to the discussion in Sects. 9.1 and 11.7, we could dispense with second-order parameters in ramified set theory $RS$. They can, however, easily be added. To do so, we allow atomic formulas of the form $t \,\varepsilon\, X$ and $t \not\varepsilon X$ where $t$ is an $\mathscr{L}_{RS}$-term and introduce the rule

$(Ax)_{RS}$ If $\{t \,\varepsilon\, X, s \not\varepsilon X\} \subseteq \Delta$ and $t^L = s^L$ then $RS(X) \vdash_\rho^\alpha \Delta$ holds true for all ordinals $\alpha$ and $\rho$.

Let $RS(X) \overset{\alpha}{\models} \Delta$ stand for $RS(X) \,|\!\frac{\alpha}{0}\, \Delta$. We obtain the following completeness theorem for the extended verification calculus.

**12.3.13 Exercise** Let $(\forall X)F(X)$ be a $\Pi^1_1$-sentence in the language of ramified set theory which contains only the parameters $\mathsf{L}_\alpha$ for $\alpha < \omega_1$. Show that $L_{\omega_1} \models (\forall X)F(X)$ iff there is an ordinal $\alpha < \omega_1$ such that $RS(X) \overset{\alpha}{\models} F(X)$.

Hint: One direction is shown straight forwardly by induction on $\alpha$ and does not need the hypothesis of countability. For the opposite direction you have to define a search tree analogous to the search tree in Definition 5.4.5.

Let us discuss the situation in which the basis semi-formal system is $L_\tau$-logic for a countable ordinal $\tau$. For $a \in L_\tau$ we put $\mathrm{rnk}_{L_\tau}(a)$ as the first ordinal at which $a$ enters $L_\tau$ and define for an $\mathcal{L}(RS_{L_\tau})$-term $a$

$$stg^{RS}_{\tau+}(a) := \begin{cases} \mathrm{rnk}_{L_\tau}(a) & \text{for } a \in \mathcal{T}^{L_\tau}_0 \\ \tau + stg^{RS}_{L_\tau}(a) & \text{otherwise.} \end{cases}$$

In $RS_{L_\tau}$, we can prove a foundation axiom, even a foundation scheme.

**12.3.14 Exercise** Show

$$RS_{L_\tau} \,|\!\frac{\alpha}{0}\, b \notin a, (\exists x \in a)(\forall y \in x)[y \notin a]$$

for $\alpha = \mathrm{rnk}(b \in a) + 5 \cdot (stg^{RS}_{\tau+}(b) + 1)$.

Hint: Use induction on $stg^{RS}_{\tau+}(b)$.

The system $RS_{L_\tau}$ represents a set universe above the urelement structure $L_\tau$. We have seen in Sect. 11.10 that there is a semi-formal system building up $L_\tau$ from scratch. We are now going to unravel the urelement structure using ramified set theory (cf. Fig. 12.1). In Lemma 11.9.3, we have shown that for every set $a \in L_\tau$ there is an $\mathcal{L}(RS_{L_\tau})$-term $t_a$ representing $a$, i.e., satisfying $t_a^L = a$. Choosing $t_a$ $<_L$-minimal, we define $a^{\tau+}$ for all $RS_{L_\tau}$-terms $a$.

**12.3.15 Definition** For $a \in \mathcal{L}(RS_{L_\tau})(X)$ put

$$a^{\tau+} := \begin{cases} t_b & \text{if } a \text{ is a name } \underline{b} \text{ for } b \in L_\tau \\ \mathsf{L}_{\tau+\alpha} & \text{if } a \text{ is a term } \mathsf{L}^{L_\tau}_\alpha \\ \{x \in \mathsf{L}_{\tau+\alpha} \mid F(x, \vec{a}^{\tau+})^{\mathsf{L}_{\tau+\alpha}}\} & \text{if } a \text{ is a term } \{x \in \mathsf{L}^{L_\tau}_\alpha \mid F(x,\vec{a})^{\mathsf{L}^{L_\tau}_\alpha}\}. \end{cases}$$

If $G$ is an $\mathcal{L}(RS_{L_\tau})$-formula $F(\vec{a})$, where $\vec{a}$ is a complete list of all occurring $\mathcal{L}(RS_{L_\tau})$-terms, we define $G^{\tau+} \;:\Leftrightarrow\; F(\vec{a}^{\tau+})$.

The next exercise follows directly from Definition 12.3.15.

**12.3.16 Exercise** Show that $a \in \mathcal{T}^{L_\tau}_\alpha$ iff $a^{\tau+} \in \mathcal{T}_{\tau+\alpha}$ holds true for all $RS_{L_\tau}$-terms $a$.

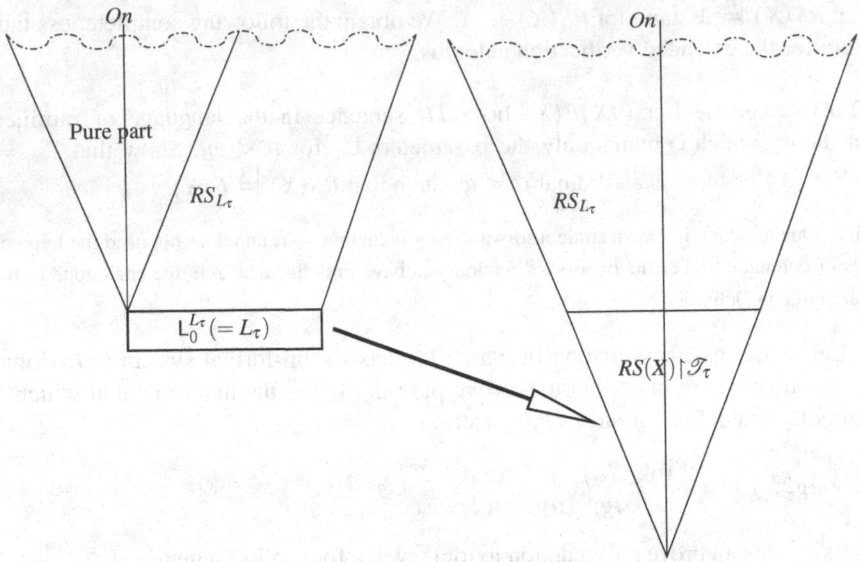


**Fig. 12.1** Unraveling $L_\tau$ into $RS(X)\restriction\mathscr{T}_\tau$

**12.3.17 Lemma** *Let $\tau$ be a countable additively indecomposable ordinal above $\omega^\omega$. Then $RS_{L_\tau} \frac{|\alpha}{\rho} \Delta$ implies $RS(X) \frac{|\tau+3\cdot\alpha}{\tau+\rho} \Delta^{\tau+}$.*

*Proof* The proof is by induction on $\alpha$. We will not show all cases but restrict ourselves to the more delicate ones.

If $RS_{L_\tau} \frac{|\alpha}{\rho} \Delta$ holds by $(Ax)_{L_\tau}$ there are $L_\tau$-terms $a$ and $b$ such that $a^{L_\tau} = b^{L_\tau}$ and $\{a \in X, b \notin X\} \subseteq \Delta$. But then we get $a^{\tau+L} = t_a^L = a^{L_\tau} = b^{L_\tau} = t_b^L = b^{\tau+L}$, and we get $RS(X) \frac{|\tau+3\cdot\alpha}{\tau+\rho} \Delta^{\tau+}$ by $(Ax)_{RS}$.

Now assume that $RS_{L_\tau} \frac{|\alpha}{\rho} \Delta$ holds by an inference $(\bigwedge)$. Then there is a formula $F \in \bigwedge$–type $\cap \Delta$ such that $RS_{L_\tau} \frac{|\alpha_G}{\rho} \Delta, G$ for all $G \in \mathrm{CS}^{\mathrm{RS}}_{L_\tau}(F)$.

If $\mathrm{CS}^{\mathrm{RS}}_{L_\tau}(F) = \emptyset$ then $F$ belongs to $Diag(L_\tau)$ and we obtain $RS(X) \frac{|\mathrm{rnk}(F)}{0} \Delta$ by Lemma 11.9.18. So assume $\mathrm{CS}^{\mathrm{RS}}_{L_\tau}(F) \neq \emptyset$.

If $F$ is a formula $s \notin s_0$, where $s_0$ is a term of stage $-1$ and $s$ a term of stage $\geq 0$, we have the premises

$$RS_{L_\tau} \frac{|\alpha_G}{\rho} \Delta, G \tag{i}$$

for all $G \in \mathrm{CS}^{\mathrm{RS}}_{L_\tau}(F)$. If $s_0$ is $L_\beta$ for some $\beta < \tau$ the premises have the form

$$RS_{L_\tau} \frac{|\alpha_t}{\rho} \Delta, s \neq t \vee t \notin L_\beta, \quad \text{hence} \quad RS_{L_\tau} \frac{|\alpha_t}{\rho} \Delta, s \neq t, t \notin L_\beta \tag{ii}$$

for all $t \in \mathcal{T}_0^{L_\tau}$. For $a \in \mathcal{T}_\beta$ we get $a^L \in L_\beta$ and thus $RS_{L_\tau} \overset{\alpha_{a^L}}{\underset{\rho}{\vdash}} \Delta, s \neq a^L$ from (ii) by Exercise 12.2.5. By induction hypothesis we thus obtain $RS(X) \overset{\tau+3\cdot\alpha_{a^L}}{\underset{\tau+\rho}{\vdash}} \Delta^{\tau+}, s^{\tau+} \neq a$ for all $a \in \mathcal{T}_\beta$. Hence $RS(X) \overset{\tau+3\cdot\alpha}{\underset{\tau+\rho}{\vdash}} \Delta^{\tau+}, s^{\tau+} \notin \mathsf{L}_\beta$.

If $s_0$ is a term $\{x \in \mathsf{L}_\beta \mid H(x)\}$ for some $\beta < \tau$ the premises have the form

$$RS_{L_\tau} \overset{\alpha_t}{\underset{\rho}{\vdash}} \Delta, s \neq t, t \notin s_0 \tag{iii}$$

for all $t \in \mathcal{T}_0^{L_\tau}$. Let $a \in \mathcal{T}_\beta$. Then $a^L \in L_\beta \subseteq L_\tau = \mathcal{T}_0^{L_\tau}$. If $L_\tau \models H(a^L)$ then $L_\tau \not\models a^L \notin s_0^L$ and thus $RS(X) \overset{\tau+3\cdot\alpha_{a^L}}{\underset{\tau+\rho}{\vdash}} \Delta^{\tau+}, s^{\tau+} \neq a, \neg H(\underline{a^L})^{\tau+}$ by (iii), Exercise 12.2.5 and the induction hypothesis. If $L_\tau \models \neg H(a^L)$ then we also obtain $L \models \neg H(a^L)^{\tau+}$ which by Corollary 11.9.19 implies $RS(X) \overset{\tau}{\underset{0}{\vdash}} \Delta^{\tau+}, s^{\tau+} \neq a, \neg H(\underline{a^L})^{\tau+}$ since $\mathrm{rnk}(\neg H(\underline{a^L})^{\tau+}) < \tau$. But $\neg H(\underline{a^L})^{\tau+} = \neg H^{\tau+}(a)$ and we have

$$RS(X) \overset{\tau+3\cdot\alpha_a+2}{\underset{\tau+\rho}{\vdash}} \Delta^{\tau+}, s^{\tau+} \neq a \vee \neg H^{\tau+}(a)$$

for all $a \in \mathcal{T}_\beta$ which implies $RS(X) \overset{\tau+3\cdot\alpha}{\underset{\tau+\rho}{\vdash}} \Delta^{\tau+}, s^{\tau+} \notin \{x \in \mathsf{L}_\beta \mid H^{\tau+}(x)\}$.

In all other cases, we get $\mathrm{CS}(F^{\tau+}) = \langle G^{\tau+} \mid G \in \mathrm{CS}^{\mathrm{RS}}_{L_\tau}(F)\rangle$ and obtain the claim directly from the induction hypothesis.

The next case is that of an $(\bigvee)$-inference. Here, we have a formula $F \in \bigvee\text{–type} \cap \Delta$ and a premise

$$RS_{L_\tau} \overset{\alpha_0}{\underset{\rho}{\vdash}} \Delta, G \tag{iv}$$

for some $G \in \mathrm{CS}^{\mathrm{RS}}_{L_\tau}(F)$. The only remarkable case is that $F$ is a formula $s \in s_0$ for a term $s_0$ of stage $-1$. If $s_0$ is $L_\beta$ for some $\beta < \tau$ then the premise has the form

$$RS_{L_\tau} \overset{\alpha_0}{\underset{\rho}{\vdash}} \Delta, s = \underline{a} \wedge \underline{a} \in L_\beta \tag{v}$$

for some $\underline{a} \in \mathcal{T}_0^{L_\tau}$. Then $\underline{a}^{\tau+} = t_a \in \mathcal{T}_\tau$. If $t_a \in \mathcal{T}_\beta$ we get $RS(X) \overset{\tau+3\cdot\alpha_0}{\underset{\tau+\rho}{\vdash}} \Delta^{\tau+}, s^{\tau+} = t_a$, hence also $RS(X) \overset{\tau+3\cdot\alpha}{\underset{\tau+\rho}{\vdash}} \Delta^{\tau+}, s^{\tau+} \in \mathsf{L}_\beta$, from (v) by $\wedge$-inversion and induction hypothesis. If $t_a \notin \mathcal{T}_\beta$ then $a = t_a^L \notin L_\beta$ and we get $RS_{L_\tau} \overset{\alpha_0}{\underset{\rho}{\vdash}} \Delta$ from (v) by $\wedge$-inversion and Exercise 12.2.5. Hence $RS(X) \overset{\tau+3\cdot\alpha}{\underset{\tau+\rho}{\vdash}} \Delta^{\tau+}, s^{\tau+} \in \mathsf{L}_\beta$ by induction hypothesis.

If $s_0 = \{x \in \mathsf{L}_\beta \mid H(x)\}$ for some $\beta < \tau$ the premise has the form

$$RS_{L_\tau} \overset{\alpha_0}{\underset{\rho}{\vdash}} \Delta, s = \underline{a} \wedge \underline{a} \in s_0 \tag{vi}$$

for some $\underline{a} \in \mathcal{T}_0^{L_\tau}$. If $L_\tau \models a \in s_0^L$ we get $(a^{\tau+})^L = a \in s_0^L$ which entails $a^{\tau+} \in \mathcal{T}_\beta$ by the minimality of $a^{\tau+}$ and $L \models H(a)^{\tau+}$. Hence $RS(X) \overset{\tau}{\underset{0}{\vdash}} \Delta^{\tau+}, H^{\tau+}(a^{\tau+})$ by Corollary 11.9.19. From (vi) we get $RS(X) \overset{\tau+3\cdot\alpha_0}{\underset{\tau+\rho}{\vdash}} \Delta^{\tau+}, s^{\tau+} = a^{\tau+}$ by $\wedge$-inversion and induction hypothesis. Hence

$$RS(X) \overset{\tau+3\cdot\alpha_0+1}{\underset{\tau+\rho}{\vdash}} \Delta^{\tau+}, s^{\tau+} = a^{\tau+} \wedge H^{\tau+}(a^{\tau+})$$

by a inference $(\bigwedge)$ which entails $RS(X) \mathrel{|\!\frac{\tau+3\cdot\alpha}{\tau+\rho}} \Delta^{\tau+}, s^{\tau+} \in \{x \in \mathsf{L}_\beta \mid H^{\tau+}(x)\}$. If $\mathsf{L} \not\models a \in s_0^L$ we get $RS_{L_\tau} \mathrel{|\!\frac{\alpha_0}{\rho}} \Delta$ from (vi) and Exercise 12.2.5 which immediately implies $RS(X) \mathrel{|\!\frac{\tau+3\cdot\alpha}{\tau+\rho}} \Delta^{\tau+}, s^{\tau+} \in \{x \in \mathsf{L}_\beta \mid H^{\tau+}(x)\}$ by the inductive hypothesis.

In the remaining cases we again have $\mathrm{CS}(F^{\tau+}) = \langle G^{\tau+} \mid G \in \mathrm{CS}^{\mathrm{RS}}_{L_\tau}(F)\rangle$ and obtain the claim easily from the induction hypothesis.

The last case to consider is a cut. But here we just observe that $\mathrm{rnk}(F^{\tau+}) \leq \tau + \mathrm{rnk}^{RS}_{L_\tau}(F)$ and apply the induction hypothesis. □

## 12.4 Asymmetric Interpretations

The key tool in using ramified set theory in the ordinal analysis of predicative theories is a technique which is known as *asymmetric interpretation.* An interpretation fixes the ranges of set quantifiers. An interpretation is asymmetric if existential and universal quantifiers get different ranges. Our first application of asymmetric interpretations is the reduction of the admissible extension $T^+$ of a theory to the the basis theory $T$ in the framework of $\mathbb{S}$-logic.

**12.4.1 Definition** Let $F$ be a formula in the language $\mathcal{L}(\mathbb{S}^{\mathsf{U}})$. By $F^{(\alpha,\beta)}$ we denote the $\mathcal{L}(RS_{\mathbb{S}})$-formula which is obtained from $F$ by replacing all quantifiers $(\mathsf{Q}x \in \mathsf{U})[\cdots x \cdots]$ by $(\mathsf{Q}x \in \mathsf{L}_0)[\cdots x \cdots]$, all unbounded quantifiers $(\forall x)[\cdots x \cdots]$ by $(\forall x \in \mathsf{L}_\alpha)[\cdots x \cdots]$ and all quantifiers $(\exists x)[\cdots x \cdots]$ by $(\exists x \in \mathsf{L}_\beta)[\cdots x \cdots]$. I.e., unbounded quantifiers are interpreted asymmetrically while quantifiers ranging over $\mathsf{U}$ are interpreted by quantifiers ranging over $\mathsf{L}_0$.

If $\Delta$ is the finite set $\{F_1, \ldots, F_n\}$, we denote by $\mathcal{D}^{(\alpha,\beta)}$ the collection of all sets $\{F_1^{(\alpha_1,\beta_1)}, \ldots, F_n^{(\alpha_n,\beta_n)}\}$ with $\alpha_i \leq \alpha$ and $\beta_i \geq \beta$ for $i = 1, \ldots, n$.

If $A$ is a formula $(\forall x_1) \ldots (\forall x_n) F_{v_1,\ldots,v_n}(x_1, \ldots, x_n)$ and $\vec{u} = u_1, \ldots, u_n$ is a tuple of variables we call $F_{\vec{v}}(\vec{u})$ a specialization of $A$.

**12.4.2 Lemma** *(a) Let $\mathbb{S}$ be a chaste semi-formal system. Assume moreover that every formula in the finite set $\Lambda(\vec{u})$ of $\mathcal{L}(\mathbb{S}^{\mathsf{U}})$-formulas is a specialization of an axiom in* $\mathsf{KP}^-$ *or an identity axiom or an axiom Tran*$(\mathsf{U})$*. Let moreover $\Delta(\vec{u})$ be a set of $\mathcal{L}(\mathbb{S}^{\mathsf{U}})$-formulas. Then $\mathbb{S}^{\mathsf{U}} \mathrel{|\!\frac{\alpha}{0}} \neg\Lambda(\vec{u}), \Delta(\vec{u})$ implies $RS_{\mathbb{S}} \mathrel{|\!\frac{\varphi_1(\beta+3^\alpha)}{\omega\cdot(\beta+3^\alpha+1)}} \Delta'_{\vec{u}}(\vec{a})$ for all ordinals $\beta \geq 2$, all finite sets $\Delta' \in \mathcal{D}^{(\beta,\beta+3^\alpha)}$ and all tuples $\vec{a}$ of $\mathcal{L}(RS_{\mathbb{S}})$-terms of stages less than $\beta$.*

*(b) If $\Lambda(\vec{u})$ contains also specializations of axiom* (Found) *and $\mathbb{S}$ is $L_\tau$-logic for some countable ordinal $\tau$ the claim remains true for $\beta \geq \tau$.*

*Proof* The proof is by induction on $\alpha$. Choose $\beta$ and let $\sigma := \beta + 3^\alpha$ and $\rho := \omega\cdot(\sigma+1)$ and pick $\Delta' \in \mathcal{D}^{(\beta,\sigma)}$.

Assume first that the critical formula(s) of the last inference (J) belong(s) to $\Delta(\vec{u})$. The claim follows trivially if (J) is an axiom $(Ax)_{\mathfrak{M}}$ or an axiom $(Ax)_L$ belonging to

$\mathscr{L}(\mathbb{S})$. If (J) is an axiom $(Ax)_L$ not belonging to $\mathscr{L}(\mathbb{S})$ then there is an atomic formula $A(u_i, u_j)$ such that $\{\neg A(u_i, u_j), A(u_i, u_j)\} \subseteq \Delta(\vec{u})$ and we obtain $RS_{\mathbb{S}} \frac{|\delta}{\rho} \Delta'(\vec{a})$ for some ordinal $\delta < \varphi_1(\beta)$ by (Taut).[6]

If the critical formula of (J) is $(\forall x \in \mathsf{U})G(x, \vec{u})$ we have the premises

$$\mathbb{S}^{\mathsf{U}} \frac{|\alpha_t}{0} \neg\Lambda(\vec{u}), \Delta(\vec{u}), G(t, \vec{u})$$

for all $\mathscr{L}(\mathbb{S})$-terms $t$, i.e., for all $t \in \mathscr{T}_0^{\mathbb{S}}$. There are ordinals $\beta' \leq \beta$ and $\sigma' \geq \sigma$ such that $((\forall x \in \mathsf{U})G(x, \vec{a}))^{(\beta', \sigma')} \in \Delta'$. By the induction hypothesis we get

$$RS_{\mathbb{S}} \frac{|\varphi_1(\beta+3^{\alpha_t})}{\rho} \Delta'(\vec{a}), G(t, \vec{a})^{(\beta', \sigma')}$$

for all $t \in \mathscr{T}_0^{\mathbb{S}}$ and obtain $RS_{\mathbb{S}} \frac{|\varphi_1(\beta+3^{\alpha})}{\rho} \Delta'(\vec{a}), (\forall x \in \mathsf{L}_0)G(x, \vec{a})^{(\beta', \sigma')}$ by an inferences $(\bigwedge)$.

The case that the critical formula is $(\exists x \in \mathsf{U})G(x, \vec{u})$ is shown analogously.

If the critical formula $F$ is $(\forall x)G(x, \vec{u})$ then we have the premises

$$\mathbb{S}^{\mathsf{U}} \frac{|\alpha_a}{0} \neg\Lambda(\vec{u}), \Delta(\vec{u}), G(t, \vec{u})$$

for all terms $t$. Choose $t$ to be a variable which does not occur in $\vec{u}$. There are ordinals $\beta' \leq \beta$ and $\sigma' \geq \sigma$ such that $((\forall x)G(x, \vec{a}))^{(\beta', \sigma')} \in \Delta'$. By the inductive hypothesis, we then obtain

$$RS_{\mathbb{S}} \frac{|\varphi_1(\beta+3^{\alpha_a})}{\rho} \Delta'(\vec{a}), G(b, \vec{a})^{(\beta', \sigma')}$$

for all $\mathscr{L}(RS_{\mathbb{S}})$-terms $b$ of stage less than $\beta'$ and by an inference $(\bigwedge)$ we get

$$RS_{\mathbb{S}} \frac{|\varphi_1(\beta+3^{\alpha})}{\rho} \Delta'(\vec{a}), (\forall x \in \mathsf{L}_{\beta'})G(x, \vec{a})^{(\beta', \sigma')}, \text{ i.e., } RS_{\mathbb{S}} \frac{|\beta+3^{\alpha}}{\rho} \Delta'(\vec{a}).$$

Now assume that $F$ is a formula $(\exists x)G(x, \vec{u})$. Let $\beta' \leq \beta$ and $\sigma' \geq \sigma$ be ordinals such that $((\exists x)G(x, \vec{a}))^{(\beta', \sigma')} \in \Delta'$. We have the premise

$$\mathbb{S}^{\mathsf{U}} \frac{|\alpha_0}{0} \neg\Lambda(\vec{u}), \Delta(\vec{u}), G(t, \vec{u})$$

for some term $t$. If $t$ is a variable occurring in the list $\vec{u}$ we replace it by the corresponding term $b$ in the list $\vec{a}$. Otherwise $t$ is either an $\mathscr{L}(\mathbb{S})$-term. i.e., a term in $\mathscr{T}_0^{\mathbb{S}}$, the term $\mathsf{U}$, which is to replace by $\mathsf{L}_0$, or a variable not occurring in $\vec{u}$. In the latter case we replace $t$ by any $\mathscr{L}(RS_{\mathbb{S}})$-term of stage less than $\beta$ and obtain

$$RS_{\mathbb{S}} \frac{|\varphi_1(\beta+3^{\alpha_0})}{\rho} \Delta'(\vec{a}), G(b, \vec{a})^{(\beta', \sigma')}$$

by the inductive hypothesis. The claim follows by an inference $(\bigvee)$.

The remaining cases that the critical formula belongs to $\Delta(\vec{u})$ are even simpler and can be treated analogously.

Assume next that the critical formula $F$ of the last inference belongs to the set $\neg\Lambda(\vec{u})$. There are two main sub-cases. First assume that the formula $F$ has the shape

---

[6] Cf. p. 281.

$(\exists x_{k+1})\dots(\exists x_n)G(\vec{u},x_{k+1})$ such that $(\forall x_{k+2})\dots(\forall x_n)\neg G(\vec{u},v)$ is still a specialization of one of the axioms. Then, we have the premise

$$\mathbb{S}^{\mathsf{U}} \,\Big|\!\frac{\alpha_0}{0}\; (\exists x_{k+2})\dots(\exists x_n)G(\vec{u},t), \neg\Lambda(\vec{u}), \Delta(\vec{u})$$

for some $\mathscr{L}(\mathbb{S}^{\mathsf{U}})$-term $t$. If $t$ is a an $\mathscr{L}(\mathbb{S})$-term, we leave $t$ unchanged. If $t$ is $\mathsf{U}$ we replace $t$ by $\mathsf{L}_0$. If $t$ is a variable in the list $\vec{u}$, we replace $t$ by the corresponding term in the list $\vec{a}$. If $t$ is a variable not occurring in $\vec{u}$, we replace $t$ by any term of stage less than $\beta$ and obtain by the induction hypothesis

$$RS_{\mathbb{S}} \,\Big|\!\frac{\varphi_1(\beta+3^{\alpha_0})}{\rho}\; \Delta''(\vec{a})$$

for all $\Delta'' \in \mathscr{D}^{(\beta,\beta+3^{\alpha_0})} \supseteq \mathscr{D}^{(\beta,\beta+3^{\alpha})}$ and thus also $RS_{\mathbb{S}} \,\Big|\!\frac{\varphi_1(\beta+3^{\alpha})}{\rho}\; \Delta'(\vec{a})$.

Now assume that the formula(s) in the premise corresponding to the critical formula $F$ are not longer specializations of an axiom. Here, we have to distinguish cases according to the axiom which $F$ specializes.

*(Nullset)* If $\neg F$ is a specialization of (Nullset) then $F$ is $(\forall x)(\exists y)[y \in x]$ and we have a premise

$$\mathbb{S} \,\Big|\!\frac{\alpha_0}{0}\; (\exists y)[y \in w], \neg\Lambda(\vec{u}), \Delta(\vec{u})$$

for a free variable $w$ not occurring in the list $\vec{u}$. Let $b := \{x \in \mathsf{L}_\beta \,|\, x \neq x\}$. Then $b$ is a term of stage $\beta$. Applying the induction hypothesis to $\beta' := \beta+1$ we obtain

$$RS_{\mathbb{S}} \,\Big|\!\frac{\varphi_1(\beta'+3^{\alpha_0})}{\rho_0}\; (\exists y \in \mathsf{L}_{\beta'})[y \in b], \Delta'(\vec{a}).$$

By (11.14) (on p. 282) and the chasteness of $\mathbb{S}$ we get $RS_{\mathbb{S}} \,\Big|\!\frac{<\varphi_1(\beta)}{0}\; a = a$ for all terms $a$ of stage less than $\beta'$ and thus

$$RS_{\mathbb{S}} \,\Big|\!\frac{<\varphi_1(\beta)}{0}\; (\forall y \in \mathsf{L}_{\beta'})[y \notin b], \Delta'(\vec{a})$$

and we obtain the claim by cut.

*(Pair)* Let $\neg F$ is a specialization of (Pair'), i.e., $F$ is a formula $(\forall z)[u \notin z \vee v \notin z]$. Then, we have a premise

$$\mathbb{S}^{\mathsf{U}} \,\Big|\!\frac{\alpha_0}{0}\; u \notin w \vee v \notin w, \neg\Lambda(\vec{u}), \Delta(\vec{u})$$

such that $w$ does not occur in the list $\vec{u}$ while $u$ and $v$ are members of the list. Let $a_1$ and $a_2$ be the terms in the list $\vec{a}$ which correspond to $u$ and $v$, respectively. Then $b := \{a_1, a_2\}$ is an $\mathscr{L}(RS_{\mathbb{S}})$-term of stage $\leq \beta$. Let $\beta' := \beta+1$ and apply the inductive hypothesis to obtain

$$RS_{\mathbb{S}} \,\Big|\!\frac{\varphi_1(\beta'+3^{\alpha_0})}{\rho_0}\; a_1 \notin b \vee a_2 \notin b, \Delta'(\vec{a}). \tag{i}$$

By (11.14) (cf. p. 282) we obtain by a few inferences

$$RS_{\mathbb{S}} \,\Big|\!\frac{<\varphi_1(\beta)}{0}\; a_1 \in b \wedge a_2 \in b. \tag{ii}$$

Since $\mathrm{rnk}(a_1 \in b \wedge a_2 \in b) < \omega{\cdot}\beta + \omega < \rho$ we get

$$RS_{\mathbb{S}} \,\Big|\!\frac{\varphi_1(\beta+3^\alpha)}{\rho}\, \Delta'(\vec{a})$$

from (i) and (ii) by cut.

*(Union)* Now assume that $\neg F$ is a specialization of (Union'). Then $F$ is a formula $(\forall z)\big[(\exists x \in u)[\neg x \subseteq z]\big]$ and we have a premise

$$\mathbb{S}^{\mathsf{U}} \,\Big|\!\frac{\alpha_0}{0}\, (\exists x \in u)[\neg x \subseteq w], \neg\Lambda(\vec{u}), \Delta(\vec{u})$$

where $w$ is a variable not occurring in the list $\vec{u}$. Let $a_0$ be the term which corresponds to $u$. Then $a := \bigcup a_0 = \{x \in \mathsf{L}_\beta \,|\, (\exists y \in a_0)[x \in y]\}$ is an $\mathscr{L}(RS_{\mathbb{S}})$-term of stage $\beta$. Applying the inductive hypothesis for $\beta' := \beta + 1$ we get

$$RS_{\mathbb{S}} \,\Big|\!\frac{\varphi_1(\beta'+3^{\alpha_0})}{\rho}\, (\exists x \in a_0)[\neg x \subseteq a], \Delta''(\vec{a}) \tag{iii}$$

for all $\Delta'' \in \mathscr{D}^{(\beta',\beta'+3^{\alpha_0})}$. Since

$$RS_{\mathbb{S}} \,\Big|\!\frac{<\varphi_1(\beta)}{0}\, (\forall x \in a_0)[x \subseteq a], \tag{iv}$$

$\mathscr{D}^{(\beta',\beta'+3^{\alpha_0})} \supseteq \mathscr{D}^{(\beta,\sigma)}$ and $\mathrm{rnk}((\forall x \in a_0)[x \subseteq a]) < \omega{\cdot}\beta + \omega < \rho$, we get the claim from (iii) and (iv) by cut.

*(Transitive Closure)* Now, assume that (Union') has to be replaced by (TranC). Then $F$ is a formula $(\forall y)[\neg \mathit{Tran}(y) \vee \neg u \subseteq y]$ and we have a premise

$$\mathbb{S}^{\mathsf{U}} \,\Big|\!\frac{\alpha_0}{0}\, \neg \mathit{Tran}(w) \vee \neg u \subseteq w, \neg\Lambda(\vec{u}), \Delta(\vec{u})$$

where $w$ is a variable not occurring in the list $\vec{u}$. Let $a_0$ be the term which corresponds to $u$. Since $a_0$ is a term of stage less than $\beta$ we obtain

$$RS_{\mathbb{S}} \,\Big|\!\frac{<\varphi_1(\beta)}{0}\, a_0 \subseteq \mathsf{L}_\beta \tag{v}$$

by (11.20) in p. 283. Since $\mathsf{L}_\beta$ is a term of stage $\beta$ we can apply the inductive hypothesis for $\beta' := \beta + 1$ and use $\bigvee$-exportation to obtain

$$RS_{\mathbb{S}} \,\Big|\!\frac{\varphi_1(\beta'+3^{\alpha_0})}{\rho}\, \neg \mathit{Tran}(\mathsf{L}_\beta), \neg a_0 \subseteq \mathsf{L}_\beta, \Delta''(\vec{a}) \tag{vi}$$

for all $\Delta'' \in \mathscr{D}^{(\beta',\beta'+3^{\alpha_0})} \supseteq \mathscr{D}^{(\beta,\beta+3^\alpha)}$. By (11.19), we have

$$RS_{\mathbb{S}} \,\Big|\!\frac{<\varphi_1(\beta)}{0}\, \mathit{Tran}(\mathsf{L}_\beta) \tag{vii}$$

and get the claim from (v),(vi) and (vii) by cuts.

*(Separation)* Assume $\neg F$ is an instance of $\Delta_0$-separation. Then $F$ is a formula

$$(\forall z)\big[(\exists x \in z)[x \notin u \vee \neg G(x,\vec{u})] \vee (\exists x \in u)[G(x,\vec{u}) \wedge x \notin z]\big].$$

for a $\Delta_0$-formula $G(x,\vec{u})$ and we have a premise

$$\mathbb{S}^{\mathsf{U}} \,\Big|\!\frac{\alpha_0}{0}\, (\exists x \in w)[(x \notin u \vee \neg G(x,\vec{u}))] \vee (\exists x \in u)[G(x,\vec{u}) \wedge x \notin w], \neg\Lambda(\vec{u}), \Delta(\vec{u})$$

for a variable $w$ not occurring in the list $\vec{u}$. Then $b := \{x \in \mathsf{L}_\beta \mid x \in a \wedge G(x,\vec{a})\}$ is an $\mathcal{L}(RS_{\mathbb{S}})$-term of stage $\beta$. Applying the induction hypothesis to $\beta' := \beta + 1$ we obtain

$$RS_{\mathbb{S}} \left|\frac{\varphi_1(\beta'+3^{\alpha_0})}{\rho}\right. (\exists x \in b)[x \notin a \vee \neg G(x,\vec{a})] \vee (\exists x \in a)[G(x,\vec{a}) \wedge x \notin b], \Delta''(\vec{a}) \quad \text{(viii)}$$

for all $\Delta'' \in \mathscr{D}^{(\beta',\beta'+3^{\alpha_0})}$. But, as shown in the proof of Lemma 11.12.6, we get

$$RS_{\mathbb{S}} \left|\frac{<\varphi_1(\beta)}{0}\right. (\forall x \in b)[x \in a \wedge G(x,\vec{a})] \wedge (\forall x \in a)[\neg G(x,\vec{a}) \vee x \in b]. \quad \text{(ix)}$$

Since $\mathscr{D}^{(\beta,\sigma)} \subseteq \mathscr{D}^{(\beta',\beta'+3^{\alpha_0})}$ we obtain the claim cutting (ix) and (viii).

*(Collection)* The crucial case is that $\neg F$ is an instance of $\Delta_0$-collection. Then $F$ is a formula

$$\neg[(\forall x \in v)(\exists y)G(x,y,\vec{u}) \to (\exists z)(\forall x \in v)(\exists y \in z)G(x,y,\vec{u})],$$

Here, we need the asymmetric interpretation. We have the premises

$$\mathbb{S}^{\mathsf{U}} \left|\frac{\alpha_0}{0}\right. \neg\Lambda(\vec{u}), \Delta(\vec{u}), (\forall x \in v)(\exists y)G(x,y,\vec{u}) \quad \text{(x)}$$

and

$$\mathbb{S}^{\mathsf{U}} \left|\frac{\alpha_0}{0}\right. \neg\Lambda(\vec{u}), \Delta(\vec{u}), (\forall z)(\exists x \in v)(\forall y \in z)\neg G(x,y,\vec{u}). \quad \text{(xi)}$$

Let $b$ be the term corresponding to $v$ and put $\sigma_0 := \beta + 3^{\alpha_0}$. Applying the inductive hypothesis to (x) we obtain

$$RS_{\mathbb{S}} \left|\frac{\varphi_1(\beta+3^{\alpha_0})}{\rho}\right. \Delta''(\vec{a}), (\forall x \in b)(\exists y \in \mathsf{L}_{\sigma_0})G(x,y,\vec{a}) \quad \text{(xii)}$$

for all $\Delta'' \in \mathscr{D}^{(\beta,\beta+3^{\alpha_0})}$. Let $\beta' := \sigma_0 + 1$ and $\sigma' := \beta' + 3^{\alpha_0}$. Applying the inductive hypothesis to (xi) we get

$$RS_{\mathbb{S}} \left|\frac{\varphi_1(\beta'+3^{\alpha_0})}{\omega\cdot(\sigma'+1)}\right. \Delta'''(\vec{a}), (\forall z \in \mathsf{L}_{\sigma_0+1})(\exists x \in b)(\forall y \in z)\neg G(x,y,\vec{a})$$

and thus by $\forall$-inversion

$$RS \left|\frac{\varphi_1(\beta'+3^{\alpha_0})}{\omega\cdot(\sigma'+1)}\right. \Delta'''(\vec{a}), (\exists x \in b)(\forall y \in \mathsf{L}_{\sigma_0})\neg G(x,y,\vec{a}) \quad \text{(xiii)}$$

for all $\Delta''' \in \mathscr{D}^{(\beta',\sigma')}$. Since $\beta < \beta'$ and $\sigma' = \beta' + 1 = \beta + 3^{\alpha_0} + 1 + 3^{\alpha_0} \leq \beta + 3^{\alpha} = \sigma$ we get $\mathscr{D}^{(\beta',\sigma')} \cap \mathscr{D}^{(\beta,\sigma_0)} \supseteq \mathscr{D}^{(\beta,\sigma)}$. Since moreover $\omega\cdot(\sigma'+1) \leq \omega\cdot(\sigma+1)$ and $\operatorname{rnk}((\exists y \in b)(\forall z \in \mathsf{L}_{\sigma_0})\neg G(x,y,\vec{a})) \leq \omega\cdot(\sigma_0+1) < \omega\cdot\sigma + \omega$ we get the claim from (xii) and (xiii) by cut.

*(Identity)* Assume that $\neg F$ is a specialization of an identity axiom, say $\neg F$ is the formula $u = v \wedge A(u) \to A(v)$ for atomic $A(u)$. Then $F$ is $u = v \wedge A(u) \wedge \neg A(v)$ and we obtain from the premises

$$\mathbb{S}^{\mathsf{U}} \left|\frac{\alpha_0}{0}\right. \neg\Lambda(\vec{u}), \Delta(\vec{u}), u = v$$

$$\mathbb{S}^{\mathsf{U}} \left|\frac{\alpha_0}{0}\right. \neg\Lambda(\vec{u}), \Delta(\vec{u}), A(u)$$

and

$$\mathbb{S}^{\mathsf{U}} \overset{\alpha_0}{\underset{0}{\vdash}} \neg\Lambda(\vec{u}), \Delta(\vec{u}), \neg A(v).$$

By the inductive hypothesis, we then get

$$RS_{\mathbb{S}} \overset{\varphi_1(\beta+3^{\alpha_0})}{\underset{\rho}{\vdash}} \Delta'(\vec{a}), a = b$$

$$RS_{\mathbb{S}} \overset{\varphi_1(\beta+3^{\alpha_0})}{\underset{\rho}{\vdash}} \Delta'(\vec{a}), A(a)$$

and

$$RS_{\mathbb{S}} \overset{\varphi_1(\beta+3^{\alpha_0})}{\underset{\rho}{\vdash}} \Delta'(\vec{a}), \neg A(b)$$

where $a$ and $b$ are the terms corresponding to $u$ and $v$ if these variables occur in the list $\vec{u}$ or arbitrary terms of stage less than $\beta$ otherwise. Either by Theorem 11.12.2 or by the fact that $\mathbb{S}$ is chaste we get

$$RS_{\mathbb{S}} \overset{\eta}{\underset{0}{\vdash}} \Delta(\vec{a}), a \neq b, \neg A(a), A(b)$$

for some ordinal $\eta < \varphi_1(\beta + 3^{\alpha_0})$. The claim then follows by cuts.

The remaining cases of identity axioms are similar and are left as exercises.

*(Extensionality)* The case that $\neg F$ is the extensionality axiom follows immediately from (11.25) on p. 292.

*(Transivity of* $\mathsf{U}$*)* Finally, let $\neg F$ be the sentence $Tran(\mathsf{U})$. Then we have the the premise

$$\mathbb{S}^{\mathsf{U}} \overset{\alpha_0}{\underset{0}{\vdash}} \Lambda(\vec{u}), (\exists y \in t)[y \notin \mathsf{U}], \Delta(\vec{u})$$

for some $\mathscr{L}(\mathbb{S})$-term $t$. Applying the inductive hypothesis yields

$$RS_{\mathbb{S}} \overset{\varphi_1(\beta+3^{\alpha_0})}{\underset{\rho}{\vdash}} (\exists y \in t)[y \notin \mathsf{L}_0], \Delta'(\vec{a}).$$

Since $\mathbb{S}$ is chaste, we have $\mathbb{S} \overset{<\omega}{\underset{0}{\vdash}} s = s$ for every $\mathscr{L}(\mathbb{S})$-term $s$ and therefore $RS_{\mathbb{S}} \overset{<\omega}{\underset{0}{\vdash}} s \notin t \vee s \in \mathsf{L}_0$. This implies

$$RS_{\mathbb{S}} \overset{<\omega}{\underset{0}{\vdash}} (\forall y \in t)[y \in \mathsf{L}_0]$$

and we obtain the claim by (cut).

To prove part (b) assume that $\mathbb{S}$ is $L_\tau$-logic and $\neg F$ is a specialization of axiom (Found). Then $F$ is a formula

$$(\exists x)[x \in u] \wedge (\forall x \in u)(\exists y \in x)[y \in u]$$

and we have the premises

$$L_\tau^{\mathsf{U}} \overset{\alpha_0}{\underset{0}{\vdash}} (\exists x)[x \in u], \neg\Lambda(\vec{u}), \Delta(\vec{u})$$

and

$$L_\tau^{\mathsf{U}} \left|\frac{\alpha_0}{0}\right. (\forall x \in u)(\exists y \in x)[y \in u], \neg\Lambda(\vec{u}), \Delta(\vec{u}).$$

For $\sigma_0 := \beta + 3^{\alpha_0}$, we obtain by the inductive hypothesis

$$RS_{L_\tau} \left|\frac{\varphi_1(\beta+3^{\alpha_0})}{\rho}\right. (\exists x \in \mathsf{L}_{\tau_0})[x \in a], \Delta'(\vec{a}) \tag{xiv}$$

and

$$RS_{L_\tau} \left|\frac{\varphi_1(\beta+3^{\alpha_0})}{\rho}\right. (\forall x \in a)(\exists y \in x)[y \in a], \Delta'(\vec{a}). \tag{xv}$$

Since $\tau \leq \beta < \sigma_0$ we get by Exercise 12.3.14

$$RS_{L_\tau} \left|\frac{<\omega^{\sigma_0}}{0}\right. b \notin a, (\exists x \in a)(\forall y \in x)[y \notin a]] \tag{xvi}$$

for all terms $b \in \mathscr{T}_{\sigma_0}^{L_\tau}$ which implies

$$RS_{L_\tau} \left|\frac{<\varphi_1(\beta+3^{\alpha})}{0}\right. (\forall x \in \mathsf{L}_{\sigma_0})[x \notin a], (\exists x \in a)(\forall y \in x)[y \notin a] \tag{xvii}$$

by an inference $(\bigwedge)$. Since $(\forall x \in \mathsf{L}_{\sigma_0})[x \notin a]$ and $(\exists x \in a)(\forall y \in x)[y \notin a]$ have ranks less than $\rho$ we obtain

$$RS_{L_\tau} \left|\frac{\varphi_1(\beta+3^{\alpha})}{\rho}\right. \Delta'(\vec{a})$$

cutting (xvii), (xiv) and (xv). □

**12.4.3 Remark** Observe that we can extend Lemma 12.4.2 to $\mathbb{S}^{\mathsf{U}}$-derivations which may contain cuts whose cut formulas are at most $\Sigma_1$ or $\Pi_1$. Then we have the additional case that there are premises

$$\mathbb{S}^{\mathsf{U}} \left|\frac{\alpha_0}{\mu}\right. \neg\Lambda(\vec{u}), \Delta(\vec{u}), (\exists x)G(x,\vec{u}) \tag{i}$$

and

$$\mathbb{S}^{\mathsf{U}} \left|\frac{\alpha_0}{\mu}\right. \neg\Lambda(\vec{u}), \Delta(\vec{u}), (\forall x)\neg G(x,\vec{u}). \tag{ii}$$

By the induction hypothesis, we get from (i) for $\sigma_0 := \beta + 3^{\alpha_0}$

$$RS_{\mathbb{S}} \left|\frac{\varphi_1(\beta+3^{\alpha_0})}{\rho}\right. \Delta'(\vec{a}), (\exists x \in \mathsf{L}_{\sigma_0})G(x,\vec{a}) \tag{iii}$$

and, putting $\sigma' := \sigma_0 + 3^{\alpha_0}$, from (ii)

$$RS_{\mathbb{S}} \left|\frac{\varphi_1(\sigma')}{\rho}\right. \Delta'(\vec{a}), (\forall x \in \mathsf{L}_{\sigma_0})\neg G(x,\vec{a}) \tag{iv}$$

Since $\sigma' = \beta + 3^{\alpha_0} + 3^{\alpha_0} < \beta + 3^{\alpha}$, we get the claim cutting (iv) and (iii).

**12.4.4 Definition** Let $T$ be a theory in the language of $\mathscr{L}(\mathbb{S})$. By $T_{\mathbb{S}}^{+} \left|\frac{\alpha}{\rho}\right. \Delta$ we denote that there is a finite subset $\Gamma$ of $T$ and finite subset $\Lambda \subseteq \mathsf{KP}^{-} + \mathit{Tran}(\mathsf{U})$ such that $\mathbb{S}^{\mathsf{U}} \left|\frac{\alpha}{\rho}\right. \neg\Gamma^{\mathsf{U}}, \neg\Lambda, \Delta$. If $T$ is empty, we just write $\mathbb{S}^{+} \left|\frac{\alpha}{\rho}\right. \Delta$. For a theory $T'$ in the language of $\mathcal{L}(\mathbb{S}^{\mathsf{U}})$, we denote by $T^+_{\mathbb{S}} + T' \mathop{\mid\!\!\!\frac{\alpha}{\rho}} \Delta$ that there is finite subset $\Sigma$ of $T'$ such that $T^+_{\mathbb{S}} \mathop{\mid\!\!\!\frac{\alpha}{\rho}} \neg\Sigma, \Delta$. The notions $T^+_{\mathbb{S}} \mathop{\mid\!\!\!\frac{<\alpha}{<\rho}} \Delta$ are defined in the obvious way.

**12.4.5 Theorem** *(a) Let $\mathbb{S}$ be a chaste semi-formal and $\alpha$ be an $\varepsilon$-number. Then $T^+_{\mathbb{S}} \mathop{\mid\!\!\!\frac{<\alpha}{0}} \Delta^{\mathsf{U}}$ implies $T_{\mathbb{S}} \mathop{\mid\!\!\!\frac{<\varphi_\alpha(0)}{0}} \Delta$.*

*(b) If $\tau$ is a countable ordinal and $\alpha$ an $\varepsilon$-number then $T^+_{L_\tau} + (\text{Found}) \mathop{\mid\!\!\!\frac{<\alpha}{0}} \Delta^{\mathsf{U}}$ implies $L_\tau \mathop{\mid\!\!\!\frac{<\varphi_\alpha(0)}{0}} \Delta$.*

*Proof* (a) If $T^+_{\mathbb{S}} \mathop{\mid\!\!\!\frac{<\alpha}{0}} \Delta^{\mathsf{U}}$ there is an ordinal $\xi < \alpha$ and finite subsets $\Gamma \subseteq T$ and $\Lambda$ of $\mathsf{KP}^- + \mathit{Tran}(\mathsf{U})$ such that $\mathbb{S}^{\mathsf{U}} \mathop{\mid\!\!\!\frac{\xi}{0}} \neg\Lambda, \neg\Gamma^{\mathsf{U}}, \Delta^{\mathsf{U}}$ for some ordinal $\xi < \alpha$. By Lemma 12.4.2, we obtain $RS_{\mathbb{S}} \mathop{\mid\!\!\!\frac{\varphi_1(3^\xi)}{\omega\cdot 3^\xi+\omega}} \neg\Gamma^{\mathsf{L}_0}, \Delta^{\mathsf{L}_0}$. Using predicative cut-elimination, we obtain $RS_{\mathbb{S}} \mathop{\mid\!\!\!\frac{<\varphi_\alpha(0)}{0}} \neg\Gamma^{\mathsf{L}_0}, \Delta^{\mathsf{L}_0}$ and finally $T_{\mathbb{S}} \mathop{\mid\!\!\!\frac{<\varphi_\alpha(0)}{0}} \Delta$ by Exercise 12.3.9.

(b) is proved analogously using Lemma 12.4.2(b). □

## 12.5 Reduction of $T^+$ to $T$

Theorem 12.4.5 shows that $T^+$ is reducible to $T$ in the framework of the infinitary $\mathbb{S}$-logic. It was Jäger's observation that there is even a reduction in the framework of ordinary first-order logic. The basic idea is the same as in the reduction within the framework of infinitary logic. The reduction chain "$T^+ \vdash F$ *implies* $\mathbb{S}^{\mathsf{U}} \mathop{\mid\!\!\!\frac{\xi}{0}} \neg\Gamma^{\mathsf{U}}, \neg\Lambda, F^{\mathsf{U}}$ *for some finite* $\Gamma \subseteq T$ *and finite* $\Lambda \subseteq \mathsf{KP}^-$ *which, in turn, implies* $RS_{\mathbb{S}} \mathop{\mid\!\!\!\frac{\eta}{0}} \neg\Gamma^{\mathsf{U}}, F^{\mathsf{U}}$ *and this implies* $T_{\mathbb{S}} \mathop{\mid\!\!\!\frac{\eta}{0}} F$" has to be modified in such a way that $RS_{\mathbb{S}} \mathop{\mid\!\!\!\frac{\eta}{0}} \neg\Gamma, F$ allows to deduce $\mathop{\vdash}_{\mathsf{T}} \neg\Gamma, F$. In general, however there is no passage from the infinitary system $RS_{\mathbb{S}}$ to ordinary first-order logic. Jäger's idea was to use a finitary fragment $RS^k_{\mathsf{T}}$ of $RS_{\mathbb{S}}$ to make this passage feasible.

Of course we cannot start with an infinitary semi-formal system. Therefore we fix a theory $T$ and regard first-order logic – as introduced in Definition 6.3.2 – as a semi-formal system where the characteristic sequence of a formula $(\mathsf{Q}x)F_u(x)$ is $\langle F_u(t) \mid t$ is an $\mathcal{L}(T)$-term $\rangle$.[7] If $\mathcal{L}(T)$ does not include the membership relation, we add it to the language and define $\mathrm{CS}^k(s \in t) := \mathrm{CS}^k(s \notin t) := \emptyset$ for $\mathcal{L}(T)$-terms $s$ and $t$.[8] The role of $k$ will become clear in a moment. To avoid unnecessary case distinctions, we again require that the language of $T$ does not contain function symbols.

---

[7] It is obvious that the $\forall$-rule and $\exists$-rule are then obtainable from the $\bigwedge$-rule and $\bigvee$-rule, respectively.

[8] Since $s \notin t$ belongs to $\bigwedge$–type this makes $s \notin t$ an axiom.

Since pure first-order logic is in general not chaste, we require that the axioms in $T$ suffice to make the system given by $T \mathrel{\vert\!\!\frac{}{\mathsf{T}}} \Delta$ chaste. This means in particular that $T$ contains all identity axioms. We call $T$ chaste if the "semi-formal" system $T \mathrel{\vert\!\!\frac{}{\mathsf{T}}} \Delta$ is chaste.

We use the language of $T^+$ to define the finitary fragment $RS_{\mathsf{T}}^k$ of ramified set theory. For an $\mathscr{L}(T^+)$-formula $F$ let $F^n$ be the formula which is obtained from $F$ replacing the constant $U$ by $\mathsf{L}_0$ and bounding all quantifiers by a constant $\mathsf{L}_n$.

**12.5.1 Definition** Assume that $\{F_i \mid i \in \omega\}$ is an enumeration of all $\mathscr{L}(T^+)$-formulas. We define the $RS_{\mathsf{T}}^k$-terms inductively by the following clauses.

- Every $\mathscr{L}(T)$-term is an $RS_{\mathsf{T}}^k$-term of stage $-1$.
- Every constant $\mathsf{L}_n$ with $n < \omega$ is an $RS_{\mathsf{T}}^k$-term of stage $n$.
- Let $G(u, u_1, \dots, u_k)$ be an $\mathscr{L}(T^+)$-formula occurring in the list $\{F_0, \dots, F_k\}$, which contains at most the free variables $u, u_1, \dots, u_k$. If $\vec{a} := a_1, \dots, a_k$ is a tuple of $RS_{\mathsf{T}}^k$-terms of stages less than $n < \omega$ then $\{x \in \mathsf{L}_n \mid G^n(x, \vec{a})\}$ is an $RS_{\mathsf{T}}^k$-term of stage $n$.
- If $F(u_1, \dots, u_n)$ is an $\mathscr{L}(T^+)$-formula and $a_1, \dots, a_n$ is a tuple of $RS_{\mathsf{T}}^k$-terms then $F^m(a_1, \dots, a_n)$ is an $RS_{\mathsf{T}}^k$-formula for any $m < \omega$.

By $\mathscr{T}_n^k$ we denote the set of $RS_{\mathsf{T}}^k$-terms of stages less than $n$.

**12.5.2 Definition** We define simultaneously a relation $\sim$ on the $RS_{\mathsf{T}}^k$-terms and the $RS_{\mathsf{T}}^k$-formulas by the following clauses.

- $a \sim b$ if $a$ and $b$ are terms of stage $-1$.
- $\mathsf{L}_n \sim \mathsf{L}_n$ for all $n < \omega$.
- $\{x \in \mathsf{L}_m \mid F(x)^{\mathsf{L}_m}\} \sim \{x \in \mathsf{L}_n \mid G(x)^{\mathsf{L}_n}\}$ iff $m = n$ and $G(u)^{\mathsf{L}_n} \sim F(u)^{\mathsf{L}_n}$.
- $A \sim B$ iff there are a $\mathscr{L}(T^+)$-formula $F(u_1, \dots, u_n)$ and $RS_{\mathsf{T}}^k$-terms $a_1, \dots, a_n$ and $b_1, \dots, b_n$ such that $a_i \sim b_i$ for $i = 1, \dots, n$ and $A = F^m(a_1, \dots, a_n)$ and $B = F^m(b_1, \dots, b_n)$ for some finite $m$.

Let $[t]_\sim := \{s \mid s \sim t\}$ and $[F]_\sim := \{G \mid G \sim F\}$.

**12.5.3 Exercise** (a) Show that for every class $[t]_\sim$ of $RS_{\mathsf{T}}^k$-terms there is a term $\widetilde{t}(\vec{u})$ such that $[t]_\sim = \{\widetilde{t}_{u_1,\dots,u_n}(c_1, \dots, c_n) \mid c_1, \dots, c_n \in \mathscr{T}_0^k\}$.
(b) Show that for every class $[F]_\sim$ of $RS_{\mathsf{T}}^k$-formulas there is a formula $\widetilde{F}(\vec{u})$ such that $[F]_\sim = \{\widetilde{F}_{u_1,\dots,u_n}(c_1, \dots, c_n) \mid c_1, \dots, c_n \in \mathscr{T}_0^k\}$.

(c) Show that $\sim$ is an equivalence relation on the $RS^k_{\mathsf{T}}$-terms and – formulas.
(d) Conclude that the equivalence classes of $\sim$ provide a finite partition of the terms in $\mathcal{T}^k_n$ for every $n < \omega$.

Hint: Show (a) and (b) simultaneously by induction on $\mathrm{rnk}(t)$ and $\mathrm{rnk}(F)$, respectively. This is more tedious than you would expect. Observe that $\tilde{t}$ and $\tilde{F}$ are obtained from $t$ or $F$, respectively, by successively marking all occurrences of terms of stages $-1$ by free variables. Pay attention that different occurrences even of the same term have to be marked by different variables. Then use (a) and (b) to prove (c) and then (d), where you need induction on $n$.

To define the relation $RS^k_{\mathsf{T}} \overset{\alpha}{\underset{\rho}{\vdash}} \Delta$ for a finite set $\Delta$ of $RS^k_{\mathsf{T}}$-formulas we modify Definition 12.3.7 of the characteristic sequences for $RS^k_{\mathsf{T}}$-formulas replacing $\mathcal{T}^{\mathbb{S}}_\alpha$ by $\mathcal{T}^k_n$ accordingly. Let $\mathrm{CS}^k(F)$ denote the characteristic sequence in the sense of $RS^k_{\mathsf{T}}$. However, not every $RS^k_{\mathsf{T}}$-formula is in ($\bigvee$–type $\cup \bigwedge$–type ).
Let

- $\mathrm{rnk}_k(A) := 0$ for all atomic $\mathcal{L}(T)$-formulas

and define

- $\mathrm{rnk}_k(F) := \sup\{\mathrm{rnk}_k(G) \mid G \in \mathrm{CS}^k(F)\} + 1$ for $F \in$ ($\bigvee$–type $\cup \bigwedge$–type ).

Observe that $\mathrm{rnk}_k(F)$ is well-defined (cf. Exercise 12.3.8).

**12.5.4 Exercise** Show that $A \sim B$ implies $\mathrm{rnk}_k(A) = \mathrm{rnk}_k(B)$. Conclude that $\mathrm{rnk}_k(F) < \omega$ for all $RS^k_{\mathsf{T}}$-formulas $F$.

**12.5.5 Definition** We are now going to define the semi-formal system for $RS^k_{\mathsf{T}}$. Since there is no underlying structure $\mathfrak{M}$ we do not need rule $(Ax)_{\mathfrak{M}}$ but only

$(Ax)^k_L$ If $A$ is an atomic $\mathcal{L}(RS^k_{\mathsf{T}})$-formula not in ($\bigwedge$–type $\cup \bigvee$–type ) and $\{A, \neg A\} \subseteq \Delta$ then $RS^k_{\mathsf{T}} \overset{\alpha}{\underset{\rho}{\vdash}} \Delta$ for all ordinals $\alpha$ and $\rho$

and we define $RS^k_{\mathsf{T}} \overset{\alpha}{\underset{\rho}{\vdash}} \Delta$ using rules $(Ax)^k_L$, ($\bigwedge$), ($\bigvee$) and (cut).

We denote by $RS^k_T \overset{\alpha}{\underset{\rho}{\vdash}} \Delta$ that there is a finite set $\Lambda \subseteq T$ of formulas such that $RS^k_{\mathsf{T}} \overset{\alpha}{\underset{\rho}{\vdash}} \neg\Lambda^{\mathsf{L}_0}, \Delta$.

**12.5.6 Exercise** Show that $RS^k_{\mathsf{T}} \overset{\alpha}{\underset{\rho}{\vdash}} \Delta$ implies that there are finite ordinals $m$ and $r$ such that $RS^k_{\mathsf{T}} \overset{m}{\underset{r}{\vdash}} \Delta$.

Hint: Show first that $RS^k_{\mathsf{I}} \overset{\alpha}{\underset{\rho}{\vdash}} \Delta(u)$ implies $RS^k_{\mathsf{T}} \overset{\alpha}{\underset{\rho}{\vdash}} \Delta_u(t)$ for all terms $t$ of stage $-1$. Then prove the claim by induction on $\alpha$ using the fact that the characteristic sequence of any $RS^k_{\mathsf{T}}$-formula contains only finitely many equivalence classes.

As a shorthand, we write $RS_{\mathsf{T}}^k \vdash \Delta$ to denote that there are finite ordinals $m$ and $r$ such that $RS_{\mathsf{T}}^k \,|\!\frac{m}{r}\, \Delta$ and $RS_{\mathsf{T}}^k \,|\!\frac{}{0}\, \Delta$ to denote that there is a finite ordinal $m$ such that $RS_{\mathsf{T}}^k \,|\!\frac{m}{0}\, \Delta$. For a theory $T$ in the language of $RS_{\mathsf{T}}^k$ we denote by $T_{RS_{\mathsf{T}}^k} \vdash \Delta$ that there is a finite set $\Lambda \subseteq T$ such that $RS_{\mathsf{T}}^k \vdash \neg\Lambda, \Delta$. We use the same notations for $RS_T^k$.

**12.5.7 Exercise** Show that $RS_{\mathsf{T}}^k \vdash \Delta$ implies $RS_{\mathsf{T}}^k \,|\!\frac{}{0}\, \Delta$.

Hint: Adapt the Reduction Lemma (Lemma 7.3.12) to $RS_{\mathsf{T}}^k$ and then copy the proof of the Basis Elimination Lemma (Lemma 7.3.13).

**12.5.8 Exercise** Let $\Delta$ be a finite set of $\mathscr{L}(T)$-formulas. Show that $RS_{\mathsf{T}}^k \,|\!\frac{}{0}\, \Delta^{\mathsf{L}_0}$ implies $\,|\!\frac{}{\mathsf{T}}\, \Delta$. Conclude that $RS_T^k \,|\!\frac{}{0}\, \Delta^{\mathsf{L}_0}$ implies $T \,|\!\frac{}{\mathsf{T}}\, \Delta$.

Hint: Show by induction on $m$ that $RS_{\mathsf{T}}^k \,|\!\frac{m}{0}\, \Delta^{\mathsf{L}_0}$ implies $\,|\!\frac{}{\mathsf{T}}\, \Delta$. The only remarkable case is that the critical formula of the last inference in $RS_{\mathsf{T}}^k \,|\!\frac{m}{0}\, \Delta^{\mathsf{L}_0}$ is a formula $(\forall x \in \mathsf{L}_0)F(x)^{\mathsf{L}_0}$. Then you have the premises $RS_{\mathsf{T}}^k \,|\!\frac{m_0}{0}\, \Delta^{\mathsf{L}_0}, F(t)^{\mathsf{L}_0}$ for all $\mathscr{L}(T)$-terms $t$. Choose a free $\mathscr{L}(T)$-variable not occurring in $\Delta, F$ and apply the inductive hypothesis followed by a universal inference.

Now we have to check in how far the results of Sects. 11.10 and 11.12 can be transformed to $RS_T^k$. Recall our general proviso that $T$ is a chaste theory. Then we observe that in transferring the results of Sect. 11.10 and Sect. 11.12 to $RS_{\mathbb{S}}$ we only used the chasteness of $RS_{\mathbb{S}}$ but never the fact that there are infinitely many $RS_{\mathbb{S}}$-terms. Therefore, we can assume that all these properties stay correct for $RS_T^k$.

In the next step we apply the method of asymmetric interpretations to the system $RS_{\mathsf{T}}^k$. In analogy to Definition 12.4.1 let $F^{(m,s)}$ be the formula obtained from the $\mathscr{L}(T^+)$-formula by replacing $\mathsf{U}$ by $\mathsf{L}_0$ and bounding all unbounded universal quantifiers by $\mathsf{L}_m$ and all unbounded existential quantifiers by $\mathsf{L}_s$. For a finite set $\Delta = \{F_1, \ldots, F_n\}$ of $\mathscr{L}(T^+)$-formulas let

$$\mathscr{D}^{(m,s)} := \{\{F_1{}^{(m_1,s_1)}, \ldots, F_n{}^{(m_n,s_n)}\} \mid m_i \le m \wedge s_i \ge s\}.$$

**12.5.9 Exercise** Let $T$ be a chaste theory. Let moreover $\Lambda(\vec{u})$ and $\Delta(\vec{u})$ be finite sets of $\mathscr{L}(T^+)$-formulas not containing free variables other than those in the list $\vec{u}$. Assume that every formula in $\Lambda(\vec{u})$ is a specialization of an axiom in $\mathsf{KP}^-$, an identity axiom, or an axiom $Tran(\mathsf{U})$. Then $\,|\!\frac{n}{\mathsf{T}}\, \neg\Lambda(\vec{u}), \Delta(\vec{u})$ implies that there is a $k$ such that $RS_T^k \vdash \Delta'(\vec{a})$ holds true for all finite ordinals $m$, all $\Delta' \in \mathscr{D}^{(m,m+3^n)}$ and and all $RS_{\mathsf{T}}^k$-terms of stages less than $m$.

Hint: The proof by induction on $n$ is essentially the same as that of Lemma 12.4.2. Exercise 12.5.6 ensures that all derivations remain finite. To find the correct $k$ you have to secure that all comprehension formulas of specializations of $\Delta_0$-separation occurring in $\Lambda(\vec{u})$ as well as the formulas which are needed to handle specializations of (Nullset), (Pair') and (Union') or (TranC), respectively, are among the formulas in the enumeration $F_0, \dots, F_k$.

**12.5.10 Theorem** *The theory $T^+$ is a conservative extension of $T$.*

*Proof* If $T^+ \vdash F^{\mathsf{U}}$ for an $\mathcal{L}(T)$-formula $F$ we obtain finite sets $\Gamma \subseteq T$ and $\Lambda \subseteq \mathsf{KP}^- \cup \mathit{Tran}(\mathsf{U})$ such that $\underset{\mathsf{T}}{\vdash} \neg\Lambda, \neg\Gamma^{\mathsf{U}}, F^{\mathsf{U}}$. By Exercise 12.5.9 we obtain a $k$ such that $RS^k_{\mathsf{T}} \vdash \neg\Gamma^{\mathsf{L}_0}, F^{\mathsf{L}_0}$. By Exercise 12.5.7, we get $RS^k_{\mathsf{T}} \underset{0}{\vdash} \neg\Gamma^{\mathsf{L}_0}, F^{\mathsf{L}_0}$ and by Exercise 12.5.8 finally $T \underset{\mathsf{T}}{\vdash} F$. □

**12.5.11 Remark** Surprisingly Theorem 12.5.10 states that the axioms for the next admissible set do not affect the strength of the basis theory $T$. Notice, however, that this crucially depends on the fact that there is no foundation in $T^+$. The main point in obtaining this conservation result is to get rid of the additional set theoretical axioms in $T^+$. The crucial axiom in this connection is the axiom of $\Delta_0$-collection. This axiom is removed by an asymmetric interpretation. Such an interpretation, however, cannot work in the presence of a foundation or induction *scheme* because there the same formula occurs positively and negatively.

It remains to study what happens in the presence of the foundation axiom. This is prepared by the following lemma.

**12.5.12 Exercise** Show that for every $RS^k_{\mathsf{T}}$-formula $F$ there is an $\mathcal{L}(T)$-formula $F_T$ such that $RS^k_{\mathsf{T}} \vdash (\forall \vec{x} \in \mathsf{L}_0)[F(\vec{x}) \leftrightarrow F_T^{\mathsf{L}_0}(\vec{x})]$.

Hint: The key in solving Exercise 12.5.12 is Exercise 12.5.3. Recall that for all formulas in $\bigwedge$–type $\cup \bigvee$–type we have

$$\mathrm{CS}(F) = \langle G(s) \mid s \in I \rangle,$$

where $I$ is a finite index set for $RS^k_{\mathsf{T}}$-terms (in case that $F$ is a conjunction or distinction $F_0 \circ F_1$ let $G(s)$ be $F_s$ for $s \in \{0,1\}$). By Exercise 12.5.3 there is a term $\tilde{s}$ containing free variables $u_1, \dots, u_n$ such that $t \sim s$ iff there is a tuple $c_1, \dots, c_n$ of $RS^k_{\mathsf{T}}$-terms such that $t$ is a term $\tilde{s}_{u_1,\dots,u_n}(c_1, \dots, c_n)$. Now define

$$F_T := \begin{cases} F & \text{iff } F \notin (\bigvee\text{–type} \cup \bigwedge\text{–type}) \text{ or } \mathrm{CS}(F) = \emptyset \\ \bigvee\{(\exists \vec{x})G(\tilde{s}_{\vec{u}}(\vec{x}))_T \mid G(s) \in \mathrm{CS}(F)\} & \text{for } F \in \bigvee\text{–type} \\ \bigwedge\{(\forall \vec{x})G(\tilde{s}_{\vec{u}}(\vec{x}))_T \mid G(s) \in \mathrm{CS}(F)\} & \text{for } F \in \bigwedge\text{–type}. \end{cases}$$

Let (FOUND($X$)) be the scheme

$$(\exists y)[y \in X \wedge F(x)] \rightarrow (\exists z)[F(z) \wedge z \subset X \wedge (\forall y \subset z)[y \subset X \rightarrow \neg F(y)]]$$

saying that $\in$ is well-founded on $X$ for definable classes which is equivalent to the scheme

$$(\forall x \in X)[(\forall y \in x)[y \in X \rightarrow F(y)] \rightarrow F(x)] \rightarrow (\forall x \in X)F(x)$$

of $\in$-induction. Recall also the axiom (Found($X$)) (introduced on p. 263) saying that $\in$ is well-founded on $X$ for sets.

**12.5.13 Exercise** Let $F$ be an $\mathcal{L}(T)$-formula. Show $T^+ + (\mathrm{Found}(X)) \underset{\mathsf{T}}{\models} F^{\mathsf{U}}$ iff $T + (\mathrm{FOUND}(X)) \underset{\mathsf{T}}{\models} F$.

Hint: The direction from right to left is simple since $\{x \in \mathsf{U} \mid F(x)^{\mathsf{U}}\}$ is a set in $T^+$. The opposite direction is proved by extending Exercise 12.5.9 to the case that $\Lambda(\vec{u})$ may also contain a specialization of axiom (Found($X$)) under the additional hypothesis $T \underset{\mathsf{T}}{\models} (\mathrm{FOUND}(X))$. To this end you have to check

$$RS_T^k \vdash b \notin a, b \notin X, (\exists x \in a)[x \in X \wedge (\forall y \in x)[y \notin X \vee y \notin a] \tag{i}$$

for all $b \in \mathcal{T}_n^k$. This is done by induction on $n$. If $a$ and $b$ are in $\mathcal{T}_0^k$ you get (i) from $T \underset{\mathsf{T}}{\models} (\mathrm{FOUND})$. So assume that $b \in \mathcal{T}_0^k$ and $stg(a) \geq 0$. According to Exercise 12.5.12 there is an $\mathcal{L}(T)$-formula $F$ such that $RS_T^k \vdash (b \in a \wedge b \in X \leftrightarrow F^{\mathsf{L}_0}(b))$. Using $T \underset{\mathsf{T}}{\models} (\mathrm{FOUND})$ you get

$$T \underset{\mathsf{T}}{\models} \neg F(b), (\exists x)[F(x) \wedge (\forall y \in x) \neg F(y)],$$

which is (i) for $b \in \mathcal{T}_0^k$. For $n > 0$ you get

$$RS_T^k \vdash c \not\in a, c \notin X, (\exists x \in a)[x \in X \wedge (\forall y \in x)[y \notin X \vee y \not\in a]^{9} \tag{ii}$$

for all $c \in \mathcal{T}_{stg(b)}^k$ by the induction hypothesis. Use (ii) to infer

$$RS_T^k \vdash (\forall x \in b)[x \not\in a \vee x \notin X], (\exists x \in a)[x \in X \wedge (\forall y \in x)[y \notin X \vee y \not\in a] \tag{iii}$$

which implies

$$RS_T^k \vdash \begin{array}{l}(\forall x \in b)[x \not\in a \vee x \notin X] \wedge b \in X \wedge b \underline{\in} a,\\ (\exists x \in a)[x \in X \wedge (\forall y \in x)[y \notin X \vee y \not\in a], b \not\in a, b \notin X\end{array} \tag{iv}$$

by tautology. From (iv) you easily get (i) for all terms $b$.

Now you have to extend the proof of Exercise 12.5.9 by the case that $\neg F$ is a specialization of axiom (Found). Then $F$ is a formula

$$(\exists x)[x \in u \wedge x \in X] \wedge (\forall x \in u)[x \in X \rightarrow (\exists y \in x)[y \in X \wedge y \in u]$$

and you have the premises

$$\underset{\mathsf{T}}{\overset{n_0}{\models}} (\exists x)[x \in u \wedge x \in X], \neg \Lambda(\vec{u}), \Delta(\vec{u})$$

and

$$\underset{\mathsf{T}}{\overset{n_0}{\models}} (\forall x \in u)[x \notin X \vee (\exists y \in x)[y \in X \wedge y \in u], \neg \Lambda(\vec{u}), \neg(\forall x)[x \in \mathsf{U}], \Delta(\vec{u}).$$

For $m_0 := m + 3^{n_0}$ you get by the inductive hypothesis

$$RS_T^k \vdash (\exists x \in \mathsf{L}_{m_0})[x \in a \wedge x \in X], \Delta'(\vec{a}) \tag{v}$$

and

$$RS_T^k \vdash (\forall x \in a)[x \notin X \vee (\exists y \in x)[y \in X \wedge y \in a], \Delta'(\vec{a}) \tag{vi}$$

[9] For the notion $\underline{\in}$ cf. p. 280.

for $\Delta'(\vec{a}) \in \mathscr{D}^{(m,m+3^n)}$. Applying two $\bigvee$ inferences and a $\bigwedge$ inference to (i) and cutting the result with (v) and (vi) yields

$$RS_T^k \vdash \Delta'(\vec{a})$$

and you have extended Exercise 12.5.9.

Now assume $T^+ + (\mathrm{Found}(X)) \vdash F^{\mathsf{U}}$. Then conclude $RS^k_{T+(\mathrm{FOUND}(X))} \vdash F^{\mathsf{L}_0}$ by Exercise 12.5.9 and finally, use exercises 12.5.7 and 12.5.8 to get $T + (\mathrm{FOUND}(X)) \vdash F$.

## 12.6 The Theories $\mathsf{KP}_n$ and $\mathsf{KP}_n^0$

In this section we introduce the theories $\mathsf{KP}_n$ which axiomatize set universes containing $n$-admissible sets. Their ordinal analyzes are already outside of a first step into impredicativity (they need $n$-steps). However, restricting the amount of foundation which is available in these theories we obtain theories $\mathsf{KP}_n^0$ which are in the realm of predicative proof theory.

**12.6.1 Definition** Recall the group $\mathsf{Ad}$ of axioms introduced on p. 262. By $\mathsf{KP}_n$ we denote the theory in the language $\mathscr{L}(\emptyset, \mathbb{A}_0, \ldots, \mathbb{A}_n, \in, Ad)$ which comprises the axioms $\mathsf{KP} + \mathsf{Ad} + \bigwedge_{i=0}^n Ad(\mathbb{A}_i) + (\forall x)[x \notin \emptyset] + \emptyset \in \mathbb{A}_0 + \bigwedge_{i=0}^{n-1} \mathbb{A}_i \in \mathbb{A}_{i+1}$ axiomatizing that there are $n$ consecutive admissible universes which entails the existence of the admissible ordinals $\omega$ and $\omega_i^{CK}$ for $i \in \{1, \ldots, n\}$.

The canonical least model for the theory $\mathsf{KP}_n$ in the constructible hierarchy is $L_{\omega_{n+1}^{CK}}$.

The theory $\mathsf{KP}_n^0$ is obtained from $\mathsf{KP}_n$ by replacing the foundation scheme (FOUND) by axiom $(\mathrm{Found}(\mathbb{A}_0))$, the theory $\mathsf{KP}_n^{\mathsf{r}}$ is obtained from $\mathsf{KP}_n$ by replacing the foundation scheme (FOUND) by axiom (Found).

If $T$ and $T'$ are theories, we write $T \sqsubseteq T'$ iff a model $\mathfrak{M}$ of $T$ is definable within any model $\mathfrak{M}'$ of $T'$. By $T \equiv T'$ we denote that $T \sqsubseteq T'$ and $T' \sqsubseteq T$ hold true.

**12.6.2 Exercise** (a) Show that $\mathsf{KP}\omega \equiv \mathsf{KP}_0$.

(b) Show that $\mathsf{KP}\omega^0 \equiv \mathsf{KP}_0^0$.

(c) Show that $\mathsf{KP}\omega^r \equiv \mathsf{KP}_0^{\mathsf{r}}$.

(d) Show that $\mathsf{KP}\omega^r + (\mathrm{IND})_\omega \equiv \mathsf{KP}_0^{\mathsf{r}} + (\mathrm{FOUND}(\mathbb{A}_0))$.

Hints: (a) $\mathsf{KP}\omega \sqsubseteq \mathsf{KP}_0$ is obvious since you get $\omega = \mathbb{A}_0 \cap On$. For the opposite direction you have to show that the hereditarily finite sets – which are definable in $\mathsf{KP}\omega$ (cf. [4]) – are an interpretation for $\mathbb{A}_0$.

(b) Again you have to show that a model of $\mathsf{KP}_0^0$ can be constructed within a model of $\mathsf{KP}\omega^0$. This is, however, much more complicated since you do not have enough induction. But you have the existence of $\omega$ and the hereditarily finite sets can be coded into $\omega$.

(c) and (d) follow easily from (b).

To establish the connection between the theories $\mathsf{KP}_n^0$ and iterations of the admissible extension we introduce the theories $\widetilde{\mathsf{KP}}_n$. In defining the admissible extension $\widetilde{\mathsf{KP}}_n^+$ we rename the new constant $\mathsf{U}$ by $\mathbb{A}_{n+1}$ in the obvious way.

**12.6.3 Definition** Let $\widetilde{\mathsf{KP}}_{-1} := \mathsf{KP}^-$ and $\widetilde{\mathsf{KP}}_{n+1} := \widetilde{\mathsf{KP}}_n^+ + (\mathrm{Found}(\mathbb{A}_0))$. $\widetilde{\mathsf{KP}^\mathsf{r}}_{-1} = \mathsf{KP}^-$ and $\widetilde{\mathsf{KP}^\mathsf{r}}_{n+1} := \widetilde{\mathsf{KP}^\mathsf{r}_n}^+ + (\mathrm{Found})$.

**12.6.4 Exercise** Show that $\mathsf{KP}_n^0 \equiv \widetilde{\mathsf{KP}}_n$ and $\mathsf{KP}_n^\mathsf{r} \equiv \widetilde{\mathsf{KP}^\mathsf{r}_n}$ hold true for all $n \geq 0$.

Hint: Roughly speaking you get $\widetilde{\mathsf{KP}}_0 = \mathsf{KP}^- + (\mathsf{KP}^-)^{\mathbb{A}_0} + \emptyset \in \mathbb{A}_0 + Tran(\mathbb{A}_0) + (\mathrm{Found}(\mathbb{A}_0)) \equiv \mathsf{KP}^- + Ad(\mathbb{A}_0) + (\mathrm{Found}(\mathbb{A}_0)) = \mathsf{KP}_0^0$ and by induction on $n$ also

$$\begin{aligned}
\widetilde{\mathsf{KP}}_{n+1} &= \mathsf{KP}^- + (\widetilde{\mathsf{KP}}_n)^{\mathbb{A}_{n+1}} + \{\mathbb{A}_i \in \mathbb{A}_{n+1} \mid i \leq n\} + Tran(\mathbb{A}_{n+1}) + (\mathrm{Found}(\mathbb{A}_0))\\
&\equiv \mathsf{KP}^- + (\mathsf{KP}_n^0)^{\mathbb{A}_{n+1}} + \{\mathbb{A}_i \in \mathbb{A}_{n+1} \mid i \leq n\} + Tran(\mathbb{A}_{n+1}) + (\mathrm{Found}(\mathbb{A}_0))\\
&\equiv \mathsf{KP}^- + Ad(\mathbb{A}_{n+1}) + \emptyset \in \mathbb{A}_0 + \textstyle\bigwedge_{i=0}^{n} Ad(\mathbb{A}_i) + \bigwedge_{i=0}^{n+1} \mathbb{A}_i \in \mathbb{A}_{i+1} + (\mathrm{Found}(\mathbb{A}_0))\\
&\equiv \mathsf{KP}_{n+1}^0.
\end{aligned}$$

Render this sketch more precisely and apply it also to $\mathsf{KP}_n^\mathsf{r}$.

**12.6.5 Remark** Recall the semi-formal system $L_\omega$ defined according to Definition 12.2.1. The semi-formal system based on $L_\omega$ must not be confused with ramified set-theory. Formulas $s \in t$ and $s = t$ are atomic in $L_\omega$. But observe that $s^{L_\omega} \in t^{L_\omega}$ and $s^{L_\omega} = t^{L_\omega}$ are still decidable (cf. [4] II.2). All formulas in $\mathcal{L}(L_\omega)$ have finite ranks.

By a simple induction on the rank of an $\mathcal{L}(L_\omega)$-sentence we get the following lemma.

**12.6.6 Lemma** *Let $F$ be an $\mathcal{L}(L_\omega)$-sentence which is true in $L_\omega$. Then $L_\omega \,\big|\!\frac{\mathrm{rnk}(F)}{0}\, F$.*

In Sect. 11.5, we have shown that all primitive recursive functions are sets. This needed the existence of $\omega$. More generally we know that every recursive function is $\Delta_1$ definable on $L_\omega$ (cf. [4] II.2.3). Therefore there is a direct translation of arithmetical pseudo $\Pi_1^1$-sentences $F$ into pseudo $\Pi_1^1$-sentences $F^*$ of $\mathcal{L}(L_\omega)$ such that $\mathbb{N} \models (\forall X)F(X)$ implies $L_\omega \models (\forall X)[X \subseteq On \to F^*(X)]$. The role of natural numbers in $L_\omega$ is played by ordinals. We identify $F$ and $F^*$ if there is no danger of confusion.

**12.6.7 Exercise** Let $F$ be an arithmetical pseudo $\Pi_1^1$-sentence. Show that $L_\omega \models^{\alpha} F^*$ implies $\mathrm{tc}(F) \leq \alpha$. Let $\prec$ be a transitive binary relation. Infer $\mathrm{otyp}(\prec) \leq \alpha$ from $L_\omega \models^{\alpha} TI(\prec, X)^*$.

Hint: For an $L_\omega$-formula $F$ let $\underline{F}$ be the formula which is obtained from $F$ by replacing all closed terms $t$ by their evaluations $\underline{t^{L_\omega}}$. Let $\Delta, F$ be a finite set of arithmetical formulas and show first

$$L_\omega \models^{\alpha} \Delta^*, F^* \;\Rightarrow\; L_\omega \models^{\alpha} \underline{\Delta^*, F^*}$$

by induction on $\alpha$. The crucial case is that the critical formula of the last inference is $F^*$ for an atomic formula $F$. Observe that then $F^*$ is not necessarily atomic, too. However, if $F \in Diag(\mathbb{N})$ then $\underline{F^*} \in Diag(L_\omega)$ and you get $L_\omega \models^{\alpha} \underline{\Delta^*, F^*}$ with an inference $(\bigwedge)$. Now assume that $F$ is a formula $t \in X$ for a term $t$ of the shape $\underline{f}s_1 \dots s_n$ where $\underline{f}$ is a symbol for an $n$-ary primitive recursive function. Then observe that $F^*$ is the formula $(\forall y \in \omega)(\forall x_1 \in \omega)\dots(\forall x_n \in \omega)[\underline{f}x_1 \dots x_n = y \to y \in X]$. Since $L_\omega \models^{\alpha} \Delta^*, F^*$ you have the premises

$$L_\omega \models^{\alpha_s} , \Delta^*, s \notin On, (\forall x_1 \in \omega)\dots(\forall x_n \in \omega)[\underline{f}x_1 \dots x_n = s \to y \in X]$$

for all $s \in L_\omega$, which by inversion entails

$$L_\omega \models^{\alpha_s} \Delta^*, s \notin On, t_1 \notin On, \dots, t_n \notin On, \underline{f}t_1 \dots t_n \neq s, s \in X.$$

(Observe that primitive recursive functions are not elements of $L_\omega$. However, $\underline{f}t_1 \dots t_n = s$ is a sentence in the language of $L_\omega$). For $s := t^{L_\omega}$ and $t_i \in On$ all the sentences $\underline{s} \notin On$, $\underline{t_i} \notin On$, $\underline{f}\underline{t_1} \dots \underline{t_n} \neq \underline{s}$ become false and by Exercise 12.2.5 and the induction hypothesis you get $L_\omega \models^{\alpha} \underline{\Delta^*, t^{L_\omega}} \in X$. The case that $F$ is a formula $t \notin X$ is symmetrical.

In a similar way you treat the case that $F$ is a false equation $s = t$ or inequality $s \neq t$ for terms $s$ and $t$ of the form $\underline{f}s_1 \dots s_m$ and $\underline{g}t_1 \dots t_m$, respectively.

In the next step you prove that $L_\omega \models^{\alpha} \underline{\Delta^*}$ implies $\models^{\alpha} \Delta$ by a simple induction on $\alpha$.

**12.6.8 Exercise** Show that $L_\omega \models^{<\omega\cdot 2}$ (FOUND).

Hint: Show first $L_\omega \vdash_0^{2\cdot \mathrm{rnk}(F)+5\cdot(\mathrm{rnk}_{L_\omega}(F(a))+1)} \neg F(a), (\exists y)[F(y) \land (\forall x \in y)\neg F(x)]$ for all $a \in L_\omega$ by $\in$-induction on $a$. Here $\mathrm{rnk}_{L_\omega}(a)$ denotes the least stage at which $a$ enters $L_\omega$ (cf. the proofs of Lemma 11.12.7 and Theorem 11.12.8). Actually the proof here is simpler and even closer to the proof of Lemma 7.3.4.

**12.6.9 Definition** Let $L_\omega{}^{(-1)} = L_\omega$ and $L_\omega{}^{(\mathbf{n+1})} = (L_\omega{}^{(\mathbf{n})})^+$. Instead of U, we denote the new constant in the admissible extension $(L_\omega{}^{(\mathbf{n})})^+$ by $\mathbb{A}_{n+1}$.

Observe that the languages of $L_\omega{}^{(\mathbf{n})}$ and $\mathsf{KP}_n$ coincide and that all formulas in $\mathcal{L}(L_\omega{}^{(\mathbf{n})})$ have finite ranks.

**12.6.10 Definition** Let $\Phi^0(\alpha) := \alpha$ and $\Phi^{n+1}(\alpha) := \Phi^n(\varphi_\alpha(0))$.

**12.6.11 Lemma** *Let $\alpha$ be an $\varepsilon$-number and $0 \leq k \leq n$. Then $L_\omega{}^{(\mathbf{n})} \vdash_0^{<\alpha} \Delta^{\mathbb{A}_k}$ implies $L_\omega{}^{(\mathbf{k-1})} \vdash_0^{<\Phi^{n+1-k}(\alpha)} \Delta$ for all finite sets $\Delta$ of formulas in $\mathcal{L}_\omega^{\mathbf{k}}$.*

*Proof* We induct on $n$. Assume $L_\omega{}^{(\mathbf{n})} \vdash_0^{<\alpha} \Delta^{\mathbb{A}_k}$. By Theorem 12.4.5 we obtain $L_\omega{}^{(\mathbf{n-1})} \vdash_0^{<\varphi_\alpha(0)} \Delta$ for $k = n$ and $L_\omega{}^{(\mathbf{n-1})} \vdash_0^{<\varphi_\alpha(0)} \Delta^{\mathbb{A}_k}$ for $k < n$. The latter implies by the induction hypothesis $L_\omega{}^{(\mathbf{k})} \vdash_0^{<\Phi^{n-k}(\varphi_\alpha(0))} \Delta$, i.e., $L_\omega{}^{(\mathbf{k})} \vdash_0^{<\Phi^{n+1-k}(\alpha)} \Delta$. ☐

**12.6.12 Exercise** (a) Show $L_\omega{}^{(\mathbf{n})} \vdash_0^{<\omega\cdot 2}$ (FOUND($\mathbb{A}_0$)).

(b) Show $L_\omega{}^{(\mathbf{n})} \vdash_{<\omega}^{<\omega\cdot 2} \tilde{A}$ for all axioms $A$ in $\widetilde{\mathsf{KP}}_n$.

(c) Conclude that $\mathsf{KP}_n+$(FOUND($\mathbb{A}_0$)) $\vdash \Delta$ implies $L_\omega{}^{(\mathbf{n})} \vdash_{<\omega}^{<\omega\cdot 2} \Delta$.

Hint: (a) Redo the proof of Exercise 12.6.8 bearing in mind that all formulas in $\mathscr{L}(L_\omega{}^{(\mathbf{n})})$ have finite ranks.

Prove part (b) by induction on $n$. The case $n = -1$ is covered by Lemma 12.6.6. For the successor case you have $L_\omega{}^{(\mathbf{m})} \mathrel{|}\!\frac{<\omega\cdot 2}{<\omega} A$ for all axioms $A$ in $\widetilde{\mathsf{KP}}_m$ which entails $(L_\omega{}^{(\mathbf{m})})^+ \mathrel{|}\!\frac{<\omega\cdot 2}{<\omega} A^{\mathbb{A}_{m+1}}$ for all axioms $A$ in $\widetilde{\mathsf{KP}}_m$. Essentially by definition, you get $L_\omega{}^{(\mathbf{m+1})} \mathrel{|}\!\frac{<\omega}{<\omega} A$ for all axioms in $\mathsf{KP}^-$ and $Tran(\mathbb{A}_{m+1})$. You easily show $L_\omega{}^{(\mathbf{m+1})} \mathrel{|}\!\frac{<\omega\cdot 2}{<\omega} \mathbb{A}_M \in \mathbb{A}_{m+1}$ and, as a special case of part (a), finally $L_\omega{}^{(\mathbf{m+1})} \mathrel{|}\!\frac{<\omega\cdot 2}{<\omega} (\mathrm{Found}(\mathbb{A}_0))$.

For (c) combine claims (a) and (b) with Exercise 12.2.4.

**12.6.13 Theorem** *$\widetilde{\mathsf{KP}}_n \vdash \Delta^{\mathbb{A}_k}$ implies $L_\omega{}^{(\mathbf{k-1})} \mathrel{|}\!\frac{<\Phi^{n-k}(\varepsilon_0)}{0} \Delta$ for $0 \leq k \leq n$.*

*Proof* Using Exercise 12.5.13 we obtain $\widetilde{\mathsf{KP}}_{n-1} + (\mathrm{FOUND}(\mathbb{A}_0)) \vdash \Delta^{\mathbb{A}_k}$ from $\widetilde{\mathsf{KP}}_n \vdash \Delta^{\mathbb{A}_k}$. By Exercise 12.6.12 this implies $L_\omega{}^{(\mathbf{n-1})} \mathrel{|}\!\frac{<\omega\cdot 2}{<\omega} \Delta^{\mathbb{A}_k}$. Predicative cut elimination yields $L_\omega{}^{(\mathbf{n-1})} \mathrel{|}\!\frac{<\varepsilon_0}{0} \Delta^{\mathbb{A}_k}$ and Lemma 12.6.11 finally $L_\omega{}^{(\mathbf{k-1})} \mathrel{|}\!\frac{<\Phi^{n-k}(\varepsilon_0)}{0} \Delta$. □

Defining

$$\|T\|_{\Sigma^{\mathbb{A}_1}} := \min\{\alpha \mid F \text{ a } \Sigma_1\text{-formula and } L_\alpha \models F \text{ and } T \vdash F^{\mathbb{A}_1}\}$$

and

$$\|T\|_{\Pi_2^{\mathbb{A}_1}} := \min\{\alpha \mid F \text{ a } \Pi_2\text{-formula and } L_\alpha \models F \text{ and } T \vdash F^{\mathbb{A}_1}\}$$

we obtain the following theorem as a corollary of Theorem 12.6.13.

**12.6.14 Theorem** *(a) For all $n$ we get $\|\mathsf{KP}_n^0\| \leq \Phi^n(\varepsilon_0)$.*

*(b) For $n \geq 1$ we have $\|\mathsf{KP}_n^0\|_{\Sigma^{\mathbb{A}_1}} \leq \Phi^{n-1}(\varepsilon_0)$.*

*(c) For $n \geq 1$ we obtain $\|\mathsf{KP}_n^0\|_{\Pi_2^{\mathbb{A}_1}} \leq \Phi^{n-1}(\varepsilon_0)$. The stage $L_{\Phi^{n-1}(\varepsilon_0)}$ of the constructible hierarchy is therefore closed under the provably $\omega_1^{CK}$-recursive functions of $\mathsf{KP}_n^0$.*

*Proof* (a) We have to determine an upper bound for the order type of orderings on $\omega$ whose well-foundedness is provable in $\mathsf{KP}_n^0$. In $\mathsf{KP}_n^0$, we obtain $\omega$ as $\mathbb{A}_0 \cap On$. So assume that $\mathsf{KP}_n^0 \vdash TI(\prec, X)^{\mathbb{A}_0 \cap On}$. Then $\widetilde{\mathsf{KP}}_n \vdash TI(\prec, X)^{\mathbb{A}_0 \cap On}$ according to Exercise 12.6.4. By Theorem 12.6.13, we get $L_\omega \mathrel{|}\!\frac{\Phi^n(\varepsilon_0)}{0} TI(\prec, X)^{On}$ and by Exercise 12.6.7 $\mathrm{otyp}(\prec) \leq \Phi^n(\varepsilon_0)$.

(b) To prove part (b) assume $\mathsf{KP}_n^0 \vdash (\exists x \in \mathbb{A}_1) F(x)$ for a $\Delta_0$-formula $F(x)$. Thus, we get $L_\omega{}^{(\mathbf{0})} \mathrel{|}\!\frac{<\Phi^{n-1}(\varepsilon_0)}{0} (\exists x) F(x)$ by Theorem 12.6.13, i.e., $L_\omega{}^{\cup} \mathrel{|}\!\frac{\eta}{0} \neg\Lambda, (\exists x) F(x)$ for some ordinal $\eta < \Phi^{n-1}(\varepsilon_0)$ and a finite set $\Lambda \subseteq \mathsf{KP}^- + \mathsf{IDEN}$. By Lemma 12.4.2, we get $RS_{L_\omega} \mathrel{|}\!\frac{\varphi_1(3^\eta)}{\omega\cdot 3^\eta + \omega} (\exists x \in \mathsf{L}_{3^\eta}) F(x)$. Applying Lemma 12.3.17 we therefore get $RS(X) \overset{\varphi_1(3^\eta)}{\underset{\omega^{\eta+2}}{\vdash}} (\exists x \in \mathsf{L}_{\omega+3^\eta})F(x)$. Since $\omega + 3^\eta < \Phi^{n-1}(\varepsilon_0)$, we obtain $\|\mathsf{KP}_n\|_{\mathbb{A}_1} \le \Phi^{n-1}(\varepsilon_0)$.

(c) Assume $\mathsf{KP}_n^0 \vdash (\forall x \in \mathbb{A}_1)(\exists y \in \mathbb{A}_1)F(x,y)$ for a $\Delta_0$-formula $F(x,y)$. By Theorem 12.6.13 we obtain $L_\omega{}^{(\mathbf{0})} \overset{\eta}{\underset{0}{\vdash}} (\forall x)(\exists y)F(x,y)$ for some $\eta < \Phi^{n-1}(\varepsilon_0)$. For $a \in L_{\Phi^{n-1}(\varepsilon_0)}$ there is a $\beta < \Phi^{n-1}(\varepsilon_0)$ such that $a \in L_\beta$ and by Lemma 12.4.2 we get

$$RS_{L_\omega} \overset{\varphi_1(\beta+3^\eta)}{\underset{\rho}{\vdash}} (\exists y \in \mathsf{L}_{\beta+3^\eta})F(a,y).$$

Hence

$$RS_{L_\omega} \overset{<\Phi^{n-1}(\varepsilon_0)}{\underset{<\Phi^{n-1}(\varepsilon_0)}{\vdash}} (\forall x \in \mathsf{L}_{\Phi^{n-1}(\varepsilon_0)})(\exists y \in \mathsf{L}_{\Phi^{n-1}(\varepsilon_0)})F(x,y)$$

which implies by Lemma 12.3.17 $RS(X) \vdash (\forall x \in \mathsf{L}_{\Phi^{n-1}(\varepsilon_0)})(\exists y \in \mathsf{L}_{\Phi^{n-1}(\varepsilon_0)})F(x,y)$. Hence $L_{\Phi^{n-1}(\varepsilon_0)} \models (\forall x)(\exists y)F(x,y)$.

A function $f$ is provably $\omega_1^{CK}$-recursive in $\mathsf{KP}_n^0$ if there is a $\Sigma$-formula $F(x,y,z)$ such that $f(x) \simeq y \;\Leftrightarrow\; \mathsf{KP}_n^0 \vdash (\forall x \in \mathbb{A}_1)(\exists y \in \mathbb{A}_1)(\exists z \in \mathbb{A}_1)F(x,y,z)$ for a $\Delta_0$-formula $F(x,y,z)$ whose parameters are all definable in $\mathsf{KP}_n^0$. Hence $L_{\Phi^{n-1}(\varepsilon_0)} \models (\forall x)(\exists y)(\exists z)F(x,y,z)$ which shows that $L_{\Phi^{n-1}(\varepsilon_0)}$ is closed under $f$. □

**12.6.15 Remark** In $\mathsf{KP}_n$, the set $\mathbb{A}_0$ figures as the set $L_\omega$ of hereditarily finite sets and thus $\mathbb{A}_1$ figures as the first admissible set above $L_\omega$, i.e., as $L_{\omega_1^{CK}}$. The ordinal $\|\mathsf{KP}_n\|_{\Sigma^{\mathbb{A}_1}}$ corresponds therefore to the ordinal $\|\mathsf{KP}_n\|_{\Sigma^{\omega_1^{CK}}}$ mentioned in Remark 11.7.3.

The theories $\mathsf{KP}_n$ are therefore examples for theories in which the proof-theoretic ordinal and the $\|\mathsf{KP}_n\|_{\Sigma^{\omega_1^{CK}}}$-ordinals differ.

In the following theorem, we collect some of the results of this section.

**12.6.16 Theorem** $\|\mathsf{KP}\omega^0\| = \|\mathsf{KP}_0^0\| = \varepsilon_0$.

$\|\mathsf{KP}\omega^r\| = \varepsilon_0$.

$\|\mathsf{KP}\omega^r + (\mathrm{IND})_\omega\| = \|\widetilde{\mathsf{KP}_0^r}\| \le \varphi_{\varepsilon_0}(0)$.

*Proof* $\|\mathsf{KP}\omega^0\| = \|\mathsf{KP}_0^0\| \le \varepsilon_0$ follows from Exercise 12.6.2 (b) and Theorem 12.6.14.

If $\mathsf{KP}\omega^r \vdash TI(\prec, X)^{\mathbb{A}_0 \cap On}$ we get $\widetilde{\mathsf{KP}_0} + (\mathrm{Found}) \vdash TI(\prec, X)^{\mathbb{A}_0 \cap On}$ and thus $\mathsf{KP}^- + (\mathrm{FOUND}) \vdash TI(\prec, X)^{On}$ by Exercise 12.5.13. Hence $L_\omega \overset{<\omega\cdot 2}{\underset{<\omega}{\vdash}} TI(\prec, X)^{On}$ which implies $L_\omega \overset{<\varepsilon_0}{\underset{0}{\vdash}} TI(\prec, X)^{On}$ by cut elimination and thus $\mathrm{otyp}(\prec) < \varepsilon_0$ by Exercise 12.6.7. Since NT is embeddable in both theories the bound $\varepsilon_0$ is exact.

If $\mathsf{KP}\omega^r + (\mathrm{IND})_\omega \vdash TI(\prec, X)$ we get $\widetilde{\mathsf{KP}_0^r} + (\mathrm{FOUND}(\mathbb{A}_0)) \vdash TI(\prec, X)^{\mathbb{A}_0 \cap On}$. Hence, $L_\omega{}^{(\mathbf{0})} + (\mathrm{Found}) \overset{<\omega\cdot 2}{\underset{<\omega}{\vdash}} TI(\prec, X)^{\mathbb{A}_0 \cap On}$ by Exercise 12.6.12(c) which by cut-elimination implies $L_\omega{}^{(0)} + (\text{Found}) \overset{<\varepsilon_0}{\underset{0}{\vdash}} TI(\prec,X)^{\mathbb{A}_0 \cap On}$. By Theorem 12.4.5(b), we get $L_\omega \overset{<\varphi_{\varepsilon_0}(0)}{\underset{0}{\vdash}} TI(\prec,X)^{On}$ and thus $\|\mathsf{KP}\omega^r + (\mathrm{IND})_\omega\| = \|\widetilde{\mathsf{KP}^{\mathsf{r}}_0}\| \leq \varphi_{\varepsilon_0}(0)$ from Exercise 12.6.7. □

**12.6.17 Remark** $\varphi_{\varepsilon_0}(0)$ is the the proof-theoretic ordinal of the theory of $\Delta^1_1$-comprehension and also of the theory of the $\Sigma^1_1$ axiom of choice in second-order arithmetic both equipped with the induction scheme. These results are scattered in the literature of predicative proof theory (e.g. [26]). Unfortunately, there is no comprehensive monograph treating these parts of predicative proof theory systematically. Schütte in his monograph [89] only treats the his system **DA** which corresponds to a $\Delta^1_1$-comprehension rule and has proof-theoretic ordinal $\varphi_\omega(0)$. Since $\mathsf{KP}\omega$ proves $\Delta$-separation and strong $\Sigma$-replacement it is not too difficult to see that ($\Delta^1_1$-CA) and ($\Sigma^1_1$-AC) can be embedded into $\mathsf{KP}^{\mathsf{r}}_\omega + (\mathrm{IND})_\omega$ (which entails that $\varphi_{\varepsilon_0}(0)$ is the exact bound for $\mathsf{KP}\omega^r + (\mathrm{IND})_\omega$) while $(\Delta^1_1\text{-CA})_0$ and $(\Sigma^1_1\text{-AC})_0$, the versions in which the induction scheme is replaced by the induction axiom, can be embedded into $\mathsf{KP}^0_\omega$ and thus have proof-theoretic ordinals $\varepsilon_0$. The result that these theories are conservative over $\mathsf{NT}$ is originally due to Schlipf (cf. [6]).

## 12.7 The Theories $\mathsf{KPl}_0$ and $\mathsf{KPi}_0$

Recall the theories $\mathsf{KPl}_0$ and $\mathsf{KPi}_0$ introduced in Sect. 11.6. The theory $\mathsf{KPl}$ comprises the axioms of $\mathsf{BST} + \mathsf{Ad}$ together with the limit axiom

(Lim) $\quad (\forall x)(\exists y)[Ad(y) \wedge x \in y]$

which states that the universe is a union of admissible universes. Even stronger is the theory $\mathsf{KPi}$ which is $\mathsf{KP} + \mathsf{BST} + \mathsf{Ad} + (\mathrm{Lim})$ which axiomatizes an admissible universe which is also a union of admissible universes. In Exercise 11.6.3 we have seen that the set $\mathbb{A}_0$ of hereditarily finite sets is definable in $\mathsf{KPl}$, hence also in $\mathsf{KPi}$. We obtain the restrictions $\mathsf{KPl}_0$ and $\mathsf{KPi}_0$ replacing the foundation scheme (FOUND) by the axiom (Found($\mathbb{A}_0$)).

We are going to reduce $\mathsf{KPi}_0$ – and thus also $\mathsf{KPl}_0$ – to the theories $\mathsf{KP}^0_k$ via an asymmetric interpretation. To this end, we need an (asymmetric) interpretation for the formulas in $\mathscr{L}(\mathsf{KPi})$ into $\mathscr{L}_{\mathsf{KP}_k}$ which is given in the following definition.

**12.7.1 Definition** Let $F$ be an $\mathscr{L}(\in,Ad)$-formula. By $F^{(m,s)}$, we understand the $\mathscr{L}(\in,\emptyset,\mathbb{A}_1,\dots,\mathbb{A}_k,Ad)$-formula which is obtained from $F$ restricting all universal quantifiers $(\forall x)[\cdots x\cdots]$ to $(\forall x \in \mathbb{A}_m)[\cdots x\cdots]$ and all existential quantifiers $(\exists x)[\cdots x\cdots]$ to $(\exists x \in \mathbb{A}_s)[\cdots x\cdots]$. If $\Delta = \{F_1,\dots,F_n\}$ is a finite set of $\mathscr{L}(\in,Ad)$-formulas then $\mathscr{D}^{(m,s)}$ is the collection of all finite sets $\{F_1^{(m_1,s_1)},\dots,F_n^{(m_n,s_n)}\}$ with $m_i \leq m$ and $s_i \geq s$ for $i = 1,\dots n$ (cf. Definition 12.4.1) .

**12.7.2 Exercise** Let $\Delta(\vec{u})$ be a finite set of $\mathcal{L}(\in,Ad)$ formulas whose free variables occur all in the list $\vec{u}$. Show that $\mathsf{KPi}_0 \underset{\mathsf{T}}{\overset{n}{\vdash}} \Delta(\vec{u})$ implies $\mathsf{KP}_k^0 \models \vec{a} \in \mathbb{A}_i \to \bigvee \Delta'(\vec{a})$ for all $m$, all $i \leq m$, all $k \geq s \geq m+3^n$, all $\Delta' \in \mathcal{D}^{(m,s)}$ and all tuples $\vec{a}$ of $\mathcal{L}(\emptyset,\mathbb{A}_1,\ldots,\mathbb{A}_n,\in,Ad)$-terms.

Hint: This is again an asymmetric interpretation. The proof is essentially that of Lemma 12.4.2. Assume $\mathsf{KPi}_0 \underset{\mathsf{T}}{\overset{n}{\vdash}} \Delta(\vec{u})$. Fix $m$ and choose $s \geq m+3^n$ and $k \geq s$. Let $\mathfrak{M}$ be an arbitrary $\mathsf{KP}_k^0$-model and $\Phi$ an $\mathfrak{M}$-assignment. Prove

- *"If $\underset{\mathsf{T}}{\overset{n}{\vdash}} \neg\Lambda(\vec{u}),\Delta(\vec{u})$ for a finite set $\Lambda(\vec{u})$ of specializations of $\mathsf{KPi}_0$-axioms, including the identity axioms, then $\mathfrak{M} \models (\vec{a} \in \mathbb{A}_i \to \bigvee \Delta'(\vec{a}))\,[\Phi]$ holds true for all $\Delta' \in \mathcal{D}^{(m,s)}$, all $i \leq m$ and all tuples $\vec{a}$ of $\mathcal{L}(\mathsf{KP}_k^0)$-terms"*

by induction on $n$. Fix $\Delta' \in \mathcal{D}^{(m,s)}$. To simplify the notation, suppress mentioning the assignment $\Phi$. The distinction of cases is as in the proof of Lemma 12.4.2. The main cases are that the critical formula of the last inference is an axiom $\neg$(Lim) or $\neg(\Delta_0$–Coll). In case of $\neg$(Lim), you have the premise

$$\underset{\mathsf{T}}{\overset{n_0}{\vdash}} \neg\Lambda(\vec{u}),\Delta(\vec{u}),(\forall y)\neg[Ad(y) \land v \in y] \tag{i}$$

for some variable $v$. If $v$ occurs in the list $\vec{u}$ let $b$ the corresponding term. Otherwise put $b := \emptyset$. Then $\mathfrak{M} \models (\vec{a} \in \mathbb{A}_i)$ implies $b \in \mathbb{A}_i$ for all $i \leq m < m+1$. Since also $k \geq s \geq m+3^n \geq m+1+3^{n_0}$ you can apply the induction hypothesis to (i) to get

$$\mathfrak{M} \models \vec{a} \in \mathbb{A}_i \to \bigvee \Delta'(\vec{a}) \lor (\forall y \in \mathbb{A}_{m+1})\neg[Ad(y) \lor b \in y]. \tag{ii}$$

Since $\mathfrak{M} \models b \in \mathbb{A}_m \in \mathbb{A}_{m+1}$ and $\mathfrak{M} \models Ad(\mathbb{A}_m)$ you get the claim from (ii).

In case of an axiom $(\Delta_0$–Coll) you have the premise

$$\underset{\mathsf{T}}{\overset{n_0}{\vdash}} \neg\Lambda(\vec{u}),\Delta(\vec{u}),(\forall x \in v)(\exists y)F(x,y,\vec{u}) \land (\forall z)(\exists x \in v)(\forall y \in z)\neg F(x,y,\vec{u})$$

from which by $\land$-inversion you get

$$\underset{\mathsf{T}}{\overset{n_0}{\vdash}} \neg\Gamma,\Delta(\vec{u}),(\forall x \in v)(\exists y)F(x,y,\vec{u}) \tag{iii}$$

and

$$\underset{\mathsf{T}}{\overset{n_0}{\vdash}} \neg\Gamma,\Delta(\vec{u}),(\forall z)(\exists x \in v)(\forall y \in z)\neg F(x,y,\vec{u}). \tag{iv}$$

Choose $b$ as in the previous case, put $s_0 := m+3^{n_0}$ and apply the inductive hypothesis to (iii) to obtain

$$\mathfrak{M} \models \vec{a} \in \mathbb{A}_i \to \bigvee \Delta'(\vec{a}) \lor (\forall x \in b)(\exists y \in \mathbb{A}_{s_0})F(x,y,\vec{a}). \tag{v}$$

Letting $s_0 := m+3^{n_0}$, $m' := s_0+1$ and $s' := m'+3^{n_0}$ and applying the inductive hypothesis to (iv) you get

$$\mathfrak{M} \models \vec{a} \in \mathbb{A}_i \to \bigvee \Delta''(\vec{a}) \lor (\forall z \in \mathbb{A}_{s_0+1})(\exists x \in b)(\forall y \in z)\neg F(x,y,\vec{a})$$

for all $\Delta'' \in \mathcal{D}^{(m',s')} \supseteq \mathcal{D}^{(m,s)}$. Since $\mathfrak{M} \models \mathbb{A}_{s_0} \in \mathbb{A}_{s_0+1}$ this implies

$$\mathfrak{M} \models \vec{a} \in \mathbb{A}_i \to \bigvee \Delta'(\vec{a}) \lor (\exists x \in b)(\forall y \in \mathbb{A}_{s_0})\neg F(x,y,\vec{a}) \tag{vi}$$

and the claim follows from (v) and (vi).

A case which does not appear in 12.4.2 is that the critical formula is $\neg\mathsf{Ad}$. Then you have a premise

$$\Big|\frac{n_0}{\mathsf{T}}\ \Lambda(\vec{u}),\Delta(\vec{u}),Ad(v)\wedge\neg B(v)$$

or

$$\Big|\frac{n_0}{\mathsf{T}}\ \Lambda(\vec{u}),\Delta(\vec{u}),Ad(v_0)\wedge Ad(v_1)\wedge\neg B'(v)$$

where $B(v)$ is the formula $Tran(v)$ or ((Pair')$^v$ $\wedge$ (Union')$^v$ $\wedge$ ($\Delta_0$–Sep)$^v$ $\wedge$ ($\Delta_0$–Coll)$^v$) and $B'(v)$ the formula $v_0\in v_1\vee v_0=v_1\vee v_1\in v_0$. Indicating only the first case you get by $\wedge$-inversion

$$\Big|\frac{n_0}{\mathsf{T}}\ \Lambda(\vec{u}),\Delta(\vec{u}),Ad(v)$$

and

$$\Big|\frac{n_0}{\mathsf{T}}\ \Lambda(\vec{u}),\Delta(\vec{u}),\neg B(v).$$

If $v$ appears in the list $\vec{u}$ replace $v$ by $b$ accordingly. Otherwise replace $v$ by $b:=\emptyset$. Then $\mathfrak{M}\models\vec{a}\in\mathbb{A}_i$ implies $b\in\mathbb{A}_i$. By induction hypothesis you obtain

$$\mathfrak{M}\models\vec{a}\in\mathbb{A}_i\rightarrow Ad(b)\vee\bigvee\Delta'(\vec{a})\tag{vii}$$

and

$$\mathfrak{M}\models\vec{a}\in\mathbb{A}_i\rightarrow\neg B(b)\vee\bigvee\Delta'(\vec{a}).\tag{viii}$$

If $\mathfrak{M}\not\models Ad(b)$ you get the claim from (vii) and $\mathfrak{M}\models Ad(b)$ implies $\mathfrak{M}\models B(b)$ and you get the claim from (viii).

The case that the critical formula is the negation of (Found($\mathbb{A}_0$)) is again different from that in Lemma 12.4.2. The problem is that $\mathbb{A}_0$ has different definitions in $\mathsf{KPi}_0$ and $\mathsf{KP}^0_k$. To emphasize this difference let us temporarily write $\mathsf{U}_0$ for $\mathbb{A}_0$ as defined in $\mathsf{KPi}_0$. Observe, however, that by Exercise 11.6.3(c) $(\forall x\in\mathsf{U}_0)[\cdots x\cdots]$ can be expressed by $(\forall x)[(\forall u)(Ad(u)\rightarrow x\in u)\rightarrow\cdots x\cdots]$. But if $\mathfrak{M}\models x\in\mathbb{A}_0$ we always get $\mathfrak{M}\models(\forall u)[Ad(u)\rightarrow x\in u]$. So $(\forall x\in\mathsf{U}_0)[\cdots x\cdots]^{\mathbb{A}_m}$ is in $\mathfrak{M}$ equivalent to $(\forall x\in\mathbb{A}_0)[\cdots x\cdots]^{\mathbb{A}_m}$. On the other hand $(\exists x\in\mathsf{U}_0)[\cdots x\cdots]$ is a shorthand for the formula $(\exists x)(\exists u)[Ad(u)\wedge(\forall y\in u)(\neg Ad(y))\wedge x\in u\wedge\cdots x\cdots]$. For $l>0$ the set $\mathbb{A}_0$ is a witness for $u$ in $(\forall x)[x\in\mathbb{A}_0\rightarrow(\exists u\in\mathbb{A}_l)[Ad(u)\wedge(\forall y\in u)(\neg Ad(y))\wedge x\in u]$. Hence, $\mathfrak{M}\models(\exists x\in\mathsf{U}_0)[\cdots x\cdots]^{\mathbb{A}_l}$ iff $\mathfrak{M}\models(\exists x\in\mathbb{A}_0)[\cdots x\cdots]^{\mathbb{A}_l}$ for all $l>0$.

The negation of (Found($\mathbb{A}_0$)) is the formula

$$(\exists u)\big[(\forall x\in\mathsf{U}_0)[(\forall x\in y)(y\in u)\rightarrow x\in u]\wedge(\exists x\in\mathsf{U}_0)(x\notin u)\big]$$

and from the premise you get by $\bigwedge$-inversion

$$\Big|\frac{n_0}{\mathsf{T}}\ \neg\Lambda(\vec{u}),\Delta(\vec{u}),(\forall x\in\mathsf{U}_0)[(\forall x\in y)(y\in u)\rightarrow x\in u]$$

and

$$\Big|\frac{n_0}{\mathsf{T}}\ \neg\Lambda(\vec{u}),\Delta(\vec{u}),(\exists x\in\mathsf{U}_0)[x\notin u]$$

for a variable $u$. If $u$ occurs in the list $\vec{u}$ replace $u$ correspondingly by a term $b$. Otherwise, let $b$ be $\emptyset$. In any case you get $\mathfrak{M}\models\vec{a}\in\mathbb{A}_i\rightarrow b\in\mathbb{A}_i$. Using the above observations you get

$$\mathfrak{M}\models\vec{a}\in\mathbb{A}_i\rightarrow\bigvee\Delta'(\vec{a})\vee(\forall x\in\mathbb{A}_0)[(\forall y\in x)(y\in b)\rightarrow x\in b]\tag{ix}$$

and

$$\mathfrak{M}\models\vec{a}\in\mathbb{A}_i\rightarrow\bigvee\Delta'(\vec{a})\vee(\exists x\in\mathbb{A}_0)(x\in b)\tag{x}$$

by the inductive hypothesis. Since $\mathfrak{M} \models (\mathrm{Found}(\mathbb{A}_0))$ you get from (ix)

$$\mathfrak{M} \models \vec{a} \in \mathbb{A}_i \to \bigvee \Delta'(\vec{a}) \vee (\forall x \in \mathbb{A}_0)(x \in b)$$

and the claim follows together with (x).

The remaining cases follow the pattern of the proof of Lemma 12.4.2.

Since $\mathsf{KPl}_0 \subseteq \mathsf{KPi}_0$, Exercise 12.7.2 also yields a reduction of $\mathsf{KPl}_0$ to $\bigcup_{n\in\omega} \mathsf{KP}_n^0$.

**12.7.3 Theorem** *(a)* $\|\mathsf{KPl}_0\| \leq \|\mathsf{KPi}_0\| \leq \Gamma_0$.
*(b)* $\|\mathsf{KPl}_0\|_{\Sigma_1^{\mathbb{A}}} \leq \|\mathsf{KPi}_0\|_{\Sigma_1^{\mathbb{A}}} \leq \Gamma_0$.
*(c)* $\|\mathsf{KPl}_0\|_{\Pi_2^{\mathbb{A}_1}} \leq \|\mathsf{KPi}_0\|_{\Pi_2^{\mathbb{A}_1}} \leq \Gamma_0$*, i.e., the stage* $L_{\Gamma_0}$ *of the constructible hierarchy is the least model for the* $\Pi_2^{\mathbb{A}_1}$*-sentences which are provable in* $\mathsf{KPl}_0$ *or* $\mathsf{KPi}_0$ *and hence closed under all* $\omega_1^{CK}$*-recursive functions which are provably total in* $\mathsf{KPl}_0$ *or even* $\mathsf{KPi}_0$.

*Proof* (a) $\|\mathsf{KPl}_0\| \leq \|\mathsf{KPi}_0\|$ is obvious. So assume $\mathsf{KPi}_0 \vdash TI(\prec,X)^{\mathsf{U}_0}$. Then, we get $\mathsf{KP}_n^0 \vdash TI(\prec,X)^{\mathbb{A}_0}$ for some $n$ which by Theorem 12.6.14 implies $\mathrm{otyp}(\prec) \leq \Phi^n(\varepsilon_0) < \Gamma_0$.

(b) First observe that for $l > 0$ we get $\mathsf{U}_0^{\mathbb{A}_l} = \{x \mid (\forall u \in \mathbb{A}_l)[Ad(u) \to x \in u]\} = \{x \mid x \in \mathbb{A}_0\} = \mathbb{A}_0$ which implies that for a $\Sigma_1$-formula $(\exists x)F(x)$ the translation $((\exists x \in \mathsf{U}_1)F(x))^{\mathbb{A}_1}$, i.e., the formula $(\exists x)\big[(\forall u)[Ad(u) \wedge \mathsf{U}_0 \in u \to x \in u] \wedge F(x)\big]^{\mathbb{A}_1}$, becomes equivalent to $(\exists x \in \mathbb{A}_1)\big[(\forall u \in \mathbb{A}_1)[Ad(u) \wedge \mathbb{A}_0 \in u \to x \in u] \wedge F(x)\big]$, which, in turn, is equivalent to $(\exists x \in \mathbb{A}_1)F(x)$. If $\mathsf{KPi}_0 \vdash (\exists x)F(x)^{\mathsf{U}_1}$ for a $\Sigma_1$-sentence $(\exists x)F(x)$ then we get from Exercise 12.7.2 $\mathsf{KP}_n \vdash (\exists x \in \mathbb{A}_1)F(x)$ which by Theorem 12.6.14 (b) implies $(\exists x \in L_{\Phi^{n-1}(\varepsilon_0)})F(x)$ and by upwards persistency also $(\exists x \in \mathsf{L}_{\Gamma_0})F(x)$. Hence $\|\mathsf{KPl}_0\|_{\mathbb{A}_1} \leq \|\mathsf{KPi}_0\|_{\mathbb{A}_1} \leq \Gamma_0$.

Claim (c) follows from (b) in the same way as in Theorem 12.6.14. □

In Exercise 11.6.6, we have seen that the theory $(\mathsf{ATR})_0$ of autonomously iterated arithmetic comprehensions can be embedded into $\mathsf{KPl}_0$. Hence $\|(\mathsf{ATR})_0\| \leq \Gamma_0$ which shows that it is reducible to a theory which is predicative in the Feferman–Schütte sense as described in Chap. 8. On the other hand, $(\mathsf{ATR})_0$ proves that for every ordinal $\alpha$ less than $\Gamma_0$ there is a primitive recursive well-ordering of order type $\alpha$ whose well-foundedness is provable in $(\mathsf{ATR})_0$ (cf. Exercise 8.5.32). Summing up we get the following theorem.

**12.7.4 Corollary** $\|\mathsf{KPl}_0\| = \|\mathsf{KPi}_0\| = \|(\mathsf{ATR})_0\| = \Gamma_0$. *This implies that the theory* $(\mathsf{ATR})_0$ *is predicatively reducible.*

**12.7.5 Remark** The theory $(\mathsf{ATR})_0$ has been introduced by Friedman in the framework of reverse mathematics (cf. [94]). The first computation of $\|(\mathsf{ATR})_0\|$ by Friedman et al. is in [27].

The method of asymmetric interpretations is not limited to $\mathsf{KPi}_0$ but also yields results for a series of theories which lay between $\Gamma_0$ and $\psi(\varepsilon_{\Omega+1})$. The analysis of these *metapredicative* theories is mostly due to Jäger and his students and collaborators. Two selected references are [53] and [54] where further references can be found.[10]

[10] Cf. also the references at the end of the book.

# Chapter 13
# Nonmonotone Inductive Definitions

We have seen in Sect. 6.3 that the fixed-point of an inductive definition comes in stages. We used a cardinality argument to show that the hierarchy of stages of an inductive definition becomes eventually stationary. Monotonicity, however, is not really necessary in the cardinality argument. The cardinality argument only needs that the operator is inflationary, i.e., that the operator $\Phi$ satisfies the condition $X \subseteq \Phi(X)$. This means, however, not really a restriction since any operator $\Phi\colon \mathrm{Pow}(\mathbb{N}) \longrightarrow \mathrm{Pow}(\mathbb{N})$ induces an operator $\Phi'(X) := X \cup \Phi(X)$ which is apparently inflationary.

## 13.1 Nonmonotone Inductive Definitions

We start with a brief synopsis of the theory of nonmonotone inductive definitions.

**13.1.1 Definition** Let $\Phi\colon \mathrm{Pow}(\mathbb{N}) \longrightarrow \mathrm{Pow}(\mathbb{N})$ be an operator. The hierarchy of stages is defined by

$$\Phi^{\alpha} := \Phi^{<\alpha} \cup \Phi(\Phi^{<\alpha})$$

where again we abbreviate $\Phi^{<\alpha} := \bigcup_{\xi<\alpha} \Phi^{\alpha}$.

Then

$$\Phi^{<\infty} := \bigcup_{\xi \in On} \Phi^{\xi}$$

is a fixed-point of $\Phi'$. We call it the fixed-point generated by $\Phi$.

**13.1.2 Lemma** *Let $\Phi\colon \mathrm{Pow}(\mathbb{N}) \longrightarrow \mathrm{Pow}(\mathbb{N})$ be an operator. Then there is a countable ordinal $\sigma$ such that $\Phi^{<\sigma} = \Phi^{\sigma}$. For this ordinal $\sigma$, we have $\Phi^{<\infty} = \Phi^{<\sigma} = \Phi^{\sigma}$.*

*Proof* Since all sets $\Phi^{\alpha}$ are countable, we obtain by cardinality reasons an ordinal $\sigma$ such that $\Phi^{<\sigma} = \Phi^{\sigma}$. By definition, we have $\Phi^{<\sigma} \subseteq \Phi^{<\infty}$. For the opposite inclusion, we prove

W. Pohlers, *Proof Theory: The First Step into Impredicativity*, Universitext,

$$\sigma \le \tau \Rightarrow \Phi^{\tau} = \Phi^{<\sigma} \tag{i}$$

by induction on $\tau$. For $\sigma = \tau$, we obtain (i) by definition of $\sigma$. For $\sigma < \tau$ we obtain $\Phi^{<\tau} = \Phi^{<\sigma}$ by induction hypothesis. Hence

$$\Phi^{\tau} = \Phi(\Phi^{<\tau}) \cup \Phi^{<\tau} = \Phi(\Phi^{<\sigma}) \cup \Phi^{<\sigma} = \Phi^{\sigma} = \Phi^{<\sigma}. \qquad \square$$

**13.1.3 Definition** Let $|\Phi| := \min\{\sigma \mid \Phi^{\sigma} = \Phi^{<\sigma}\}$. We call $|\Phi|$ the closure ordinal of $\Phi$.

As a consequence of Lemma 13.1.2, we obtain

$$\Phi^{\infty} := \Phi(\Phi^{<\infty}) \cup \Phi^{<\infty} = \Phi(\Phi^{<|\Phi|}) \cup \Phi^{<|\Phi|} = \Phi^{|\Phi|} = \Phi^{<\infty}. \tag{13.1}$$

By a norm on a set $P$ we commonly understand a surjective mapping $f{:}P \xrightarrow{onto} \lambda \in On$.

An operator $\Phi{:}\mathrm{Pow}(\mathbb{N}) \longrightarrow \mathrm{Pow}(\mathbb{N})$ induces a mapping $|\ |_{\Phi}{:}\mathbb{N} \longrightarrow On$ defined by

$$|n|_{\Phi} := \begin{cases} \min\{\alpha \mid x \in \Phi^{\alpha}\} & \text{if } x \in \Phi^{<\infty} \\ \infty & \text{otherwise} \end{cases}$$

which satisfies the following property.

**13.1.4 Lemma** *We obtain* $|\Phi| = \{|x|_{\Phi} \mid x \in \Phi^{<\infty}\}$. *Hence* $|\ |_{\Phi}{:}\Phi^{<\infty} \xrightarrow{onto} |\Phi|$.

*Proof* Let $\sigma := |\Phi|$. By Lemma 13.1.2, we have $\Phi^{<\infty} = \Phi^{<\sigma}$. Hence $\{|x|_{\Phi} \mid x \in \Phi^{<\infty}\} \subseteq \sigma$. If $\alpha < \sigma$ then $\Phi^{<\alpha} \subsetneq \Phi^{\alpha}$. Then, there is an $x \in \Phi^{\alpha} \setminus \Phi^{<\alpha}$. But then $|x|_{\Phi} = \alpha$ for that $x$. $\square$

By Lemma 13.1.4 $|\ |_{\Phi}$ is a norm on $\Phi^{<\infty}$. We call it the *inductive norm* induced by $\Phi$.

**13.1.5 Remark** The standard example for nonmonotone operators are operators of the form $[\Gamma_0, \Gamma_1]$ where $\Gamma_0$ and $\Gamma_1$ are inductive definitions, i.e., monotone operators, and

$$[\Gamma_0, \Gamma_1](X) := \{x \in \mathbb{N} \mid x \in \Gamma_0(X) \vee [\Gamma_0(X) \subseteq X \wedge x \in \Gamma_1(X)]\}.$$

We obtain the stages of $[\Gamma_0, \Gamma_1]$ by first iterating $\Gamma_0$ until a $\Gamma_0$-closed set is obtained and then use $\Gamma_1$, then falling back to iterating $\Gamma_0$ until the next $\Gamma_0$ closed set is obtained, then using $\Gamma_1$, etc.

The origin of operators of this kind was the need of higher constructive number classes in abstract recursion theory. The first constructive number class comprises all ordinals which can be represented by recursive well-orderings on the natural numbers, i.e., the ordinals below $\omega_1^{CK}$. Kleene designed an abstract notation system $\mathcal{O}$ for the ordinals in the first constructive number class using an abstract notion of fundamental sequence. Kleene's notation system consists of a set $\mathcal{O}$ of ordinal notations together with an interpretation $|a|_{\mathcal{O}} \in On$ for $a \in \mathcal{O}$. Originally, it is defined by the clauses

- $1 \in \mathcal{O}$ and $|1|_{\mathcal{O}} := 0$,

- if $a \in \mathcal{O}$ then $2^a \in \mathcal{O}$ and $|2^a|_{\mathcal{O}} := |a|_{\mathcal{O}} + 1$

and

- if $e$ is an index for a recursive fundamental sequence, i.e., if $(\forall n \in \omega)[\{e\}^1(x) \in \mathcal{O}]$ and $|\{e\}^1(n)|_{\mathcal{O}} < |\{e\}^1(n+1)|_{\mathcal{O}}$, then $3{\cdot}5^e \in \mathcal{O}$ and $|3{\cdot}5^e|_{\mathcal{O}} := \sup_{n\in\mathbb{N}} |\{e\}^1(n)|_{\mathcal{O}}$,

where $\{e\}^1$ denotes the Kleene bracket, i.e.,$\{e\}^1(x) \simeq U(\mu y.\, T^1(e,x,y))$ (cf. Theorem 10.1.1 on p. 208).

The second constructive number class $\mathcal{O}_2$ is obtained by allowing fundamental sequences of length $\omega_1^{CK}$, i.e., by the limit clause

- if $(\forall x \in \mathcal{O})[\{e\}^1(x) \in \mathcal{O}_2]$ and $|x|_{\mathcal{O}} < |y|_{\mathcal{O}} \Rightarrow |\{e\}^1(x)|_{\mathcal{O}_2} < |\{e\}^1(y)|_{\mathcal{O}_2}$ then $3^2{\cdot}5^e \in \mathcal{O}_2$ and $|3^2{\cdot}5^e|_{\mathcal{O}_2} := \sup_{x\in\mathcal{O}} |\{e\}^1(x)|_{\mathcal{O}_2}$.

Iterating this procedure, we obtain all finite constructive number classes. By $\omega_n^{CK}$, we denote the order type of the $n$th constructive number class.

It is not too difficult to see that Kleene's $\mathcal{O}$ can be defined by a positive arithmetical inductive definition. This implies that the order type of the ordinals denoted in Kleene's $\mathcal{O}$ cannot exceed $\omega_1^{CK}$ (as we defined it in this book). Again, it is not too difficult to see that the order type is in fact equal to $\omega_1^{CK}$. The order type of the ordinals in $\mathcal{O}_2$ therefore exceeds $\omega_1^{CK}$ which implies that already $\mathcal{O}_2$ is unattainable by a positive arithmetical inductive definition but needs an inductive definition in which $\mathcal{O}$ occurs negatively as given above.

This indicates that nonmonotone inductive definitions are more powerful than monotone inductive definitions. We will not pursue the abstract theory further. A careful study of the connections between large cardinals, their constructible counterparts in the constructible hierarchy $L$ and the closure ordinals of nonmonotone inductive definitions is due to Aczel and Richter [83].[1]

From a proof-theoretical standpoint only definable nonmonotone operators (cf. Definition 6.4.1) can be interesting. Following Richter–Aczel, we define

$$[\mathscr{F}_1,\dots,\mathscr{F}_n] := \{[\Gamma_1,\dots,\Gamma_n] \mid \Gamma_i \text{ is positively } \mathscr{F}_i\text{-definable}\},$$

where $\mathscr{F}_1,\dots,\mathscr{F}_n$ are complexity classes. The closure ordinal of the class $[\Pi_1^0,\Pi_1^0]$ is already a recursive Mahlo ordinal and thus widely outside the reach of a first step into impredicativity. But Richter and Aczel prove in [83] that $|\Pi_1^0| = |\Sigma_2^0| = \omega_1^{CK}$, a result which goes back to Gandy. An ordinal analysis of a theory for $\Pi_1^0$-definable nonmonotone inductive definitions should therefore still be in the realm of this book. We will give its ordinal analysis as an application of the ordinal analysis of $(\Pi_2\text{–REF})$.

[1] One of the most influential papers for ordinally informative proof theory.

## 13.2 Prewellorderings

The axiomatization of a theory of fixed-points $\Phi$ of $\Pi^0_1$-definable operators is not as simple as the axiomatization of positive inductive definitions. In the latter, the fixed-point could be easily described as the least $\Phi$-closed set. To define fixed-points of nonmonotone operators, we need ordinals which are, however, not available in the language of arithmetic. The solution is given by the stage comparison relations for inductive definitions which are defined by

$$\begin{aligned} x \preceq_\Phi y \ &:\Leftrightarrow \ (\exists\alpha)[x \in \Phi^\alpha \wedge y \notin \Phi^{<\alpha}] \\ x \prec_\Phi y \ &:\Leftrightarrow \ (\exists\alpha)[x \in \Phi^\alpha \wedge y \notin \Phi^{\alpha}]. \end{aligned} \tag{13.2}$$

Our aim is to axiomatize $\Phi^{<\infty}$ via the stage comparison relations. Therefore, we have to study them more profoundly. We start with two observations.

$$x \preceq_\Phi y \ \Leftrightarrow \ x \in \Phi^{<\infty} \wedge [y \notin \Phi^{<\infty} \vee \neg(y \prec_\Phi x)] \tag{13.3}$$

and

$$x \prec_\Phi y \ \Leftrightarrow \ x \in \Phi^{<\infty} \wedge [y \notin \Phi^{<\infty} \vee \neg(y \preceq_\Phi x)]. \tag{13.4}$$

To prove the direction from left to right in (13.3) let $x \preceq_\Phi y$. Then there is an $\alpha$ such that $x \in \Phi^\alpha$ and $y \notin \Phi^{<\alpha}$. Hence $x \in \Phi^{<\infty}$ and we are done if $y \notin \Phi^{<\infty}$. If $y \in \Phi^\beta$, then $\alpha \leq \beta$ which implies $x \in \Phi^\beta$. Hence $\neg(y \prec_\Phi x)$. For the opposite inclusion assume $x \in \Phi^{<\infty}$. If $y \notin \Phi^{<\infty}$ then $x \preceq_\Phi y$ is immediate. Otherwise, $y \in \Phi^\alpha$ implies $x \in \Phi^\alpha$ for all ordinals $\alpha$. For $\alpha := |x|_\Phi$, we therefore obtain $y \notin \Phi^{<\alpha}$ but $x \in \Phi^\alpha$. Hence $x \preceq_\Phi y$ and we have (13.3).

Equation (13.4) is proved similarly. From the left-hand side hypothesis, we obtain an ordinal $\alpha$ such that $x \in \Phi^\alpha$ but $y \notin \Phi^\alpha$. This implies $x \in \Phi^{<\infty}$. If $y \in \Phi^\beta$ then $\alpha < \beta$ which implies $x \in \Phi^{<\beta}$. Hence $\neg(y \preceq_\Phi x)$. For the opposite implication assume $x \in \Phi^{<\infty}$. If $y \notin \Phi^{<\infty}$, we obviously have $x \prec_\Phi y$. Otherwise, we obtain $x \in \Phi^{<\alpha}$ whenever $y \in \Phi^\alpha$. For $\alpha := |x|_\Phi$, we therefore have $x \in \Phi^\alpha$ but $y \notin \Phi^\alpha$. Hence $x \prec_\Phi y$ and we are done with (13.4), too.

The agreement $\alpha < \infty$ for all ordinals $\alpha$ allows for a simple characterization of the relations $\preceq_\Phi$ and $\prec_\Phi$. We obtain

$$\{x \mid x \preceq_\Phi y\} = \Phi^{|y|_\Phi} \tag{13.5}$$

and

$$\{x \mid x \prec_\Phi y\} = \Phi^{<|y|_\Phi}. \tag{13.6}$$

To prove (13.5) let $x \preceq_\Phi y$. Then there is an $\alpha$ such that $x \in \Phi^\alpha$ and $y \notin \Phi^{<\alpha}$. If $y \notin \Phi^{<\infty}$ then $|y|_\Phi = \infty$ which immediately implies $x \in \Phi^{<\infty} \subseteq \Phi^{|y|_\Phi}$. Otherwise we have $y \in \Phi^{|y|_\Phi}$ which implies $\alpha \leq |y|_\Phi$. Hence, $x \in \Phi^{|y|_\Phi}$ and we have $\{x \mid x \preceq_\Phi y\} \subseteq \Phi^{|y|_\Phi}$ in any case. For the opposite inclusion let $x \in \Phi^{|y|_\Phi}$. If $|y|_\Phi < \infty$ we obtain $y \notin \Phi^{<|y|_\Phi}$ and thus $x \preceq_\Phi y$. If $|y|_\Phi = \infty$, we obtain $x \in \Phi^{<\infty}$ by (13.1) and $y \notin \Phi^{<\alpha}$ for all $\alpha$. Hence $x \preceq_\Phi y$.

To check (13.6) let $x \prec_\Phi y$. Then there is an $\alpha$ such that $x \in \Phi^\alpha$ and $y \notin \Phi^\alpha$. Hence $x \in \Phi^{<\infty}$. If $|y|_\Phi < \infty$, we have $y \in \Phi^{|y|_\Phi}$ which implies $\alpha < |y|_\Phi$ and thus $x \in \Phi^{<|y|_\Phi}$. Hence $\{x \mid x \prec_\Phi y\} \subseteq \Phi^{<|y|_\Phi}$. For the opposite inclusion let $x \in \Phi^{<|y|_\Phi}$. Then there is an $\alpha < |y|_\Phi$ such that $x \in \Phi^\alpha$. Since $y \notin \Phi^{<|y|_\Phi}$, we get $y \notin \Phi^\alpha$. Hence $x \prec_\Phi y$. □

As a consequence of (13.6) and (13.5) we obtain

$$x \preceq_\Phi y \;\Leftrightarrow\; x \prec_\Phi y \lor x \in \Phi(\{z \mid z \prec_\Phi y\}). \tag{13.7}$$

This is immediate because for $\alpha := |y|_\Phi$ we have

$$\{x \mid x \preceq_\Phi y\} = \Phi^\alpha = \Phi^{<\alpha} \cup \Phi(\Phi^{<\alpha}) = \{x \mid x \prec_\Phi y\} \cup \Phi(\{x \mid x \prec_\Phi y\}).$$

The relations $\prec_\Phi$ and $\preceq_\Phi$ are in some sense the unique relations which satisfy (13.7). To make precise in which sense we introduce the notion of a prewellordering.

**13.2.1 Definition** Let $P$ be a set and $(\preceq, \prec)$ pair of relations. The triple $(P, \preceq, \prec)$ is a *prewellordering* if there is a function $f\colon P \xrightarrow{onto} \lambda \in On$ such that

$$x \preceq y \;\Leftrightarrow\; x \in P \land [y \in P \Rightarrow f(x) \leq f(y)]$$

and

$$x \prec y \;\Leftrightarrow\; x \in P \land [y \in P \Rightarrow f(x) < f(y)].$$

We call $f$ the *associated norm* of $(P, \preceq, \prec)$.

**13.2.2 Theorem** *(Prewellordering Theorem) The triple $(\Phi^{<\infty}, \preceq_\Phi, \prec_\Phi)$ is the uniquely determined prewellordering which satisfies*

$$(FP_\Phi) \qquad x \preceq_\Phi y \;\Leftrightarrow\; x \prec_\Phi y \lor x \in \Phi(\{z \mid z \prec_\Phi y\}).$$

*Proof* From (13.6) and (13.5), we obtain $x \preceq_\Phi y \Leftrightarrow x \in \Phi^{<\infty} \land [y \notin \Phi^{<\infty} \lor |x|_\Phi \leq |y|_\Phi]$ and $x \prec_\Phi y \;\Leftrightarrow\; x \in \Phi^{<\infty} \land [y \notin \Phi^{<\infty} \lor |x|_\Phi < |y|_\Phi]$. So $(\Phi^{<\infty}, \preceq_\Phi, \prec_\Phi)$ is a prewellordering with associated norm $|y|_\Phi$. By (13.7), we know that it also satisfies $(FP_\Phi)$. To show the uniqueness of $(\Phi^{<\infty}, \preceq_\Phi, \prec_\Phi)$ let $(P, \preceq, \prec)$ be any prewellordering satisfying $(FP_\Phi)$. Let moreover $f\colon P \longrightarrow \lambda \in On$ be its associated norm. By induction on $f(y)$ we show

$$y \in P \;\Rightarrow\; \Phi^{f(y)} = \{z \mid z \preceq y\} \tag{i}$$

From the induction hypothesis we obtain

$$\Phi^{<f(y)} = \bigcup_{x \in P} \{z \mid z \preceq x \land f(x) < f(y)\} = \{z \mid z \prec y\}. \tag{ii}$$

Hence $\Phi^{f(y)} = \Phi^{<f(y)} \cup \Phi(\Phi^{<f(y)}) = \{z \mid z \prec y\} \cup \Phi(\{z \mid z \prec y\}) = \{z \mid z \preceq y\}$. From (i) and $y \preceq y$ for $y \in P$, we obtain

$$y \in P \;\Rightarrow\; y \in \Phi^{f(y)} \text{ hence } |y|_\Phi \leq f(y). \tag{iii}$$

If we assume $|y|_\Phi < f(y)$ then we obtain a $z \in P$ such that $|y|_\Phi = f(z)$ because $f$ is onto. Hence $y \in \Phi^{f(z)}$ which by (i) implies $y \preceq z$, i.e., $f(y) \leq f(z) < f(y)$, a contradiction.

Therefore we have

$$y \in P \Rightarrow f(y) = |y|_\Phi. \tag{iv}$$

Next, we show

$$\Phi^{|x|_\Phi} = \{z \,|\, z \preceq x\} \tag{v}$$

by induction on $|x|_\Phi$. By the induction hypothesis and (iv) we obtain $\Phi^{<|x|_\Phi} = \{z \,|\, z \prec x\}$. Hence $\Phi^{|x|_\Phi} = \Phi^{<|x|_\Phi} \cup \Phi(\Phi^{<|x|_\Phi}) = \{z \,|\, z \prec x\} \cup \Phi(\{z \,|\, z \prec x\}) = \{z \,|\, z \preceq x\}$.

Since $x \in \Phi^{|x|_\Phi}$, we get from (v) $x \preceq x$ and thus $x \in P$ which proves $\Phi^{<\infty} \subseteq P$. Together with (i) we therefore get $P = \Phi^{<\infty}$. By (iv) we know that the associated norms coincide. Hence $(P, \preceq, \prec) = (\Phi^{<\infty}, \preceq_\Phi, \prec_\Phi)$. □

The definition of a prewellordering still depends on its associated norm function which needs ordinals in its definition. To get rid of the explicit need of ordinals we develop a description of prewellorderings which only uses the language of arithmetic.

**13.2.3 Theorem** *Let $P \subseteq \mathbb{N}$ be a set and $(\preceq, \prec)$ be a pair of transitive order relations. The triple $(P, \preceq, \prec)$ is a prewellordering if and only if it satisfies the following conditions*

*(PWO1)* $x \preceq y \;\Leftrightarrow\; x \in P \wedge [y \notin P \vee \neg(y \prec x)]$.

*(PWO2)* $x \prec y \;\Leftrightarrow\; x \preceq y \wedge \neg(y \preceq x)$.

*(PWO3) The relation $\prec$ is well-founded.*

*Proof* Assume that $(P, \preceq, \prec)$ is a prewellordering with associated norm $f$. Then $x \prec y \;\Leftrightarrow\; f(x) < f(y)$ holds for $x, y \in P$ and we obtain property (PWO1) since

$$\begin{aligned} x \preceq y \;&\Leftrightarrow\; x \in P \wedge [y \notin P \vee f(x) \leq f(y)] \\ &\Leftrightarrow\; x \in P \wedge [y \notin P \vee \neg(f(y) < f(x))] \\ &\Leftrightarrow\; x \in P \wedge [y \notin P \vee \neg(y \prec x)]. \end{aligned} \tag{i}$$

We obtain property (PWO2) because

$$\begin{aligned} x \preceq y \wedge \neg(y \preceq x) \;&\Leftrightarrow\; x \in P \wedge [y \notin P \vee f(x) \leq f(y)] \\ &\qquad \wedge (y \notin P \vee [x \in P \wedge f(x) < f(y)]) \\ &\Leftrightarrow\; x \in P \wedge [y \notin P \vee f(x) < f(y)] \\ &\Leftrightarrow\; x \prec y. \end{aligned}$$

Since $f$ is a norm every infinite descending $\prec$ sequence will induce an infinite descending sequence in the ordinals. The relation $\prec$ is therefore well-founded.

Now assume properties (PWO1) – (PWO3). For $x \in P$ let $f(x) := \mathrm{otyp}_{\prec}(x)$. We show that $(P, \preceq, \prec)$ is a prewellordering with associated norm $f$. First, we observe

$$\begin{aligned} x \in P \wedge y \notin P \;&\Rightarrow\; x \preceq y \wedge \neg(y \preceq x) \\ &\Rightarrow\; x \prec y \end{aligned} \tag{ii}$$

by (PWO1) and (PWO2). By the definition of the order type, we obtain

$$x \in P \wedge y \in P \;\Rightarrow\; (x \prec y \;\Leftrightarrow\; f(x) < f(y)). \tag{iii}$$

Again by (PWO1) and (PWO2), we have

$$x \prec y \;\Rightarrow\; x \in P. \tag{iv}$$

Pulling together (ii), (iii), and (iv) we obtain

$$x \prec y \;\Leftrightarrow\; x \in P \wedge [y \in P \Rightarrow f(x) < f(y)]. \tag{v}$$

From (v) and (PWO1) we obtain

$$\begin{aligned} x \preceq y \;&\Leftrightarrow\; x \in P \wedge [y \notin P \vee \neg(y \prec x)] \\ &\Leftrightarrow\; x \in P \wedge [y \notin P \vee \neg(f(y) < f(x))] \\ &\Leftrightarrow\; x \in P \wedge [y \in P \Rightarrow f(x) \leq f(y)]. \end{aligned} \tag{vi}$$

By (v) and (vi), we see that $(P, \preceq, \prec)$ is a prewellordering with associated norm $\lambda x \,.\, \mathrm{otyp}_{\prec}(x)$. □

The next observation is that the set $P$ is completely determined by a prewellordering $(P, \preceq, \prec)$.

**13.2.4 Lemma** *Let $(P, \preceq, \prec)$ be a triple satisfying (PWO1)–(PWO3). Then*

$$P = \{x \,|\, x \preceq x\} =: D_{\preceq}. \tag{13.8}$$

*We call $D_{\preceq}$ the* diagonalization *of $\preceq$.*

*Proof* From $x \preceq x$, we obtain $x \in P$ from (PWO1). Conversely, we obtain from $\neg(x \preceq x)$ by (PWO1) $x \notin P \vee [x \in P \wedge x \prec x]$. Since $\prec$ is well-founded, we have $\neg(x \prec x)$. Hence $x \notin P$. □

**13.2.5 Lemma** *Let $(\preceq, \prec)$ be transitive orderings satisfying (PWO1)–(PWO3). Then we obtain*

$$x \prec y \preceq z \;\Rightarrow\; x \prec z \tag{13.9}$$

*and*

$$x \preceq y \prec z \;\Rightarrow\; x \prec z. \tag{13.10}$$

*Proof* For (13.9) check $x \prec y \preceq z \;\Rightarrow\; x \preceq z$ by transitivity of $\preceq$ and (PWO2). The assumption $z \preceq x$ leads to $z \preceq x \preceq y$, i.e., $z \preceq y$ contradicting $y \prec z$.

For (13.10), we again derive $x \preceq z$ from the hypotheses. The assumption $z \preceq x$ implies $z \preceq x \preceq y$ contradicting $y \prec z$. □

**13.2.6 Theorem** *Let $\Phi_F$ be a definable operator, say $\Phi_F(X) = \{x \in \mathbb{N} \mid F(X,x)\}$ and $(\preceq_F, \prec_F)$ a pair of transitive binary relations satisfying the following properties:*

*(FIX)* $\quad x \preceq_F y \;\Leftrightarrow\; x \prec_F y \vee F(\{z \mid z \prec_F y\}, x)$

*(PWO1)* $\quad x \preceq_F y \;\Leftrightarrow\; x \preceq_F x \wedge [y \preceq_F y \rightarrow \neg(y \prec_F x)]$

*(PWO2)* $\quad x \prec_F y \;\Leftrightarrow\; x \preceq_F y \wedge \neg(y \preceq_F x)$

*(PWO3)* $\quad \prec_F$ *is well-founded.*

*Let $D_F := D_{\preceq_F} = \{x \mid x \preceq_F x\}$. Then $(D_F, \preceq_F, \prec_F)$ is a prewellordering and therefore $D_F = \Phi_F^{<\infty}$.*

*Proof* By Theorem 13.2.3, we obtain that $(D_F, \preceq_F, \prec_F)$ is a prewellordering. Since $x \preceq_F y \;\Leftrightarrow\; x \prec_F y \vee F(\{z \mid z \prec_F y\}, x)$, we obtain $D_F = \Phi_F^{\leq\infty}$ by the Prewellordering Theorem (Theorem 13.2.2). □

**13.2.7 Remark** Observe that Theorem 13.2.6 implies that the conditions (FIX) and (PWO1) – (PWO2) already guarantee that the order type of $\prec_F$ is large enough to turn the diagonalization $D_F$ into the fixed point of the operator $\Phi'_F$. This is a consequence of the Prewellordering Theorem. We can, however, see this more directly also. First, we observe that $x \notin D_F$ implies $D_F = \{y \mid y \prec_F x\}$. This holds because $x \notin D_F$, i.e., $\neg(x \preceq_F x)$, implies $y \preceq_F x \;\Leftrightarrow\; y \prec_F x$ for any $y$ by (PWO2) which in turn implies $y \prec_F x \;\Leftrightarrow\; y \preceq_F y$ by (PWO1). Trivially, we have $D_F \subseteq \Phi'_F(D_F)$. Now assume $F(D_F, x)$ and $x \notin D_F$. Then $D_F = \{y \mid y \prec_F x\}$ and $F(D_F, x)$, implies $x \preceq_F x$ by (FIX) which contradicts $x \notin D_F$.

## 13.3 The Theory $(\Pi^0_1\text{–FXP})_0$

The aim of this section is to introduce the theory $(\Pi^0_1\text{–FXP})_0$ which axiomatizes the existence of fixed-points for $\Pi^0_1$-definable operators. The language is based on the language of second-order arithmetic. We introduce some abbreviations which are needed to formalize the results of the earlier sections.

For a set $X$ let

$$Rel(X) \;:\Leftrightarrow\; (\forall x \varepsilon X)[Seq(x) \wedge lh(x) = 2]$$

denote that $X$ codes a binary relation. For simplicity, we mostly write $x\,X\,y$ instead of $\langle x, y\rangle \in X$. By

$$Tran(X) \;:\Leftrightarrow\; Rel(X) \land (\forall x)(\forall y)(\forall z)[x\,X\,y \land y\,X\,z \to x\,X\,z]$$

we express that $X$ codes a transitive binary relation. Let

$$\mathit{field}(X) := \{x \mid (\exists y)[x\,X\,y \lor y\,X\,x]\}$$

denote the field and

$$Wf(X) \;:\Leftrightarrow\; (\forall Y)\big[Y \subseteq \mathit{field}(X) \land (\exists x)(x\,\varepsilon\,Y) \to (\exists x)[x\,\varepsilon\,Y \land (\forall y)[y\,X\,x \to y \notin Y]]\big]$$

express the well-foundedness of the relation coded by $X$. By

$$X_x := \{y \mid \langle x,y\rangle\,\varepsilon\,X\}$$

we denote the $x$-slice of $X$. The formula

$$PO(X) \;:\Leftrightarrow\; (\forall x)[x\,\varepsilon\,X \leftrightarrow Seq(x) \land lh(x) = 2 \land ((x)_0 = 0 \lor (x)_0 = 1) \land Tran(X_0) \land Tran(X_1)]$$

formalizes that $X$ codes two transitive orderings. If $PO(X)$ let $\preceq_X := X_0$ and $\prec_X := X_1$. We define

$$PRO(X) \;:\Leftrightarrow\; PO(X) \land (\forall x)(\forall y)\big[[x \preceq_X y \leftrightarrow x \preceq_X x \land (y \preceq_X y \to \neg(y \prec_X x))] \land [(x \prec_X y) \leftrightarrow x \preceq_X y \land \neg(y \preceq_X x)]\big].$$

The formula $PRO(X)$ formalizes properties (PWO1) and (PWO2) to express that $(D_{\preceq_X}, \preceq_X, \prec_X)$ is a pre-ordering. By

$$PRWO(X) \;:\Leftrightarrow\; PRO(X) \land Wf(\prec_X),$$

we denote that $(D_{\preceq_X}, \preceq_X, \prec_X)$ is a prewellordering. Finally, we use Theorem 13.2.6 to express that the diagonalization of a prewellordering represents the fixed-point of an operator defined by a formula $F(X,x)$. We define

$$FXP_F(X) \;:\Leftrightarrow\; PRWO(X) \land (\forall x)(\forall y)[x \preceq_X y \leftrightarrow x \prec_X y \lor F(\{z \mid z \prec_X y\}, x)].$$

It follows from Theorem 13.2.6, that the set satisfying $FXP_F(X)$ is uniquely determined by the formula $F$. For $FXP_F(X)$ we, therefore introduce the notations $\preceq_F$ for $\preceq_X$, $\prec_F$ for $\prec_X$ and $\Phi^F$ for $D_{\preceq_X}$.

Now we have all the material to define $(\Pi^0_1\text{–FXP})_0$.

**13.3.1 Definition** The theory $(\mathsf{ACA})_0$ is a second-order theory[2] in the language of arithmetic which comprises:

- All axioms of NT.
- The axiom (Ind)[2] of Mathematical Induction.

---

[2] In the weak sense of Sect. 4.6.

- The scheme of arithmetical comprehension

  $(\Delta_0^1\text{–}CA) \quad (\exists X)(\forall x)[x \,\varepsilon\, X \leftrightarrow F(x)]$

  where $F(x)$ is an arbitrary first-order formula.

The theory $(\Pi_1^0\text{–FXP})_0$ is a second-order theory whose nonlogical axioms comprise all axioms of $(\mathsf{ACA})_0$ and

- for every $\Pi_1^0$-formula $F(X,x)$, in which $X$ is the only set variable, the axiom

  $(FXP(F)) \quad (\exists X)[FXP_F(X)].$

**13.3.2 Exercise** Show that $(\Pi_1^0\text{–FXP})_0 \vdash (\exists! X)[FXP_F(X)]$.

## 13.4 $\mathsf{ID}_1$ as Sub-Theory of $(\Pi_1^0\text{–FXP})_0$

To obtain a lower bound for the proof-theoretical ordinal of $(\Pi_1^0\text{–FXP})_0$ we want to show that $\mathsf{ID}_1$ is a sub-theory of $(\Pi_1^0\text{–FXP})_0$. The easiest way to obtain that is to use the fact that all positive-inductively definable relations are $\Pi_1^1$-definable. Then, we use the $\omega$-Completeness Theorem (Theorem 5.4.9) in the form that a $\Pi_1^1$-relation is valid if and only if the associated search tree is well-founded. By Theorem 6.5.5, we know that the well-foundedness of a primitive recursively definable tree is expressible by a relation which is positive-inductively definable by a $\Pi_1^0$-formula. Finally we only have to check that $(\Pi_1^0\text{–FXP})_0$ comprises also all positive $\Pi_1^0$-definable fixed-points.

**13.4.1 Definition** The theory $(\Pi_1^1\text{–CA})_0^-$ is a second-order theory whose nonlogical axioms are:

- all axioms of $(\mathsf{ACA})_0$;
- the scheme of parameter free $\Pi_1^1$-comprehension

  $(\Pi_1^1\text{–}CA)^- \quad (\exists X)(\forall x)[x \,\varepsilon\, X \leftrightarrow F(x)]$

  where $F(x)$ is a $\Pi_1^1$-formula which must not contain free set parameters.

**13.4.2 Theorem** *The theory* $\mathsf{ID}_1$ *is embeddable into* $(\Pi_1^1\text{–CA})_0^-$.

*Proof* We first have to embed the language. Let $F(X,\vec{x})$ be an $X$-positive arithmetical formula which contains only the shown variables free. We define

$$\mathscr{I}(F) := \{\vec{x} \mid (\forall X)\big[(\forall \vec{y})[F(X,\vec{y}) \to \vec{y} \,\varepsilon\, X] \;\to\; \vec{x} \,\varepsilon\, X\big]\}.$$

Then $\mathscr{I}(F)$ is a set by $(\Pi^1_1\text{–}CA)^-$. Replacing all occurrences of $I_F$ in the language $\mathscr{L}(\mathsf{ID})$ by $\mathscr{I}(F)$ we obtain $\mathscr{L}(\mathsf{ID})$ as a sub-language of second-order arithmetic. If $F(x)$ is a formula of $\mathscr{L}(\mathsf{ID})$ then $\{x \mid F(x)\}$ is arithmetical in finitely many sets $\mathscr{I}(F_i)$ and, since we have arithmetical comprehension with set parameters, also a set in $(\Pi^1_1\text{–CA})^-_0$. Any instance of the scheme of Mathematical Induction in the language $\mathscr{L}(\mathsf{ID})$ is therefore a specialization of the axiom of Mathematical Induction in $(\Pi^1_1\text{–CA})^-_0$. All defining axioms for primitive recursive functions are of course also axioms of $(\Pi^1_1\text{–CA})^-_0$. So it remains to show that $\mathsf{ID}^1_1$ and $\mathsf{ID}^2_1$ are derivable in $(\Pi^1_1\text{–CA})^-_0$. To prove first $\mathsf{ID}^2_1$ let $F(X,\vec{x})$ be an $X$-positive arithmetical formula. Put $\mathfrak{M}_F(X) \;:\Leftrightarrow\; (\forall \vec{x})[F(X,\vec{x}) \to \vec{x} \,\varepsilon\, X]$. By definition of $\mathscr{I}(F)$, we obtain

$$\vec{x} \,\varepsilon\, \mathscr{I}(F) \leftrightarrow (\forall X)[\mathfrak{M}_F(X) \to \vec{x} \,\varepsilon\, X]. \tag{i}$$

If $G(\vec{x})$ is an $\mathscr{L}(\mathsf{ID})$-formula then $S := \{\vec{x} \mid G(\vec{x})\}$ is a set in $(\Pi^1_1\text{–CA})^-_0$. Therefore (i) implies

$$\mathfrak{M}_F(S) \to \vec{x} \in \mathscr{I}(F) \to \vec{x} \,\varepsilon\, S \tag{ii}$$

and therefore

$$(\forall \vec{y})[F(S,\vec{y}) \to \vec{y} \,\varepsilon\, S] \to (\forall \vec{x})[\vec{x} \,\varepsilon\, \mathscr{I}(F) \to \vec{x} \,\varepsilon\, S].$$

This is the translation of $\mathsf{ID}^2_1$. To prove also the translation of $\mathsf{ID}^1_1$ we obtain from (i)

$$\mathfrak{M}_F(X) \quad\to\quad (\forall \vec{x})[\vec{x} \,\varepsilon\, \mathscr{I}(F) \to \vec{x} \,\varepsilon\, X]. \tag{iii}$$

By the $X$-positivity of $F(X,\vec{x})$ we obtain from (iii)

$$\mathfrak{M}_F(X) \to F(\mathscr{I}(F),\vec{x}) \to F(X,\vec{x}) \tag{iv}$$

for any tuple $\vec{x}$ of free variables. But

$$\mathfrak{M}_F(X) \to F(X,\vec{x}) \to \vec{x} \,\varepsilon\, X \tag{v}$$

holds true by definition of $\mathfrak{M}_F(X)$ and from (v) and (iv), we obtain

$$F(\mathscr{I}(F),\vec{x}) \to \mathfrak{M}_F(X) \to \vec{x} \,\varepsilon\, X \tag{vi}$$

which entails

$$F(\mathscr{I}(F),\vec{x}) \to (\forall X)[\mathfrak{M}_F(X) \to \vec{x} \,\varepsilon\, X],$$

i.e.,

$$F(\mathscr{I}(F),\vec{x}) \to \vec{x} \in \mathscr{I}(F). \tag{vii}$$

The universal closure of (vii) is axiom $\mathsf{ID}^1_1$. □

We check next that positive inductive $\Pi^0_1$-definitions are obtainable in $(\Pi^0_1\text{–FXP})_0$. Let $F(X,x)$ be a $\Pi^0_1$-formula. Then, we have by $FXP(F)$ a prewellordering $X$ satisfying $FXP_F(X)$. To simplify notations, we denote the the prewellordering by $(\preceq_F, \prec_F)$ and put

$$\Phi^x_F := \{y \mid y \preceq_F x\} \text{ and } \Phi^{<x}_F := \{y \mid y \prec_F x\}$$

as well as

$$\Phi^F := \{x \mid x \preceq_F x\}.$$

If $F(X,x)$ is an $X$-positive $\Pi^0_1$-formula, we obtain

$$F(\Phi^F, x) \;\rightarrow\; x \,\varepsilon\, \Phi^F \tag{13.11}$$

and

$$(\forall y)[F(\{z \mid G(z)\}, y) \rightarrow G(y)] \;\rightarrow\; \Phi^F \subseteq \{z \mid G(z)\}. \tag{13.12}$$

To prove (13.11), assume $F(\Phi^F, x)$ and $x \not\varepsilon\, \Phi^F$. Then $\Phi^F = \Phi^{<x}_F$. But $F(\Phi^{<x}_F, x)$ implies $x \preceq_F x$. Hence $x \,\varepsilon\, \Phi^F$ contradicting our assumption.

To prove (13.12), we show

$$(\forall y)[F(\{z \mid G(z)\}, y) \rightarrow G(y)] \;\rightarrow\; \Phi^x_F \subseteq \{z \mid G(z)\}$$

by induction on $\prec_F$. By induction hypothesis, we have

$$(\forall y)[F(\{z \mid G(z)\}, y) \rightarrow G(y)] \;\rightarrow\; \Phi^{<x}_F \subseteq \{z \mid G(z)\}$$

which by $X$-positivity of $F(X,x)$ implies

$$(\forall y)[F(\{z \mid G(z)\}, y) \rightarrow G(y)] \;\rightarrow\; (\forall x)[F(\Phi^{<x}_F, x) \rightarrow F(\{z \mid G(z)\}, x)].$$

Hence

$$(\forall y)[F(\{z \mid G(z)\}, y) \rightarrow G(y)] \;\rightarrow\; \Phi^x_F \subseteq \{z \mid G(z)\}$$

by $FXP(F)$. Since this holds for all $x$, we obtain finally

$$(\forall y)[F(\{z \mid G(z)\}, y) \rightarrow G(y)] \;\rightarrow\; \Phi^F \subseteq \{z \mid G(z)\}.$$

**13.4.3 Theorem** *The theory* $(\Pi^1_1\text{–CA})^-_0$ *is a sub-theory of* $(\Pi^0_1\text{–FXP})_0$.

*Proof* We have only to show that parameter free $\Pi^1_1$-comprehension is provable in $(\Pi^0_1\text{–FXP})_0$. Let $(\forall X)F(X,x)$ be a $\Pi^1_1$-formula without further free set parameters. Inspecting the proofs of the Syntactical and Semantical Main Lemma (Lemmas 5.4.7 and 5.4.8) we see that these proofs are easily formalizable in $(\Pi^0_1\text{–FXP})_0$ if we replace induction on otyp$(s)$ by bar induction. According to both lemmas we get $(\forall X)[F(X,n)]$ if and only if $S^{\omega}_{\{F(X,n)\}}$ is well-founded. The search tree $S^{\omega}_{\{F(X,n)\}}$ is defined by a primitive recursive formula. The formula $G(Y,y,n) \;:\Leftrightarrow\; (\forall z)[y^\frown\langle z\rangle \in S^{\omega}_{\{F(X,n)\}} \rightarrow y^\frown\langle z\rangle \,\varepsilon\, Y]$ is therefore a $Y$-positive $\Pi^0_1$-formula. According to Theorem 6.5.5 – which is provable in $(\Pi^0_1\text{–FXP})_0$ – the search tree $S^{\omega}_{\{F(X,n)\}}$ is well-founded if and only if $\langle\,\rangle \,\varepsilon\, \Phi^{G(n)}$. Hence

$$(\Pi^0_1\text{–FXP})_0 \vdash (\forall X)[F(X,n)] \leftrightarrow \langle\,\rangle \,\varepsilon\, \Phi^{G(n)}. \tag{i}$$

By arithmetical comprehension, we obtain $\{y \mid \langle\,\rangle \,\varepsilon\, \Phi^{G(y)}\}$ as a set which shows

$$(\Pi^0_1\text{–FXP})_0 \vdash (\exists Z)(\forall y)[y \,\varepsilon\, Z \leftrightarrow (\forall X)[F(X,y)]]. \qquad \square$$

Summing up, we obtain the following theorem.

**13.4.4 Theorem** *The theory* $\mathsf{ID}_1$ *is a sub-theory of* $(\Pi^0_1\text{–FXP})_0$. *This implies that not only* $\Pi^0_1$*-definable but any arithmetically definable positive inductive definition is obtainable in* $(\Pi^0_1\text{–FXP})_0$.

As another consequence we obtain

**13.4.5 Theorem** $\psi(\varepsilon_{\Omega+1}) = \|\mathsf{ID}_1\| \leq \|(\Pi^1_1\text{–CA})^-_0\| \leq \|(\Pi^0_1\text{–FXP})_0\|$.

## 13.5 The Upper Bound for $\|(\Pi^0_1\text{–FXP})_0\|$

In Sect. 11.5, we have shown that the first-order language of arithmetic can be regarded as a sublanguage of set theory. We cannot directly extend that to the second-order language of $(\Pi^0_1\text{–FXP})_0$ because we know that the second-order variables of the theory $(\Pi^0_1\text{–FXP})_0$, which contains $(\Pi^1_1\text{–CA})^-_0$ as sub-theory, range over a domain which contains $\Pi^1_1$-sets. The attempt to translate second-order variables in the language of arithmetic by quantifiers ranging over subsets of $\omega$ must fail because in theories like $\mathsf{KP}\omega$ or $(\Pi_2\text{–REF})$, which have $L_{\omega_1^{CK}}$ as least constructible model, these quantifiers range over sets in $\mathrm{Pow}(\omega) \cap L_{\omega_1^{CK}}$, a narrower class than the class of $\Pi^1_1$-definable sets which corresponds to the class of sets which are $\Sigma_1$-definable over $L_{\omega_1^{CK}}$.

Therefore, we have to extend $(\Pi_2\text{–REF})$ to a second-order theory $(\Pi_2\text{–REF})^2$. Second-order variables, again denoted by capital Latin letters, are supposed to range over classes.

**13.5.1 Definition** The theory $(\Pi_2\text{–REF})^2$ is a second-order theory which comprises the axioms $\mathsf{BST}$ of Basic Set Theory in which the scheme of foundation is replaced by the axiom

$$(\text{Found})^2 \qquad (\forall X)\big[X \neq \emptyset \rightarrow (\exists x \in X)[(\forall y \in x)(y \notin X)]\big]$$

together with the comprehension scheme

(CS) $(\exists X)(\forall y)[y \in X \leftrightarrow F(y)]$, where $F(y)$ is a first-order formula not containing $X$,

and the scheme *$(\Pi_2$–REF$)$*.

The second-order variables in $(\Pi_2\text{–REF})^2$ range over classes which are first-order definable. Therefore, every model of $(\Pi_2\text{–REF})$ can easily be extended to a model of $(\Pi_2\text{–REF})^2$ which shows that $(\Pi_2\text{–REF})^2$ is a conservative extension of $(\Pi_2\text{–REF})$. The theories $(\Pi_2\text{–REF})$ and $(\Pi_2\text{–REF})^2$ have therefore the same $\Sigma$-ordinal.

We obtain a translation of the language of second-order arithmetic into the language of second-order set theory by translating first-order quantifiers $(\mathsf{Q}x)[\cdots]$ into

$(\mathsf{Q}x \in \omega)[\cdots]$ and second-order quantifiers $(\forall X)[\cdots]$ into $(\forall X)[X \subseteq \omega \to \cdots]$ and $(\exists X)[\cdots]$ into $(\exists X)[X \subseteq \omega \wedge \cdots]$. If we want to emphasize the difference between formulas in the language of arithmetic and formulas in the language of set theory, we denote the translation of an arithmetical formula $F$ by $F^\omega$, but to simplify notations we mostly identify $F$ and $F^\omega$. It will be clear from the context (or inessential) whether we mean the arithmetical formula or its translation into the language of set theory.

For any arithmetical formula $F(X,x,i_1,\ldots,i_n)$ we get by Lemma 11.5.2 an $\Sigma$-function symbol $I_F$ such that $(\Pi_2\text{–REF})$ proves

$$I_F(\alpha,i_1,\ldots,i_n) = \{x \in \omega \,|\, F(I_F^{<\alpha},x,i_1,\ldots,i_n)\} \cup I_F^{<\alpha} \tag{13.13}$$

for

$$I_F^{<\alpha} := \{z \in \omega \,|\, (\exists \beta < \alpha)[z \in I_F(\beta,i_1,\ldots,i_n)]\}.$$

Put

$$I_F^{\alpha} := \{x \in \omega \,|\, (\exists \xi \leq \alpha)[x \in I_F(\xi,i_1,\ldots,i_n)]\}$$

and define

$$x \preceq_F y \;:\Leftrightarrow\; (\exists \alpha)[x \in I_F^{\alpha} \wedge y \notin I_F^{<\alpha}]$$

as well as

$$x \prec_F y \;:\Leftrightarrow\; (\exists \alpha)[x \in I_F^{\alpha} \wedge y \notin I_F^{\alpha}].$$

The pair $(\preceq_F, \prec_F)$ are the stage comparison relations defined in (13.2) on p. 336. Let

$$\begin{aligned} X_F := \{x \in \omega \mid\ & Seq(x) \wedge lh(x) = 2 \wedge Seq((x)_1) \wedge lh((x)_1) = 2 \wedge (x)_0 \leq 1 \wedge \\ & ((x)_0 = 0 \to (x)_{1,0} \preceq_F (x)_{1,1}) \wedge ((x)_0 = 1 \to (x)_{1,0} \prec_F (x)_{1,1})\}. \end{aligned}$$

We have shown in Sect. 11.5 that all primitive recursive functions on $\omega$ are definable in $\mathsf{KP}\omega$ and thus also in $(\Pi_2\text{–REF})^2$. Therefore, we obtain $X_F$ as a class in $(\Pi_2\text{–REF})^2$. We observe

$$x \preceq_F x \leftrightarrow (\exists \alpha)[x \in I_F^{\alpha}] \leftrightarrow (\exists \alpha)[x \in I_F(\alpha,i_1,\ldots,i_n)]$$

and define

$$I_F^{<\infty} := \{x \in \omega \,|\, (\exists \alpha)[x \in I_F(\alpha,i_1,\ldots,i_n)]\} = \{x \in \omega \,|\, x \preceq_F x\}. \tag{13.14}$$

We have to show that $(\Pi_2\text{–REF})^2$ is strong enough to prove that $\preceq_F$ and $\prec_F$ fulfill properties (PWO1) – (PWO3) and (FIX) of Theorem 13.2.3. It is obvious that $\prec_F$ and $\preceq_F$ are transitive binary relations. It is also obvious that $\prec_F$ is well-founded. So, we have property (PWO3). Moreover, we can prove

$$\begin{aligned} x \preceq_F y &\leftrightarrow (\exists \alpha)[x \in I_F^{\alpha} \wedge (\forall \beta)[y \in I_F^{\beta} \to \alpha \leq \beta]] \\ &\leftrightarrow x \in I_F^{<\infty} \wedge (y \in I_F^{<\infty} \to \neg(y \prec_F x)) \end{aligned}$$

which is (PWO1). In addition, we prove

$$\begin{aligned} x \prec_F y &\leftrightarrow (\exists\alpha)[x \in I_F^{\alpha} \wedge y \notin I_F^{<\alpha}] \wedge (\forall\beta)[y \in I_F^{\beta} \to x \in I_F^{<\beta}] \\ &\leftrightarrow x \preceq_F y \wedge \neg(y \preceq_F x) \end{aligned}$$

which is (PWO2). It remains to check (FIX). Clearly, we can define

$$|x|_F := \min\{\alpha \mid x \in I_F^{\alpha}\}$$

for $x \in I_F^{<\infty}$ and obtain $I_F^{|x|_F} = \{y \mid y \preceq_F x\}$ as well as $I_F^{<|x|_F} = \{y \mid y \prec_F x\}$. For $x \in I_F^{<\infty}$ we therefore obtain by (13.13)

$$\begin{aligned} x \preceq_F y &\leftrightarrow x \in I_F^{|y|_F} \leftrightarrow x \in I_F^{<|y|_F} \vee F(I_F^{<|y|_F}, x) \\ &\leftrightarrow x \prec_F y \vee F(\{z \mid z \prec y\}, x). \end{aligned} \tag{13.15}$$

To check (FIX) completely we have to get rid of the hypothesis $x \in I_F^{<\infty}$. However, the cardinality argument which we used in the proof of Lemma 13.1.2 is not available in $(\Pi_2\text{–REF})^2$. We need some other argument and this is the point at which we have to exploit the fact that $F$ is $\Pi_1^0$. This new argument needs a few preparations.

**13.5.2 Lemma** *Let $A(X,x,i_1,\dots,i_n)$ be a formula which contains at most quantifiers which are bounded by ordinals $< \omega$. Then*

$$A(I_F^{<\infty}, x, i_1, \dots, i_n) \leftrightarrow (\forall\alpha)(\exists\beta)[\alpha \le \beta \wedge A(I_F^{<\beta}, x, i_1, \dots, i_n)].$$

*Proof* Let $\tilde{A}(X^+, X^-)$ be the formula $A(X,x,i_1,\dots,i_n)$, where we have separated the positive and negative occurrences of $X$ and suppressed the mentioning of the parameters $x, i_1, \dots, i_n$. Then we have

$$\xi \ge \eta \;\Rightarrow\; \tilde{A}(X^+, I_F^{<\xi}) \to \tilde{A}(X^+, I_F^{<\eta}) \tag{13.16}$$

and

$$\xi \le \eta \;\Rightarrow\; \tilde{A}(I_F^{<\xi}, X^-) \to \tilde{A}(I_A^{<\eta}, X^-). \tag{13.17}$$

First we prove

$$(\forall\alpha)(\exists\beta)[\alpha \le \beta \wedge \tilde{A}(I_F^{<\infty}, I_F^{<\beta})] \to \tilde{A}(I_F^{<\infty}, I_F^{<\infty}). \tag{i}$$

Assume that for all $\alpha$ there is a $\beta \ge \alpha$ and $\tilde{A}(I_F^{<\infty}, I_F^{<\beta})$. We show $\tilde{A}(I_F^{<\infty}, I_F^{<\infty})$ by induction on the complexity of the formula $\tilde{A}(X^+, X^-)$. The claim is trivial if $X^-$ does not occur in $\tilde{A}(X^+, X^-)$.

If $\tilde{A}(X^+, X^-)$ is a formula $s \notin X^-$ then

$$s \notin I_F^{<\beta} \to s \notin I_F^{<\infty}$$

holds by definition of $I_F^{<\infty}$.

Next let $\tilde{A}(X^+, X^-)$ be a formula $\tilde{A}_1(X^+, X^-) \wedge \tilde{A}_2(X^+, X^-)$. From the hypothesis $(\forall\alpha)(\exists\beta)[\alpha \le \beta \wedge \tilde{A}(I_F^{<\infty}, I_F^{<\beta})]$ we obtain $(\forall\alpha)(\exists\beta)[\alpha \le \beta \wedge \tilde{A}_i(I_F^{<\infty}, I_F^{<\beta})]$ for $i \in \{1,2\}$. Hence $\tilde{A}_1(I_A^{<\infty}, I_F^{<\infty}) \wedge \tilde{A}(I_F^{<\infty}, I_F^{<\infty})$ by the induction hypothesis.

Now assume that $\tilde{A}(X^+,X^-)$ is a formula $\tilde{A}_1(X^+,X^-) \vee \tilde{A}_2(X^+,X^-)$. Then we get from $(\forall\alpha)(\exists\beta)[\alpha \le \beta \wedge \tilde{A}(I_F^{<\infty}, I_F^{<\beta})]$ that $(\forall\alpha)(\exists\beta)[\alpha \le \beta \wedge \tilde{A}_i(I_F^{<\infty}, I_F^{<\beta})]$ holds true for at last one $i \in \{1,2\}$ because otherwise there would exist an $\alpha_1$ and an $\alpha_2$ such that $\tilde{A}_i(I_F^{<\infty}, I_F^{<\beta})$ would be false for all $\beta \ge \alpha_i$. But then the disjunction $\tilde{A}_1(I_F^{<\infty}, I_F^{<\beta}) \vee A_1(I_F^{<\infty}, I_F^{<\beta})$ would be false for all $\beta \ge \max\{\alpha_1, \alpha_2\}$. By the induction hypothesis, we therefore obtain $\tilde{A}_i(I_F^{<\infty}, I_F^{<\infty})$ for at least one $i \in \{1,2\}$ which entails $\tilde{A}(I_F^{<\infty}, I_F^{<\infty})$.

In case that $\tilde{A}(X^+,X^-)$ is a formula $(\forall y \in n)\tilde{A}_0(X^+,X^-,y)$ we obtain from the hypothesis $(\forall\alpha)(\exists\beta)[\alpha \le \beta \wedge \tilde{A}(I_F^{<\infty}, I_F^{<\beta})]$ that $(\forall\alpha)(\exists\beta)\tilde{A}_0(I_F^{<\infty}, I_F^{<\beta}, y)$ holds true for all $y$ less than $n$. From the induction hypothesis, we therefore obtain $\tilde{A}_0(I_F^{<\infty}, I_F^{<\infty}, y)$ for all $y < n$ and this implies $\tilde{A}(I_F^{<\infty}, I_F^{<\infty})$.

If $\tilde{A}(X^+,X^-)$ is a formula $(\exists y \in n)\tilde{A}_0(X^+,X^-,y)$ we obtain from the hypothesis that there is a $y < n$ such that $(\forall\alpha)(\exists\beta)[\alpha \le \beta \wedge \tilde{A}_0(I_F^{<\infty}, I_F^{<\beta}, y)]$ holds true. Otherwise for every $y < n$ there would exists an $\alpha_y$ such that $\tilde{A}(I_F^{<\infty}, I_F^{<\beta}, y)$ becomes false for all $\beta \ge \alpha_y$ and $\tilde{A}(I_F^{<\infty}, I_F^{<\beta})$ would be false for all $\beta \ge \max\{\alpha_y \,|\, y < n\}$. From the induction hypothesis we therefore obtain that $\tilde{A}_0(I_F^{<\infty}, I_F^{<\infty}, y)$ holds true for some $y < n$ and this implies $\tilde{A}(I_F^{<\infty}, I_F^{<\infty})$.

As a consequence of (13.17) and (i) we have

$$(\forall\alpha)(\exists\beta)[\alpha \le \beta \wedge \tilde{A}(I_F^{<\beta}, I_F^{<\beta})] \to \tilde{A}(I_F^{<\infty}, I_F^{<\infty}). \tag{ii}$$

For the opposite direction we prove

$$\tilde{A}(I_F^{<\infty}, I_F^{<\infty}) \to (\exists\beta)\tilde{A}(I_F^{<\beta}, I_F^{<\infty}) \tag{iii}$$

by induction of the complexity of the formula $\tilde{A}(X^+,X^-)$. The claim is trivial if $X^+$ does not occur in $\tilde{A}(X^+,X^-)$. If $\tilde{A}(X^+,X^-)$ is a formula $s \in X^+$ then we obtain the claim from

$$s \in I_F^{<\infty} \to (\exists\xi)[s \in I_F^{<\xi}]. \tag{iv}$$

If we assume that $\tilde{A}(X^+,X^-)$ is a formula $\tilde{A}_1(X^+,X^-) \wedge \tilde{A}_2(X^+,X^-)$ or a formula $\tilde{A}_1(X^+,X^-) \vee \tilde{A}_2(X^+,X^-)$ we have the induction hypotheses

$$\tilde{A}_i(I_F^{<\infty}, I_F^{<\infty}) \to (\exists\beta_i)\tilde{A}_i(I_F^{<\beta}, I_F^{<\infty})$$

for $i = 1,2$ or $i \in \{1,2\}$ and obtain by (13.17) $\tilde{A}(I_F^{<\infty}, I_F^{<\infty}) \to (\exists\beta)[\tilde{A}_1(I_F^{<\beta}, I_F^{<\infty})]$ with $\beta := \max\{\beta_1, \beta_2\}$ or $\beta := \beta_i$ as witness.

If $\tilde{A}(X^+,X^-)$ is a formula $(\exists y \in n)\tilde{A}_0(X^+,X^-,y)$ then there is a $y < n$ and by induction hypothesis a $\beta$ such that $\tilde{A}_0(I_F^{<\beta}, I_F^{<\infty}, y)$. Hence $(\exists\beta)\tilde{A}(I_F^{<\beta}, I_F^{<\infty})$.

If $\tilde{A}(X^+,X^-)$ is a formula $(\forall y \in n)\tilde{A}_0(X^+,X^-,y)$ then there is for every $y < n$ by induction hypothesis a $\beta_y$ such that $\tilde{A}(I_F^{<\beta_y}, I_F^{<\infty}, y)$ and, by putting $\beta := \max\{\beta_0, \ldots, \beta_{n-1}\}$, we obtain $(\exists\beta)(\forall y \in n)\tilde{A}_0(I_F^{<\beta}, I_F^{<\infty}, y)$. This finishes the proof of (iii). But (iii) together with (13.16) and (13.17) show

$$\tilde{A}(I_F^{<\infty}, I_F^{<\infty}) \to (\forall\alpha)(\exists\beta)[\alpha \le \beta \wedge \tilde{A}(I_F^{<\beta}, I_F^{<\beta})]. \tag{v}$$

From (v) and (ii) we finally obtain

$$\tilde{A}(I_F^{<\infty}, I_F^{<\infty}) \leftrightarrow (\forall\alpha)(\exists\beta)[\alpha \le \beta \wedge \tilde{A}(I_F^{<\beta}, I_F^{<\beta})].$$ □

As a corollary of the proof of Lemma 13.5.2, we obtain its relativization to limit ordinals.

**13.5.3 Lemma** *Let $A(X,x,i_1,\dots,i_n)$ be a formula which contains at most quantifiers which are bounded by ordinals $< \omega$. Then*

$$A(I_F^{<\gamma}, x, i_1, \dots, i_n) \leftrightarrow (\forall\alpha < \gamma)(\exists\beta < \gamma)[\alpha \le \beta \wedge A(I_A^{<\beta}, x, i_1, \dots, i_n)]$$

*for all limit ordinals $\gamma$.*

*Proof* The proof is a simple relativization of the proof of Lemma 13.5.2. □

**13.5.4 Lemma** *Let $F(X,y)$ be the $\Pi^0_1$-formula $(\forall x)A(X,x,y)$ in the language of arithmetic. Then*

$$(\Pi_2\text{–REF})^2 \vdash (\forall y \in \omega)\big[(\forall x \in \omega)A(I_F^{<\infty}, x, y) \to (\exists\gamma)(\forall x \in \omega)A(I_F^{<\gamma}, x, y)\big].$$

*Proof* Assume

$$(\forall x \in \omega)A(I_F^{<\infty}, x, y). \tag{i}$$

By Lemma 13.5.2 we obtain

$$(\forall x \in \omega)(\forall\alpha)(\exists\beta)[\alpha \le \beta \wedge A(I_F^{<\beta}, x, y)]. \tag{ii}$$

The formula $A(I_F^{<\beta}, x, y)$ is $\Delta_0$ and by $\Sigma$-reflection, we obtain

$$(\forall\alpha)(\exists b)(\forall x \in \omega)(\exists\beta \in b)[\alpha \le \beta \wedge A(I_F^{<\beta}, x, y)]. \tag{iii}$$

From (iii) and the fact that $(\Pi_2\text{–REF})^2$ proves $(\forall\gamma)(\exists\xi)[\gamma < \xi]$ we get by *($\Pi_2$–REF)* a transitive nonvoid set $z$ such that

$$(\forall\alpha \in z)(\exists b \in z)(\forall x \in \omega)(\exists\beta \in b)[\alpha \le \beta \wedge A(I_F^{<\beta}, x, y) \\ \wedge (\forall\gamma \in z)(\exists\xi \in z)[\gamma < \xi]]. \tag{iv}$$

Then $\gamma := \sup(z \cap On) \in Lim$ and

$$(\forall x \in \omega)(\forall\alpha < \gamma)(\exists\beta < \gamma)[\alpha \le \beta \wedge A(I_F^{<\beta}, x, y)]. \tag{v}$$

From (v) and Lemma 13.5.3 we obtain

$$(\forall x \in \omega)A(I_F^{<\gamma}, x, y).$$ □

**13.5.5 Corollary** *For any $\Pi^0_1$-formula $F(X,x)$ in the language of arithmetic we obtain*

$$(\Pi_2\text{–REF})^2 \vdash (\forall x \in \omega)[F(I_F^{<\infty}, x) \rightarrow x \in I_F^{<\infty}].$$

*Proof* We obtain $F(I_F^{<\infty}, x) \rightarrow (\exists \alpha)F(I_F^{<\alpha}, x)$ according to Lemma 13.5.4. By (13.14) this implies $x \in I_F^{<\infty}$. □

Putting

$$I_F^{\infty} := \{x \in \omega \,|\, F(I_F^{<\infty}, x)\} \cup I_F^{<\infty}$$

we obtain

$$I_F^{<\infty} = I_F^{\infty} \tag{13.18}$$

which corresponds to (13.1) with $\Phi(X) := \{x \in \omega \,|\, F(X,x)\}$.

Together with (13.15) we get from Corollary 13.5.5 that $\prec_F$ and $\preceq_F$ fulfill (FIX). By already shown facts, we know that (PWO1) – (PWO3) are satisfied, too, and we obtain

$$(\Pi_2\text{–REF})^2 \vdash (\exists X)[X \subseteq \omega \wedge FXP_F(X)]. \tag{13.19}$$

**13.5.6 Theorem** *Let $F$ be a sentence in the second-order language of arithmetic such that $(\Pi^0_1\text{–FXP})_0 \vdash F$. Then $(\Pi_2\text{–REF})^2 \vdash F^\omega$.*

*Proof* It suffices to show that all axioms of $(\Pi^0_1\text{–FXP})_0$ are provable in $(\Pi_2\text{–REF})^2$. We have already seen in Sect. 11.5 that all axioms of NT, except the axiom of Mathematical Induction, are provable in KP$\omega$ and thence also in $(\Pi_2\text{–REF})^2$. The axiom (Found)$^2$ is apparently equivalent to

$$(\forall X)\big[(\forall x)[(\forall y \in x)(y \in X) \rightarrow x \in X] \rightarrow (\forall x)(x \in X)\big]. \tag{i}$$

Pick $X \subseteq \omega$, such that $0 \in X$ and $y \in X$ implies $y+1 \in X$ and $x \in \omega$ such that $(\forall y \in x)[y \in X]$. If $x = 0$, we get $x \in X$ by hypothesis. Otherwise, there is a $y \in \omega$ such that $x = y+1$ which implies $y \in x$ and therefore $y \in X$ by choice of $x$. By choice of $X$, we get $x = y+1 \in X$. From (i) we then obtain $(\forall x \in \omega)[x \in X]$ and have shown

$$(\forall X)[X \subseteq \omega \wedge 0 \in X \wedge (\forall y \in \omega)[y \in X \rightarrow y+1 \in X] \rightarrow (\forall x \in \omega)[x \in X]]$$

which is the translation of axiom (Ind)$^2$. The translation of the scheme ($\Pi^1_0$-CA) of arithmetical comprehension is covered by the the scheme (CS) of class comprehension. Finally, we have seen in (13.19) that $(\Pi_2\text{–REF})^2$ also proves $FXP(F)^\omega$. □

To obtain the upper bound for the proof-theoretical ordinal of $(\Pi^0_1\text{–FXP})_0$ we adopt Definition 11.7.1 to the theory $(\Pi^0_1\text{–FXP})_0$ and define the ordinal

$$\sigma_{\Pi^0_1}^{(\Pi^0_1\text{–FXP})_0} := \sup\{|n|_F + 1 \,|\, F(X,x) \text{ is a } \Pi^0_1\text{-formula and } (\Pi^0_1\text{–FXP})_0 \vdash \underline{n} \,\varepsilon\, \Phi^F\}.$$

Using the fact that every positive inductive definition can already be defined by a $\Pi^0_1$-definable positive inductive definition – a fact which we showed in the proof of Theorem 13.4.4 – we apparently have

$$\kappa^{(\Pi^0_1\text{–FXP})_0} \leq \sigma^{(\Pi^0_1\text{–FXP})_0}_{\Pi^0_1}.$$

If $(\Pi^0_1\text{–FXP})_0 \vdash \underline{n}\ \varepsilon\ \Phi^F$ we obtain $(\Pi_2\text{–REF})^2 \vdash (\exists\alpha)[\underline{n} \in I_F(\alpha)]$ by Theorem 13.5.6 and therefore $|n|_F < \|(\Pi_2\text{–REF})^2\|_\Sigma$. Hence, $\sigma^{(\Pi^0_1\text{–FXP})_0}_{\Pi^0_1} \leq \|(\Pi_2\text{–REF})^2\|_\Sigma$ and we obtain

$$\|(\Pi^0_1\text{–FXP})_0\| = \kappa^{(\Pi^0_1\text{–FXP})_0} \leq \sigma^{(\Pi^0_1\text{–FXP})_0}_{\Pi^0_1} \leq \|(\Pi_2\text{–REF})^2\|_\Sigma = \|(\Pi_2\text{–REF})\| = \psi(\varepsilon_{\Omega+1}). \tag{13.20}$$

Summing up, we have

**13.5.7 Theorem** $\|(\Pi^0_1\text{–FXP})_0\| = \|(\Pi^1_1\text{–CA})^-_0\| = \|\mathsf{ID}_1\| = \psi(\varepsilon_{\Omega+1})$

*Proof* By Theorem 13.4.5 and (13.20) we have $\psi(\varepsilon_{\Omega+1}) = \|\mathsf{ID}_1\| \leq \|(\Pi^1_1\text{–CA})^-_0\| \leq \|(\Pi^0_1\text{–FXP})_0\| = \kappa^{(\Pi^0_1\text{–FXP})_0} \leq \sigma^{(\Pi^0_1\text{–FXP})_0}_{\Pi^0_1} \leq \psi(\varepsilon_{\Omega+1})$. □

# Chapter 14
# Epilogue

The leitmotif of this book was to pursue and extend Gentzen's work in proof theory. Gentzen's research was guided by Hilbert's program. Therefore, it is likely that already his work in pure logic was inspired by the observation that the consistency of a formal system could only be obtained by scrutinizing deviation free derivations. According to Theorem 7.5.4, it suffices to find a single formula which is underivable within a formal system to establish its consistency. The obvious strategy for finding such a formula is to show that an atomic formula, which is not among the axioms of the formal system, cannot be the end-formula of any derivation. This, however, is only obvious if the derivation is deviation free.

The natural first step in pursuing this strategy is therefore to develop a system of pure logic which is deviation free. Thus, it is likely that similar considerations lead to Gentzen's papers [29] and [30] which contain his Hauptsatz (corresponding to Theorems 4.5.1 and 4.5.2 in this book). However, this result turned out to be still insufficient to establish the consistency of pure number theory. Here the presence of nonlogical axioms, especially the axiom of Mathematical Induction, spoils the cut-elimination procedure.

Later Gentzen overcame these obstacles replacing the induction axiom by an induction rule and subsequently eliminating deviations, not in all derivations, but at least in the end-piece ("Endstück") of derivations of quantifier free sentences in a formal system of Peano Arithmetic (cf. [30] and [32]). In this way, he could show that the empty sequent is not derivable which entailed the consistency of pure number theory. So his results were certainly meant as a step toward a performance of Hilbert's program.

However, we believe that Gentzen was aware that his work meant more. This belief is corroborated by his paper [31] in which he showed the unprovability of induction up to $\varepsilon_0$ in Peano arithmetic without referring to Gödel's incompleteness theorem, i.e., without referring to consistency. Simultaneously, he proved that this bound is exact.

Roughly speaking Gentzen showed that all applications of rules of Mathematical Induction occurring in the end-piece of a derivation of an atomic formula can be unraveled into a series of cuts, which are eliminated afterwards. This unraveling procedure is not fully visible in Gentzen's original papers, where he works with finite derivations. It becomes, however, completely plain when turning to infinitary systems as they have been systematically used by Schütte. There it becomes apparent that unraveling a formal derivation from the axioms of number theory leads to a cut free derivation of length below $\varepsilon_0$, i.e., – in our terminology – to a verification of length below $\varepsilon_0$, showing that the truth complexities of the provable formulas of number theory are below $\varepsilon_0$. This unraveling process, i.e., the process of resolving abstract axioms into an iteration of elementary principles, is the central issue of infinitary proof theory.

This aspect of infinitary proof theory becomes even more transparent in the ordinal analysis of subsystems of set theory. There, we can study how "complicated" axioms, e.g., the reflection axiom, are resolved into the verification calculus for the constructible hierarchy. The verification calculus for the constructible hierarchy has the advantage over the verification calculus for $\Pi^1_1$-sentences that it needs no "deep" theorem, like the $\omega$-completeness theorem (Theorem 5.4.9), to prove its completeness and also no "deep" theorem, as the Boundedness theorem (Theorem 6.6.10), to read off the information kept in a verification.[1] The definition of the constructible hierarchy is locally predicative. It is a well-known fact that the notion of definability can even be replaced by a finite set of "Gödel" functions, very elementary set theoretic functions, which suffice to generate the stage $L_{\alpha+1}$ from $L_\alpha$. The passage from $L_\alpha$ to $L_{\alpha+1}$ is therefore even more elementary than displayed it in this book.[2] Resolving formal proofs into the verification calculus for the constructible hierarchy therefore means a reduction of abstract concepts to a (transfinite) iteration of very elementary principles.

So in some respect, we moved far away from Hilbert's original programme of a finitist consistency proof. Nonetheless ordinal analysis, as performed in this book, has a strong connection to another aspect of Hilbert's program. One of the issues of his program concerned the elimination of ideal objects (cf. [43]). Actual[3] statements (in Hilbert's setting) are statements which are directly verifiable. He compares them to the sort of physical statements which are verifiable by experiments.[4] Statements of this type are $\Sigma^0_1$-statements with parameters, i.e., essentially $\Pi^0_2$-sentences. If a $\Pi^0_2$-statement is true we can verify every instance of it. In Sect. 10, we have shown in the paradigmatic example of the theory NT, that, in case that there is an ordinal analysis of the theory which proves the $\Pi^0_2$-sentences, we even have an upper bound for the number of steps which we need for the verification.[5] We have, however, also

[1] Compare Theorem 11.9.9 and its consequences in Sect. 11.9 to the $\omega$-completeness and Boundedness Theorem.

[2] Using Gödel functions instead of definability in the verification calculus would have made the verification calculus unnecessarily complicated.

[3] In [43] Hilbert is talking (in German) about "reale Aussagen". In order to not confuse real objects with real numbers, we opted to translate that by "actual" objects and "actual" statements.

[4] Cf. [43]" ...– sowie in meiner Beweistheorie nur die realen Aussagen unmittelbar einer Verifikation fähig sind." i.e., "...– as well as my prooftheory only allows a direct verification of actual statements."

[5] This is also true for stronger systems (cf. [10]).

shown in Sect. 10, that the description of this upper bound in terms of iterations of an "elementary" function needs transfinitely many steps.

In this respect Hilbert's "dream" of obtaining a *justification of the infinite by finite reasoning* failed. Eliminating ideal objects in a proof of an actual statement needs in general infinitely many steps. In some other respects, however, Hilbert's idea can be generalized. Instead of only regarding natural numbers as actual objects, we may also count infinite ordinals among the actual objects, if they can be coded by an effectively decidable well-ordering on the natural numbers, i.e., we may regard all ordinals below $\omega_1^{CK}$ as actual objects. In this generalized setting the actual statements are represented by $\Sigma_1^{L_{\omega_1^{CK}}}$ formulas with parameters. An example for an ideal object then is the ordinal $\omega_1^{CK}$ – or rather the stage $L_{\omega_1^{CK}}$ in the constructible hierarchy – whose defining axiom is the scheme of $\Pi_2$-reflection. Our ordinal analysis of $(\Pi_2\text{–REF})$ then corresponds exactly to an elimination of ideal objects in a proof of an actual statement. In this interpretation the hypothesis in the collapsing theorem (Theorem 11.11.2) – that $\Delta$ has to be a set of $\Sigma_1$-sentences – is therefore not only of technical nature.

The ordinal analysis of subsystems of second ordinal arithmetic carries of course the same flavor. There we computed the truth complexities of the provable $\Pi_1^1$-sentences of a theory. Since a $\Pi_1^1$-sentence in the language of arithmetic corresponds to an $\Sigma_1^{L_{\omega_1^{CK}}}$-sentence in the language of set theory, we again computed an upper bound for the length of a verification of (generalized) actual statements.

Using this generalized interpretation of actual objects, we could even show that Hilbert's idea of justifying the infinite by finite reasoning, i.e., actual reasoning in our generalization, gains some substance. As shown in Sect. 9.7 the "ideal ordinal" $\Omega$, can be alternatively interpreted by an ordinal below $\omega_1^{CK}$, i.e., by an actual object.[6]

Leaving aside these speculations about Hilbert's program, ordinal analysis has also a genuinely mathematical aspect which we could not fully handle in this book. As shown in Sect. 9.6.1, we can extract an ordinal notation system from the iterations of the (least) controlling operator. In this sense an ordinal analysis not only computes the proof-theoretic ordinal $\|T\|$ of a theory $T$ but also provides us with a notation system for the ordinals below $\|T\|$. According to Definition 10.3.1 and the discussion in Sect. 10.6, an ordinal notation system induces a subrecursive hierarchy based on some "operator", i.e., on a strictly increasing number theoretic function, which eventually majorizes the Skolem functions for the provable $\Pi_2^0$-sentences of $T$. Moreover, our discussion in Sect. 10.6 has also shown that this hierarchy is pretty independent of the choice of the base function.

Most combinatorial principles lead to $\Pi_2^0$-sentences. A famous example stems from Ramsey Theory. For a set $M$ denote by $[M]^k := \{S \mid S \subseteq M \wedge \overline{\overline{S}} = k\}$ the set of subsets of size $k$. Let $c$ be a cardinal. A $c$ coloring of $[M]^k$ is a map $P\colon [M]^k \longrightarrow c$. A subset $H \subseteq M$ is homogeneous if $P\restriction[H]^k$ is constant, i.e., if every $k$-sized subset of $H$ gets the same color.

---

6 This follows also already from Sect. 9.6.1, by which we obtain that the order type of the ordinals in the set $\mathscr{B}_{\varepsilon_{\Omega+1}}(\emptyset) \cap \varepsilon_{\Omega+1}$ is below $\omega_1^{CK}$.

The finite version of Ramsey's Theorem states that for all natural numbers $k$, $c$, and $n$ you can find a natural number $R$ such that every finite set $M$ of cardinality $\geq R$ and any $c$ coloring of $[M]^k$ there is a homogeneous subset $H \subseteq M$ of cardinality $\geq n$. This can easily be formalized as an $\Pi^0_2$-sentence. The Ramsey number $R$ then depends on $k$, $c$, and $n$, i.e., $R = R(k,c,n)$. Erdös and Rado showed that this function has a growth rate in terms of exponential towers. A size which lies within the Skolem functions of the provable $\Pi^0_2$-sentences of NT. In fact, the finite Ramsey Theorem is a theorem of Peano arithmetic.

Due to an observation by Paris and Harrington (cf. [66]) the situation changes dramatically if we require that the homogeneous set is relatively large, i.e., that $\min H \leq \overline{\overline{H}}$. The then emerging Paris–Harrington Ramsey function gets the growth rate of $H_{\varepsilon_0}$ (cf. Sect. 10.6). Since $H_{\varepsilon_0}$ majorizes all Skolem functions of the provable $\Pi^0_2$-sentences of NT this shows that the Paris–Harrington version of the finite Ramsey theorem cannot be provable within NT.[7]

As indicated in Chap. 10, in the example of the theory NT, the Skolem functions for the provable $\Pi^0_2$-sentences of a theory are governed by the subrecursive hierarchy generated by the notation system for its proof-theoretical ordinal. The combinatorial content of theory is therefore reflected by the notation system generated by its ordinal analysis. This holds still true for stronger theories (cf. e.g. [11]). This example may illuminate the importance of ordinal analysis. There are, however, other examples of applications of ordinal analysis. One should be aware that the analysis presented in this book is still quite rough. We only looked at the truth complexities of the provable $\Pi^1_1$-sentences. A finer analysis would also take into account the derivations itself and not only their ordinal heights. An example of such a finer analysis is in [12]. A broader discussion about the importance of ordinal analysis is Rathjen's paper [80].

In this book, we restricted ourselves to the first step into impredicativity. The further steps are in general more complicated. The reason for that is that, we needed only one collapsing step and could afterwards argue mainly semantically (e.g., using semantical cut-elimination (Theorem 9.3.1)). The analysis of stronger theories need more, in general infinitely many, collapsing steps. Although this meant a big obstacle in the beginning of impredicative proof theory, most of the then occurring problems are solved today. So it was for more pedagogical reasons that we restricted ourselves to the first step into impredicativity. Nevertheless, we hope that we will soon be able to present a continuation of this book which includes the further steps.

---

[7] The requirement $\min H \leq \overline{\overline{H}}$ of relative largeness can be specified to $f(\min H) \leq \overline{\overline{H}}$ for an arithmetical function $f$. The Paris–Harrington result holds for $f = Id$. It is an extremely interesting problem to classify the "threshold" function $f$ which is on the edge between provability and unprovability in NT. This is a research project of Weiermann, who already obtained a series of spectacular results also regarding other combinatorial principles. A survey is published in [109].

# References

1. Ackermann, W.: Begründung des 'Tertium non datur' mittels der Hilbertschen Theorie der Widerspruchsfreiheit. Mathematische Annalen **93** (1924)
2. Arai, T.: Variations on a theme by Weiermann. Journal of Symbolic Logic **63**, 897–925 (1998)
3. Arai, T.: Proof theory for theories of ordinals I: Recursively Mahlo ordinals. Annals of Pure and Applied Logic **122**, 163–208 (2003)
4. Barwise, J.: Admissible Sets and Structures. Perspectives in Mathematical Logic. Springer-Verlag, Berlin/Heidelberg/New York (1975)
5. Barwise, J. (ed.): Handbook of Mathematical Logic. North-Holland Publishing Company, Amsterdam (1977)
6. Barwise, J., Schlipf, J.S.: On recursively saturated models of arithmetic. In: D.H. Saracino, V.B. Weispfenning (eds.) Model Theory and Algebra, no. 498 in Lecture Notes in Mathematics, pp. 42–55. Springer-Verlag, Heidelberg/New York (1975)
7. Beckmann, A.: Dynamic ordinal analysis. Archive for Mathematical Logic **42**, 303–334 (2003)
8. Beckmann, A., Pohlers, W.: Application of cut–free infinitary derivations to generalized recursion theory. Annals of Pure and Applied Logic **94**, 1–19 (1998)
9. Bernays, P.: Hilberts Untersuchungen über die Grundlagen der Mathematik. In: D. Hilbert (ed.) Gesammelte Abhandlungen, vol. III, pp. 196–216. Springer-Verlag, Berlin (1935)
10. Blankertz, B., Weiermann, A.: How to characterize provably total functions by the Buchholz operator method. No. 6 in Lecture Notes in Logic. Springer-Verlag, Heidelberg/New York (1996)
11. Buchholz, W.: An independence result for $\Pi^1_1$-CA + BI. Annals of Pure and Applied Logic **33**, 131–155 (1987)
12. Buchholz, W.: Notation systems for infinitary derivations. Archive for Mathematical Logic **30**, 277–296 (1991)
13. Buchholz, W.: A simplified version of local predicativity. In: P. Aczel, H. Simmons, S.S. Wainer (eds.) Proof Theory, pp. 115–147. Cambridge University Press, Cambridge (1992)
14. Buchholz, W., Cichon, E.A., Weiermann, A.: A uniform approach to fundamental sequences and hierarchies. Mathematical Logic Quarterly **40**, 273–286 (1994)
15. Buchholz, W., Feferman, S., Pohlers, W., Sieg, W. (eds.): Iterated Inductive Definitions and Subsystems of Analysis: Recent Proof-Theoretical Studies. No. 897 in Lecture Notes in Mathematics. Springer-Verlag, Heidelberg/New York (1981)
16. Buchholz, W., Schütte, K.: Syntaktische Abgrenzungen von formalen Systemen der $\Pi^1_1$-Analysis und $\Delta^1_2$-Analysis. Bayerische Akademie der Wissenschaften, Sitzungsberichte 1980 pp. 1–35 (1981)
17. Buchholz, W., Schütte, K.: Proof Theory of Impredicative Subsystems of Analysis. No. 2 in Studies in Proof Theory, Monographs. Bibliopolis, Naples (1988)

18. Buss, S.R. (ed.): Handbook of Proof Theory. Studies in Logic and the Foundations of Mathematics. North-Holland Publishing Company (1998)
19. Buss, S.R., Hájek, P., Pudlák, P. (eds.): Logic Colloquium '98, no. 13 in Lecture Notes in Logic. Association for Symbolic Logic, Natik, Massachusetts (2000)
20. Dedekind, R.: Gesammelte mathematische Werke, vol. III. Friedr. Vieweg & Sohn, Wiesbaden (1932)
21. Devlin, K.J.: Constructibility. Perspectives in Mathematical Logic. Springer-Verlag, Heidelberg/New York (1984)
22. Erdös, P., Rado, R.: Combinatorial theorems on classifications of subsets of a given set. Proceedings of the London Mathematical Society **28**, 417–439 (1952)
23. Feferman, S.: Systems of predicative analysis. Journal of Symbolic Logic **29**, 1–30 (1964)
24. Feferman, S.: Formal theories for transfinite iteration of generalized inductive definitions and some subsystems of analysis. In: Kino et al. [60], pp. 303–326
25. Friedman, H.M.: Iterated inductive definitions and $\Sigma^1_2$-AC. In: Kino et al. [60], pp. 435–442
26. Friedman, H.M.: Subsystems of set theory and analysis. Ph.D. thesis, MIT Press, Boston (1987)
27. Friedman, H.M., McAloon, K., Simpson, S.G.: A finite combinatorical principle which is equivalent to the 1-consistency of predicative analysis. In: G. Metakides (ed.) Patras Logic Symposium, no. 109 in Studies in Logic and the Foundations of Mathematics, pp. 197–230. North-Holland Publishing Company, Amsterdam (1982)
28. Gentzen, G.: Untersuchungen über das logische Schließen I. Mathematische Zeitschrift **39**, 176–210 (1934)
29. Gentzen, G.: Untersuchungen über das logische Schließen II. Mathematische Zeitschrift **39**, 405–431 (1935)
30. Gentzen, G.: Die Widerspruchsfreiheit der reinen Zahlentheorie. Mathematische Annalen **112**, 493–565 (1936)
31. Gentzen, G.: Beweisbarkeit und Unbeweisbarkeit von Anfangsfällen der transfiniten Induktion in der reinen Zahlentheorie. Mathematische Annalen **119**, 140–161 (1943)
32. Gentzen, G.: Der erste Widerspruchsfreiheitsbeweis für die klassische Zahlentheorie. Archiv für Mathematische Logik und Grundlagenforschung **16**, 97–118 (1974)
33. Girard, J.Y.: Une extension de l'interpretation de Gödel a l'analyse et son application a l'elimination des coupures dans l'analyse et la theorie des types. In: J.E. Fenstad (ed.) Proceedings of the 2nd Scandinavian Logic Symposium, no. 63 in Studies in Logic and the Foundations of Mathematics, pp. 63–92. North-Holland Publishing Company, Amsterdam (1971)
34. Girard, J.Y.: $\Pi^1_2$-logic I. Dilators. Annals of Mathematical Logic **21**, 75–219 (1981)
35. Girard, J.Y.: Proof Theory and Logical Complexity, vol. 1. Bibliopolis, Naples(1987)
36. Gödel, K.: Die Vollständigkeit der Axiome des logischen Funktionenkalküls. Monatshefte für Mathematik und Physik **37**, 349–360 (1930)
37. Gödel, K.: Über formal unentscheidbare Sätze der 'Prinzipia Mathematica' und verwandter Systeme. Monatshefte für Mathematik und Physik **38**, 173–198 (1931)
38. Gödel, K.: Über eine bisher noch nicht benützte Erweiterung des finiten Standpunktes. Dialectica **12**, 280–287 (1958)
39. Herbrand, J.: Recherches sur la theorie de la demonstration. Societe des Science et des Lettres Varsovic, Science Mathematiques et Physiques **33**, 128 (1930)
40. Herbrand, J.: Sur la non-contradiction de l'arithmétique. Journal für die reine und angewandte Mathematik **166** (1930)
41. Hilbert, D.: Neubegründung der Mathematik. Abhandlungen aus dem Math. Seminar d. Hamb. Univ. **I**, 157–122 (1922)
42. Hilbert, D.: Über das Unendliche. Mathematische Annalen **95**, 161–190 (1926)
43. Hilbert, D.: Die Grundlagen der Mathematik. Vortrag gehalten auf Einladung des Mathematischen Seminars im Juli 1927 in Hamburg. Hamburger Mathematische Einzelschriften **5. Heft**, 1–21 (1928)
44. Hilbert, D.: Die Grundlegung der elementaren Zahlenlehre. Mathematische Annalen **104**, 485–494 (1931)

45. Hinman, P.G.: Recursion-Theoretic Hierarchies. Perspectives in Mathematical Logic. Springer-Verlag, Heidelberg/New York (1978)
46. Howard, W.A.: Assignment of ordinals to terms for primitive recursive functionals of finite type. In: Kino et al. [60], pp. 443–458
47. Howard, W.A.: A system of abstract constructive ordinals. Journal of Symbolic Logic **37**, 355–374 (1972)
48. Jäger, G.: Die konstruktible Hierarchie als Hilfsmittel zur beweistheoretischen Untersuchung von Teilsystemen der Mengenlehre und Analysis. Dissertation, Ludwig-Maximilians-Universität, Munich (1979)
49. Jäger, G.: A well ordering proof for Feferman's theory $T_0$. Archiv für Mathematische Logik und Grundlagenforschung **23**, 65–77 (1983)
50. Jäger, G.: The strength of admissibility without foundation. Journal of Symbolic Logic **49**, 867–879 (1984)
51. Jäger, G.: A version of Kripke-Platek set theory which is conservative over Peano arithmetic. Zeitschrift für Mathematische Logik und Grundlagen der Mathematik **30**, 3–9 (1984)
52. Jäger, G.: Theories for Admissible Sets. A Unifying Approach to Proof Theory. No. 2 in Studies in Proof Theory, Lecture Notes. Bibliopolis, Naples (1986)
53. Jäger, G.: Metapredicative and explicit mahlo: a proof-theoretic perspective. In: R. Cori, A. Razborov, S. Todorcevic, C. Wood (eds.) Logic Colloquium '00, *Lecture Notes in Logic*, vol. 19, pp. 272–293. AK Peters, Wellesley, MA (2005)
54. Jäger, G.: Reflections on reflections in explicit mathematics. Annals of Pure and Applied Logic **66** (2005)
55. Jäger, G., Pohlers, W.: Eine beweistheoretische Untersuchung von $(\Delta^1_2\text{-CA}) + (\text{BI})$ und verwandter Systeme. Bayerische Akademie der Wissenschaften, Sitzungsberichte 1982 pp. 1–28 (1983)
56. Jäger, G., Schütte, K.: Eine syntaktische Abgrenzung der $(\Delta^1_1\text{-CA})$-Analysis. Bayerische Akademie der Wissenschaften, Sitzungsberichte 1979 pp. 15–34 (1979)
57. Jäger, G., Strahm, T.: Upper bounds for metapredicative mahlo in explicit mathematics and admissible set theory. Journal of Symbolic Logic **66(2)**, 935–958 (2001)
58. Jech, T.J.: The Axiom of Choice. No. 75 in Studies in Logic and the Foundations of Mathematics. North-Holland Publishing Company, Amsterdam (1973)
59. Jech, T.J.: Set Theory. Academic Press, New York (1978)
60. Kino, A., Myhill, J., Vesley, R.E. (eds.): Intuitionism and Proof Theory, Studies in Logic and the Foundations of Mathematics. North-Holland Publishing Company, Amsterdam (1970)
61. Kreisel, G.: Generalized inductive defintions. Stanford Report, mimeographed **section III** (1963)
62. Kreisel, G.: Notes concerning the elements of proof theory. Lecture Notes, University of Calfornia, Los Angeles, Los Angeles, CA (1967, 1968)
63. Lorenzen, P.: Algebraische und logistische Untersuchungen über freie Verbände. Journal of Symbolic Logic **16**, 81–106 (1951)
64. Mostowski, A.: On $\omega$–models which are not $\beta$–models. Fundamenta Mathematicae **65**, 83–93 (1969)
65. Neumann, J.v.: Zur Hilbertschen Beweistheorie. Mathematische Annalen **26** (1927)
66. Paris, J.B., Harrington, L.: A mathematical incompleteness in Peano arithmetic. In: Barwise [5], pp. 1133–1142
67. Peano, G.: Arithmetices principia, nova methodo exposita. Bocca, Torino (1889)
68. Pohlers, W.: An upper bound for the provability of transfinite induction. In: J. Diller, G.H. Müller (eds.) $\models$ ISILC Proof Theory Symposium, no. 500 in Lecture Notes in Mathematics, pp. 271–289. Springer-Verlag, Heidelberg/New York (1975)
69. Pohlers, W.: Ordinals connected with formal theories for transfinitely iterated inductive definitions. Journal of Symbolic Logic **43**, 161–182 (1978)
70. Pohlers, W.: Cut-elimination for impredicative infinitary systems I. Ordinal analysis for $\text{ID}_1$. Archiv für Mathematische Logik und Grundlagenforschung **21**, 113–129 (1981)
71. Pohlers, W.: Proof-theoretical analysis of $\text{ID}_\nu$ by the method of local predicativity. In: Buchholz et al. [15], pp. 261–357

72. Pohlers, W.: Cut elimination for impredicative infinitary systems II. Ordinal analysis for iterated inductive definitions. Archiv für Mathematische Logik und Grundlagenforschung **22**, 69–87 (1982)
73. Pohlers, W.: Proof Theory. An Introduction. No. 1407 in Lecture Notes in Mathematics. Springer-Verlag, Berlin/Heidelberg/New York (1989)
74. Pohlers, W.: Subsystems of set theory and second order number theory. In: Buss [18], pp. 209–335
75. Prawitz, D.: Hauptsatz for higher order logic. Journal of Symbolic Logic **33**, 452–457 (1968)
76. Probst, D.: On the relationship between fixed points and iteration in admissible set theory without foundation. Archive for Mathematical Logic **44**(5), 561–580 (2005)
77. Rathjen, M.: Proof-theoretic analysis of KPM. Archive for Mathematical Logic **30**, 377–403 (1991)
78. Rathjen, M.: Eine Ordinalzahlanalyse der $\Pi_3$-Reflexion. Habilitationsschrift, Westfälische Wilhelms-Universität, Münster (1992)
79. Rathjen, M.: Proof theory of reflection. Annals of Pure and Applied Logic **68**, 181–224 (1994)
80. Rathjen, M.: The realm of ordinal analysis. In: S. Cooper, J. Truss (eds.) Sets and Proofs, pp. 219–79. Cambridge University Press (1999)
81. Rathjen, M.: An ordinal analyis of parameter free $\Pi^1_2$-comprehension. Archive for Mathematical Logic **48/3**, 263–362 (2005)
82. Rathjen, M.: An ordinal analyis of stability. Archive for Mathematical Logic **48/2**, 1–62 (2005)
83. Richter, W.H., Aczel, P.: Inductive definitions and reflecting properties of admissible ordinals. In: J.E. Fenstad, P.G. Hinman (eds.) Generalized Recursion Theory I, no. 79 in Studies in Logic and the Foundations of Mathematics, pp. 301–381. North-Holland Publishing Company, Amsterdam (1974)
84. Rogers jun., H.: Theory of Recursive Functions and Effective Computability. McGraw-Hill Book Company, New York (1967)
85. Schütte, K.: Beweistheoretische Untersuchungen der verzweigten Analysis. Mathematische Annalen **124**, 123–147 (1952)
86. Schütte, K.: Syntactical and semantical properties of simple type theory. Journal of Symbolic Logic **25**, 305–326 (1960)
87. Schütte, K.: Eine Grenze für die Beweisbarkeit der transfiniten Induktion in der verzweigten Typenlogik. Archiv für Mathematische Logik und Grundlagenforschung **7**, 45–60 (1965)
88. Schütte, K.: Predicative well-orderings. In: J.N. Crossley, M.A.E. Dummett (eds.) Formal Systems and Recursive Functions, Studies in Logic and the Foundations of Mathematics, pp. 280–303. North-Holland Publishing Company, Amsterdam (1965)
89. Schütte, K.: Proof Theory. No. 225 in Grundlehren der mathematischen Wissenschaften. Springer-Verlag, Heidelberg/New York (1977)
90. Schwichtenberg, H.: Proof theory: Some applications of cut-elimination. In: Barwise [5], pp. 867–895
91. Schwichtenberg, H., Troelstra, A.: Basic Proof Theory. No. 43 in Cambridge Tracts in Theoretical Computer Science. Cambridge University Press, Cambridge, UK (1996)
92. Shoenfield, J.R.: Mathematical Logic. Addison-Wesley Publishing Company, Reading (1967)
93. Sieg, W.: Hilbert's Programs: 1917–1922. Bulletin of Symbolic Logic **5**, 1–44 (1999)
94. Simpson, S.G.: Subsystems of Second Order Arithmetic. Springer-Verlag, Berlin/ Heidelberg/ New York (1999)
95. Spector, C.: Provably recursive functionals of analysis: a consistency proof of analysis by an extension of principles formulated in current intuitionistic mathematics. In: J.C.E. Dekker (ed.) Recursive Function Theory, no. 5 in Proceedings of Symposia in Pure Mathematics, pp. 1–27. American Mathematical Society, Providence (1962)
96. Szabo, M.E. (ed.): The Collected Papers of Gerhard Gentzen. North-Holland Publishing Company, Amsterdam (1969)

97. Tait, W.W.: A non constructive proof of gentzen's hauptsatz for second order predicate logic. Bulletin of the American Mathematical Society **66**, 980–983 (1966)
98. Tait, W.W.: Intensional interpretatons of functionals of finite type. Journal of Symbolic Logic **32/2**, 198–212 (1967)
99. Tait, W.W.: Applications of the cut elemination theorem to some subsystems of classical analysis. In: Kino et al. [60], pp. 475–488
100. Takahashi, M.o.: A proof of cut-elimination in simple type theory. Journal of the Mathematical Society of Japan **19**, 399–410 (1967)
101. Takeuti, G.: On a generalized logic calculus. Japanese Journal of Mathematics **24**, 149–156 (1953)
102. Takeuti, G.: Consistency proofs of subsystems of classical analysis. Annals of Mathematics **86**, 299–348 (1967)
103. Takeuti, G.: Proof Theory, 2. edn. No. 81 in Studies in Logic and the Foundations of Mathematics. North-Holland Publishing Company, Amsterdam (1987)
104. Takeuti, G., Yasugi, M.: Reflection principles of subsystems of analysis. In: H.A. Schmidt, K. Schütte, H.J. Thiele (eds.) Contributions to Mathematical Logic, Studies in Logic and the Foundations of Mathematics, pp. 255–273. North-Holland Publishing Company, Amsterdam (1968)
105. Takeuti, G., Yasugi, M.: The ordinals of the systems of second order arithmetic with the provably $\Delta^1_2$-comprehension axiom and with the $\Delta^1_2$-comprehension axiom respectively. Japanese Journal of Mathematics **41**, 1–67 (1973)
106. Weiermann, A.: How to characterize provably total functions by local predicativity. Journal of Symbolic Logic **61**, 52–69 (1996)
107. Weiermann, A.: Sometimes slow growing is fast growing. Annals of Pure and Applied Logic **83**, 199–223 (1997)
108. Weiermann, A.: What makes a (pointwise) subrecursive hierarchy slow growing? In: B. Cooper, J.K. Truss (eds.) Sets and Proofs, no. 258 in London Mathematical Society Lecture Notes Series, pp. 403–423. Cambridge University Press, Cambridge (1999)
109. Weiermann, A.: Analytic combinatorics, proof-theoretic ordinals and phase transitions for independence results. Annals of Pure and Applied Logic **136**, 189–218 (2005)
110. Weyl, H.: Das Kontinuum. Veit & Co., Leipzig (1918)
111. Weyl, H.: Der Circulus vitiosus in der heutigen Begründung der Mathematik. Jahresbericht der Deutschen Mathematiker-Vereinigung pp. 85–92 (1919)
112. Weyl, H.: Über die neue Grundlagenkrise der Mathematik. Mathematische Zeitschrift **10**, 39–79 (1921)
113. Woodin, W.H.: The continuum hypothesis, Part I. Notices of the AMS **48/6**, 567–576 (2001)
114. Woodin, W.H.: The continuum hypothesis, Part II. Notices of the AMS **48/7**, 681–690 (2001)
115. Zucker, J.I.: Iterated inductive definitions, trees, and ordinals. In: A.S. Troelstra (ed.) Metamathematical Investigation of Intuitionistic Arithmetic and Analysis, no. 344 in Lecture Notes in Mathematics, pp. 392–453. Springer-Verlag, Heidelberg/New York (1973)

# Index

## Notations

**Key–words**

# Universitext

*Aguilar, M.; Gitler, S.; Prieto, C.:* Algebraic Topology from a Homotopical Viewpoint
*Ahlswede, R.; Blinovsky, V.:* Lectures on Advances in Combinatorics
*Aksoy, A.; Khamsi, M. A.:* Methods in Fixed Point Theory
*Alevras, D.; Padberg M. W.:* Linear Optimization and Extensions
*Andersson, M.:* Topics in Complex Analysis
*Aoki, M.:* State Space Modeling of Time Series
*Arnold, V. I.:* Lectures on Partial Differential Equations
*Arnold, V. I.; Cooke, R.:* Ordinary Differential Equations
*Audin, M.:* Geometry
*Aupetit, B.:* A Primer on Spectral Theory
*Bachem, A.; Kern, W.:* Linear Programming Duality
*Bachmann, G.; Narici, L.; Beckenstein, E.:* Fourier and Wavelet Analysis
*Badescu, L.:* Algebraic Surfaces
*Balakrishnan, R.; Ranganathan, K.:* A Textbook of Graph Theory
*Balser, W.:* Formal Power Series and Linear Systems of Meromorphic Ordinary Differential Equations
*Bapat, R.B.:* Linear Algebra and Linear Models
*Benedetti, R.; Petronio, C.:* Lectures on Hyperbolic Geometry
*Benth, F. E.:* Option Theory with Stochastic Analysis
*Berberian, S. K.:* Fundamentals of Real Analysis
*Berger, M.:* Geometry I, and II
*Bhattacharya, R; Waymire, E. C.:* A Basic Course in Probability Theory
*Bliedtner, J.; Hansen, W.:* Potential Theory
*Blowey, J. F.; Coleman, J. P.; Craig, A. W. (Eds.):* Theory and Numerics of Differential Equations
*Blowey, J. F.; Craig, A.; Shardlow, T. (Eds.):* Frontiers in Numerical Analysis, Durham 2002, and Durham 2004
*Blyth, T. S.:* Lattices and Ordered Algebraic Structures
*Börger, E.; Grädel, E.; Gurevich, Y.:* The Classical Decision Problem
*Böttcher, A; Silbermann, B.:* Introduction to Large Truncated Toeplitz Matrices
*Boltyanski, V.; Martini, H.; Soltan, P. S.:* Excursions into Combinatorial Geometry
*Boltyanskii, V. G.; Efremovich, V. A.:* Intuitive Combinatorial Topology
*Bonnans, J. F.; Gilbert, J. C.; Lemarchal, C.; Sagastizbal, C. A.:* Numerical Optimization
*Booss, B.; Bleecker, D. D.:* Topology and Analysis
*Borkar, V. S.:* Probability Theory
*Brides/Vita:* Techniques of Constructive Analysis
*Bruiner, J. H.:* The 1-2-3 of Modular Forms
*Brunt B. van:* The Calculus of Variations
*Bühlmann, H.; Gisler, A.:* A Course in Credibility Theory and its Applications
*Carleson, L.; Gamelin, T. W.:* Complex Dynamics
*Cecil, T. E.:* Lie Sphere Geometry: With Applications of Submanifolds
*Chae, S. B.:* Lebesgue Integration
*Chandrasekharan, K.:* Classical Fourier Transform
*Charlap, L. S.:* Bieberbach Groups and Flat Manifolds
*Chern, S.:* Complex Manifolds without Potential Theory
*Chorin, A. J.; Marsden, J. E.:* Mathematical Introduction to Fluid Mechanics
*Cohn, H.:* A Classical Invitation to Algebraic Numbers and Class Fields
*Curtis, M. L.:* Abstract Linear Algebra
*Curtis, M. L.:* Matrix Groups
*Cyganowski, S.; Kloeden, P.; Ombach, J.:* From Elementary Probability to Stochastic Differential Equations with MAPLE
*Da Prato, G.:* An Introduction to Infinite Dimensional Analysis
*Dalen, D. van:* Logic and Structure
*Das, A.:* The Special Theory of Relativity: A Mathematical Exposition

*Debarre, O.:* Higher-Dimensional Algebraic Geometry

*Deitmar, A.:* A First Course in Harmonic Analysis

*Demazure, M.:* Bifurcations and Catastrophes

*Devlin, K. J.:* Fundamentals of Contemporary Set Theory

*DiBenedetto, E.:* Degenerate Parabolic Equations

*Diener, F.; Diener, M. (Eds.):* Nonstandard Analysis in Practice

*Dimca, A.:* Sheaves in Topology

*Dimca, A.:* Singularities and Topology of Hypersurfaces

*DoCarmo, M. P.:* Differential Forms and Applications

*Duistermaat, J. J.; Kolk, J. A. C.:* Lie Groups

*Dumortier.:* Qualitative Theory of Planar Differential Systems

*Dundas, B. I.; Levine, M.; Østvaer, P. A.; Röndip, O.; Voevodsky, V.:* Motivic Homotopy Theory

*Edwards, R. E.:* A Formal Background to Higher Mathematics Ia, and Ib

*Edwards, R. E.:* A Formal Background to Higher Mathematics IIa, and IIb

*Emery, M.:* Stochastic Calculus in Manifolds

*Emmanouil, I.:* Idempotent Matrices over Complex Group Algebras

*Endler, O.:* Valuation Theory

*Engel, K.-J.; Nagel, R.:* A Short Course on Operator Semigroups

*Erez, B.:* Galois Modules in Arithmetic

*Everest, G.; Ward, T.:* Heights of Polynomials and Entropy in Algebraic Dynamics

*Farenick, D. R.:* Algebras of Linear Transformations

*Foulds, L. R.:* Graph Theory Applications

*Franke, J.; Härdle, W.; Hafner, C. M.:* Statistics of Financial Markets: An Introduction

*Frauenthal, J. C.:* Mathematical Modeling in Epidemiology

*Freitag, E.; Busam, R.:* Complex Analysis

*Friedman, R.:* Algebraic Surfaces and Holomorphic Vector Bundles

*Fuks, D. B.; Rokhlin, V. A.:* Beginner's Course in Topology

*Fuhrmann, P. A.:* A Polynomial Approach to Linear Algebra

*Gallot, S.; Hulin, D.; Lafontaine, J.:* Riemannian Geometry

*Gardiner, C. F.:* A First Course in Group Theory

*Gårding, L.; Tambour, T.:* Algebra for Computer Science

*Godbillon, C.:* Dynamical Systems on Surfaces

*Godement, R.:* Analysis I, and II

*Goldblatt, R.:* Orthogonality and Spacetime Geometry

*Gouvêa, F. Q.:* $p$-Adic Numbers

*Gross, M. et al.:* Calabi-Yau Manifolds and Related Geometries

*Grossman, C.; Roos, H.-G.; Stynes, M:* Numerical Treatment of Partial Differential Equations

*Gustafson, K. E.; Rao, D. K. M.:* Numerical Range. The Field of Values of Linear Operators and Matrices

*Gustafson, S. J.; Sigal, I. M.:* Mathematical Concepts of Quantum Mechanics

*Hahn, A. J.:* Quadratic Algebras, Clifford Algebras, and Arithmetic Witt Groups

*Hájek, P.; Havránek, T.:* Mechanizing Hypothesis Formation

*Heinonen, J.:* Lectures on Analysis on Metric Spaces

*Hlawka, E.; Schoißengeier, J.; Taschner, R.:* Geometric and Analytic Number Theory

*Holmgren, R. A.:* A First Course in Discrete Dynamical Systems

*Howe, R., Tan, E. Ch.:* Non-Abelian Harmonic Analysis

*Howes, N. R.:* Modern Analysis and Topology

*Hsieh, P.-F.; Sibuya, Y. (Eds.):* Basic Theory of Ordinary Differential Equations

*Humi, M., Miller, W.:* Second Course in Ordinary Differential Equations for Scientists and Engineers

*Hurwitz, A.; Kritikos, N.:* Lectures on Number Theory

*Huybrechts, D.:* Complex Geometry: An Introduction

*Isaev, A.:* Introduction to Mathematical Methods in Bioinformatics

*Istas, J.:* Mathematical Modeling for the Life Sciences

*Iversen, B.:* Cohomology of Sheaves

*Jacod, J.; Protter, P.:* Probability Essentials

*Jennings, G. A.:* Modern Geometry with Applications

*Jones, A.; Morris, S. A.; Pearson, K. R.:* Abstract Algebra and Famous Inpossibilities

*Jost, J.:* Compact Riemann Surfaces

*Jost, J.:* Dynamical Systems. Examples of Complex Behaviour

*Jost, J.:* Postmodern Analysis

*Jost, J.:* Riemannian Geometry and Geometric Analysis

*Kac, V.; Cheung, P.:* Quantum Calculus

*Kannan, R.; Krueger, C. K.:* Advanced Analysis on the Real Line

*Kelly, P.; Matthews, G.:* The Non-Euclidean Hyperbolic Plane

*Kempf, G.:* Complex Abelian Varieties and Theta Functions

*Kitchens, B. P.:* Symbolic Dynamics

*Klenke, A.:* Probability Theory

*Kloeden, P.; Ombach, J.; Cyganowski, S.:* From Elementary Probability to Stochastic Differential Equations with MAPLE

*Kloeden, P. E.; Platen; E.; Schurz, H.:* Numerical Solution of SDE Through Computer Experiments

*Koralov, L. B.; Sinai, Ya. G.:* Theory of Probability and Random Processes. 2[nd] edition

*Kostrikin, A. I.:* Introduction to Algebra

*Krasnoselskii, M. A.; Pokrovskii, A. V.:* Systems with Hysteresis

*Kuo, H.-H.:* Introduction to Stochastic Integration

*Kurzweil, H.; Stellmacher, B.:* The Theory of Finite Groups. An Introduction

*Kyprianou, A. E.:* Introductory Lectures on Fluctuations of Lévy Processes with Applications

*Lang, S.:* Introduction to Differentiable Manifolds

*Lefebvre, M.:* Applied Stochastic Processes

*Lorenz, F.:* Algebra I: Fields and Galois Theory

*Lorenz, F.:* Algebra II: Fields with Structure, Algebras and Advanced Topics

*Luecking, D. H., Rubel, L. A.:* Complex Analysis. A Functional Analysis Approach

*Ma, Zhi-Ming; Roeckner, M.:* Introduction to the Theory of (non-symmetric) Dirichlet Forms

*Mac Lane, S.; Moerdijk, I.:* Sheaves in Geometry and Logic

*Marcus, D. A.:* Number Fields

*Martinez, A.:* An Introduction to Semiclassical and Microlocal Analysis

*Matoušek, J.:* Using the Borsuk-Ulam Theorem

*Matsuki, K.:* Introduction to the Mori Program

*Mazzola, G.; Milmeister G.; Weissman J.:* Comprehensive Mathematics for Computer Scientists 1

*Mazzola, G.; Milmeister G.; Weissman J.:* Comprehensive Mathematics for Computer Scientists 2

*Mc Carthy, P. J.:* Introduction to Arithmetical Functions

*McCrimmon, K.:* A Taste of Jordan Algebras

*Meyer, R. M.:* Essential Mathematics for Applied Field

*Meyer-Nieberg, P.:* Banach Lattices

*Mikosch, T.:* Non-Life Insurance Mathematics

*Mines, R.; Richman, F.; Ruitenburg, W.:* A Course in Constructive Algebra

*Moise, E. E.:* Introductory Problem Courses in Analysis and Topology

*Montesinos-Amilibia, J. M.:* Classical Tessellations and Three Manifolds

*Morris, P.:* Introduction to Game Theory

*Mortveit, H.; Reidys, C.:* An Introduction to Sequential Dynamical Systems

*Nicolaescu, L.:* An Invitation to Morse Theory